Contents

X   Contents

Engineering a Lightweight and Efficient Local Search SAT Solver

Adrian Balint and Uwe Schöning[⊠]

Institute of Theoretical Computer Science, Ulm University,
89069 Ulm, Germany
{adrian.balint,uwe.schoening}@uni-ulm.de

Abstract. One important category of SAT solver implementations use stochastic local search (SLS, for short). These solvers try to find a satisfying assignment for the input Boolean formula (mostly, required to be in CNF) by modifying the (mostly randomly chosen) initial assignment by bit flips until a satisfying assignment is possibly reached. Usually such SLS type algorithms proceed in a greedy fashion by increasing the number of satisfied clauses until some local optimum is reached. Trying to find its way out of such local optima typically requires the use of randomness. We present an easy, straightforward SLS type SAT solver, called probSAT, which uses just one simple strategy being based on biased probabilistic flips. Within an extensive empirical study we evaluate the current state-of-the-art solvers on a wide range of SAT problems, and show that our approach is able to exceed the performance of other solving techniques.

1 Introduction

The SAT problem is one of the most studied $\mathcal{NP}$-complete problems in computer science. One reason is the wide range of SAT's practical applications ranging from hardware verification to planning and scheduling. Given a propositional formula in CNF with variables $\{x_1, \ldots, x_n\}$ the SAT-problem consists in finding an assignment for the variables such that all clauses are satisfied.

Stochastic local search (SLS) solvers operate on complete assignments and try to find a solution by flipping variables according to a given heuristic. Most SLS solvers are based on the following scheme: Initially, a random assignment is chosen. If the formula is satisfied by the assignment the solution is found. If not, a variable is chosen according to a (possibly probabilistic) variable selection heuristic, which is further called *pickVar*. The heuristics mostly depend on some score, which counts the number of satisfied/unsatisfied clauses, as well as other aspects like the "age" of variables, and others. It was believed that a good flip heuristic should be designed in a very sophisticated way to obtain a really efficient solver. We show in the following that it is worth to "come back to the roots" since a very elementary and (as we think) elegant design principle

© Springer International Publishing AG 2016
L. Kliemann and P. Sanders (Eds.): Algorithm Engineering, LNCS 9220, pp. 1–18, 2016.
DOI: 10.1007/978-3-319-49487-6_1

for the *pickVar* heuristic just based on probability distributions will do the job extraordinary well.

It is especially popular (and successful) to pick the flip variable from an unsatisfied clause. This is called *focused* local search in [14]. In each round, the selected variable is flipped and the process starts over again until a solution is eventually found.

Most important for the flip heuristic seems to be the *score* of an assignment, i.e. the number of satisfied clauses. Considering the process of flipping one variable, we get the *relative score change* produced by a candidate variable for flipping as: (*score after flipping* minus *score before flipping*) which is equal to *make* minus *break*. Here *make* means the number of newly satisfied clauses which come about by flipping the variable, and *break* means the number of clauses which become false by flipping the respective variable. To be more precise, we will denote $make(x, \alpha)$ and $break(x, \alpha)$ as functions of the respective flip variable x and the actual assignment α (before flipping). Notice that in case of focused flipping mentioned above the value of *make* is always at least 1.

Most of the SLS solvers so far, if not all, follow the strategy that whenever the score improves by flipping a certain variable from an unsatisfied clause, they will indeed flip this variable without referring to probabilistic decisions. Only if no improvement is possible as is the case in local minima, a probabilistic strategy is performed. The winner of the SAT Competition 2011 category random SAT, Sparrow, mainly follows this strategy but when it comes to a probabilistic strategy it uses a probability distribution function [2]. The probability distribution in Sparrow is defined as an exponential function of the score value. In this chapter we analyze several simple SLS solvers which are based only on probability distributions.

2   The New Algorithm Paradigm

We propose a new class of solvers here, called probSAT, which base their probability distributions for selecting the next flip variable solely on the make and break values, but not necessarily on the value of the $score = make - break$, as it was the case in Sparrow. Our experiments indicate that the influence of make should be kept rather weak – it is even reasonable to ignore make completely, like in implementations of WalkSAT [13]. The role of make and break in these SLS-type algorithms should be seen in a new light. The new type of algorithm presented here can also be applied for general constraint satisfaction problems and works as follows.

Algorithm 1. ProbSAT

> **Input**   : Formula F, $maxTries$, $maxFlips$
> **Output**: satisfying assignment α or UNKNOWN
> 1 **for** $i = 1$ *to* $maxTries$ **do**
> 2 $\alpha \leftarrow$ randomly generated assignment
> 3   **for** $j = 1$ *to* $maxFlips$ **do**
> 4   **if** *(α is model for F)* **then**
> 5   return α
> 6 $C_u \leftarrow$ randomly selected unsatisfied clause
> 7   **for** x *in* C_u **do**
> 8   compute $f(x, \alpha)$
> 9 $var \leftarrow$ random variable x according to probability $\frac{f(x,\alpha)}{\sum_{z \in C_u} f(z,\alpha)}$
> 10 $\alpha \leftarrow$ flip(var) in α
> 11 return UNKNOWN;

The idea here is that the function f should give a high value to variable x if flipping x seems to be advantageous, and a low value otherwise. Using f the probability distribution for the potential flip variables is calculated. The flip probability for x is proportional to $f(x, \alpha)$. Letting f be a constant function leads in the k-SAT case to the probabilities $(\frac{1}{k}, \ldots, \frac{1}{k})$ morphing the probSAT algorithm to the random walk algorithm that is theoretically analyzed in [15]. In all our experiments with various functions f we made f depend on $break(x, \alpha)$ and possibly on $make(x, \alpha)$, and no other properties of x and α nor the history of previous search course. In the following we analyze experimentally the effect of several functions to be plugged in for f.

2.1   An Exponential Function

First we considered an exponential decay, 2-parameter function:

$$f(x, \alpha) = \frac{(c_m)^{make(x,\alpha)}}{(c_b)^{break(x,\alpha)}}$$

The parameters of the function are c_b and c_m. Because of the exponential functions used here (think of $c^x = e^{\frac{1}{T}x}$) this is reminiscence of the way Metropolis-like algorithms (see [17]) select a variable. Also, this is similar to the Softmax probabilistic decision for actions in reinforcement learning [19]. We call this the *exp-algorithm*. The separation into the two base constants c_m and c_b will allow us to find out whether there is a different influence of the make and the break value – and there is one, indeed.

It seems reasonable to try to maximize *make* and to minimize *break*. Therefore, we expect $c_m > 1$ and $c_b > 1$ to be good choices for these parameters. Actually, one might expect that c_m should be identical to c_b such that the above formula simplifies to $c^{make-break} = c^{score}$ for an appropriate parameter c.

To get a picture on how the performance of the solver varies for different values of c_m and c_b, we have done a uniform sampling of $c_b \in [1.0, 4.0]$ and of $c_m \in [0.1, 2.0]$ for this exponential function and of $c_m \in [-1.0, 1.0]$ for the polynomial function (see below). We have then run the solver with the different parameter settings on a set of randomly generated 3-SAT instances with 1000 variables at a clause to variable ratio of 4.26. The cutoff limit was set to 10 s. As a performance measure we use PAR10: penalized average runtime, where a timeout of the solver is penalized with 10·(cutoff limit). A parameter setting where the solver is not able to solve anything has a PAR10 value of 100 in our case.

In the case of 3-SAT a very good choice of the parameters is $c_b > 1$ (as expected) and $c_m < 1$ (totally unexpected), for example, $c_b = 3.6$ and $c_m = 0.5$ (see Fig. 1 left upper diagram and the survey in Table 1) with small variation

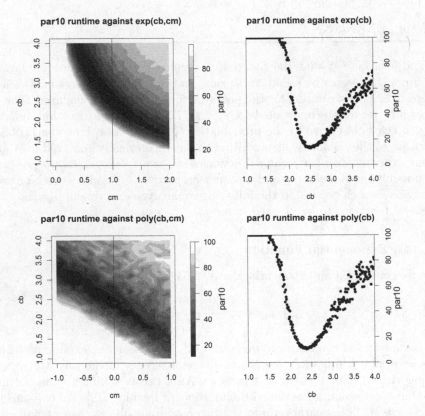

Fig. 1. Parameter space performance plot: The left plots show the performance of different combinations of c_b and c_m for the exponential (upper left corner) and the polynomial (lower left corner) functions. The darker the area the better the runtime of the solver with that parameter settings. The right plots show the performance variation if we ignore the make values (correspond to the cut in the left plots) by setting $c_m = 1$ for the exponential function and $c_m = 0$ for the polynomial function.

depending on the considered set of benchmarks. In the interval $c_m \in [0.3, 1.8]$ the optimal choice of parameters can be described by the hyperbola-like function $(c_b - 1.3) \cdot c_m = 1.1$. Almost optimal results were also obtained if c_m is set to 1 (and c_b to 2.5), see Fig. 1, both upper diagrams. In other words, the value of make is not taken into account in this case.

As mentioned, it turns out that the influence of make is rather weak, therefore it is reasonable, and still leads to very good algorithms – also because the implementation is simpler and has less overhead – if we ignore the make value completely and consider the one-parameter function:

$$f(x, \alpha) = (c_b)^{-break(x,\alpha)}$$

We call this the *break-only-exp-algorithm*.

2.2   A Polynomial Function

Our experiments showed that the exponential decay in probability with growing break value might be too strong in the case of 3-SAT. The above formulas have an exponential decay in probability comparing different (say) break values. The relative decay is the same when we compare $break = 0$ with $break = 1$, and when we compare, say, $break = 5$ with $break = 6$. A "smoother" function for high values would be a polynomial decay function. This led us to consider the following, 2-parameter function ($\epsilon = 1$ in all experiments):

$$f(x, \alpha) = \frac{(make(x, \alpha))^{c_m}}{(\epsilon + break(x, \alpha))^{c_b}}$$

We call this the *poly-algorithm*. The best parameters in case of 3-SAT turned out to be $c_m = -0.8$ (notice the minus sign!) and $c_b = 3.1$ (See Fig. 1, lower part). In the interval $c_m \in [-1.0, 1.0]$ the optimal choice of parameters can be described by the linear function $c_b + 0.9 c_m = 2.3$. Without harm one can set $c_m = 0$, and then take $c_b = 2.3$, and thus ignore the make value completely.

Ignoring the make value (i.e. setting $c_m = 0$) brings us to the function

$$f(x, \alpha) = (\epsilon + break(x, \alpha))^{-c_b}$$

We call this the *break-only-poly-algorithm*.

2.3   Some Remarks

As mentioned above, in both cases, the exp- and the poly-algorithm, it was a good choice to ignore the make value completely (by setting $c_m = 1$ in the exp-algorithm, or by setting $c_m = 0$ in the poly-algorithm). This corresponds to the vertical lines in Fig. 1, left diagrams. But nevertheless, the optimal choice in both cases, was to set $c_m = 0.5$ and $c_b = 3.6$ in the case of the exp-algorithm (and similarly for the poly-algorithm.) We have $\frac{0.5^{make}}{3.6^{break}} \approx 3.6^{-(break+make/2)}$.

This can be interpreted as follows: instead of the usual $score = make - break$ a better score measure is $-(break + make/2)$.

The value of c_b determines the greediness of the algorithm. We concentrate on c_b in this discussion since it seems to be the more important parameter. The higher the value of c_b, the more greedy is the algorithm. A low value of c_b (in the extreme, $c_b = 1$ in the exp-algorithm) morphs the algorithm to a random walk algorithm with flip probabilities $(\frac{1}{k}, \ldots \frac{1}{k})$ like the one considered in [15]. Examining Fig. 1, almost a phase-transition can be observed. If c_b falls under some critical value, like 2.0, the expected run time increases tremendously. Turning towards the other side of the scale, increasing the value of c_b, i.e. making the algorithm more greedy, also degrades the performance but not with such an abrupt rise of the running time as in the other case. These observations have also been made empirically by Hoos in [9], where he proposed to approximate the noise value from above, rather from below.

3   Experimental Analysis of the Functions

To determine the performance of our probability distribution based solver we have designed a wide variety of experiments. In the first part of our experiments we try to determine good settings for the parameters c_b and c_m by means of automatic configuration procedures. In the second part we will compare our solver to other state-of-the-art solvers.

3.1   The Benchmark Formulae

All random instances used in our settings are uniform random k-SAT problems generated with different clause to variable ratios, denoted with r. The class of random 3-SAT problems is the best studied class of random problems and because of this reason we have four different sets of 3-SAT instances.

1. 3sat1k [21]: 10^3 variables at $r = 4.26$ (500 instances)
2. 3sat10k [21]: 10^4 variables at $r = 4.2$ (500 instances)
3. 3satComp[1]: all large 3-SAT instances from the SAT Competition 2011 category random with variables range $2 \cdot 10^3 \ldots 5 \cdot 10^4$ at $r = 4.2$ (100 instances)
4. 3satExtreme: $10^5 \ldots 5 \cdot 10^5$ variables at $r = 4.2$ (180 instances)

The 5-SAT and 7-SAT problems used in our experiments come from [21]: 5sat500 (500 variables at $r = 20$) and 7sat90 (90 variables at $r = 85$). The 3sat1k, 3sat10k, 5sat500 and 7sat90 instance classes are divided into two equal sized classes called train and test. The train set is used to determine good parameters for c_b and c_m and the second class is used to report the performance. Further we also include the set of satisfiable random and crafted instances from the SAT Competition 2011.

---

[1] www.satcompetition.org.

3.2   Good Parameter Setting

The problem that every solver designer is confronted with is the determination of good parameters for its solvers. We have avoided to accomplish this task by manual tuning but instead have used an automatic procedure.

As our parameter search space is relatively small, we have opted to use a modified version of the iterated F-Race [5] configurator, which we have implemented in Java. The idea of F-race is relatively simple: good configurations should be evaluated more often than poor ones which should be dropped as soon as possible. F-Race uses a familywise Friedman test (see Test 25 in [18] for more details about the test) to check if there is a significant performance difference between solver configurations. The test is conducted every time the solvers have run on an instance. If the test is positive, poor configurations are dropped, and only the good ones are further evaluated. The configurator ends when the number of solvers left in the race is less than 2 times the number of parameters or if there are no more instances to evaluate on.

To determine good values for c_b and c_m we have run our modified version of F-Race on the training sets 3sat1k, 3sat10k, 5sat500 and 7sat90. The cutoff time for the solvers were set to 10 s for 3sat1k and to 100 s for the rest. The best parameter values returned by this procedure are listed in Table 1. Values for the class of 3sat1k problems were also included, because the preliminary analysis of the parameter search space was done on this class. The best parameter of the break-only-exp-algorithm for k-SAT can be roughly described by the formula $c_b = k^{0.8}$.

Table 1. Parameter setting for c_b and c_m: Each cell represents a good setting for c_b and c_m dependent on the function used by the solver. Parameter values close to these values have similar good performance.

	3sat1k	3sat10k	5sat500	7sat90
$exp(c_b, c_m)$	3.6 0.5	3.97 0.3	3.1 1.3	3.2 1.4
$poly(c_b, c_m)$	3.1 −0.8	2.86 −0.81	-	-
$exp(c_b)$	2.50	2.33	3.6	4.4
$poly(c_b)$	2.38	2.16	-	-

4   Empirical Evaluation

In the second part of our experiments we compare the performance of our solvers to that of the SAT Competition 2011 winners and also to WalkSAT [13]. An additional comparison to a survey propagation algorithm will show how far our probSAT local search solver can get.

Soft- and Hardware. The solvers were run on a part of the bwGrid clusters [8] (Intel Harpertown quad-core CPUs with 2.83 GHz and 8 GByte RAM). The operating system was Scientific Linux. All experiments were conducted with EDACC, a platform that distributes solver execution on clusters [1].

The Competitors. The WalkSAT solver is implemented within our own code basis. We use our own implementation and not the original code (version 48) provided by Henry Kautz[2], because our implementation is approximately 1.35 times faster[3].

We have used version 1.4 of the survey propagation solver provided by Zecchina[4], which was changed to be DIMACS conform. For all other solvers we have used the binaries from the SAT Competition 2011[5].

Parameter Settings of Competitors. Sparrow is highly tuned on our target set of instances and incorporates optimal settings for each set within its code. WalkSAT [13] has only one single parameter, the walk probability wp. In case of 3-SAT we took the optimal values for $wp = 0.567$ which have been established in an extensive analysis in [11]. Because we could not find any settings for 5-SAT and 7-SAT problems we have run our modified version of F-Race to find good settings. For 5sat500 the configurator reported $wp = 0.25$ and for 7sat90 $wp = 0.1$. The survey propagation solver was evaluated with the default settings reported in [6] (fixing 5 % of the variables per step).

Results. We have evaluated our solvers and the competitors on the test set of the instance sets 3sat1k, 3sat10k, 5sat500 and 7sat90 (note that the training set was used only for finding good parameters for the solvers). The parameter setting for c_b and c_m are those from Table 1 (in case of 3-SAT we have always used the parameters for 3sat10k). The results of the evaluations are listed in Table 2.

On the 3-SAT instances, the polynomial function yields the overall best performance. On the 3-SAT competition set all of our solver variants exhibited the most stable performance, being able to solve all problems within cutoff time. The survey propagation solver has problems with the 3sat10k and the 3sat-Comp problems (probably because of the relatively small number of variables). The good performance of the break-only-poly-solver remains surprisingly good even on the 3satExtreme set where the number of variables reaches $5 \cdot 10^5$ (ten times larger than that from the SAT Competition 2011). From the class of SLS solvers it exhibits the best performance on this set and is only approx. 2 times slower than survey propagation. Note that a value of $c_b = 2.165$ for the break-only-poly solver further improved the runtime of the solver by approximately 30 % on the 3satExtreme set.

---

[2] http://www.cs.rochester.edu/u/kautz/walksat/.

[3] The latest version 50 of WalkSAT has been significantly improved, but was not available at the time we have performed the experiments.

[4] http://users.ictp.it/~zecchina/SP/.

[5] http://www.cril.univ-artois.fr/SAT11/solvers/SAT2011-static-binaries.tar.gz.

Table 2. Evaluation results: Each cell represents the PAR10 (Penalized average runtime with penalization factor 10 - every unsuccessful run is penalized with 10 times the maximum runtime.) runtime and the number of successful runs for the solvers on the given instance set. Results are highlighted if the solver succeeded in solving all instances within the cutoff time, or if it has the best PAR10 runtime. Cutoff times are 600 s for 3sat10k, 5sat500 and 7sat90 and 5000 s for the rest. The blank cells indicates that we have no parameter setting worth evaluating.

	3sat10k	3satComp	3satExtreme	5sat500	7sat90
$exp(c_b, c_m)$	46.6	93.84	-	12.49	201.68
	(998)	**(500)**		(10^3)	(974)
$poly(c_b, c_m)$	46.65	76.81	-	-	-
	996	**(500)**			
$exp(c_b)$	53.02	126.59	-	**7.84**	134.06
	(997)	**(500)**		(10^3)	(984)
$poly(c_b)$	**22.80**	**54.37**	1121.34	-	-
	(1000)	**(500)**	**(180)**		
Sparrow	199.78	498.05	47419	9.52	**14.94**
	(973)	(498)	(10)	(10^3)	(10^3)
WalkSAT	61.74	172.21	1751.77	14.71	69.34
	(995)	(499)	(178)	(10^3)	(994)
sp 1.4	3146.17	18515.79	**599.01**	5856	6000
	(116)	(63)	**(180)**	(6)	(0)

On the 5-SAT instances the exponential break-only-exp solver yields the best performance being able to beat even Sparrow, which was the best solver for 5-SAT within the SAT Competition 2011. On the 7-SAT instances though the performance of our solvers is relatively poor. We observed a very strong variance of the run times on this set and it was relatively hard for the configurator to cope with such high variances.

Overall the performance of our simple probability based solvers reaches state-of-the-art performance and can even get into problem size regions where only survey propagation could catch ground.

Scaling Behavior with the Number of Variables n. Experiments show that the survey propagation algorithm scales linearly with n on formulas generated near the threshold ratio. The same seems to hold for WalkSAT with optimal noise as the results in [11] show. The 3satExtreme instance set contains very large instances with varying $n \in \{10^5 \dots 5 \cdot 10^5\}$. To analyze the scaling behavior of probSAT in the break-only-poly variant we have computed for each run the number of flips per variable performed by the solver until a solution was found. The number of flips per variable remains constant at about $2 \cdot 10^3$ independent of n. The same holds for WalkSAT, though WalkSAT seems to have a slightly larger variance of the runtimes.

Results on the SAT Competition 2011 Satisfiable Random Set. We have compiled an adaptive version of probSAT and of WalkSAT, that first checks the size of the clauses (i.e. k) and then sets the parameters accordingly (like Sparrow does). We have ran these solvers on the complete satisfiable instances set from the SAT Competition 2011 random category along with all other competition winning solvers from this category: Sparrow2011, sattime2011 and EagleUP. Cutoff time was set to 5000 s. We report only the results on the large set, as the medium set was completely solved by all solvers and the solvers had a median runtime under one second. As can be seen from the results of the cactus plot in Fig. 2, the adaptive version of probSAT would have been able to win the competition. Interestingly is to see that the adaptive version of WalkSAT would have ranked third.

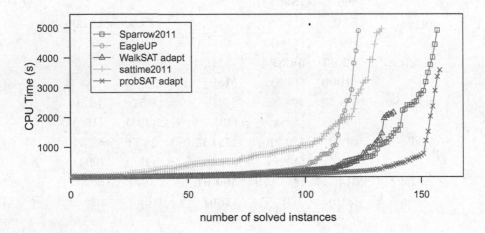

Fig. 2. Results on the "large" set of the SAT Competition 2011 random instances represented as a cactus plot. The x-axis represents the number of problems a solver was able to solve ordered by runtime; the y-axis is the runtime. The lower a curve (low runtimes) and the more it gets to the right (more problems solved) the better the solver.

Results on the SAT Competition 2011 Satisfiable Crafted Set. We have also run the different solvers on the satisfiable instances from the crafted set of SAT Competition 2011 (with a cutoff time of 5000 s). The results are listed in Table 3. We have also included the results of the best three complete solvers from the crafted category. probSAT and WalkSAT performed best in their 7-SAT break-only configuration solving 81 respectively 101 instances. The performance of WalkSAT could not be improved by changing the walk probability. probSAT though exhibited better performance with $c_b = 7$ and a switch to the polynomial break-only scheme, being then able to solve 93 instances. With such a high c_b value (very greedy) the probability of getting stuck in local minima is very high. By adding a static restart strategy after $2 \cdot 10^4$ flips per variable probSAT was then able to solve 99 instances (as listed in the table).

Table 3. Results on the crafted satisfiable instances: Each cell reports the number of solved instances within the cutoff time (5000 s). The first line shows the results on the original instances and the second on the preprocessed instances.

	Sattime	Sparrow	WalkSAT	probSAT	MPhaseSAT (complete)	clasp (complete)
Crafted	**107**	104	101	99	93	81
Crafted pre	86	97	**101**	95	98	80

The high greediness level needed for WalkSAT and probSAT to solve the crafted instances indicates that this instances might be more similar to the 7-SAT instances (generally to higher k-SAT). A confirmation of this conjecture is that Sparrow with fixed parameters for 7-SAT instances could solve 103 instances vs. 104 in the default setting (which adapts the parameters according to the maximum clause length found in the problem). We suppose that improving SLS solvers for random instances with large clause length would also yield improvements for non random instances.

To check whether the performance of SLS solvers can be improved by preprocessing the instances first, we have run the preprocessor of lingeling [4], which incorporates all main preprocessing techniques, to simplify the instances. The results unluckily show the contrary of what would have been expected (see Table 3). None of the SLS solvers could benefit from the preprocessing step, solving equal or less instances. These results motivated the analysis of preprocessing techniques in more detail, which was performed in [3]. It turns out that bounded variable elimination, which performs variable elimination through resolution rules up to certain bound is a good preprocessing technique for SLS solvers and can indeed improve the performance of SLS solvers.

Results on the SAT Challenge 2012 Random Set. We have submitted the probSAT solver (the adaptive version) to the SAT Challenge 2012 random satisfiable category. The results of the best performing solvers can be seen as a cactus plot in Fig. 3. probSAT was the second best solver on these instances been only outperformed by CCAsat.

While the difference to all other competitors is significant in terms of a Mann-Whitney-U test, the difference to CCAsat is not.

Results on the SAT Competition 2013 Satisfiable Random Set. We have also submitted an improved version of probSAT to the SAT Competition 2013 to the Random Satisfiable category. The implementation of probSAT was improved with respect to parameters, data structure and work flow.

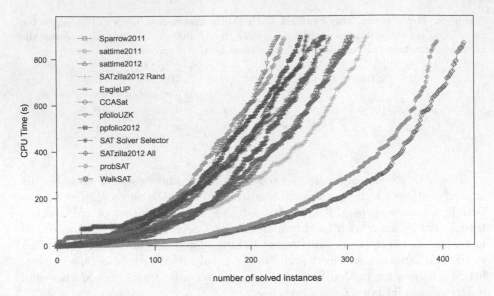

Fig. 3. Results of the best performing solvers on the SAT Challenge 2012 random instances as a cactus plot. For details about cactus plot see Fig. 2.

The parameters of probSAT have been set as follows:

k	fct	cb	ϵ
3	poly	2.06	0.9
4	exp	2.85	-
5	exp	3.7	-
6	exp	5.1	-
≥ 7	exp	5.4	-

where k is the size of the longest clause found in the problem during parsing. These parameter values have been determined in different configuration experiments.

All array data structures where ended by a sentinel[6] (i.e. the last element in the array is the stop value; in our case we have used 0). All for-loops have been changed into while-loops that have no counter but only a sentinel check, allowing us to save several memory dereferences and variables. As most of the operations performed by SLS solvers are loops over some small sized arrays, this optimization turns out to improve the performance of the solver between 10 %–25 % (dependent on the instances).

---

[6] We would like to thank Armin Biere for this suggestion.

Compared to the original version the version submitted to the competition is not selecting an unsatisfied clause randomly but will iterate through the set of unsatisfied clauses with the flip counter (i.e. instead of `c=rand() modulo numUnsat` we use `c=flipCounter modulo numUnsat`). This scheme will reduce the probability of undoing a change right in the next step. This small change seems to improve in some cases the stagnation behavior of the solver giving it a further boost[7].

To measure the isolated effect of the different changes we have performed a small experiment on the 3sat10k instance set. We start with the version that was submitted to the SAT Challenge 2012 with new parameters (sc12(1)), then we add the code optimizations (sc12(2)) and finally we remove the random selection of a false clause (sc13). A further version was added to this evaluation that does not cache the break values, but computes them on the fly. This version is denoted with (nc) in the table and was analyzed only after the competition. The results of the evaluation are listed in Table 4.

Table 4. The results of the evaluation of different implementation variants of the probSAT solver on the 3sat10k instance set. The last column shows the speed up with respect to the last row. Time is measured in seconds.

	Solver	Total CPU time	Average CPU time	Median CPU time	Average speedup
1	probSAT sc13 (nc)	4356.0729	17.4243	7.886	2.01x
2	probSAT sc13	4696.9674	18.7879	8.499	1.86x
3	probSAT sc12(2)	7632.1326	30.5285	10.695	1.15x
4	probSAT sc12(1)	8781.8255	35.1273	12.489	-

The code optimizations yielded an average speedup of 15 %, while the removal of random clause selection is further improving the performance by around 70 %. Further adding on the fly computation of the break values yields a twofold speedup compared to the original version with new parameters.

probSAT sc13 was submitted to SAT Competition 2013[8]. The results of the best performing solvers submitted to SAT Competition 2013 can be seen as a cactus plot in Fig. 4. probSAT is able to outperform all its competitors. The instances used in SAT Competition 2013 contained randomly generated instances on the phase transition point for $k = 3, \ldots, 7$ and also a small set of huge instances (in terms of number of variables). The last were intended to test the robustness of the solvers. probSAT turns out to be a very robust solver, being able to solve many of the huge instances 18 out of the 26 that have been solved by some solver (out of a total of 36). From the set of phase transition instances

---

[7] This might also be the case for the WalkSAT solver.

[8] The code was compiled with the Intel®Compiler 12.0 with the following parameters: *-O3 -xhost -static -unroll-aggressive -opt-prefetch -fast*.

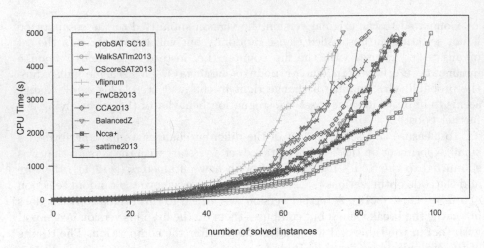

Fig. 4. Results of the best performing solvers on the SAT Competition 2013 random satisfiable instances.

probSAT solved 81 out of 109 that could be solved by any solver. Altogether this shows that the solving approach (and the parameter settings) used by probSAT has an overall good performance.

5   Comparison with WalkSAT

In principle, WalkSAT [13] also uses a certain pattern of probabilities for flipping one of the variables within a non-satisfied clause. But the probability distribution does not depend on a single continuous function f as in our algorithms described above, but uses some non-continuous if-then-else decisions as described in [13].

In Table 5 we compare the flipping probabilities in WalkSAT (setting the wp parameter i.e. the noise value to $wp = 0.567$) with the break-only-poly-algorithm (with $c_b = 2.06$ and $\epsilon = 0.9$) using several examples of *break* values combinations that might occur within a 3-CNF clause.

Even though the probabilities look very similar, we think that the small differences renders our approach to be more robust. Further, probSAT has the PAC property [10, p. 153]. In each step every variable has a probability greater zero to be picked for flipping. This is though not the case for WalkSAT. A variable occurring in a clause where an other variable has a score of zero can not be chosen for flipping. There is no published example for which WalkSAT gets trapped in cycles. Though, during a talk given by Donald Knuth in Trento at the SAT Conference in 2012 where he presented details about his implementation of WalkSAT, he mentioned that Bram Cohen, the designer of WalkSAT, has provided such an example.

Table 5. Probability comparison of WalkSAT and probSAT: The first columns show some possible *break* value combinations that occur within a clause in a 3-SAT formula during the search. For the different solvers considered here the probabilities for each of the 3 variables to be flipped are listed.

Breaks			WalkSAT			Break-only-poly		
0	0	0	0.33	0.33	0.33	0.33	0.33	0.33
0	0	1	0.5	0.5	0	0.45	0.45	0.10
0	1	1	1.0	0	0	0.70	0.15	0.15
0	1	2	1.0	0	0	0.76	0.16	0.07
0	2	2	1.0	0	0	0.85	0.07	0.07
1	1	1	0.33	0.33	0.33	0.33	0.33	0.33
1	1	2	0.41	0.41	0.17	0.41	0.41	0.18
1	2	2	0.53	0.23	0.23	0.54	0.23	0.22
1	2	3	0.53	0.23	0.23	0.61	0.25	0.14

6   Implementation Variations

In the previous sections we have compared the solvers based on their runtime. As a consequence the efficiency of the implementation plays a crucial role and the best available implementation should be taken for comparison. Another possible comparison measure is the number of flips the solver needs to perform to find a solution. From a practical point of view this is not optimal. The number of flips per second (denoted with $flips/sec$) is a key measure of SLS solvers when it comes to compare algorithm implementations or two different similar algorithms. In this Section we would like to address this problem by comparing two different implementations of probSAT and WalkSAT on a set of very large 3-SAT problems.

All efficient implementations of SLS solvers are computing the scores of variables from scratch only within the initialization phase. During the search of the solver, the scores are only updated. This is possible because only the score of variables can change that are in the neighborhood of the variable being flipped. This method is also known as *caching* (the scores of the variables are being cached) in [10, p. 273] or *incremental* approach in [7].

The other method would be to compute the score of variables on the fly before taking them into consideration for flipping. This method is called *non-caching* or *non-incremental* approach. In case of probSAT or WalkSAT only the score of variables from one single clause has to be computed as opposed to other solvers where all variables from all unsatisfied clauses are taken into consideration for flipping.

We have implemented two different versions of probSAT and WalkSAT within the same code basis (i.e. the solvers are identical with exception of the *pickVar* method), one that uses caching and one that does not. We have evaluated the

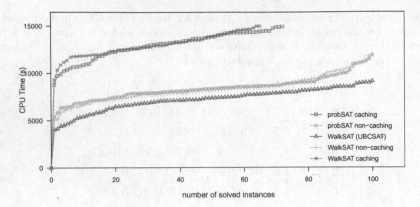

Fig. 5. Comparison of the different implementation variants of probSAT and WalkSAT on extreme large 3-SAT problems (within the same code basis), with and without caching of the break values. We also evaluate the best known WalkSAT implementation (non-caching) from UBCSAT as a reference.

four different solvers on a set of 100 randomly generated 3-SAT problems with 10^6 variables and a ratio of 4.2. The results can be seen in Fig. 5.

Within the time limit of $1.5 \cdot 10^4$ s only the variants not using caching were able to solve all problems. The implementation with caching solved only 72 (probSAT) respectively 65 instances (WalkSAT). Note that all solvers started with the same seed (i.e. they perform search on the exactly same search trajectory). The difference between the different implementations in terms of performance can be explained by the different number of *flips/sec*. While the version with caching performs around $1.4 \cdot 10^5$ flips/sec the version without caching is able to perform around $2.2 \cdot 10^5$ flips/sec. This explains the difference in runtime between the two different implementations. Similar findings have also been observed in [20, p. 27] and in [7].

The advantage of non-caching decreases with increasing k (for random generated k-SAT problems) and becomes even a disadvantage for 5-SAT problems and upwards. As a consequence the latest version of probSAT uses caching for 3-SAT problems and non-caching for the other types of problems.

7   Conclusion and Future Work

We introduced a simple algorithmic design principle for a SLS solver which does its job without heuristics and "tricks". It just relies on the concept of probability distribution and focused search. It is though flexible enough to allow plugging in various functions f which guide the search.

Using this concept we were able to discover a non-symmetry regarding the importance of the *break* and *make* values: the *break* value is the more important one; one can even do without the *make* value completely.

We have systematically used an automatic configurator to find the best parameters and to visualize the mutual dependency and impact of the parameters.

Furthermore, we observe a large variation regarding the running times even on the same input formula. Therefore the issue of introducing an optimally chosen restart point arises. Some initial experiments show that performing restarts, even after a relatively short period of flips (e.g. $20\,n$) gives favorable results on hard instances. It seems that the probability distribution of the number of flips until a solution is found, shows some strong heavy tail behavior (cf. [12, 16]).

Finally, a theoretical analysis of the Markov chain convergence and speed of convergence underlying this algorithm would be most desirable, extending the results in [15].

Acknowledgments. We would like to thank the BWGrid [8] project for providing the computational resources. This project was funded by the Deutsche Forschungsgemeinschaft (DFG) under the number SCHO 302/9-1. We thank Daniel Diepold and Simon Gerber for implementing the F-race configurator and providing different analysis tools within the EDACC framework. We would also like to thank Andreas Fröhlich for fruitful discussions on this topic and Armin Biere for helpful suggestions regarding code optimizations.

References

1. Balint, A., Diepold, D., Gall, D., Gerber, S., Kapler, G., Retz, R.: EDACC - an advanced platform for the experiment design, administration and analysis of empirical algorithms. In: Coello, C.A.C. (ed.) LION 2011. LNCS, vol. 6683, pp. 586–599. Springer, Heidelberg (2011). doi:10.1007/978-3-642-25566-3_46
2. Balint, A., Fröhlich, A.: Improving stochastic local search for SAT with a new probability distribution. In: Strichman, O., Szeider, S. (eds.) SAT 2010. LNCS, vol. 6175, pp. 10–15. Springer, Heidelberg (2010). doi:10.1007/978-3-642-14186-7_3
3. Balint, A., Manthey, N.: Analysis of preprocessing techniques and their utility for CDCL and SLS solver. In: Proceedings of POS2013 (2013)
4. Biere, A.: Lingeling and friends at the SAT competition 2011. Technical report, FMV Reports Series, Institute for Formal Models and Verification, Johannes Kepler University, Altenbergerstr. 69, 4040 Linz, Austria (2011)
5. Birattari, M., Yuan, Z., Balaprakash, P., Stützle, T.: F-Race and iterated F-Race: an overview. In: Bartz-Beielstein, T., Chiarandini, M., Paquete, L., Preuss, M. (eds.) Experimental Methods for the Analysis of Optimization Algorithms, pp. 311–336. Springer, Heidelberg (2010). http://dx.doi.org/10.1007/978-3-642-02538-9_13
6. Braunstein, A., Mézard, M., Zecchina, R.: Survey propagation: an algorithm for satisfiability. Random Structures & Algorithms **27**(2), 201–226 (2005)
7. Fukunaga, A.: Efficient implementations of SAT local search. In: Seventh International Conference on Theory and Applications of Satisfiability Testing (SAT 2004), pp. 287–292 (2004, this volume)
8. bwGRiD(http://www.bwgrid.de/): Member of the German D-Grid initiative, funded by the Ministry of Education and Research (Bundesministeriumfür Bildung und Forschung) and the Ministry for Science, Research and Arts Baden-Württemberg (Ministerium für Wissenschaft, Forschung und Kunst Baden-Württemberg). Techical report, Universities of Baden-Württemberg (2007-2010)

9. Hoos, H.H.: An adaptive noise mechanism for WalkSAT. In: Proceedings of the Eighteenth National Conference in Artificial Intelligence (AAAI 2002), pp. 655–660 (2002)

10. Hoos, H.H., Stützle, T.: Stochastic Local Search: Foundations and Applications. Morgan Kaufmann, San Francisco (2005)

11. Kroc, L., Sabharwal, A., Selman, B.: An empirical study of optimal noise and runtime distributions in local search. In: Strichman, O., Szeider, S. (eds.) SAT 2010. LNCS, vol. 6175, pp. 346–351. Springer, Heidelberg (2010). doi:10.1007/978-3-642-14186-7_31

12. Luby, M., Sinclair, A., Zuckerman, D.: Optimal speedup of Las Vegas algorithms. In: ISTCS, pp. 128–133 (1993). http://dblp.uni-trier.de/db/conf/istcs/istcs1993.html#LubySZ93

13. McAllester, D., Selman, B., Kautz, H.: Evidence for invariants in local search. In: Proceedings of the Fourteenth National Conference on Artificial Intelligence (AAAI 1997), pp. 321–326 (1997)

14. Papadimitriou, C.H.: On selecting a satisfying truth assignment. In: Proceedings of the 32nd Annual Symposium on Foundations of Computer Science (FOCS 1991), pp. 163–169 (1991)

15. Schöning, U.: A probabilistic algorithm for k-SAT and constraint satisfaction problems. In: Proceedings of the Fourtieth Annual Symposium on Foundations of Computer Science (FOCS 1999), p. 410 (1999)

16. Schöning, U.: Principles of stochastic local search. In: Akl, S.G., Calude, C.S., Dinneen, M.J., Rozenberg, G., Wareham, H.T. (eds.) UC 2007. LNCS, vol. 4618, pp. 178–187. Springer, Heidelberg (2007). doi:10.1007/978-3-540-73554-0_17

17. Seitz, S., Alava, M., Orponen, P.: Focused local search for random 3-satisfiability. CoRR abs/cond-mat/0501707 (2005)

18. Sheskin, D.J.: Handbook of Parametric and Nonparametric Statistical Procedures, 4th edn. Chapman & Hall/CRC, Boca Raton (2007)

19. Sutton, R.S., Barto, A.G.: Reinforcement Learning: An Introduction. MIT Press, Cambridge (1998). http://www.cs.ualberta.ca/%7Esutton/book/ebook/the-book.html

20. Tompkins, D.A.D.: Dynamic local search for SAT: design, insights and analysis. Ph.D. thesis, University of British Columbia, October 2010

21. Tompkins, D.A.D., Balint, A., Hoos, H.H.: Captain jack: new variable selection heuristics in local search for SAT. In: Sakallah, K.A., Simon, L. (eds.) SAT 2011. LNCS, vol. 6695, pp. 302–316. Springer, Heidelberg (2011). doi:10.1007/978-3-642-21581-0_24

Route Planning in Transportation Networks

Hannah Bast[1], Daniel Delling[2], Andrew Goldberg[3],
Matthias Müller-Hannemann[4], Thomas Pajor[5(✉)], Peter Sanders[6],
Dorothea Wagner[6], and Renato F. Werneck[3]

[1] University of Freiburg, Freiburg im Breisgau, Germany
bast@informatik.uni-freiburg.de
[2] Apple Inc., Cupertino, USA
ddelling@apple.com
[3] Amazon, Seattle, USA
{andgold,werneck}@amazon.com
[4] Martin-Luther-Universität Halle-Wittenberg, Halle, Germany
muellerh@informatik.uni-halle.de
[5] Microsoft Research, Mountain View, USA
microsoft@tpajor.com
[6] Karlsruhe Institute of Technology, Karlsruhe, Germany
{sanders,dorothea.wagner}@kit.edu

Abstract. We survey recent advances in algorithms for route planning in transportation networks. For road networks, we show that one can compute driving directions in milliseconds or less even at continental scale. A variety of techniques provide different trade-offs between preprocessing effort, space requirements, and query time. Some algorithms can answer queries in a fraction of a microsecond, while others can deal efficiently with real-time traffic. Journey planning on public transportation systems, although conceptually similar, is a significantly harder problem due to its inherent time-dependent and multicriteria nature. Although exact algorithms are fast enough for interactive queries on metropolitan transit systems, dealing with continent-sized instances requires simplifications or heavy preprocessing. The multimodal route planning problem, which seeks journeys combining schedule-based transportation (buses, trains) with unrestricted modes (walking, driving), is even harder, relying on approximate solutions even for metropolitan inputs.

1 Introduction

This survey is an introduction to the state of the art in the area of practical algorithms for routing in transportation networks. Although a thorough survey by Delling et al. [94] has appeared fairly recently, it has become outdated due to significant developments in the last half-decade. For example, for continent-sized

---

This work was mostly done while the authors Daniel Delling, Andrew Goldberg, and Renato F. Werneck were at Microsoft Research Silicon Valley.

© Springer International Publishing AG 2016
L. Kliemann and P. Sanders (Eds.): Algorithm Engineering, LNCS 9220, pp. 19–80, 2016.
DOI: 10.1007/978-3-319-49487-6_2

road networks, newly-developed algorithms can answer queries in a few hundred nanoseconds; others can incorporate current traffic information in under a second on a commodity server; and many new applications can now be dealt with efficiently. While Delling et al. focused mostly on road networks, this survey has a broader scope, also including schedule-based public transportation networks as well as multimodal scenarios (combining schedule-based and unrestricted modes).

Section 2 considers shortest path algorithms for static networks; although it focuses on methods that work well on road networks, they can be applied to arbitrary graphs. Section 3 then considers the relative performance of these algorithms on real road networks, as well as how they can deal with other transportation applications. Despite recent advances in routing in road networks, there is still no "best" solution for the problems we study, since solution methods must be evaluated according to different measures. They provide different trade-offs in terms of query times, preprocessing effort, space usage, and robustness to input changes, among other factors. While solution quality was an important factor when comparing early algorithms, it is no longer an issue: as we shall see, all current state-of-the-art algorithms find provably exact solutions. In this survey, we focus on algorithms that are not clearly dominated by others. We also discuss approaches that were close to the dominance frontier when they were first developed, and influenced subsequent algorithms.

Section 4 considers algorithms for journey planning on schedule-based public transportation systems (consisting of buses, trains, and trams, for example), which is quite different from routing in road networks. Public transit systems have a time-dependent component, so we must consider multiple criteria for meaningful results, and known preprocessing techniques are not nearly as effective. Approximations are thus sometimes still necessary to achieve acceptable performance. Advances in this area have been no less remarkable, however: in a few milliseconds, it is now possible to find good journeys within public transportation systems at a very large scale.

Section 5 then considers a true *multimodal* scenario, which combines schedule-based means of transportation with less restricted ones, such as walking and cycling. This problem is significantly harder than its individual components, but reasonable solutions can still be found.

A distinguishing feature of the methods we discuss in this survey is that they quickly made real-life impact, addressing problems that need to be solved by interactive systems at a large scale. This demand facilitated technology transfer from research prototypes to practice. As our concluding remarks (Sect. 6) will explain, several algorithms we discuss have found their way into mainstream production systems serving millions of users on a daily basis.

This survey considers research published until January 2015. We refer to the final (journal) version of a result, citing conference publications only if a journal version is not yet available. The reader should keep in mind that the journal publications we cite often report on work that first appeared (at a conference) much earlier.

2 Shortest Paths Algorithms

Let $G = (V, A)$ be a (directed) graph with a set V of vertices and a set A of arcs. Each arc $(u, v) \in A$ has an associated nonnegative *length* $\ell(u, v)$. The length of a path is the sum of its arc lengths. In the *point-to-point shortest path problem*, one is given as input the graph G, a source $s \in V$, and a target $t \in V$, and must compute the length of the shortest path from s to t in G. This is also denoted as $\mathrm{dist}(s, t)$, the distance between s and t. The *one-to-all* problem is to compute the distances from a given vertex s to all vertices of the graph. The *all-to-one* problem is to find the distances from all vertices to s. The *many-to-many* problem is as follows: given a set S of sources and a set T of targets, find the distances $\mathrm{dist}(s, t)$ for all $s \in S$, $t \in T$. For $S = T = V$ we have the *all pairs shortest path* problem.

In addition to the distances, many applications need to find the corresponding shortest paths. An *out-shortest path tree* is a compact representation of one-to-all shortest paths from the root r. (Likewise, the in-shortest path tree represents the all-to-one paths.) For each vertex $u \in V$, the path from r to u in the tree is the shortest path.

In this section, we focus on the basic point-to-point shortest path problem under the basic *server model*. We assume that all data fits in RAM. However, locality matters, and algorithms with fewer cache misses run faster. For some algorithms, we consider multi-core and machine-tailored implementations. In our model, preprocessing may be performed on a more powerful machine than queries (e.g., a machine with more memory). While preprocessing may take a long time (e.g., hours), queries need to be fast enough for interactive applications.

In this section, we first discuss basic techniques, then those using preprocessing. Since all methods discussed could in principle be applied to arbitrary graphs, we keep the description as general as possible. For intuition, however, it pays to keep road networks in mind, considering that they were the motivating application for most approaches we consider. We will explicitly consider road networks, including precise performance numbers, in Sect. 3.

2.1 Basic Techniques

The standard solution to the one-to-all shortest path problem is Dijkstra's algorithm [108]. It maintains a priority queue Q of vertices ordered by (tentative) distances from s. The algorithm initializes all distances to infinity, except $\mathrm{dist}(s, s) = 0$, and adds s to Q. In each iteration, it extracts a vertex u with minimum distance from Q and *scans* it, i.e., looks at all arcs $a = (u, v) \in A$ incident to u. For each such arc, it determines the distance to v via arc a by computing $\mathrm{dist}(s, u) + \ell(a)$. If this value improves $\mathrm{dist}(s, v)$, the algorithm performs an arc *relaxation*: it updates $\mathrm{dist}(s, v)$ and adds vertex v with key $\mathrm{dist}(s, v)$ to the priority queue Q. Dijkstra's algorithm has the *label-setting* property: once a vertex $u \in V$ is scanned (settled), its distance value $\mathrm{dist}(s, u)$ is correct. Therefore, for point-to-point queries, the algorithm may stop as soon as it scans the

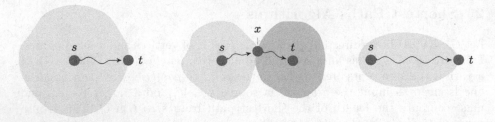

Fig. 1. Schematic search spaces of Dijkstra's algorithm (left), bidirectional search (middle), and the A* algorithm (right).

target t. We refer to the set of vertices $S \subseteq V$ scanned by the algorithm as its *search space*. See Fig. 1 for an illustration.

The running time of Dijkstra's algorithm depends on the priority queue used. The running time is $\mathcal{O}((|V| + |A|) \log |V|)$ with binary heaps [254], improving to $\mathcal{O}(|A| + |V| \log |V|)$ with Fibonacci heaps [129]. For arbitrary (non-integral) costs, generalized versions of binary heaps (such as 4-heaps or 8-heaps) tend to work best in practice [61]. If all arc costs are integers in the range $[0, C]$, multilevel buckets [103] yield a running time of $\mathcal{O}(|A| + |V| \sqrt{\log C})$ [8,62] and work well in practice. For the average case, one can get an $\mathcal{O}(|V| + |A|)$ (linear) time bound [147,192]. Thorup [244] has improved the theoretical worst-case bound of Dijkstra's algorithm to $O(|A| + |V| \log \log \min\{|V|, C\})$, but the required data structure is rather involved and unlikely to be faster in practice.

In practice, one can reduce the search space using *bidirectional search* [67], which simultaneously runs a forward search from s and a backward search from t. The algorithm may stop as soon as the intersection of their search spaces provably contains a vertex x on the shortest path from s to t. For road networks, bidirectional search visits roughly half as many vertices as the unidirectional approach.

An alternative method for computing shortest paths is the Bellman-Ford algorithm [46,127,198]. It uses no priority queue. Instead, it works in rounds, each scanning all vertices whose distance labels have improved. A simple FIFO queue can be used to keep track of vertices to scan next. It is a *label-correcting* algorithm, since each vertex may be scanned multiple times. Although it runs in $\mathcal{O}(|V| |A|)$ time in the worst case, it is often much faster, making it competitive with Dijkstra's algorithm in some scenarios. In addition, it works on graphs with negative edge weights.

Finally, the Floyd-Warshall algorithm [126] computes distances between *all* pairs of vertices in $\Theta(|V|^3)$ time. For sufficiently dense graphs, this is faster than $|V|$ calls to Dijkstra's algorithm.

2.2 Goal-Directed Techniques

Dijkstra's algorithm scans all vertices with distances smaller than $\mathrm{dist}(s, t)$. Goal-directed techniques, in contrast, aim to "guide" the search toward the target by avoiding the scans of vertices that are not in the direction of t. They either exploit the (geometric) embedding of the network or properties of the graph itself, such as the structure of shortest path trees toward (compact) regions of the graph.

A* Search. A classic goal-directed shortest path algorithm is A* search [156]. It uses a potential function $\pi\colon V \to \mathbb{R}$ on the vertices, which is a *lower bound* on the distance $\mathrm{dist}(u,t)$ from u to t. It then runs a modified version of Dijkstra's algorithm in which the priority of a vertex u is set to $\mathrm{dist}(s,u) + \pi(u)$. This causes vertices that are closer to the target t to be scanned earlier during the algorithm. See Fig. 1. In particular, if π were an *exact* lower bound ($\pi(u) = \mathrm{dist}(u,t)$), *only* vertices along shortest s–t paths would be scanned. More vertices may be visited in general but, as long as the potential function is *feasible* (i.e., if $\ell(v,w) - \pi(v) + \pi(w) \geq 0$ for $(v,w) \in E$), an s–t query can stop with the correct answer as soon as it is about to scan the target vertex t.

The algorithm can be made bidirectional, but some care is required to ensure correctness. A standard approach is to ensure that the forward and backward potential functions are consistent. In particular, one can combine two arbitrary feasible functions π_f and π_r into consistent potentials by using $(\pi_f - \pi_r)/2$ for the forward search and $(\pi_r - \pi_f)/2$ for the backward search [163]. Another approach, which leads to similar results in practice, is to change the stopping criterion instead of making the two functions consistent [148,166,216,220].

In road networks with travel time metric, one can use the geographical distance [217,237] between u and t divided by the maximum travel speed (that occurs in the network) as the potential function. Unfortunately, the corresponding bounds are poor, and the performance gain is small or non-existent [148]. In practice, the algorithm can be accelerated using more aggressive bounds (for example, a smaller denominator), but correctness is no longer guaranteed. In practice, even when minimizing travel distances in road networks, A* with geographical distance bound performs poorly compared to other modern methods.

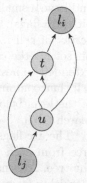

Fig. 2. Triangle inequalities for ALT.

One can obtain much better lower bounds (and preserve correctness) with the *ALT* (*A*, landmarks, and triangle inequality*) algorithm [148]. During a preprocessing phase, it picks a small set $L \subseteq V$ of *landmarks* and stores the distances between them and all vertices in the graph. During an s–t query, it uses triangle inequalities involving the landmarks to compute a valid lower bound on $\mathrm{dist}(u,t)$ for any vertex u. More precisely, for any landmark l_i, both $\mathrm{dist}(u,t) \geq \mathrm{dist}(u,l_i) - \mathrm{dist}(t,l_i)$ and $\mathrm{dist}(u,t) \geq \mathrm{dist}(l_i,t) - \mathrm{dist}(l_i,u)$ hold. If several landmarks are available, one can take the maximum overall bound. See Fig. 2 for an illustration. The corresponding potential function is feasible [148].

The quality of the lower bounds (and thus query performance) depends on which vertices are chosen as landmarks during preprocessing. In road networks, picking well-spaced landmarks close to the boundary of the graph leads to the best results, with acceptable query times on average [112,150]. For a small (but noticeable) fraction of the queries, however, speedups relative to bidirectional Dijkstra are minor.

Geometric Containers. Another goal-directed method is *Geometric Containers.* It precomputes, for each arc $a = (u, v) \in A$, an arc label $L(a)$ that encodes the set V_a of vertices to which a shortest path from u begins with the arc a. Instead of storing V_a explicitly, $L(a)$ approximates this set by using geometric information (i.e., the coordinates) of the vertices in V_a. During a query, if the target vertex t is not in $L(a)$, the search can safely be pruned at a. Schulz et al. [235] approximate the set V_a by an angular sector (centered at u) that covers all vertices in V_a. Wagner et al. [251] consider other geometric containers, such as ellipses and the convex hull, and conclude that bounding boxes perform consistently well. For graphs with no geometric information, one can use graph layout algorithms and then create the containers [55,250]. A disadvantage of Geometric Containers is that its preprocessing essentially requires an all-pairs shortest path computation, which is costly.

Arc Flags. The *Arc Flags* approach [157,178] is somewhat similar to Geometric Containers, but does not use geometry. During preprocessing, it partitions the graph into K cells that are roughly *balanced* (have similar number of vertices) and have a small number of boundary vertices. Each arc maintains a vector of K bits (arc flags), where the i-th bit is set if the arc lies on a shortest path to some vertex of cell i. The search algorithm then prunes arcs which do not have the bit set for the cell containing t. For better query performance, arc flags can be extended to nested multilevel partitions [197]. Whenever the search reaches the cell that contains t, it starts evaluating arc flags with respect to the (finer) cells of the level below. This approach works best in combination with bidirectional search [157].

The arc flags for a cell i are computed by growing a backward shortest path tree from each boundary vertex (of cell i), setting the i-th flag for all arcs of the tree. Alternatively, one can compute arc flags by running a label-correcting algorithm from all boundary vertices simultaneously [157]. To reduce preprocessing space, one can use a compression scheme that flips some flags from zero to one [58], which preserves correctness. As Sect. 3 will show, Arc Flags currently have the fastest query times among purely goal-directed methods for road networks. Although high preprocessing times (of several hours) have long been a drawback of Arc Flags, the recent PHAST algorithm (cf. Sect. 2.7) can make this method more competitive with other techniques [75].

Precomputed Cluster Distances. Another goal-directed technique is *Precomputed Cluster Distances* (PCD) [188]. Like Arc Flags, it is based on a (preferably balanced) partition $\mathcal{C} = (C_1, \ldots, C_K)$ with K cells (or clusters). The preprocessing algorithm computes the shortest path distances between all pairs of cells.

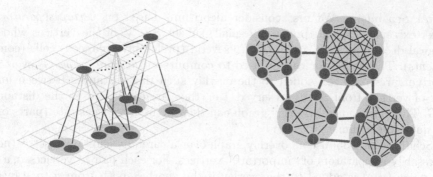

Fig. 3. Left: Multilevel overlay graph with two levels. The dots depict separator vertices in the lower and upper level. Right: Overlay graph constructed from arc separators. Each cell contains a full clique between its boundary vertices, and cut arcs are thicker.

The query algorithm is a pruned version of Dijkstra's algorithm. For any vertex u visited by the search, a valid lower bound on its distance to the target is $\text{dist}(s, u) + \text{dist}(C(u), C(t)) + \text{dist}(v, t)$, where $C(u)$ is the cell containing u and v is the boundary vertex of $C(t)$ that is closest to t. If this bound exceeds the best current upper bound on $\text{dist}(s, t)$, the search is pruned. For road networks, PCD has similar query times to ALT, but requires less space.

Compressed Path Databases. The Compressed Path Databases (CPD) [52,53] method implicitly stores all-pairs shortest path information so that shortest paths can be quickly retrieved during queries. Each vertex $u \in V$ maintains a label $L(u)$ that stores the *first move* (the arc incident to u) of the shortest path toward *every* other vertex v of the graph. A query from s simply scans $L(u)$ for t, finding the first arc (s, u) of the shortest path (to t); it then recurses on u until it reaches t. Explicitly storing the first arc of every shortest path (in $\Theta(|V|^2)$ space) would be prohibitive. Instead, Botea and Harabor [53] propose a lossless data compression scheme that groups vertices that share the same first move (out of u) into nonoverlapping geometric rectangles, which are then stored with u. Further optimizations include storing the most frequent first move as a default and using more sophisticated compression techniques. This leads to fast queries, but space consumption can be quite large; the method is thus dominated by other approaches. CPD can be seen as an evolution of the *Spatially Induced Linkage Cognizance* (SILC) algorithm [228], and both can be seen as stronger versions of Geometric Containers.

2.3 Separator-Based Techniques

Planar graphs have small (and efficiently-computable) separators [181]. Although road networks are not planar (think of tunnels or overpasses), they have been observed to have small separators as well [79,123,227]. This fact is exploited by the methods in this section.

Vertex Separators. We first consider algorithms based on *vertex separators.* A vertex separator is a (preferably small) subset $S \subset V$ of the vertices whose removal decomposes the graph G into several (preferably balanced) cells (components). This separator can be used to compute an *overlay graph* G' over S. Shortcut arcs [249] are added to the overlay such that distances between *any* pair of vertices from S are preserved, i.e., they are equivalent to the distance in G. The much smaller overlay graph can then be used to accelerate (parts of) the query algorithm.

Schulz et al. [235] use an overlay graph over a carefully chosen subset S (not necessarily a separator) of "important" vertices. For each pair of vertices $u, v \in S$, an arc (u, v) is added to the overlay if the shortest path from u to v in G does not contain any other vertex w from S. This approach can be further extended [160, 236] to multilevel hierarchies. In addition to arcs between separator vertices of the same level, the overlay contains, for each cell on level i, arcs between the confining level i separator vertices and the interior level $(i-1)$ separator vertices. See Fig. 3 for an illustration.

Other variants of this approach offer different trade-offs by adding many more shortcuts to the graph during preprocessing, sometimes across different levels [151, 164]. In particular *High-Performance Multilevel Routing* (HPML) [83] substantially reduces query times but significantly increases the total space usage and preprocessing time. A similar approach, based on path separators for planar graphs, was proposed by Thorup [245] and implemented by Muller and Zachariasen [205]. It works reasonably well to find approximate shortest paths on undirected, planarized versions of road networks.

Arc Separators. The second class of algorithms we consider uses *arc separators* to build the overlay graphs. In a first step, one computes a partition $\mathcal{C} = (C_1, \ldots, C_k)$ of the vertices into balanced cells while attempting to minimize the number of cut arcs (which connect boundary vertices of different cells). Shortcuts are then added to preserve the distances between the boundary vertices within each cell.

An early version of this approach is the *Hierarchical MulTi* (HiTi) method [165]. It builds an overlay graph containing all boundary vertices and all cut arcs. In addition, for each pair u, v of boundary vertices in C_i, HiTi adds to the overlay a shortcut (u, v) representing the shortest path from u to v in G restricted to C_i. The query algorithm then (implicitly) runs Dijkstra's algorithm on the subgraph induced by the cells containing s and t plus the overlay. This approach can be extended to use nested multilevel partitions. HiTi has only been tested on grid graphs [165], leading to modest speedups. See also Fig. 3.

The recent *Customizable Route Planning* (CRP) [76, 78] algorithm uses a similar approach, but is specifically engineered to meet the requirements of real-world systems operating on road networks. In particular, it can handle turn costs and is optimized for fast updates of the cost function (metric). Moreover, it uses PUNCH [79], a graph partitioning algorithm tailored to road networks. Finally, CRP splits preprocessing in two phases: metric-independent preprocessing and customization. The first phase computes, besides the multilevel partition, the topology of the overlays, which are represented as matrices in contiguous memory for efficiency. Note that the partition does not depend on the cost function. The second phase (which takes the cost function as input) computes the costs of the clique arcs by processing the cells in bottom-up fashion and in parallel. To process a cell, it suffices to run Dijkstra's algorithm from each boundary vertex, but the second phase is even faster using the Bellman-Ford algorithm paired with (metric-independent) contraction [100] (cf. Sect. 2.4), at the cost of increased space usage. Further acceleration is possible using GPUs [87]. Queries are bidirectional searches in the overlay graph, as in HiTi.

2.4   Hierarchical Techniques

Hierarchical methods aim to exploit the inherent hierarchy of road networks. Sufficiently long shortest paths eventually converge to a small arterial network of important roads, such as highways. Intuitively, once the query algorithm is far from the source and target, it suffices to only scan vertices of this subnetwork. In fact, using input-defined road categories in this way is a popular heuristic [115,158], though there is no guarantee that it will find exact shortest paths. Fu et al. [130] give an overview of early approaches using this technique. Since the algorithms we discuss must find exact shortest paths, their correctness must not rely on unverifiable properties such as input classifications. Instead, they use the preprocessing phase to compute the importance of vertices or arcs according to the actual shortest path structure.

Contraction Hierarchies. An important approach to exploiting the hierarchy is to use *shortcuts*. Intuitively, one would like to augment G with shortcuts that could be used by long-distance queries to skip over "unimportant" vertices.

The *Contraction Hierarchies* (CH) algorithm, proposed by Geisberger et al. [142], implements this idea by repeatedly executing a *vertex contraction* operation. To contract a vertex v, it is (temporarily) removed from G, and a shortcut is created between each pair u, w of neighboring vertices if the shortest path from u to w is unique and contains v. During preprocessing, CH (heuristically) orders the vertices by "importance" and contracts them from least to most important.

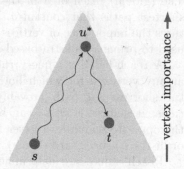

Fig. 4. Illustrating a Contraction Hierarchies query.

The query stage runs a bidirectional search from s and t on G augmented by the shortcuts computed during preprocessing, but only visits arcs leading to vertices of higher ranks (importance). See Fig. 4 for an illustration. Let $d_s(u)$ and $d_t(u)$ be the corresponding distance labels obtained by these *upward* searches (set to ∞ for vertices that are not visited). It is easy to show that $d_s(u) \geq \mathrm{dist}(s, u)$ and $d_t(u) \geq \mathrm{dist}(u, t)$; equality is not guaranteed due to pruning. Nevertheless, Geisberger et al. [142] prove that the highest-ranked vertex u^* on the original s–t path will be visited by both searches, and that both its labels will be exact, i. e., $d_s(u^*) = \text{dist}(s, u^*)$ and $d_t(u^*) = \text{dist}(u^*, t)$. Therefore, among all vertices u visited by both searches, the one minimizing $d_s(u) + d_t(u)$ represents the shortest path. Note that, since u^* is not necessarily the first vertex that is scanned by both searches, they cannot stop as soon as they meet.

Query times depend on the vertex order. During preprocessing, the vertex order is usually determined online and bottom-up. The overall (heuristic) goal is to minimize the number of edges added during preprocessing. One typically selects the vertex to be contracted next by considering a combination of several factors, including the net number of shortcuts added and the number of nearby vertices already contracted [142, 168]. Better vertex orders can be obtained by combining the bottom-up algorithm with (more expensive) top-down offline algorithms that explicitly classify vertices hitting many shortest paths as more important [5, 77]. Since road networks have very small separators [79], one can use nested dissection to obtain reasonably good orders that work for any length function [100, 107]. Approximate CH has been considered as a way to accommodate networks with less inherent hierarchy [143].

CH is actually a successor of Highway Hierarchies [225] and Highway Node Routing [234], which are based on similar ideas. CH is not only faster, but also conceptually simpler. This simplicity has made it quite versatile, serving as a building block not only for other point-to-point algorithms [4, 15, 40, 100], but also for extended queries (cf. Sect. 2.7) and applications (cf. Sect. 3.2).

Reach. An earlier hierarchical approach is *Reach* [154]. Reach is a centrality measure on vertices. Let P be a shortest s–t path that contains vertex u. The reach $r(u, P)$ of u with respect to P is defined as $\min\{\text{dist}(s, u), \text{dist}(u, t)\}$. The (global) reach of u in the graph G is the maximum reach of u over *all* shortest paths that contain u. Like other centrality measures [54], reach captures the importance of vertices in the graph, with the advantage that it can be used to prune a Dijkstra-based search.

A reach-based s–t query runs Dijkstra's algorithm, but prunes the search at any vertex u for which both $\text{dist}(s, u) > r(u)$ and $\text{dist}(u, t) > r(u)$ hold; the shortest s–t path provably does not contain u. To check these conditions, it suffices [149] to run bidirectional searches, each using the radius of the opposite search as a lower bound on $\text{dist}(u, t)$ (during the forward search) or $\text{dist}(s, u)$ (backward search).

Reach values are determined during the preprocessing stage. Computing exact reaches requires computing shortest paths for all pairs of vertices, which is too expensive on large road networks. But the query is still correct if $r(u)$ represents only an upper bound on the reach of u. Gutman [154] has shown that such bounds can be obtained faster by computing partial shortest path trees. Goldberg et al. [149] have shown that adding shortcuts to the graph effectively reduces the reaches of most vertices, drastically speeding up both queries and preprocessing and making the algorithm practical for continent-sized networks.

2.5   Bounded-Hop Techniques

The idea behind bounded-hop techniques is to precompute distances between pairs of vertices, implicitly adding "virtual shortcuts" to the graph. Queries can then return the length of a virtual path with very few hops. Furthermore, they use only the precomputed distances between pairs of vertices, and not the input graph. A naïve approach is to use single-hop paths, i. e., precompute the distances among *all* pairs of vertices $u, v \in V$. A single table lookup then suffices to retrieve the shortest distance. While the recent PHAST algorithm [75] has made precomputing all-pairs shortest paths feasible, storing all $\Theta(|V|^2)$ distances is prohibitive already for medium-sized road networks. As we will see in this section, considering paths with slightly more hops (two or three) leads to algorithms with much more reasonable trade-offs.

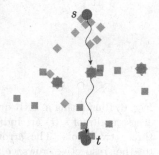

Fig. 5. Illustrating hub labels of vertices s (diamonds) and t (squares).

Labeling Algorithms. We first consider *labeling algorithms* [215]. During preprocessing, a *label* $L(u)$ is computed for each vertex u of the graph, such that, for any pair u, v of vertices, the distance $\text{dist}(u, v)$ can be determined by only looking at the labels $L(u)$ and $L(v)$. A natural special case of this approach is *Hub Labeling* (HL) [64,135], in which the label $L(u)$ associated with vertex u consists of a set of vertices (the *hubs* of u), together with their distances from u. These labels are chosen such that they obey the *cover property*: for any pair (s, t) of vertices, $L(s) \cap L(t)$ must contain at least one vertex on the shortest s–t path. Then, the distance $\text{dist}(s, t)$ can be determined in linear (in the label size) time by evaluating $\text{dist}(s, t) = \min\{\text{dist}(s, u) + \text{dist}(u, t) \mid u \in L(s) \text{ and } u \in L(t)\}$. See Fig. 5 for an illustration. For directed graphs, the label associated with u is actually split in two: the forward label $L_f(u)$ has distances from u *to* the hubs, while the backward label $L_b(u)$ has distances *from* the hubs to u; the shortest s–t path has a hub in $L_f(s) \cap L_b(t)$.

Although the required average label size can be $\Theta(|V|)$ in general [135], it can be significantly smaller for some graph classes. For road networks, Abraham et al. [4] have shown that one can obtain good results by defining the label of vertex u as the (upward) search space of a CH query from u (with suboptimal entries removed). In general, any vertex ordering fully defines a labeling [5], and an ordering can be converted into the corresponding labeling efficiently [5,12]. The CH-induced order works well for road networks. For even smaller labels, one can pick the most important vertices greedily, based on how many shortest paths they hit [5]. A sampling version of this greedy approach works efficiently for a wide range of graph classes [77].

Note that, if labels are sorted by hub ID, a query consists of a linear sweep over two arrays, as in mergesort. Not only is this approach very simple, but it also has an almost perfect locality of access. With careful engineering, one does not even have to look at all the hubs in a label [4]. As a result, HL has the fastest known queries for road networks, taking roughly the time needed for five accesses to main memory (see Sect. 3.1). One drawback is space usage, which, although not prohibitive, is significantly higher than for competing methods. By combining common substructures that appear in multiple labels, *Hub Label Compression* (HLC) [82] (see also [77]) reduces space usage by an order of magnitude, at the expense of higher query times.

Distance Table

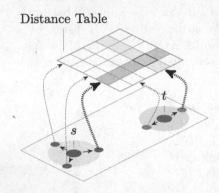

Fig. 6. Illustrating a TNR query. The access nodes of s (t) are indicated by three (two) dots. The arrows point to the respective rows/columns of the distance table. The highlighted entries correspond to the access nodes which minimize the combined s–t distance.

Transit Node Routing. The *Transit Node Routing* (TNR) [15, 28, 30, 224] technique uses distance tables on a subset of the vertices. During preprocessing, it selects a small set $T \subseteq V$ of *transit nodes* and computes all pairwise distances between them. From those, it computes, for each vertex $u \in V \setminus T$, a relevant set of *access nodes* $A(u) \subseteq T$. A transit node $v \in T$ is an access node of u if there is a shortest path P from u in G such that v is the first transit node contained in P. In addition to the vertex itself, preprocessing also stores the distances between u and its access nodes.

An s–t query uses the distance table to select the path that minimizes the combined s–$a(s)$–$a(t)$–t distance, where $a(s) \in A(s)$ and $a(t) \in A(t)$ are access nodes. Note that the result is incorrect if the shortest path does not contain a vertex from T.

To account for such cases, a *locality filter* decides whether the query might be local (i. e., does not contain a vertex from T). In that case, a fallback shortest path algorithm (typically CH) is run to compute the correct distance. Note that TNR is still correct even if the locality filter occasionally misclassifies a global query as local. See Fig. 6 for an illustration of a TNR query. Interestingly, global TNR queries (which use the distance tables) tend to be faster than local ones (which perform graph searches). To accelerate local queries, TNR can be extended to multiple (hierarchical) layers of transit (and access) nodes [28, 224].

The choice of the transit node set is crucial to the performance of the algorithm. A natural approach is to select vertex separators or boundary vertices of arc separators as transit nodes. In particular, using grid-based separators yields natural locality filters and works well enough in practice for road networks [28]. (Although an optimized preprocessing routine for this grid-based approach was later shown to have a flaw that could potentially result in suboptimal queries [257], the version with slower preprocessing reported in [28] is correct and achieves the same query times.)

For better performance [3, 15, 142, 224], one can pick as transit nodes vertices that are classified as important by a hierarchical speedup technique (such as CH).

Locality filters are less straightforward in such cases: although one can still use geographical distances [142, 224], a graph-based approach considering the Voronoi regions [189] induced by transit nodes tends to be significantly more accurate [15]. A theoretically justified TNR variant [3] also picks important vertices as transit nodes and has a natural graph-based locality filter, but is impractical for large networks.

Pruned Highway Labeling. The Pruned Highway Labeling (PHL) [11] algorithm can be seen as a hybrid between pure labeling and transit nodes. Its preprocessing routine decomposes the input into disjoint shortest paths, then computes a label for each vertex v containing the distance from v to vertices in a small subset of such paths. The labels are such that any shortest s–t path can be expressed as s–u–w–t, where u–w is a subpath of a path P that belongs to the labels of s and t. Queries are thus similar to HL, finding the lowest-cost intersecting path. For efficient preprocessing, the algorithm uses the pruned labeling technique [12]. Although this method has some similarity with Thorup's distance oracle for planar graphs [245], it does not require planarity. PHL has only been evaluated on undirected graphs, however.

2.6   Combinations

Since the individual techniques described so far exploit different graph properties, they can often be combined for additional speedups. This section describes such hybrid algorithms. In particular, early results [161, 235] considered the combination of Geometric Containers, multilevel overlay graphs, and (Euclidean-based) A* on transportation networks, resulting in speedups of one or two orders of magnitude over Dijkstra's algorithm.

More recent studies have focused on combining hierarchical methods (such as CH or Reach) with fast goal-directed techniques (such as ALT or Arc Flags). For instance, the *REAL* algorithm combines Reach and ALT [149]. A basic combination is straightforward: one simply runs an ALT query with additional pruning by reach (using the ALT lower bounds themselves for reach evaluations). A more sophisticated variant uses *reach-aware landmarks*: landmarks and their distances are only precomputed for vertices with high reach values. This saves space (only a small fraction of the graph needs to store landmark distances), but requires two-stage queries (goal direction is only used when the search is far enough from both source and target).

A similar space-saving approach is used by *Core-ALT* [40, 88]. It first computes an overlay graph for the *core graph*, a (small) subset (e.g., 1 %) of vertices (which remain after "unimportant" ones are contracted), then computes landmarks for the core vertices only. Queries then work in two stages: first plain bidirectional search, then ALT is applied when the search is restricted to the core. The (earlier) *HH** approach [95] is similar, but uses Highway Hierarchies [225] to determine the core.

Another approach with two-phase queries is *ReachFlags* [40]. During pre-processing, it first computes (approximate) reach values for all vertices in G, then extracts the subgraph H induced by all vertices whose reach value exceeds a certain threshold. Arc flags are then only computed for H, to be used in the second phase of the query.

The *SHARC* algorithm [39] combines the computation of shortcuts with multilevel arc flags. The preprocessing algorithm first determines a partition of the graph and then computes shortcuts and arc flags in turn. Shortcuts are obtained by contracting unimportant vertices with the restriction that shortcuts never span different cells of the partition. The algorithm then computes arc flags such that, for each cell C, the query uses a shortcut arc if and only if the target vertex is not in C. Space usage can be reduced with various compression techniques [58]. Note that SHARC is unidirectional and hierarchical: arc flags not only guide the search toward the target, but also vertically across the hierarchy. This is useful when the backward search is not well defined, as in time-dependent route planning (discussed in Sect. 2.7).

Combining CH with Arc Flags results in the *CHASE* algorithm [40]. During preprocessing, a regular contraction hierarchy is computed and the search graph that includes all shortcuts is assembled. The algorithm then extracts the subgraph H induced by the top k vertices according to the contraction order. Bidirectional arc flags (and the partition) are finally computed on the restricted subgraph H. Queries then run in two phases. Since computing arc flags was somewhat slow, k was originally set to a small fraction (about 5 %) of the total number $|V|$ of vertices [40]. More recently, Delling et al. showed that PHAST (see Sect. 2.7) can compute arc flags fast enough to allow k to be set to $|V|$, making CHASE queries much simpler (single-pass), as well as faster [75].

Finally, Bauer et al. [40] combine Transit Node Routing with Arc Flags to obtain the TNR+AF algorithm. Recall that the bottleneck of the TNR query is performing the table lookups between pairs of access nodes from $A(s)$ and $A(t)$. To reduce the number of lookups, TNR+AF's preprocessing decomposes the set of transit nodes T into k cells. For each vertex s and access node $u \in A(s)$, it stores a k-bit vector, with bit i indicating whether there exists a shortest path from s to cell i through u. A query then only considers the access nodes from s that have their bits set with respect to the cells of $A(t)$. A similar pruning is done at the target.

2.7   Extensions

In various applications, one is often interested in more than just the length of the shortest path between two points in a static network. Most importantly, one should also be able to retrieve the shortest path itself. Moreover, many of the techniques considered so far can be adapted to compute batched shortest paths (such as distance tables), to more realistic scenarios (such as dynamic networks), or to deal with multiple objective functions. In the following, we briefly discuss each of these extensions.

Path Retrieval. Our descriptions so far have focused on finding only the *length* of the shortest path. The algorithms we described can easily be augmented to provide the actual list of edges or vertices on the path. For techniques that do not use shortcuts (such as Dijkstra's algorithm, A* search, or Arc Flags), one can simply maintain a parent pointer for each vertex v, updating it whenever the distance label of v changes. When shortcuts are present (such as in CH, SHARC, or CRP), this approach gives only a *compact* representation of the shortest path (in terms of shortcuts). The shortcuts then need to be unpacked. If each shortcut is the concatenation of two other arcs (or shortcuts), as in CH, storing the middle vertex [142] of each shortcut allows for an efficient (linear-time) recursive unpacking of all shortcuts on the output path. If shortcuts are built from multiple arcs (as for CRP or SHARC), one can either store the entire sequence for each shortcut [225] or run a local (bidirectional) Dijkstra search from its endpoints [78]. These two techniques can be used for bounded-hop algorithms as well.

Batched Shortest Paths. Some applications require computing multiple paths at once. For example, advanced logistics applications may need to compute all distances between a source set S and a target set T. This can be trivially done with $|S| \cdot |T|$ point-to-point shortest-path computations. Using a hierarchical speedup technique (such as CH), this can be done in time comparable to $\mathcal{O}(|S| + |T|)$ point-to-point queries in practice, which is much faster. First, one runs a backward upward search from each $t_i \in T$; for each vertex u scanned during the search from t_i, one stores its distance label $d_{t_i}(u)$ in a bucket $\beta(u)$. Then, one runs a forward upward search from each $s_j \in S$. Whenever such a search scans a vertex v with a non-empty bucket, one searches the bucket and checks whether $d_{s_j}(v) + d_{t_i}(v)$ improves the best distance seen so far between s_j and t_i. This *bucket-based approach* was introduced for Highway Hierarchies [172], but can be used with any other hierarchical speedup technique (such as CH) and even with hub labels [81]. When the bucket-based approach is combined with a separator-based technique (such as CRP), it is enough to keep buckets only for the boundary vertices [99]. Note that this approach can be used to compute one-to-many or many-to-many distances.

Some applications require one-to-all computations, i. e., finding the distances from a source vertex s to all other vertices in the graph. For this problem, Dijkstra's algorithm is optimal in the sense that it visits each edge exactly once, and hence runs in essentially linear time [147]. However, Dijkstra's algorithm has bad locality and is hard to parallelize, especially for sparse graphs [186,193]. PHAST [75] builds on CH to improve this. The idea is to split the search in two phases. The first is a forward upward search from s, and the second runs a linear scan over the shortcut-enriched graph, with distance values propagated from more to less important vertices. Since the instruction flow of the second phase is (almost) independent of the source, it can be engineered to exploit parallelism and improve locality. In road networks, PHAST can be more than an order of magnitude faster than Dijkstra's algorithm, even if run sequentially, and can be further accelerated using multiple cores and even GPUs. This approach can also be extended to the *one-to-many problem*, i. e., computing distances from a source to a subset of predefined targets [81]. Similar techniques can also be applied with graph separators (instead of CH), yielding comparable query times but with faster (metric-dependent) preprocessing [113].

Dynamic Networks. Transportation networks tend to be dynamic, with unpredictable delays, traffic, or closures. If one assumes that the modified network is stable for the foreseeable future, the obvious approach for speedup techniques to deal with this is to rerun the preprocessing algorithm. Although this ensures queries are as fast as in the static scenario, it can be quite costly. As a result, four other approaches have been considered.

It is often possible to just "repair" the preprocessed data instead of rebuilding it from scratch. This approach has been tried for various techniques, including Geometric Containers [251], ALT [96], Arc Flags [66], and CH [142,234], with varying degrees of success. For CH, for example, one must keep track of dependencies between shortcuts, partially rerunning the contraction as needed. Changes that affect less important vertices can be dealt with faster.

Another approach is to adapt the query algorithms to work around the "wrong" parts of the preprocessing phase. In particular, ALT is resilient to increases in arc costs (due to traffic, for example): queries remain correct with the original preprocessing, though query times may increase [96]. Less trivially, CH queries can also be modified to deal with dynamic changes to the network [142,234] by allowing the search to bypass affected shortcuts by going "down" the hierarchy. This is useful when queries are infrequent relative to updates.

A third approach is to make the preprocessing stage completely metric-independent, shifting all metric-dependent work to the query phase. Funke et al. [131] generalize the multilevel overlay graph approach to encode *all* k-hop paths (for small k) in an overlay graph. Under the assumption that edge costs are defined by a small number of physical parameters (as in simplified road networks) this allows setting the edge costs at query time, though queries become significantly slower.

For more practical queries, the fourth approach splits the preprocessing phase into metric-independent and metric-dependent stages. The metric-independent phase takes as input only the network topology, which is fairly stable. When edge costs change (which happens often), only the (much cheaper) metric-dependent stage must be rerun, partially or in full. This concept can again be used for various techniques, with ALT, CH, and CRP being the most prominent. For ALT, one can keep the landmarks, and just recompute the distances to them [96,112]. For CH, one can keep the ordering, and just rerun contraction [107,142]. For CRP, one can keep the partitioning and the overlay topology, and just recompute the shortcut lengths using a combination of contraction and graph searches [78]. Since the contraction is metric-independent, one can precompute and store the sequence of contraction operations and reexecute them efficiently whenever edge lengths change [78,87]. The same approach can be used for CH with metric-independent orders [107].

Time-Dependence. In real transportation networks, the best route often depends on the departure time in a predictable way [102]. For example, certain roads are consistently congested during rush hours, and certain buses or trains run with different frequencies during the day. When one is interested in the earliest possible arrival given a specified departure time (or, symmetrically, the latest departure), one can model this as the *time-dependent* shortest path problem, which assigns travel time functions to (some of) the edges, representing how long it takes to traverse them at each time of the day. Dijkstra's algorithm still works [65] as long as later departures cannot lead to earlier arrivals; this *non-overtaking* property is often called first-in first-out (FIFO). Although one must deal with functions instead of scalars, the theoretical running time of Dijkstra-based algorithms can still be bounded [71,128]. Moreover, many of the techniques described so far work in this scenario, including bidirectional ALT [88,207], CH [32], or SHARC [72]. Recently, Kontogiannis and Zaroliagis [175] have introduced a theoretical (approximate) distance oracle with sublinear running time. Other scenarios (besides FIFO with no waiting at vertices) have been studied [69,70,208,209], but they are less relevant for transportation networks.

There are some challenges, however. In particular, bidirectional search becomes more complicated (since the time of arrival is not known), requiring changes to the backward search [32,207]. Another challenge is that shortcuts become more space-consuming (they must model a more complicated travel time function), motivating compression techniques that do not sacrifice correctness, as demonstrated for SHARC [58] or CH [32]. Batched shortest paths can be computed in such networks efficiently as well [141].

Time-dependent networks motivate some elaborate (but still natural) queries, such as finding the best departure time in order to minimize the total time in transit. Such queries can be dealt with by *range searches*, which compute the travel time function between two points. There exist Dijkstra-based algorithms [71] for this problem, and most speedup techniques can be adapted to deal with this as well [32,72].

Unfortunately, even a slight deviation from the travel time model, where total cost is a linear combination of travel time and a constant cost offset, makes the problem NP-hard [9,33]. However, a heuristic adaptation of time-dependent CH shows negligible errors in practice [33].

Multiple Objective Functions. Another natural extension is to consider multiple cost functions. For example, certain vehicle types cannot use all segments of the transportation network. One can either adapt the preprocessing such that these *edge restrictions* can be applied during query time [140], or perform a metric update for each vehicle type.

Also, the search request can be more flexible. For example, one may be willing to take a more scenic route even if the trip is slightly longer. This can be dealt with by performing a multicriteria search. In such a search, two paths are incomparable if neither is better than the other in all criteria. The goal is to find a *Pareto set*, i.e., a maximum set of incomparable paths. Such sets of shortest paths can be computed by extensions of Dijkstra's algorithm; see [117] for a survey on multicriteria combinatorial optimization. More specifically, the *Multicriteria Label-Setting* (MLS) algorithm [155,187,196,243] extends Dijkstra's algorithm by keeping, for each vertex, a *bag* of nondominated labels. Each label is represented as a tuple, with one entry per optimization criterion. The priority queue maintains labels instead of vertices, typically ordered lexicographically. In each iteration, it extracts the minimum label L and scans the incident arcs $a = (u, v)$ of the vertex u associated with L. It does so by adding the cost of a to L and then merging L into the bag of v, eliminating possibly dominated labels on the fly. In contrast, the *Multi-Label-Correcting* (MLC) algorithm [68,98] considers the whole bag of nondominated labels associated with u at once when scanning the vertex u. Hence, individual labels of u may be scanned multiple times during one execution of the algorithm.

Both MLS and MLC are fast enough as long as the Pareto sets are small [109,204]. Unfortunately, Pareto sets may contain exponentially many solutions, even for the restricted case of two optimization criteria [155], which makes it hard to achieve large speedups [47,97]. To reduce the size of Pareto sets, one can relax domination. In particular, $(1 + \varepsilon)$-Pareto sets have provable polynomial size [212] and can be computed efficiently [182,246,253]. Moreover, large Pareto sets open up a potential for parallelization that is not present for a single objective function [124,222].

A reasonable alternative [138] to multicriteria search is to optimize a linear combination $\alpha c_1 + (1 - \alpha)c_2$ of two criteria (c_1, c_2), with the parameter α set at query time. Moreover, it is possible to efficiently compute the values of α where the path actually changes. Funke and Storandt [133] show that CH can handle such functions with polynomial preprocessing effort, even with more than two criteria.

2.8 Theoretical Results

Most of the algorithms mentioned so far were developed with practical performance in mind. Almost all methods we surveyed are exact: they provably find the exact shortest path. Their performance (in terms of both preprocessing and queries), however, varies significantly with the input graph. Most algorithms work well for real road networks, but are hardly faster than Dijkstra's algorithm on some other graph classes. This section discusses theoretical work that helps understand why the algorithms perform well and what their limitations are.

Most of the algorithms considered have some degree of freedom during preprocessing (such as which partition, which vertex order, or which landmarks to choose). An obvious question is whether one could efficiently determine the best such choices for a particular input so as to minimize the query search space (a natural proxy for query times). Bauer et al. [36] have determined that finding optimal landmarks for ALT is NP-hard. The same holds for Arc Flags (with respect to the partition), SHARC (with respect to the shortcuts), Multilevel Overlay Graphs (with respect to the separator), Contraction Hierarchies (with respect to the vertex order), and Hub Labels (with respect to the hubs) [252]. In fact, minimizing the number of shortcuts for CH is APX-hard [36,194]. For SHARC, however, a greedy factor-k approximation algorithm exists [38]. Deciding which k shortcuts (for fixed k) to add to a graph in order to minimize the SHARC search space is also NP-hard [38]. Bauer et al. [35] also analyze the preprocessing of Arc Flags in more detail and on restricted graph classes, such as paths, trees, and cycles, and show that finding an optimal partition is NP-hard even for binary trees.

Besides complexity, theoretical performance bounds for query algorithms, which aim to explain their excellent practical performance, have also been considered. Proving better running time bounds than those of Dijkstra's algorithm is unlikely for general graphs; in fact, there are inputs for which most algorithms are ineffective. That said, one can prove nontrivial bounds for specific graph classes. In particular, various authors [37,194] have independently observed a natural relationship between CH and the notions of filled graphs [214] and elimination trees [232]. For planar graphs, one can use nested dissection [180] to build a CH order leading to $\mathcal{O}(|V|\log|V|)$ shortcuts [37,194]. More generally, for minor-closed graph classes with balanced $\mathcal{O}(\sqrt{|V|})$-separators, the search space is bounded by $\mathcal{O}(\sqrt{|V|})$ [37]. Similarly, on graphs with treewidth k, the search space of CH is bounded by $\mathcal{O}(k\log|V|)$ [37].

Road networks have motivated a large amount of theoretical work on algorithms for planar graphs. In particular, it is known that planar graphs have separators of size $\mathcal{O}(\sqrt{|V|})$ [180,181]. Although road networks are not strictly planar, they do have small separators [79,123], so theoretically efficient algorithms for planar graphs are likely to also perform well in road networks. Sommer [238] surveys several approximate methods with various trade-offs. In practice, the observed performance of most speedup techniques is much better on actual road networks than on arbitrary planar graphs (even grids). A theoretical explanation of this discrepancy thus requires a formalization of some property related to key features of real road networks.

One such graph property is *Highway Dimension*, proposed by Abraham et al. [3] (see also [1,7]). Roughly speaking, a graph has highway dimension h if, at any scale r, one can hit all shortest paths of length at least r by a hitting set S that is *locally sparse*, in the sense that any ball of radius r has at most h elements from S. Based on previous experimental observations [30], the authors [7] conjecture that road networks have small highway dimension. Based on this notion, they establish bounds on the performance of (theoretically justified versions of) various speedup techniques in terms of h and the graph diameter D, assuming the graph is undirected and that edge lengths are integral. More precisely, after running a polynomial-time preprocessing routine, which adds $\mathcal{O}(h\log h\log D)$ shortcuts to G, Reach and CH run in $\mathcal{O}((h\log h\log D)^2)$ time. Moreover, they also show that HL runs in $\mathcal{O}(h\log h\log D)$ time and long-range TNR queries take $\mathcal{O}(h^2)$ time. In addition, Abraham et al. [3] show that a graph with highway dimension h has doubling dimension $\log(h+1)$, and Kleinberg et al. [171] show that landmark-based triangulation yields good bounds for most pairs of vertices of graphs with small doubling dimension. This gives insight into the good performance of ALT in road networks.

The notion of highway dimension is an interesting application of the scientific method. It was originally used to explain the good observed performance of CH, Reach, and TNR, and ended up predicting that HL (which had not been implemented yet) would have good performance in practice.

Generative models for road networks have also been proposed and analyzed. Abraham et al. [3,7] propose a model that captures some of the properties of road networks and generates graphs with provably small highway dimension. Bauer e al. [42] show experimentally that several speedup techniques are indeed effective on graphs generated according to this model, as well as according to a new model based on Voronoi diagrams. Models with a more geometric flavor have been proposed by Eppstein and Goodrich [123] and by Eisenstat [118].

Besides these results, Rice and Tsotras [220] analyze the A* algorithm and obtain bounds on the search space size that depend on the underestimation error of the potential function. Also, maintaining and updating multilevel overlay graphs have been theoretically analyzed in [57]. For Transit Node Routing, Eisner and Funke [120] propose instance-based lower bounds on the size of the transit node set. For labeling algorithms, bounds on the label size for different graph classes are given by Gavoille et al. [135]. Approximation algorithms to compute small labels have also been studied [16,64,80]; although they can find slightly better labels than faster heuristics [5,77], their running time is prohibitive [80].

Because the focus of this work is on algorithm engineering, we refrain from going into more detail about the available theoretical work. Instead, we refer the interested reader to overview articles with a more theoretical emphasis, such as those by Sommer [238], Zwick [262], and Gavoille and Peleg [134].

3   Route Planning in Road Networks

In this section, we experimentally evaluate how the techniques discussed so far perform in road networks. Moreover, we discuss applications of some of the techniques, as well as alternative settings such as databases or mobile devices.

3.1   Experimental Results

Our experimental analysis considers carefully engineered implementations, which is very important when comparing running times. They are written in C++ with custom-built data structures. Graphs are represented as adjacency arrays [190], and priority queues are typically binary heaps, 4-heaps, or multilevel buckets. As most arcs in road networks are bidirectional, state-of-the-art implementations use edge compression [233]: each road segment is stored at both of its endpoints, and each occurrence has two flags indicating whether the segment should be considered as an incoming and/or outgoing arc. This representation is compact and allows efficient iterations over incoming and outgoing arcs.

We give data for two models. The *simplified model* ignores turn restrictions and penalties, while the *realistic model* includes the turn information [255]. There are two common approaches to deal with turns. The *arc-based representation* [59] blows up the graph so that roads become vertices and feasible turns become arcs. In contrast, the *compact representation* [76, 144] keeps intersections as vertices, but with associated *turn tables*. One can save space by sharing turn tables among many vertices, since the number of intersection types in a road network is rather limited. Most speedup techniques can be used as is for the arc-based representation, but may need modification to work on the compact model.

Most experimental studies are restricted to the simplified model. Since some algorithms are more sensitive to how turns are modeled than others, it is hard to extrapolate these results to more realistic networks. We therefore consider experimental results for each model separately.

Simplified Model. An important driving force behind the research on speedup techniques for Dijkstra's algorithm was its application to road networks. A key aspect for the success of this research effort was the availability of continent-sized benchmark instances. The most widely used instance has been the road network of Western Europe from PTV AG, with 18.0 million vertices and 42.5 million directed arcs. Besides ferries (for which the traversal time was given), it has 13 road categories. Category i has been assigned an average speed of $10i$ km/h. This synthetic assignment is consistent with more realistic proprietary data [78, 82]. Another popular (and slightly bigger) instance, representing the TIGER/USA road network, is undirected and misses several important road segments [6]. Although the inputs use the simplified model, they allowed researchers from various groups to run their algorithms on the same instance, comparing their performance. In particular, both instances were tested during the DIMACS Challenge on Shortest Paths [101].

Figure 7 succinctly represents the performance of previously published implementations of various point-to-point algorithms on the Western Europe instance, using travel time as the cost function. For each method, the plot relates its preprocessing and average query times. Queries compute the length of the shortest path (but not its actual list of edges) between sources and targets picked uniformly at random from the full graph. For readability, space consumption (a third important quality measure) is not explicitly represented.[1] We reproduce the numbers reported by Bauer et al. [40] for Reach, HH, HNR, ALT, (bidirectional) Arc Flags, REAL, HH*, SHARC, CALT, CHASE, ReachFlags and TNR+AF. For CHASE and Arc Flags, we also consider variants with quicker PHAST-based preprocessing [75]. In addition, we consider the recent ALT implementation by Efentakis and Pfoser [112]. Moreover, we report results for several variants of TNR [15, 40], Hub Labels [5, 82], HPML [83], Contraction Hierarchies (CH) [142], and Customizable Contraction Hierarchies (CCH) [107]. CRP (and the corresponding PUNCH) figures [78] use a more realistic graph

---

[1] The reader is referred to Sommer [238] for a similar plot (which inspired ours) relating query times to preprocessing space.

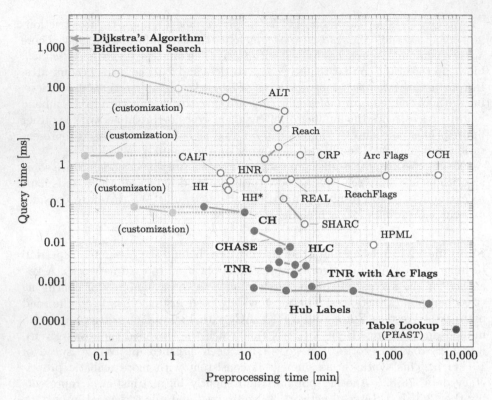

Fig. 7. Preprocessing and average query time performance for algorithms with available experimental data on the road network of Western Europe, using travel times as edge weights. Connecting lines indicate different trade-offs for the same algorithm. The figure is inspired by [238].

model that includes turn costs. For reference, the plot includes unidirectional and bidirectional implementations of Dijkstra's algorithm using a 4-heap. (Note that one can obtain a 20 % improvement when using a multilevel bucket queue [147].) Finally, the table-lookup figure is based on the time of a single memory access in our reference machine and the precomputation time of $|V|$ shortest path trees using PHAST [75]. Note that a machine with more than one petabyte of RAM (as required by this algorithm) would likely have slower memory access times.

Times in the plot are on a single core of an Intel X5680 3.33 GHz CPU, a mainstream server at the time of writing. Several of the algorithms in the plot were originally run on this machine [5,75,78,82]; for the remaining, we divide by the following scaling factors: 2.322 for [40,83], 2.698 for [142], 1.568 for [15], 0.837 for [107], and 0.797 for [112]. These were obtained from a benchmark (developed for this survey) that measures the time of computing several shortest path trees on the publicly available USA road network with travel times [101]. For the machines we did not have access to, we asked the authors to run the benchmark for us [112]. The benchmark is available from http://algo.iti.kit.edu/~pajor/survey/, and we encourage future works to use it as a base to compare (sequential) running times with existing approaches.

The figure shows that there is no best technique. To stress this point, techniques with at least one implementation belonging to the Pareto set (considering preprocessing time, query time, and space usage) are drawn as solid circles; hollow entries are dominated. The Pareto set is quite large, with various methods allowing for a wide range of space-time trade-offs. Moreover, as we shall see when examining more realistic models, these three are not the only important criteria for real-world applications.

Table 1. Performance of various speedup techniques on Western Europe. Column *source* indicates the implementation tested for this survey.

Algorithm	Impl. source	Data structures		Queries	
		Space [GiB]	Time [h:m]	Scanned vertices	Time [μs]
Dijkstra	[75]	0.4	–	9 326 696	2 195 080
Bidir. Dijkstra	[75]	0.4	–	4 914 804	1 205 660
CRP	[78]	0.9	1:00	2 766	1 650
Arc Flags	[75]	0.6	0:20	2 646	408
CH	[78]	0.4	0:05	280	110
CHASE	[75]	0.6	0:30	28	5.76
HLC	[82]	1.8	0:50	–	2.55
TNR	[15]	2.5	0:22	–	2.09
TNR+AF	[40]	5.4	1:24	–	0.70
HL	[82]	18.8	0:37	–	0.56
HL-∞	[5]	17.7	60:00	–	0.25
table lookup	[75]	1 208 358.7	145:30	–	0.06

Table 1 has additional details about the methods in the Pareto set, including two versions of Dijkstra's algorithm, one Dijkstra-based hierarchical technique (CH), three non-graph-based algorithms (TNR, HL, HLC), and two combinations (CHASE and TNR+AF). For reference, the table also includes a goal-directed technique (Arc Flags) and a separator-based algorithm (CRP), even though they are dominated by other methods. All algorithms were rerun for this survey on the reference machine (Intel X5680 3.33 GHz CPU), except those based on TNR, for which we report scaled results. All runs are single-threaded for this experiment, but note that all preprocessing algorithms could be accelerated using multiple cores (and, in some cases, even GPUs) [75,144].

For each method, Table 1 reports the total amount of space required by all data structures (including the graph, if needed, but excluding extra information needed for path unpacking), the total preprocessing time, the number of vertices scanned by an average query (where applicable) and the average query time. Once again, queries consist of pairs of vertices picked uniformly at random. We note that all methods tested can be parametrized (typically within a relatively narrow band) to achieve different trade-offs between query time, preprocessing time, and space. For simplicity, we pick a single "reasonable" set of parameters for each method. The only exception is HL-∞, which achieves the fastest reported query times but whose preprocessing is unreasonably slow.

Observe that algorithms based on any one of the approaches considered in Sect. 2 can answer queries in milliseconds or less. Separator-based (CRP), hierarchical (CH), and goal-directed (Arc Flags) methods do not use much more space than Dijkstra's algorithm, but are three to four orders of magnitude faster. By combining hierarchy-based pruning and goal direction, CHASE improves query times by yet another order of magnitude, visiting little more than the shortest path itself. Finally, when a higher space overhead is acceptable, non-graph-based methods can be more than a million times faster than the baseline. In particular, HL-∞ is only 5 times slower than the trivial table-lookup method, where a query consists of a single access to main memory. Note that the table-lookup method itself is impractical, since it would require more than one petabyte of RAM.

The experiments reported so far consider only random queries, which tend to be long-range. In a real system, however, most queries tend to be local. For that reason, Sanders and Schultes [223] introduced a methodology based on *Dijkstra ranks*. When running Dijkstra's algorithm from a vertex s, the rank of a vertex u is the order in which it is taken from the priority queue. By evaluating pairs of vertices for Dijkstra ranks $2^1, 2^2, \ldots, 2^{\lfloor \log |V| \rfloor}$ for some randomly chosen sources, all types (local, mid-range, global) of queries are evaluated. Figure 8 reports the median running times for all techniques from Table 1 (except TNR+AF, for which such numbers have never been published) for 1 000 random sources and Dijkstra ranks $\geq 2^6$. As expected, algorithms based on graph searches (including Dijkstra, CH, CRP, and Arc Flags) are faster for local queries. This is not true for bounded-hop algorithms. For TNR, in particular, local queries must actually use a (significantly slower) graph-based approach. HL is more uniform overall because it never uses a graph.

Realistic Setting. Although useful, the results shown in Table 1 do not capture all features that are important for real-world systems. First, systems providing actual driving directions must account for turn costs and restrictions, which the simplified graph model ignores. Second, systems must often support multiple metrics (cost functions), such as shortest distances, avoid U-turns, avoid/prefer freeways, or avoid ferries; metric-specific data structures should therefore be as small as possible. Third, query times should be robust to the choice of cost functions: the system should not time out if an unfriendly cost function is chosen. Finally, one should be able to incorporate a new cost function quickly to account for current traffic conditions (or even user preferences).

CH has the fastest preprocessing among the algorithms in Table 1 and its queries are fast enough for interactive applications. Its performance degrades

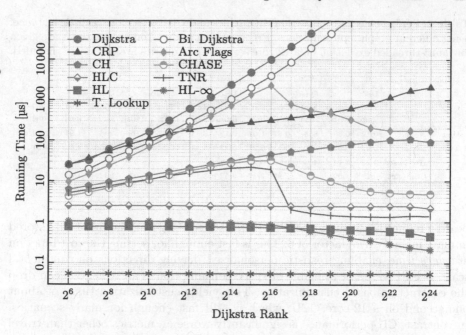

Fig. 8. Performance of speedup techniques for various Dijkstra ranks.

under realistic constraints [78], however. In contrast, CRP was developed with these constraints in mind. As explained in Sect. 2.3, it splits its preprocessing phase in two: although the initial metric-independent phase is relatively slow (as shown in Table 1), only the subsequent (and fast) metric-dependent customization phase must be rerun to incorporate a new metric. Moreover, since CRP is based on edge separators, its performance is (almost) independent of the cost function.

Table 2 (reproduced from [78]) compares CH and CRP with and without turn costs, as well as for travel distances. The instance tested is the same in Table 1, augmented by turn costs (set to 100 seconds for U-turns and zero otherwise). This simple change makes it almost as hard as fully realistic (proprietary) map data used in production systems [78]. The table reports metric-independent preprocessing and metric-dependent customization separately; "DS" refers to the data structures shared by all metrics, while "CUSTOM" refers to the additional space and time required by each individual metric. Unlike in Table 1, space consumption also includes data structures used for path unpacking. For queries, we report the time to get just the length of the shortest path (*dist*), as well as the total time to retrieve both the length and the full path (*path*). Moreover, preprocessing (and customization) times refer to multi-threaded executions on 12 cores; queries are still sequential.

As the table shows, CRP query times are very robust to the cost function and the presence of turns. Also, a new cost function can be applied in roughly 370 ms, fast enough to even support user-specific cost functions. Customization times can

Table 2. Performance of Contraction Hierarchies and CRP on a more realistic instance, using different graph representations. Preprocessing and customization times are given for multi-threaded execution on a 12-core server, while queries are run single-threaded.

Metric	Turn info	CH					CRP						
		DS		Queries			DS		Custom		Queries		
		Time [h:m]	Space [GiB]	Nmb. scans	Dist [ms]	Path [ms]	Time [h:m]	Space [GiB]	Time [s]	Space [GiB]	Nmb. scans	Dist. [ms]	Path [ms]
Dist	None	0:12	0.68	858	0.87	1.07	0:12	3.11	0.37	0.07	2942	1.91	2.49
Time	None	0:02	0.60	280	0.11	0.21	0:12	3.11	0.37	0.07	2766	1.65	1.81
	Arc-based	0:23	3.14	404	0.20	0.30	–	–	–	–	–	–	–
	Compact	0:29	1.09	1998	2.27	2.37	0:12	3.11	0.37	0.07	3049	1.67	1.85

be even reduced to 36 ms with GPUs [87], also reducing the amount of data stored in main memory by a factor of 6. This is fast enough for setting the cost function at *query time*, enabling realistic personalized driving directions on continental scale. If GPUs are not available or space consumption is an issue, one can drop the contraction-based customization. This yields customization times of about one second on a 12-core CPU, which is still fast enough for many scenarios. In contrast, CH performance is significantly worse on metrics other than travel times without turn costs.

We stress that not all applications have the same requirements. If only good estimates on travel times (and not actual paths) are needed, ignoring turn costs and restrictions is acceptable. In particular, ranking POIs according to travel times (but ignoring turn costs) already gives much better results than ranking based on geographic distances. Moreover, we note that CH has fast queries even with fully realistic turn costs. If space (for the expanded graph) is not an issue, it can still provide a viable solution to the static problem; the same holds for related methods such as HL and HLC [82]. For more dynamic scenarios, CH preprocessing can be made parallel [144] or even distributed [168]; even if run sequentially, it is fast enough for large metropolitan areas.

3.2 Applications

As discussed in Sect. 2.7, many speedup techniques can handle more than plain point-to-point shortest path computations. In particular, hierarchical techniques such as CH or CRP tend to be quite versatile, with many established extensions.

Some applications may involve more than one path between a source and a target. For example, one may want to show the user several "reasonable" paths (in addition to the shortest one) [60]. In general, these alternative paths should be short, smooth, and significantly different from the shortest path (and other alternatives). Such paths can either be computed directly as the concatenation of partial shortest paths [6,60,78,173,184] or compactly represented as a small graph [17,174,213]. A related problem is to compute a *corridor* [86] of paths between source and target, which allows deviations from the best route (while driving) to be handled without recomputing the entire path.

These robust routes can be useful in mobile scenarios with limited connectivity. Another useful tool to reduce communication overhead in such cases is route compression [31].

Extensions that deal with nontrivial cost functions have also been considered. In particular, one can extend CH to handle flexible arc restrictions [140] (such as height or weight limitations) or even multiple criteria [133,138] (such as optimizing costs and travel time). Minimizing the energy consumption of electric vehicles [43,44,122,152,240,241] is another nontrivial application, since batteries are recharged when the car is going downhill. Similarly, optimal cycling routes must take additional constraints (such as the amount of uphill cycling) into account [239].

The ability of computing many (batched) shortest paths fast enables interesting new applications. By quickly analyzing multiple candidate shortest paths, one can efficiently match GPS traces to road segments [119,121]. Traffic simulations also benefit from acceleration techniques [183], since they must consider the likely routes taken by *all* drivers in a network. Another application is route prediction [177]: one can estimate where a vehicle is (likely) headed by measuring how good its current location is as a via point towards each candidate destination. Fast routing engines allow more locations to be evaluated more frequently, leading to better predictions [2,121,162,176]. Planning placement of charging stations can also benefit from fast routing algorithms [132]. Another important application is *ride sharing* [2,110,139], in which one must match a ride request with the available offer in a large system, typically by minimizing drivers' detours.

Finally, batched shortest-path computations enable a wide range of point-of-interest queries [2,99,114,119,137,179,221,260]. Typical examples include finding the closest restaurant to a given location, picking the best post office to stop on the way home, or finding the best meeting point for a group of friends. Typically using the bucket-based approach (cf. Sect. 2.7), fast routing engines allow POIs to be ranked according to network-based cost functions (such as travel time) rather than geographic distances. This is crucial for accuracy in areas with natural (or man-made) obstacles, such as mountains, rivers, or rail tracks. Note that more elaborate POI queries must consider concatenations of shortest paths. One can handle these efficiently using an extension of the bucket-based approach that indexes pairs of vertices instead of individual ones [2,99].

3.3 Alternative Settings

So far, we have assumed that shortest path computations take place on a standard server with enough main memory to hold the input graph and the auxiliary data. In practice, however, it is often necessary to run (parts of) the routing algorithm in other settings, such as mobile devices, clusters, or databases. Many of the methods we discuss can be adapted to such scenarios.

Of particular interest are mobile devices, which typically are slower and (most importantly) have much less available RAM. This has motivated external memory implementation of various speedup techniques, such as ALT [150], CH [226], and time-dependent CH [167]. CH in particular is quite practical, supporting interactive queries by compressing the routing data structures and optimizing their access patterns.

Relational databases are another important setting in practice, since they allow users to formulate complex queries on the data in SQL, a popular and expressive declarative query language [230].

Unfortunately, the table-based computational model makes it hard (and inefficient) to implement basic data structures such as graphs or even priority queues. Although some distance oracles based on geometric information could be implemented on a database [229], they are approximate and very expensive in terms of time and space, limiting their applicability to small instances. A better solution is to use HL, whose queries can very easily be expressed in SQL, allowing interactive applications based on shortest path computations entirely within a relational database [2].

For some advanced scenarios, such as time-dependent networks, the preprocessing effort increases quite a lot compared to the time-independent scenario. One possible solution is to run the preprocessing in a distributed fashion. One can achieve an almost linear speedup as the number of machine increases, for both CH [168] and CRP [116].

4   Journey Planning in Public Transit Networks

This section considers journey planning in (schedule-based) public transit networks. In this scenario, the input is given by a timetable. Roughly speaking, a timetable consists of a set of stops (such as bus stops or train platforms), a set of routes (such as bus or train lines), and a set of trips. Trips correspond to individual vehicles that visit the stops along a certain route at a specific time of the day. Trips can be further subdivided into sequences of elementary connections, each given as a pair of (origin/destination) stops and (departure/arrival) times between which the vehicle travels without stopping. In addition, footpaths model walking connections (transfers) between nearby stops.

A key difference to road networks is that public transit networks are inherently *time-dependent*, since certain segments of the network can only be traversed at specific, discrete points in time. As such, the first challenge concerns modeling the timetable appropriately in order to enable the computation of journeys, i.e., sequences of trips one can take within a transportation network. While in road networks computing a single shortest path (typically the quickest journey) is often sufficient, in public transit networks it is important to solve more involved problems, often taking several optimization criteria into account. Section 4.1 will address such modeling issues.

Accelerating queries for efficient journey planning is a long-standing problem [45, 235, 247, 248]. A large number of algorithms have been developed not only to answer basic queries fast, but also to deal with extended scenarios that incorporate delays, compute robust journeys, or optimize additional criteria, such as monetary cost.

4.1 Modeling

The first challenge is to model the timetable in order to enable algorithms that compute optimal journeys. Since the shortest-path problem is well understood in the literature, it seems natural to build a graph $G = (V, A)$ from the timetable such that shortest paths in G correspond to optimal journeys. This section reviews the two main approaches to do so (*time-expanded* and *time-dependent*), as well as the common types of problems one is interested to solve. For a more detailed overview of these topics, we refer the reader to an overview article by Müller-Hannemann et al. [203].

Time-Expanded Model. Based on the fact that a timetable consists of time-dependent *events* (e.g., a vehicle departing at a stop) that happen at *discrete* points in time, the idea of the *time-expanded* model is to build a space-time graph (often also called an event graph) [211] that "unrolls" time. Roughly speaking, the model creates a vertex for every event of the timetable and uses arcs to connect subsequent events in the direction of time flow. A basic version of the model [196, 235] contains a vertex for every departure and arrival event, with consecutive departure and arrival events connected by *connection* (or *travel*) arcs. To enable transfers between vehicles, all vertices at the same stop are (linearly, in chronological order) interlinked by *transfer* (or *waiting*) arcs. Müller-Hannemann and Weihe [204] extend the model to distinguish trains (to optimize the number of transfers taken during queries) by subdividing each connection arc by a new vertex, and then interlinking the vertices of each trip (in order of travel). Pyrga et al. [218, 219] and Müller-Hannemann and Schnee [200] extend the time-expanded model to incorporate *minimum change times* (given by the input) that are required as buffer when changing trips at a station. Their *realistic* model introduces an additional *transfer vertex* per departure event, and

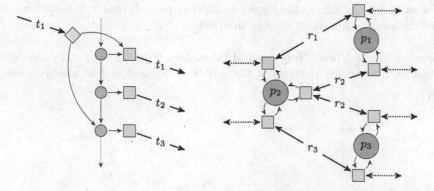

Fig. 9. Realistic time-expanded (left) and time-dependent (right) models. Different vertex types are highlighted by shape: diamond (arrival), circle (transfer) and square (departure) for the left figure; and circle (stop) and square (route) for the right figure. Connection arcs in the time-expanded model are annotated with its trips t_i, and route arcs in the time-dependent model with its routes r_i.

connects each arrival vertex to the first transfer vertex that obeys the minimum change time constraints. See Fig. 9 for an illustration. If there is a footpath from stop p_i to stop p_j, then for each arrival event at stop p_i one adds an arc to the earliest reachable transfer vertex at p_j. This model has been further engineered [90] to reduce the number of arcs that are explored "redundantly" during queries.

A timetable is usually valid for a certain period of time (up to one year). Since the timetables of different days of the year are quite similar, a space-saving technique (*compressed model*) is to consider events modulo their traffic days [202, 219].

Time-Dependent Model. The main disadvantage of the time-expanded model is that the resulting graphs are quite large [218]. For smaller graphs, the time-dependent approach (see Sect. 2.7) has been considered by Brodal and Jacob [56]. In their model, vertices correspond to stops, and an arc is added from u to v if there is at least one elementary connection serving the corresponding stops in this order. Precise departure and arrival times are encoded by the travel time function associated with the arc (u, v). Fig. 10 shows the typical shape of a travel time function: each filled circle represents an elementary connection; the line segments (with slope -1) reflect not only the travel time, but also the waiting time until the next departure. Pyrga et al. [219] further extended this basic model to enable minimum change times by creating, for each stop p and each route that serves p, a dedicated *route vertex*. Route vertices at p are connected to a common *stop vertex* by arcs with constant cost depicting the minimum change time of p. Trips are distributed among *route arcs* that connect the subsequent route vertices of a route, as shown in Fig. 9. They also consider a model that allows arbitrary minimum change times between pairs of routes within each stop [219]. Footpaths connecting nearby stops are naturally integrated into the time-dependent model [109]. For some applications, one may merge route vertices of the same stop as long as they never connect trips such that a transfer between them violates the minimum change time [85].

Frequency-Based Model. In real-world timetables trips often operate according to specific frequencies at times of the day. For instance, a bus may run every

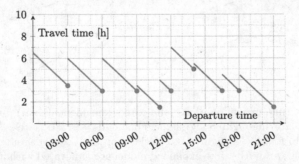

Fig. 10. Travel time function on an arc.

5 min during rush hour, and every 10 min otherwise. Bast and Storandt [27] exploit this fact in the *frequency-based model*: as in the time-dependent approach, vertices correspond to stops, and an arc between a pair of stops (u, v) is added if there is at least one elementary connection from u to v. However, instead of storing the departures of an arc explicitly, those with coinciding travel times are compressed into a set of tuples consisting of an initial departure time τ_{dep}, a time interval Δ, and a frequency f. The corresponding original departures can thus be reconstructed by computing each $\tau_{\text{dep}} + fi$ for those $i \in \mathbb{Z}_{\geq 0}$ that satisfy $\tau_{\text{dep}} + fi \leq \tau_{\text{dep}} + \Delta$. Bast and Storandt compute these tuples by covering the set of departure times by a small set of overlapping arithmetic progressions, then discarding duplicate entries (occurring after decompression) at query time [27].

Problem Variants. Most research on road networks has focused on computing the shortest path according to a given cost function (typically travel times). For public transit networks, in contrast, there is a variety of natural problem formulations.

The simplest variant is the *earliest arrival problem*. Given a source stop p_s, a target stop p_t, and a departure time τ, it asks for a journey that departs p_s no earlier than τ and arrives at p_t as early as possible. A related variant is the *range* (or *profile*) *problem* [206], which replaces the departure time by a time range (e. g. 8–10 am, or the whole day). This problem asks for a set of journeys of minimum travel time that depart within that range.

Both the earliest arrival and the range problems only consider (arrival or travel) time as optimization criterion. In public-transit networks, however, other criteria (such as the number of transfers) are just as important, which leads to the *multicriteria problem* [204]. Given source and target stops p_s, p_t and a departure time τ as input, it asks for a (maximal) Pareto set $\mathcal{J}$ of nondominating journeys with respect to the optimization criteria considered. A journey J_1 is said to dominate journey J_2 if J_1 is better than or equal to J_2 in all criteria. Further variants of the problem relax or strengthen these domination rules [200].

4.2 Algorithms Without Preprocessing

This section discusses algorithms that can answer queries without a preprocessing phase, which makes them a good fit for dynamic scenarios that include delays, route changes, or train cancellations. We group the algorithms by the problems they are meant to solve.

Earliest Arrival Problem. Earliest arrival queries on the time-expanded model can be answered in a straightforward way by Dijkstra's algorithm [235], in short TED (time-expanded Dijkstra). It is initialized with the vertex that corresponds to the earliest event of the source stop p_s that occurs after τ (in the realistic model, a transfer vertex must be selected). The first scanned vertex associated with the target stop p_t then represents the earliest arrival s–t journey. In the compressed time-expanded model, slight modifications to Dijkstra's algorithm are necessary because an event vertex may appear several times on the optimal shortest path (namely for different consecutive days). One possible solution is to use a bag of labels for each vertex as in the multicriteria variants described below. Another solution is described in Pyrga et al. [219].

On time-dependent graphs, Dijkstra's algorithm can be augmented to compute shortest paths [65,111], as long as the cost functions are nonnegative and FIFO [208,209]. The only modification is that, when the algorithm scans an arc (u, v), the arc cost is evaluated at time $\tau + \text{dist}(s, u)$. Note that the algorithm retains the label-setting property, i. e., each vertex is scanned at most once. In the time-dependent public transit model, the query is run from the stop vertex corresponding to p_s and the algorithm may stop as soon as it extracts p_t from the priority queue. The algorithm is called TDD (time-dependent Dijkstra).

Another approach is to exploit the fact that the time-expanded graph is directed and acyclic. (Note that overnight connections can be handled by unrolling the timetable for several consecutive periods.) By scanning vertices in topological order, arbitrary queries can be answered in linear time. This simple and well-known observation has been applied for journey planning by Mellouli and Suhl [191], for example. While this idea saves the relatively expensive priority queue operations of Dijkstra's algorithm, one can do even better by not maintaining the graph structure explicitly, thus improving locality and cache efficiency. The recently developed *Connection Scan Algorithm* (CSA) [105] organizes the elementary connections of the timetable in a single array, sorted by departure time. The query then only scans this array once, which is very efficient in practice. Note that CSA requires footpaths in the input to be closed under transitivity to ensure correctness.

Range Problem. The range problem can be solved on the time-dependent model by variants of Dijkstra's algorithm. The first variant [68,206] maintains, at each vertex u, a travel-time function (instead of a scalar label) representing the optimal travel times from s to u for the considered time range. Whenever the algorithm relaxes an arc (u, v), it first *links* the full travel-time function associated with u to the (time-dependent) cost function of the arc (u, v), resulting in a function that represents the times to travel from s to v via u. This function is then *merged* into the (tentative) travel time function associated with v, which corresponds to taking the element-wise minimum of the two functions. The algorithm loses the label-setting property, since travel time functions cannot be totally ordered. As a result the algorithm may reinsert vertices into the priority queue whenever it finds a journey that improves the travel time function of an already scanned vertex.

Another algorithm [34] exploits the fact that trips depart at discrete points in time, which helps to avoid redundant work when propagating travel time functions. When it relaxes an arc, it does not consider the full function, but each of its encoded connections individually. It then only propagates the parts of the function that have improved.

The *Self-Pruning Connection Setting* algorithm (SPCS) [85] is based on the observation that *any* optimal journey from s to t has to start with one of the trips departing from s. It therefore runs, for each such trip, Dijkstra's algorithm from s

at its respective departure time. SPCS performs these runs simultaneously using a shared priority queue whose entries are ordered by arrival time. Whenever the algorithm scans a vertex u, it checks if u has been already scanned for an associated (departing) trip with a *later* departure time (at s), in which case it prunes u. Moreover, SPCS can be parallelized by assigning different subsets of departing trips from s to different CPU cores.

Bast and Storandt [27] propose an extension of Dijkstra's algorithm that operates on the (compressed) frequency-based model directly. It maintains with every vertex u a set of tuples consisting of a time interval, a frequency, and the travel time. Hence, a single tuple may represent multiple optimal journeys, each departing within the tuple's time interval. Whenever the algorithm relaxes an arc (u, v), it first extends the tuples from the bag at u with the ones stored at the arc (u, v) in the compressed graph. The resulting tentative bag of tuples (representing all optimal journeys to v via u) is then *merged* into the bag of tuples associated with v. The main challenge of this algorithm is efficiently merging tuples with incompatible frequencies and time intervals [27].

Finally, the Connection Scan Algorithm has been extended to the range problem [105]. It uses the same array of connections, ordered by departure time, as for earliest arrival queries. It still suffices to scan this array once, even to obtain optimal journeys to all stops of the network.

Multicriteria Problem. Although Pareto sets can contain exponentially many solutions (see Sect. 2.7), they are often much smaller for public transit route planning, since common optimization criteria are positively correlated. For example, for the case of optimizing earliest arrival time and number of transfers, the *Layered Dijkstra* (LD) algorithm [56,219] is efficient. Given an upper bound K on the number of transfers, it (implicitly) copies the timetable graph into K layers, rewiring transfer arcs to point to the next higher level. It then suffices to run a time-dependent (single criterion) Dijkstra query from the lowest level to obtain Pareto sets.

In the time-expanded model, Müller-Hannemann and Schnee [200] consider the Multicriteria Label-Setting (MLS) algorithm (cf. Sect. 2.7) to optimize arrival time, ticket cost, and number of transfers. In the time-dependent model, Pyrga et al. [219] compute Pareto sets of journeys for arrival time and number of transfers. Disser et al. [109] propose three optimizations to MLS that reduce the number of queue operations: hopping reduction, label forwarding, and dominance by early results (or *target pruning*). Bast and Storandt [27] extend the frequency-based range query algorithm to also include number of transfers as criterion.

A different approach is *RAPTOR* (Round-bAsed Public Transit Optimized Router) [92]. It is explicitly developed for public transit networks and its basic version optimizes arrival time and the number of transfers taken. Instead of using a graph, it organizes the input as a few simple arrays of trips and routes. Essentially, RAPTOR is a dynamic program: it works in rounds, with round i computing earliest arrival times for journeys that consist of exactly i transfers. Each round takes as input the stops whose arrival time improved in the previous round (for the first round this is only the source stop). It then *scans* the routes served by these stops. To scan route r, RAPTOR traverses its stops in order of travel, keeping track of the earliest possible trip (of r) that can be taken. This trip may improve the tentative arrival times at subsequent stops of route r. Note that RAPTOR scans each route at most once per round, which is very efficient in practice (even faster than Dijkstra's algorithm with a single criterion). Moreover, RAPTOR can be parallelized by distributing non-conflicting routes to different CPU cores. It can also be extended to handle range queries (rRAPTOR) and additional optimization criteria (McRAPTOR). Note that, like CSA, RAPTOR also requires footpaths in the input to be closed under transitivity.

Trip-Based Routing [256] accelerates RAPTOR by executing a BFS-like search on a network of trips and precomputed *sensible* transfers.

4.3   Speedup Techniques

This section presents an overview of preprocessing-based speedup techniques for journey planning in public transit networks. A natural (and popular) approach is to adapt methods that are effective on road networks (see Fig. 7). Unfortunately, the speedups observed in public transit networks are several orders of magnitude lower than in road networks. This is to some extent explained by the quite different structural properties of public transit and road networks [22]. For example, the neighborhood of a stop can be much larger than the number of road segments incident to an intersection. Even more important is the effect of the inherent time-dependency of public transit networks. Thus, developing efficient preprocessing-based methods for public transit remains a challenge.

Some road network methods were tested on public transit graphs without performing realistic queries (i. e., according to one of the problems from Sect. 4.1). Instead, such studies simply perform point-to-point queries on public-transit graphs. In particular, Holzer et al. [161] evaluate basic combinations of bidirectional search, goal directed search, and Geometric Containers on a simple stop graph (with average travel times). Bauer et al. [41] also evaluated bidirectional search, ALT, Arc Flags, Reach, REAL, Highway Hierarchies, and SHARC on time-expanded graphs. Core-ALT, CHASE, and Contraction Hierarchies have also been evaluated on time-expanded graphs [40].

A* Search. On public transit networks, basic A* search has been applied to the time-dependent model [109,219]. In the context of multicriteria optimization, Disser et al. [109] determine lower bounds for each vertex u to the target stop p_t (before the query) by running a backward search (from p_t) using the (constant) lower bounds of the travel time functions as arc cost.

ALT. The (unidirectional) ALT [148] algorithm has been adapted to both the time-expanded [90] and the time-dependent [207] models for computing earliest arrival queries. In both cases, landmark selection and distance precomputation is performed on an auxiliary stop graph, in which vertices correspond to stops and an arc is added between two stops p_i, p_j if there is an elementary connection from p_i to p_j in the input. Arc costs are lower bounds on the travel time between their endpoints.

Geometric Containers. Geometric containers [235,251] have been extensively tested on the time-expanded model for computing earliest arrival queries. In fact, they were developed in the context of this model. As mentioned in Sect. 2, bounding boxes perform best [251].

Arc Flags and SHARC. Delling et al. [90] have adapted Arc Flags [157,178] to the time-expanded model as follows. First, they compute a partition on the stop graph (defined as in ALT). Then, for each boundary stop p of cell C, and each of its arrival vertices, a backward search is performed on the time-expanded graph. The authors observe that public transit networks have many paths of equal length between the same pair of vertices [90], making the choice of tie-breaking rules important. Furthermore, Delling et al. [90] combine Arc Flags, ALT, and a technique called *Node Blocking*, which avoids exploring multiple arcs from the same route.

SHARC, which combines Arc Flags with shortcuts [39], has been tested on the time-dependent model with earliest arrival queries by Delling [72]. Moreover, Arc Flags with shortcuts for the Multi-Label-Setting algorithm (MLS) have been considered for computing full (i. e., using strict domination) Pareto sets using arrival time and number of transfers as criteria [47]. In time-dependent graphs, a flag must be set if its arc appears on a shortest path toward the corresponding cell at least once during the time horizon [72]. For better performance, one can use different sets of flags for different time periods (e. g., every two hours). The resulting total speedup is still below 15, from which it is concluded that "accelerating time-dependent multicriteria timetable information is harder than expected" [47]. Slight additional speedups can be obtained if one restricts the search space to only those solutions in the Pareto set for which the travel time is within an interval defined by the earliest arrival time and some upper bound. Berger et al. [49] observed that in such a scenario optimal substructure in combination with lower travel time bounds can be exploited and yield additional pruning during search. It is worth noting that this method does not require any preprocessing and is therefore well-suited for a dynamic scenario.

Overlay Graphs. To accelerate earliest arrival queries, Schulz et al. [235] compute single-level overlays between "important" hub stations in the time-expanded model, with importance values given as input. More precisely, given a subset of important stations, the overlay graph consists of *all* vertices (events) that are associated with these stations. Edges in the overlay are computed such that distances between any pair of vertices (events) are preserved. Extending this approach to overlay graphs over multiple levels of hub stations (selected by importance or degree) results in speedups of about 11 [236].

Separator-based techniques. Strasser and Wagner [242] combine the Connection Scan Algorithm [105] with ideas of customizable route planning (CRP) [78] resulting in the Accelerated Connection Scan Algorithm (ACSA). It is designed for both earliest arrival and range queries. ACSA first computes a multilevel partition of stops, minimizing the number of elementary connections with endpoints in different cells. Then, it precomputes for each cell the partial journeys (transit connections) that cross the respective cell. For queries, the algorithm essentially runs CSA restricted to the elementary connections of the cells containing the source or target stops, as well as transit connections of other (higher-level) cells. As shown in Sect. 4.5, it achieves excellent query and preprocessing times on country-sized instances.

Contraction Hierarchies. The Contraction Hierarchies algorithm [142] has been adapted to the realistic time-dependent model with minimum change times for computing earliest arrival and range queries [136]. It turns out that simply applying the algorithm to the route model graph results in too many shortcuts to be practical. Therefore, contraction is performed on a condensed graph that contains only a single vertex per stop. Minimum change times are then ensured by the query algorithm, which must maintain multiple labels per vertex.

Transfer Patterns. A speedup technique specifically developed for public transit networks is called *Transfer Patterns* [24]. It is based on the observation that many optimal journeys share the same transfer pattern, defined as the sequence of stops where a transfer occurs. Conceptually, these transfer patterns are precomputed using range queries for all pairs of stops and departure times. At query time, a query graph is built as the union of the transfer patterns between the source and target stops. The arcs in the query graph represent direct connections between stops (without transfers), and can be evaluated very fast. Dijkstra's algorithm (or MLS) is then applied to this much smaller query graph.

If precomputing transfer patterns between *all* pairs of stops is too expensive, one may resort to the following two-level approach. It first selects a subset of (important) hub stops. From the hubs, global transfer patterns are precomputed to all other stops. For the non-hubs, local transfer patterns are computed only towards relevant hub stops. This approach is similar to TNR, but the idea is applied asymmetrically: transfer patterns are computed from all stops to the hub stops, and from the hub stops to everywhere. If preprocessing is still impractical, one can restrict the local transfer patterns to at most three legs (two transfers). Although this restriction is heuristic, the algorithm still almost always finds the optimal solution in practice, since journeys requiring more than two transfers to reach a hub station are rare [24].

TRANSIT. Finally, Transit Node Routing [28,30,224] has been adapted to public transit journey planning in [14]. Preprocessing of the resulting *TRANSIT* algorithm uses the (small) stop graph to determine a set of transit nodes (with a similar method as in [28]), between which it maintains a distance table that contains sets of journeys with minimal travel time (over the day). Each stop p maintains, in addition, a set of access nodes $A(p)$, which is computed on the time-expanded graph by running local searches from each departure event of p toward the transit stops. The query then uses the access nodes of p_s and p_t and the distance table to resolve global requests. For local requests, it runs goal-directed A* search. Queries are slower than for Transfer Patterns.

4.4 Extended Scenarios

Besides computing journeys according to one of the problems from Sect. 4.1, extended scenarios (such as incorporating delays) have been studied as well.

Uncertainty and Delays. Trains, buses and other means of transport are often prone to delays in the real world. Thus, handling delays (and other sources of uncertainty) is an important aspect of a practical journey planning system. Firmani et al. [125] recently presented a case study for the public transport network of the metropolitan area of Rome. They provide strong evidence that computing journeys according to the published timetable often fails to deliver optimal or even high-quality solutions. However, incorporating real-time GPS location data of vehicles into the journey planning algorithm helps improve the journey quality (e. g., in terms of the experienced delay) [13, 84].

Müller-Hannemann and Schnee [201] consider the online problem where delays, train cancellations, and extra trains arrive as a continuous stream of information. They present an approach which quickly updates the time-expanded model to enable queries according to current conditions. Delling et al. [74] also discuss updating the time-dependent model and compare the required effort with the time-expanded model. Cionini et al. [63] propose a new graph-based model which is tailored to handle dynamic updates, and they experimentally show its effectiveness in terms of both query and update times. Berger et al. [48] propose a realistic stochastic model that predicts how delays propagate through the network. In particular, this model is evaluated using real (delay) data from Deutsche Bahn. Bast et al. [25] study the robustness of Transfer Patterns with respect to delays. They show that the transfer patterns computed for a scenario without any delays give optimal results for 99 % of queries, even when large and area-wide (random) delays are injected into the networks.

Disser et al. [109] and Delling et al. [93] study the computation of *reliable* journeys via multicriteria optimization. The reliability of a transfer is defined as a function of the available buffer time for the transfer. Roughly speaking, the larger the buffer time, the more likely it is that the transfer will be successful. According to this notion, transfers with a high chance of success are still considered reliable even if there is no backup alternative in case they fail.

To address this issue, Dibbelt et al. [105] minimize the *expected arrival time* (with respect to a simple model for the probability that a transfer breaks). Instead of journeys, their method (which is based on the CSA algorithm) outputs a *decision graph* representing optimal instructions to the user at each point of their journey, including cases in which a connecting trip is missed. Interestingly, minimizing the expected arrival time implicitly helps minimizing the number of transfers, since each "unnecessary" transfer introduces additional uncertainty, hurting the expected arrival time.

Finally, Goerigk et al. [146] study the computation of *robust* journeys, considering both strict robustness (i. e., computing journeys that are always feasible for a given set of delay scenarios) and light robustness (i. e., computing journeys that are most reliable when given some extra slack time). While strict robustness turns out to be too conservative in practice, the notion of light robustness seems more promising. *Recoverable robust* journeys (which can always be updated when delays occur) have recently been considered in [145]. A different, new robustness concept has been proposed by Böhmová et al. [51]. In order to propose solutions that are robust for typical delays, past observations of real traffic situations are used. Roughly speaking, a route is more robust the better it has performed in the past under different scenarios.

Night Trains. Gunkel et al. [153] have considered the computation of overnight train journeys, whose optimization goals are quite different from regular "daytime" journeys. From a customer's point of view, the primary objective is usually to have a reasonably long sleeping period. Moreover, arriving too early in the morning at the destination is often not desired. Gunkel et al. present two approaches to compute overnight journeys. The first approach explicitly enumerates all overnight trains (which are given by the input) and computes, for each such train, the optimal feeding connections. The second approach runs multicriteria search with sleeping time as a maximization criterion.

Fares. Müller-Hannemann and Schnee [199] have analyzed several pricing schemes, integrating them as an optimization criterion (cost) into MOTIS, a multicriteria search algorithm that works on the time-expanded model. In general, however, optimizing exact monetary cost is a challenging problem, since real-world pricing schemes are hard to capture by a mathematical model [199].

Delling et al. [92] consider computing Pareto sets of journeys that optimize fare zones with the McRAPTOR algorithm. Instead of using (monetary) cost as an optimization criterion directly, they compute all nondominated journeys that traverse different combinations of fare zones, which can then be evaluated by cost in a quick postprocessing step.

Guidebook Routing. Bast and Storandt [26] introduce *Guidebook Routing*, where the user specifies only source and target stops, but neither a day nor a time of departure. The desired answer is then a set of routes, each of which is given by a sequence of train or bus numbers and transfer stations. For example, an answer may read like *take bus number 11 towards the bus stop at X, then change to bus number 13 or 14 (whichever comes first) and continue to the bus stop at Y.* Guidebook routes can be computed by first running a multicriteria range query, and then extracting from the union of all Pareto-optimal time-dependent paths a subset of routes composed by arcs which are most frequently used. The Transfer Patterns algorithm lends itself particularly well to the computation of such guidebook routes. For practical guidebook routes (excluding "exotic" connections at particular times), the preprocessing space and query times of Transfer Patterns can be reduced by a factor of 4 to 5.

4.5   Experiments and Comparison

This section compares the performance of some of the journey planning algorithms discussed in this section. As in road networks, all algorithms have been carefully implemented in C++ using mostly custom-built data structures.

Table 3 summarizes the results. Running times are obtained from a sequential execution on one core of a dual 8-core Intel Xeon E5-2670 machine clocked at 2.6 GHz with 64 GiB of DDR3-1600 RAM. The exceptions are Transfer Patterns and Contraction Hierarchies, for which we reproduce the values reported in the original publication (obtained on a comparable machine).

For each algorithm, we report the instance on which it has been evaluated, as well as its total number of elementary connections (a proxy for size) and the number of consecutive days covered by the connections. Unfortunately, realistic benchmark data of country scale (or larger) has not been widely available to the research community. Some metropolitan transit agencies have recently started making their timetable data publicly available, mostly using the General Transit Feed format[2]. Still, research groups often interpret the data differently, making it hard to compare the performance of different algorithms. The largest metropolitan instance currently available is the full transit network of London[3]. It contains approximately 21 thousand stops, 2.2 thousand routes, 133 thousand trips, 46 thousand footpaths, and 5.1 million elementary connections for one full day. We therefore use this instance for the evaluation of most algorithms. The instances representing Germany and long-distance trains in Europe are generated in a similar way, but from proprietary data.

The table also contains the preprocessing time (where applicable), the average number of label comparisons per stop, the average number of journeys computed by the algorithm, and its running time in milliseconds. Note that the number of journeys can be below 1 because some stops are unreachable for certain late departure times. References indicate the publications from which the figures are taken (which may differ from the first publication); TED was run by the authors for this survey. (Our TED implementation uses a single-level bucket queue [104] and stops as soon as a vertex of the target stop has been extracted.) The columns labeled "criteria" indicate whether the algorithm minimizes arrival time (arr), number of transfers (tran), fare zones (fare), reliability (rel), and whether it computes range queries (rng) over the full timetable period of 1, 2, or 7 days. Methods with multiple criteria compute Pareto sets.

Among algorithms without preprocessing, we observe that those that do not use a graph (RAPTOR and CSA) are consistently faster than their graph-based counterparts. Moreover, running Dijkstra's algorithm on the time-expanded graph model (TED) is significantly slower than running it on the time-dependent graph model (TDD), since time-expanded graphs are much larger. For earliest arrival queries on metropolitan areas, CSA is the fastest algorithm without preprocessing, but preprocessing-based methods (such as Transfer Patterns) can

---

[2] https://developers.google.com/transit/gtfs/.
[3] http://data.london.gov.uk/.

Table 3. Performance of various public transit algorithms on random queries. For each algorithm, the table indicates the implementation tested (which may not be the publication introducing the algorithm), the instance it was tested on, its total number of elementary connections (in millions) as well as the number of consecutive days they cover. A "p" indicates that the timetable is periodic (with a period of one day). The table then shows the criteria that are optimized (a subset of arrival times, transfers, full range, fares, and reliability), followed by total preprocessing time, average number of comparisons per stop, average number of journeys in the Pareto set, and average query times in milliseconds. Missing entries either do not apply (–) or are well-defined but not available (n/a).

Algorithm	Impl.	INPUT			CRITERIA					QUERY			
		Name	Conn. [10^6]	Dy.	Arr.	Tran.	Rng.	Fare	Rel.	Prep [h]	Comp./ stop	jn	Time [ms]
TED		London	5.1	1	●	○	○	○	○	–	50.6	0.9	44.8
TDD	[93]	London	5.1	1	●	○	○	○	○	–	7.4	0.9	11.0
CH	[136]	Europe (lng)	1.7	p	●	○	○	○	○	< 0.1	< 0.1	n/a	0.3
CSA	[105]	London	4.9	1	●	○	○	○	○	–	26.6	n/a	2.0
ACSA	[242]	Germany	46.2	2	●	○	○	○	○	–	n/a	n/a	8.7
T. Patterns	[27]	Germany	90.4	7	●	○	○	○	○	541	–	1.0	0.4
LD	[93]	London	5.1	1	●	●	○	○	○	–	15.6	1.8	28.7
MLS	[93]	London	5.1	1	●	●	○	○	○	–	23.7	1.8	50.0
RAPTOR	[93]	London	5.1	1	●	●	○	○	○	–	10.9	1.8	5.4
T. Patterns	[27]	Germany	90.4	7	●	●	○	○	○	566	–	2.0	0.8
CH	[136]	Europe (lng)	1.7	p	●	○	●	○	○	< 0.1	< 0.1	n/a	3.7
SPCS	[105]	London	4.9	1	●	○	●	○	○	–	372.5	98.2	843.0
CSA	[105]	London	4.9	1	●	○	●	○	○	–	436.9	98.2	161.0
ACSA	[242]	Germany	46.2	2	●	○	●	○	○	8	n/a	n/a	171.0
T. Patterns	[27]	Germany	90.4	7	●	○	●	○	○	541	–	121.2	22.0
rRAPTOR	[105]	London	4.9	1	●	●	●	○	○	–	1634.0	203.4	922.0
CSA	[105]	London	4.9	1	●	●	●	○	○	–	3824.9	203.4	466.0
T. Patterns	[27]	Germany	90.4	7	●	●	●	○	○	566	–	226.0	39.6
MLS	[93]	London	5.1	1	●	●	○	●	○	–	818.2	8.8	304.2
McRAPTOR	[93]	London	5.1	1	●	●	○	●	○	–	277.5	8.8	100.9
MLS	[93]	London	5.1	1	●	●	○	○	●	–	286.6	4.7	239.8
McRAPTOR	[93]	London	5.1	1	●	●	○	○	●	–	89.6	4.7	71.9

be even faster. For longer-range transit networks, preprocessing-based methods scale very well. CH takes 210 seconds to preprocess the long-distance train connections of Europe, while ACSA takes 8 hours to preprocess the full transit network of Germany. Transfer Patterns takes over 60 times longer to preprocess (a full week of) the full transit network of Germany, but has considerably lower query times.

For multicriteria queries, RAPTOR is about an order of magnitude faster than Dijkstra-based approaches like LD and MLS. RAPTOR is twice as fast as TDD, while computing twice as many journeys on average. Adding further criteria (such as fares and reliability) to MLS and RAPTOR increases the Pareto set, but performance is still reasonable for metropolitan-sized networks. Thanks to preprocessing, Transfer Patterns has the fastest queries overall, by more than an order of magnitude. Note that in public transit networks the optimization criteria are often positively correlated (such as arrival time and number of transfers), which keeps the Pareto sets at a manageable size. Still, as the number of criteria increases, exact real-time queries become harder to achieve.

The reported figures for Transfer Patterns are based on preprocessing leveraging the frequency-based model with traffic days compression, which makes quadratic (in the number of stops) preprocessing effort feasible. Consequently, hub stops and the three-leg heuristic are not required, and the algorithm is guaranteed to find the optimal solution. The data produced by the preprocessing is shown to be robust against large and area-wide delays, resulting in much less than 1 % of suboptimal journeys [25] (not shown in the table).

For range queries, preprocessing-based techniques (CH, ACSA, Transfer Patterns) scale better than CSA or SPCS. For full multicriteria range queries (considering transfers), Transfer Patterns is by far the fastest method, thanks to preprocessing. Among search-based methods, CSA is faster than rRAPTOR by a factor of two, although it does twice the amount of work in terms of label comparisons. Note, however, that while CSA cannot scale to smaller time ranges by design [105], the performance of rRAPTOR depends linearly on the number of journeys departing within the time range [92]. For example, for 2-hour range queries rRAPTOR computes 15.9 journeys taking only 61.3 ms on average [93] (not reported in the table). Guidebook routes covering about 80 % of the optimal results (for the full period) can be computed in a fraction of a millisecond [26].

5 Multimodal Journey Planning

We now consider journey planning in a multimodal scenario. Here, the general problem is to compute journeys that *reasonably* combine different modes of transportation by a *holistic* algorithmic approach. That is, not only does an algorithm consider each mode of transportation in isolation, but it also optimizes the choice (and sequence) of transportation modes in some integrated way. Transportation modes that are typically considered include (unrestricted) walking, (unrestricted) car travel, (local and long-distance) public transit, flight networks, and rental bicycle schemes. We emphasize that our definition of "multimodal" requires some diversity from the transportation modes, i. e., both unrestricted and schedule-based variants should be considered by the algorithm. For example, journeys that only use buses, trams, or trains are not truly multimodal (according to our definition), since these transportation modes can be represented as a single public transit schedule and dealt with by algorithms from Sect. 4.

In fact, considering modal transfers explicitly by the algorithm is crucial in practice, since the solutions it computes must be *feasible*, excluding sequences of transportation modes that are impossible for the user to take (such as a private car between train rides). Ideally, even user preferences should be respected. For example, some users may prefer taxis over public transit at certain parts of the journey, while others may not.

A general approach to obtain a multimodal network is to first build an individual graph for each transportation mode, then merge them into a single multimodal graph with *link arcs* (or vertices) added to enable modal transfers [89, 210, 258]. Typical examples [89, 210] model car travel and walking as time-independent (static) graphs, public transit networks using the realistic time-dependent model [219], and flight networks using a dedicated flight model [91]. Beyond that, Kirchler et al. [169, 170] compute multimodal journeys in which car travel is modeled as a time-dependent network in order to incorporate historic data on rush hours and traffic congestion (see Sect. 2.7 for details).

Overview. The remainder of this section discusses three different approaches to the multimodal problem. The first (Sect. 5.1) considers a combined cost function of travel time with some penalties to account for modal transfers. The second approach (Sect. 5.2) uses the label-constrained shortest path problem to obtain journeys that explicitly include (or exclude) certain sequences of transportation modes. The final approach (Sect. 5.3) computes Pareto sets of multimodal journeys using a carefully chosen set of optimization criteria that aims to provide diverse (regarding the transportation modes) alternative journeys.

5.1   Combining Costs

To aim for journeys that reasonably combine different transport modes, one may use penalties in the objective function of the algorithm. These penalties are often considered as a linear combination with the primary optimization goal (typically travel time). Examples for this approach include Aifadopoulou et al. [10], who present a linear program that computes multimodal journeys. The TRANSIT algorithm [14] also uses a linear utility function and incorporates travel time, ticket cost, and "inconvenience" of transfers. Finally, Modesti and Sciomachen [195] consider a combined network of unrestricted walking, unrestricted car travel, and public transit, in which journeys are optimized according to a linear combination of several criteria, such as cost and travel time. Moreover, their utility function incorporates user preferences on the transportation modes.

5.2   Label-Constrained Shortest Paths

The *label-constrained shortest paths* [21] approach computes journeys that explicitly obey certain constraints on the modes of transportation. It defines an alphabet Σ of modes of transportation and labels each arc of the graph by the appropriate symbol from Σ. Then, given a language L over Σ as additional input to the query, any journey (path) must obey the constraints imposed by the language L, i. e., the concatenation of the labels along the path must satisfy L. The problem of computing *shortest* label-constrained paths is tractable for *regular* languages [21], which suffice to model reasonable transport mode constraints in multimodal journey planning [18, 20]. Even restricted classes of regular languages can be useful, such as those that impose a hierarchy of transport modes [50,89,169,170,210,258] or Kleene languages that can only globally exclude (and include) certain transport modes [140].

Barrett et al. [21] have proven that the label-constrained shortest path problem is solvable in deterministic polynomial time. The corresponding algorithm, called *label-constrained shortest path problem Dijkstra* (LCSPP-D), first builds a product network G of the input (the multimodal graph) and the (possibly non-deterministic) finite automaton that accepts the regular language L. For given source and target vertices s, t (referring to the original input), the algorithm determines origin and destination sets of product vertices from G, containing those product vertices that refer to s/t and an initial/final state of the automaton. Dijkstra's algorithm is then run on G between these two sets of product vertices. In a follow-up experimental study, Barrett et al. [20] evaluate this algorithm using linear regular languages, a special case.

Basic speedup techniques, such as bidirectional search [67], A* [156], and heuristic A* [237] have been evaluated in the context of multimodal journey planning in [159] and [19]. Also, Pajor [210] combines the LCSPP-D algorithm with time-dependent Dijkstra [65] to compute multimodal journeys that contain a time-dependent subnetwork. He also adapts and analyzes bidirectional search [67], ALT [148], Arc Flags [157,178], and shortcuts [249] with respect to LCSPP.

Access-Node Routing. The *Access-Node Routing* (ANR) [89] algorithm is a speedup technique for the label-constrained shortest path problem (LCSPP). It handles *hierarchical languages*, which allow constraints such as restricting walking and car travel to the beginning and end of the journey. It works similarly to Transit Node Routing [28–30,224] and precomputes for each vertex u of the road (walking and car) network its relevant set of entry (and exit) points (*access nodes*) to the public transit and flight networks. More precisely, for any shortest path P originating from vertex u (of the road network) that also uses the public transit network, the first vertex v of the public transit network on P must be an access node of u. The query may skip over the road network by running a multi-source multi-target algorithm on the (much smaller) transit network between the access nodes of s and t, returning the journey with earliest combined arrival time.

The *Core-Based ANR* [89] method further reduces preprocessing space and time by combining ANR with contraction. As in Core-ALT [40,88], it precomputes access nodes only for road vertices in a much smaller core (overlay) graph. The query algorithm first (quickly) determines the relevant core vertices of s and t (i.e., those covering the branches of the shortest path trees rooted at s and t), then runs a multi-source multi-target ANR query between them.

Access-Node Routing has been evaluated on multimodal networks of intercontinental size that include walking, car travel, public transit, and flights. Queries run in milliseconds, but preprocessing time strongly depends on the density of the public transit and flight networks [89]. Moreover, since the regular language is used during preprocessing, it can no longer be specified at query time without loss of optimality.

State-Dependent ALT. Another multimodal speedup technique for LCSPP is *State-Dependent ALT* (SDALT) [170]. It augments the ALT algorithm [148] to overcome the fact that lower bounds from a vertex u may depend strongly on the current state q of the automaton (expressing the regular language) with which u is scanned. SDALT thus uses the automaton to precompute state-dependent distances, providing lower bound values per vertex *and* state. For even better query performance, SDALT can be extended to use more aggressive (and potentially incorrect) bounds to guide the search toward the target, relying on a label-correcting algorithm (which may scan vertices multiple times) to preserve correctness [169]. SDALT has been evaluated [169,170] on a realistic multimodal network covering the Île-de-France area (containing Paris) incorporating rental and private bicycles, public transit, walking, and a time-dependent road network for car travel. The resulting speedups are close to 30. Note that SDALT, like ANR, also predetermines the regular language constraints during preprocessing.

Contraction Hierarchies. Finally, Dibbelt et al. [106] have adapted Contraction Hierarchies [142] to LCSPP, handling arbitrary mode *sequence* constraints. The resulting User-Constrained Contraction Hierarchies (UCCH) algorithm works by (independently) only contracting vertices whose incident arcs belong to the same modal subnetwork. All other vertices are kept uncontracted. The query algorithm runs in two phases. The first runs a regular CH query in the subnetworks given as initial or final transport modes of the sequence constraints until the uncontracted *core graph* is reached. Between these entry and exit vertices, the second phase then runs a regular LCSPP-Dijkstra restricted to the (much smaller) core graph. Query performance of UCCH is comparable to Access-Node Routing, but with significantly less preprocessing time and space. Also, in contrast to ANR, UCCH also handles arbitrary mode sequence constraints at query time.

5.3   Multicriteria Optimization

While label constraints are useful to define feasible journeys, computing the (single) shortest label-constrained path has two important drawbacks. First, in order to define the constraints, users must know the characteristics of the particular transportation network; second, alternative journeys that combine the available transportation modes differently are not computed. To obtain a set of diverse alternatives, multicriteria optimization has been considered.

The criteria optimized by these methods usually include arrival time and, for each mode of transportation, some mode-dependent optimization criterion [23, 73]. The resulting Pareto sets will thus contain journeys with different usage of the available transportation modes, from which users can choose their favorites.

Delling et al. [73] consider networks of metropolitan scale and use the following criteria as proxies for "convenience": number of transfers in public transit, walking duration for the pedestrian network, and monetary cost for taxis. They observe that simply applying the MLS algorithm [155,187,196,243] to a comprehensive multimodal graph turns out to be slow, even when partial contraction is applied to the road and pedestrian networks, as in UCCH [106]. To get better query performance, they extend RAPTOR [92] to the multimodal scenario, which results in the *multimodal multicriteria RAPTOR* algorithm (MCR) [73]. Like RAPTOR, MCR operates in rounds (one per transfer) and computes Pareto sets of optimal journeys with exactly i transfers in round i. It does so by running, in each round, a dedicated subalgorithm (RAPTOR for public transit; MLS for walking and taxi) which obtains journeys with the respective transport mode as their last leg.

Since with increasing number of optimization criteria the resulting Pareto sets tend to get very large, Delling et al. identify the most significant journeys in a quick postprocessing step by a scoring method based on fuzzy logic [259]. For faster queries, MCR-based heuristics (which relax domination during the algorithm) successfully find the most significant journeys while avoiding the computation of insignificant ones in the first place.

Bast et al. [23] use MLS with contraction to compute multimodal multicriteria journeys at a metropolitan scale. To identify the significant journeys of the Pareto set, they propose a method called *Types aNd Thresholds* (TNT). The method is based on a set of simple *axioms* that summarize what most users would consider as unreasonable multimodal paths. For example, if one is willing to take the car for a large fraction of the trip, one might as well take it for the whole trip. Three types of reasonable trips are deduced from the axioms: (1) only car, (2) arbitrarily much transit and walking with no car, and (3) arbitrarily much transit with little or no walking and car. With a concrete threshold for "little" (such as 10 min), the rules can then be applied to filter the reasonable journeys. As in [73], filtering can be applied during the algorithm to prune the search space and reduce query time. The resulting sets are fairly robust with respect to the choice of threshold.

6   Final Remarks

The last decade has seen astonishing progress in the performance of shortest path algorithms on transportation networks. For routing in road networks, in particular, modern algorithms can be up to seven orders of magnitude faster than standard solutions. Successful approaches exploit different properties of road networks that make them easier to deal with than general graphs, such as goal direction, a strong hierarchical structure, and the existence of small separators. Although some early acceleration techniques relied heavily on geometry (road networks are after all embedded on the surface of the Earth), no current state-of-the-art algorithm makes explicit use of vertex coordinates (see Table 1). While one still sees the occasional development (and publication) of geometry-based algorithms they are consistently dominated by established techniques. In particular, the recent Arterial Hierarchies [261] algorithm is compared to CH (which has slightly slower queries), but not to other previously published techniques (such as CHASE, HL, and TNR) that would easily dominate it. This shows that results in this rapidly-evolving area are often slow to reach some communities; we hope this survey will help improve this state of affairs.

Note that experiments on real data are very important, as properties of production data are not always accurately captured by simplified models and folklore assumptions. For example, the common belief that an algorithm can be augmented to include turn penalties without significant loss in performance turned out to be wrong for CH [76].

Another important lesson from recent developments is that careful engineering is essential to unleash the full computational power of modern computer architectures. Algorithms such as CRP, CSA, HL, PHAST, and RAPTOR, for example, achieve much of their good performance by carefully exploiting locality of reference and parallelism (at the level of instructions, cores, and even GPUs).

The ultimate validation of several of the approaches described here is that they have found their way into systems that serve millions of users every day. Several authors of papers cited in this survey have worked on routing-related projects for companies like Apple, Esri, Google, MapBox, Microsoft, Nokia, PTV, TeleNav, TomTom, and Yandex. Although companies tend to be secretive about the actual algorithms they use, in some cases this is public knowledge. TomTom uses a variant of Arc Flags with shortcuts to perform time-dependent queries [231]. Microsoft's Bing Maps[4] use CRP for routing in road networks. OSRM [185], a popular route planning engine using OpenStreetMap data, uses CH for queries. The Transfer Patterns [24] algorithm has been in use for public-transit journey planning on Google Maps[5] since 2010. RAPTOR is currently in use by OpenTripPlanner[6].

These recent successes do not mean that all problems in this area are solved. The ultimate goal, a worldwide multimodal journey planner, has not yet been reached. Systems like Rome2Rio[7] provide a simplified first step, but a more useful system would take into account real-time traffic and transit information, historic patterns, schedule constraints, and monetary costs. Moreover, all these elements should be combined in a personalized manner. Solving such a general problem efficiently seems beyond the reach of current algorithms. Given the recent pace of progress, however, a solution may be closer than expected.

---

[4] http://www.bing.com/blogs/site_blogs/b/maps/archive/2012/01/05/
bing-maps-new-routing-engine.aspx.
[5] http://www.google.com/transit.
[6] http://opentripplanner.com.
[7] http://www.rome2rio.com.

References

1. Abraham, I., Delling, D., Fiat, A., Goldberg, A.V., Werneck, R.F.: VC-dimension and shortest path algorithms. In: Aceto, L., Henzinger, M., Sgall, J. (eds.) ICALP 2011. LNCS, vol. 6755, pp. 690–699. Springer, Heidelberg (2011). doi:10.1007/978-3-642-22006-7_58
2. Abraham, I., Delling, D., Fiat, A., Goldberg, A.V., Werneck, R.F.: HLDB: Location-based services in databases. In: Proceedings of the 20th ACM SIGSPA-TIAL International Symposium on Advances in Geographic Information Systems (GIS 2012), pp. 339–348. ACM Press 2012. Best Paper Award
3. Abraham, I., Delling, D., Fiat, A., Goldberg, A.V., Werneck, R.F.: Highway dimension and provably efficient shortest path algorithms. Technical report MSR-TR-2013-91, Microsoft Research (2013)
4. Abraham, I., Delling, D., Goldberg, A.V., Werneck, R.F.: A hub-based labeling algorithm for shortest paths in road networks. In: Pardalos, P.M., Rebennack, S. (eds.) SEA 2011. LNCS, vol. 6630, pp. 230–241. Springer, Heidelberg (2011). doi:10.1007/978-3-642-20662-7_20
5. Abraham, I., Delling, D., Goldberg, A.V., Werneck, R.F.: Hierarchical hub label-ings for shortest paths. In: Epstein, L., Ferragina, P. (eds.) ESA 2012. LNCS, vol. 7501, pp. 24–35. Springer, Heidelberg (2012). doi:10.1007/978-3-642-33090-2_4
6. Abraham, I., Delling, D., Goldberg, A.V., Werneck, R.F.: Alternative routes in road networks. ACM J. Exp. Algorithm. **18**(1), 1–17 (2013)
7. Abraham, I., Fiat, A., Goldberg, A.V., Werneck, R.F.: Highway dimension, short-est paths, and provably efficient algorithms. In: Proceedings of the 21st Annual ACM-SIAM Symposium on Discrete Algorithms (SODA 2010), pp. 782–793. SIAM (2010)
8. Ahuja, R.K., Mehlhorn, K., Orlin, J.B., Tarjan, R.: Faster algorithms for the shortest path problem. J. ACM **37**(2), 213–223 (1990)
9. Ahuja, R.K., Orlin, J.B., Pallottino, S., Scutellà, M.G.: Dynamic shortest paths minimizing travel times and costs. Networks **41**(4), 197–205 (2003)
10. Aifadopoulou, G., Ziliaskopoulos, A., Chrisohoou, E.: Multiobjective optimum path algorithm for passenger pretrip planning in multimodal transportation net-works. J. Transp. Res. Board **2032**(1), 26–34 (2007). doi:10.3141/2032-04
11. Akiba, T., Iwata, Y., Kawarabayashi, K., Kawata, Y.: Fast shortest-path distance queries on road networks by pruned highway labeling. In: Proceedings of the 16th Meeting on Algorithm Engineering and Experiments (ALENEX 2014), pp. 147–154. SIAM (2014)
12. Akiba, T., Iwata, Y., Yoshida,Y.: Fast exact shortest-path distance queries on large networks by pruned landmark labeling. In: Proceedings of the 2013 ACM SIGMOD International Conference on Management of Data (SIGMOD 2013), pp. 349–360. ACM Press (2013)
13. Allulli, L., Italiano, G.F., Santaroni, F.: Exploiting GPS data in public transport journey planners. In: Gudmundsson, J., Katajainen, J. (eds.) SEA 2014. LNCS, vol. 8504, pp. 295–306. Springer, Heidelberg (2014). doi:10.1007/978-3-319-07959-2_25
14. Antsfeld, L., Walsh, T.: Finding multi-criteria optimal paths in multi-modal pub-lic transportation networks using the transit algorithm. In: Proceedings of the 19th ITS World Congress (2012)
15. Arz, J., Luxen, D., Sanders, P.: Transit node routing reconsidered. In: Bonifaci, V., Demetrescu, C., Marchetti-Spaccamela, A. (eds.) SEA 2013. LNCS, vol. 7933, pp. 55–66. Springer, Heidelberg (2013). doi:10.1007/978-3-642-38527-8_7

16. Babenko, M., Goldberg, A.V., Gupta, A., Nagarajan, V.: Algorithms for hub label optimization. In: Fomin, F.V., Freivalds, R., Kwiatkowska, M., Peleg, D. (eds.) ICALP 2013. LNCS, vol. 7965, pp. 69–80. Springer, Heidelberg (2013). doi:10.1007/978-3-642-39206-1_7

17. Bader, R., Dees, J., Geisberger, R., Sanders, P.: Alternative route graphs in road networks. In: Marchetti-Spaccamela, A., Segal, M. (eds.) TAPAS 2011. LNCS, vol. 6595, pp. 21–32. Springer, Heidelberg (2011). doi:10.1007/978-3-642-19754-3_5

18. Barrett, C., Bisset, K., Holzer, M., Konjevod, G., Marathe, M., Wagner, D.: Engineering label-constrained shortest-path algorithms. In: Fleischer, R., Xu, J. (eds.) AAIM 2008. LNCS, vol. 5034, pp. 27–37. Springer, Heidelberg (2008). doi:10.1007/978-3-540-68880-8_5

19. Barrett, C., Bisset, K., Holzer, M., Konjevod, G., Marathe, M.V., Wagner, D.: Engineering label-constrained shortest-path algorithms. In: The Shortest Path Problem: Ninth DIMACS Implementation Challenge, DIMACS Book, vol. 74, pp. 309–319. American Mathematical Society (2009)

20. Barrett, C., Bisset, K., Jacob, R., Konjevod, G., Marathe, M.: Classical and contemporary shortest path problems in road networks: implementation and experimental analysis of the TRANSIMS router. In: Möhring, R., Raman, R. (eds.) ESA 2002. LNCS, vol. 2461, pp. 126–138. Springer, Heidelberg (2002). doi:10.1007/3-540-45749-6_15

21. Barrett, C., Jacob, R., Marathe, M.V.: Formal-language-constrained path problems. SIAM J. Comput. **30**(3), 809–837 (2000)

22. Bast, H.: Car or public transport—two worlds. In: Albers, S., Alt, H., Näher, S. (eds.) Efficient Algorithms. LNCS, vol. 5760, pp. 355–367. Springer, Heidelberg (2009). doi:10.1007/978-3-642-03456-5_24

23. Bast, H., Brodesser, M., Storandt, S.: Result diversity for multi-modal route planning. In: Proceedings of the 13th Workshop on Algorithmic Approaches for Transportation Modeling, Optimization, and Systems (ATMOS 2013), OpenAccess Series in Informatics (OASIcs), pp. 123–136 (2013)

24. Bast, H., Carlsson, E., Eigenwillig, A., Geisberger, R., Harrelson, C., Raychev, V., Viger, F.: Fast routing in very large public transportation networks using transfer patterns. In: Berg, M., Meyer, U. (eds.) ESA 2010. LNCS, vol. 6346, pp. 290–301. Springer, Heidelberg (2010). doi:10.1007/978-3-642-15775-2_25

25. Bast, H., Sternisko, J., Storandt, S.: Delay-robustness of transfer patterns in public transportation route planning. In: Proceedings of the 13th Workshop on Algorithmic Approaches for Transportation Modeling, Optimization, and Systems (ATMOS 2013), OpenAccess Series in Informatics (OASIcs), pp. 42–54 (2013)

26. Bast, H., Storandt, S.: Flow-based guidebook routing. In: Proceedings of the 16th Meeting on Algorithm Engineering and Experiments (ALENEX 2014), pp. 155–165. SIAM (2014)

27. Bast, H., Storandt, S.: Frequency-based search for public transit. In: Proceedings of the 22nd ACM SIGSPATIAL International Conference on Advances in Geographic Information Systems, pp. 13–22. ACM Press, November 2014

28. Bast, H., Funke, S., Matijevic, D.: Ultrafast shortest-path queries via transit nodes. In: The Shortest Path Problem: Ninth DIMACS Implementation Challenge, DIMACS Book, vol. 74, pp. 175–192. American Mathematical Society (2009)

29. Bast, H., Funke, S., Matijevic, D., Sanders, P., Schultes, D.: In transit to constant shortest-path queries in road networks. In: Proceedings of the 9th Workshop on Algorithm Engineering and Experiments (ALENEX 2007), pp. 46–59. SIAM (2007)

30. Bast, H., Funke, S., Sanders, P., Schultes, D.: Fast routing in road networks with transit nodes. Science **316**(5824), 566 (2007)
31. Batz, G.V., Geisberger, R., Luxen, D., Sanders, P., Zubkov, R.: Efficient route compression for hybrid route planning. In: Even, G., Rawitz, D. (eds.) MedAlg 2012. LNCS, vol. 7659, pp. 93–107. Springer, Heidelberg (2012). doi:10.1007/978-3-642-34862-4_7
32. Batz, G.V., Geisberger, R., Sanders, P., Vetter, C.: Minimum time-dependent travel times with contraction hierarchies. ACM J. Exp. Algorithm. **18**(1.4), 1–43 (2013)
33. Batz, G.V., Sanders, P.: Time-dependent route planning with generalized objective functions. In: Epstein, L., Ferragina, P. (eds.) ESA 2012. LNCS, vol. 7501, pp. 169–180. Springer, Heidelberg (2012). doi:10.1007/978-3-642-33090-2_16
34. Bauer, A.: Multimodal profile queries. Bachelor thesis, Karlsruhe Institute of Technology, May 2012
35. Bauer, R., Baum, M., Rutter, I., Wagner, D.: On the complexity of partitioning graphs for arc-flags. J. Graph Algorithms Appl. **17**(3), 265–299 (2013)
36. Bauer, R., Columbus, T., Katz, B., Krug, M., Wagner, D.: Preprocessing speed-up techniques is hard. In: Calamoneri, T., Diaz, J. (eds.) CIAC 2010. LNCS, vol. 6078, pp. 359–370. Springer, Heidelberg (2010). doi:10.1007/978-3-642-13073-1_32
37. Bauer, R., Columbus, T., Rutter, I., Wagner, D.: Search-space size in contraction hierarchies. In: Fomin, F.V., Freivalds, R., Kwiatkowska, M., Peleg, D. (eds.) ICALP 2013. LNCS, vol. 7965, pp. 93–104. Springer, Heidelberg (2013). doi:10.1007/978-3-642-39206-1_9
38. Bauer, R., D'Angelo, G., Delling, D., Schumm, A., Wagner, D.: The shortcut problem - complexity and algorithms. J. Graph Algorithms Appl. **16**(2), 447–481 (2012)
39. Bauer, R., Delling, D.: SHARC: Fast and robust unidirectional routing. ACM J. Exp. Algorithm. **14**(2.4), 1–29 (2009). Special Section on Selected Papers from ALENEX 2008
40. Bauer, R., Delling, D., Sanders, P., Schieferdecker, D., Schultes, D., Wagner, D.: Combining hierarchical, goal-directed speed-up techniques for Dijkstra's algorithm. ACM J. Exp. Algorithm. **15**(2.3), 1–31 (2010). Special Section devoted to WEA 2008
41. Bauer, R., Delling, D., Wagner, D.: Experimental study on speed-up techniques for timetable information systems. Networks **57**(1), 38–52 (2011)
42. Bauer, R., Krug, M., Meinert, S., Wagner, D.: Synthetic road networks. In: Chen, B. (ed.) AAIM 2010. LNCS, vol. 6124, pp. 46–57. Springer, Heidelberg (2010). doi:10.1007/978-3-642-14355-7_6
43. Baum, M., Dibbelt, J., Hübschle-Schneider, L., Pajor, T., Wagner, D.: Speed-consumption tradeoff for electric vehicle route planning. In: Proceedings of the 14th Workshop on Algorithmic Approaches for Transportation Modeling, Optimization, and Systems (ATMOS 2014), OpenAccess Series in Informatics (OASIcs), pp. 138–151 (2014)
44. Baum, M., Dibbelt, J., Pajor, T., Wagner, D.: Energy-optimal routes for electric vehicles. In: Proceedings of the 21st ACM SIGSPATIAL International Conference on Advances in Geographic Information Systems, pp. 54–63. ACM Press (2013)
45. Baumann, N., Schmidt, R.: Buxtehude-Garmisch in 6 Sekunden. die elektronische Fahrplanauskunft (EFA) der Deutschen Bundesbahn. Zeitschrift für aktuelle Verkehrsfragen **10**, 929–931 (1988)

46. Bellman, R.: On a routing problem. Q. Appl. Math. **16**, 87–90 (1958)
47. Berger, A., Delling, D., Gebhardt, A., Müller-Hannemann, M.: Accelerating time-dependent multi-criteria timetable information is harder than expected. In: Proceedings of the 9th Workshop on Algorithmic Approaches for Transportation Modeling, Optimization, and Systems (ATMOS 2009), OpenAccess Series in Informatics (OASIcs) (2009)
48. Berger, A., Gebhardt, A., Müller-Hannemann, M., Ostrowski, M.: Stochastic delay prediction in large train networks. In: Proceedings of the 11th Workshop on Algorithmic Approaches for Transportation Modeling, Optimization, and Systems (ATMOS 2011), OpenAccess Series in Informatics (OASIcs), vol. 20, pp. 100–111 (2011)
49. Berger, A., Grimmer, M., Müller-Hannemann, M.: Fully dynamic speed-up techniques for multi-criteria shortest path searches in time-dependent networks. In: Festa, P. (ed.) SEA 2010. LNCS, vol. 6049, pp. 35–46. Springer, Heidelberg (2010). doi:10.1007/978-3-642-13193-6_4
50. Bielli, M., Boulmakoul, A., Mouncif, H.: Object modeling and path computation for multimodal travel systems. Eur. J. Oper. Res. **175**(3), 1705–1730 (2006)
51. Böhmová, K., Mihalák, M., Pröger, T., Šrámek, R., Widmayer, P.: Robust routing in urban public transportation: how to find reliable journeys based on past observations. In: Proceedings of the 13th Workshop on Algorithmic Approaches for Transportation Modeling, Optimization, and Systems (ATMOS 2013), OpenAccess Series in Informatics (OASIcs), pp. 27–41 (2013)
52. Botea, A.: Ultra-fast optimal pathfinding without runtime search. In: Proceedings of the Seventh AAAI Conference on Artificial Intelligence and Interactive Digital Entertainment (AIIDE 2011), pp. 122–127. AAAI Press (2011)
53. Botea, A., Harabor, D.: Path planning with compressed all-pairs shortest paths data. In: Proceedings of the 23rd International Conference on Automated Planning and Scheduling, AAAI Press (2013)
54. Brandes, U., Erlebach, T.: Network Analysis: Methodological Foundations. Theoretical Computer Science and General Issues, vol. 3418. Springer, Heidelberg (2005)
55. Brandes, U., Schulz, F., Wagner, D., Willhalm, T.: Travel planning with self-made maps. In: Buchsbaum, A.L., Snoeyink, J. (eds.) ALENEX 2001. LNCS, vol. 2153, pp. 132–144. Springer, Heidelberg (2001). doi:10.1007/3-540-44808-X_10
56. Brodal, G., Jacob, R.: Time-dependent networks as models to achieve fast exact time-table queries. In: Proceedings of the 3rd Workshop on Algorithmic Methods and Models for Optimization of Railways (ATMOS 2003), Electronic Notes in Theoretical Computer Science, vol. 92, pp. 3–15 (2004)
57. Bruera, F., Cicerone, S., D'Angelo, G., Di Stefano, G., Frigioni, D.: Dynamic multi-level overlay graphs for shortest paths. Math. Comput. Sci. **1**(4), 709–736 (2008)
58. Brunel, E., Delling, D., Gemsa, A., Wagner, D.: Space-efficient SHARC-routing. In: Festa, P. (ed.) SEA 2010. LNCS, vol. 6049, pp. 47–58. Springer, Heidelberg (2010). doi:10.1007/978-3-642-13193-6_5
59. Caldwell, T.: On finding minimum routes in a network with turn penalties. Commun. ACM **4**(2), 107–108 (1961)
60. Cambridge Vehicle Information Technology Ltd. Choice routing (2005). http://www.camvit.com
61. Cherkassky, B.V., Goldberg, A.V., Radzik, T.: Shortest paths algorithms. Math. Programm. Ser. A **73**, 129–174 (1996)

62. Cherkassky, B.V., Goldberg, A.V., Silverstein, C.: Buckets, heaps, lists, and monotone priority queues. In: Proceedings of the 8th Annual ACM-SIAM Symposium on Discrete Algorithms (SODA 1997), pp. 83–92. IEEE Computer Society Press (1997)
63. Cionini, A., D'Angelo, G., D'Emidio, M., Frigioni, D., Giannakopoulou, K., Paraskevopoulos, A.: Engineering graph-based models for dynamic timetable information systems. In: Proceedings of the 14th Workshop on Algorithmic Approaches for Transportation Modeling, Optimization, and Systems (ATMOS 2014), OpenAccess Series in Informatics (OASIcs), pp. 46–61 (2014)
64. Cohen, E., Halperin, E., Kaplan, H., Zwick, U.: Reachability and distance queries via 2-hop labels. SIAM J. Comput. **32**(5), 1338–1355 (2003)
65. Cooke, K., Halsey, E.: The shortest route through a network with time-dependent intermodal transit times. J. Math. Anal. Appl. **14**(3), 493–498 (1966)
66. D'Angelo, G., D'Emidio, M., Frigioni, D., Vitale, C.: Fully dynamic maintenance of arc-flags in road networks. In: Klasing, R. (ed.) SEA 2012. LNCS, vol. 7276, pp. 135–147. Springer, Heidelberg (2012). doi:10.1007/978-3-642-30850-5_13
67. George, B.D.: Linear Programming and Extensions. Princeton University Press, Princeton (1962)
68. Dean, B.C.: Continuous-time dynamic shortest path algorithms. Master's thesis, Massachusetts Institute of Technology (1999)
69. Dean, B.C.: Algorithms for minimum-cost paths in time-dependent networks with waiting policies. Networks **44**(1), 41–46 (2004)
70. Dean, B.C.: Shortest paths in FIFO time-dependent networks: theory and algorithms. Technical report, Massachusetts Institute Of Technology (2004)
71. Dehne, F., Omran, M.T., Sack, J.-R.: Shortest paths in time-dependent FIFO networks. Algorithmica **62**, 416–435 (2012)
72. Delling, D.: Time-dependent SHARC-routing. Algorithmica **60**(1), 60–94 (2011)
73. Delling, D., Dibbelt, J., Pajor, T., Wagner, D., Werneck, R.F.: Computing multimodal journeys in practice. In: Bonifaci, V., Demetrescu, C., Marchetti-Spaccamela, A. (eds.) SEA 2013. LNCS, vol. 7933, pp. 260–271. Springer, Heidelberg (2013). doi:10.1007/978-3-642-38527-8_24
74. Delling, D., Giannakopoulou, K., Wagner, D., Zaroliagis, C.: Timetable information updating in case of delays: modeling issues. Technical report 133, Arrival Technical report (2008)
75. Delling, D., Goldberg, A.V., Nowatzyk, A., Werneck, R.F.: PHAST: Hardware-accelerated shortest path trees. J. Parallel Distrib. Comput. **73**(7), 940–952 (2013)
76. Delling, D., Goldberg, A.V., Pajor, T., Werneck, R.F.: Customizable route planning. In: Pardalos, P.M., Rebennack, S. (eds.) SEA 2011. LNCS, vol. 6630, pp. 376–387. Springer, Heidelberg (2011). doi:10.1007/978-3-642-20662-7_32
77. Delling, D., Goldberg, A.V., Pajor, T., Werneck, R.F.: Robust distance queries on massive networks. In: Schulz, A.S., Wagner, D. (eds.) ESA 2014. LNCS, vol. 8737, pp. 321–333. Springer, Heidelberg (2014). doi:10.1007/978-3-662-44777-2_27
78. Delling, D., Goldberg, A.V., Pajor, T., Werneck, R.F.: Customizable route planning in road networks. Transp. Sci. (2015)
79. Delling, D., Goldberg, A.V., Razenshteyn, I., Werneck, R.F.: Graph partitioning with natural cuts. In: 25th International Parallel and Distributed Processing Symposium (IPDPS 2011), pp. 1135–1146. IEEE Computer Society (2011)
80. Delling, D., Goldberg, A.V., Savchenko, R., Werneck, R.F.: Hub labels: theory and practice. In: Gudmundsson, J., Katajainen, J. (eds.) SEA 2014. LNCS, vol. 8504, pp. 259–270. Springer, Heidelberg (2014). doi:10.1007/978-3-319-07959-2_22

81. Delling, D., Goldberg, A.V., Werneck, R.F.: Faster batched shortest paths in road networks. In: Proceedings of the 11th Workshop on Algorithmic Approaches for Transportation Modeling, Optimization, and Systems (ATMOS 2011), OpenAccess Series in Informatics (OASIcs), vol. 20, pp. 52–63 (2011)

82. Delling, D., Goldberg, A.V., Werneck, R.F.: Hub label compression. In: Bonifaci, V., Demetrescu, C., Marchetti-Spaccamela, A. (eds.) SEA 2013. LNCS, vol. 7933, pp. 18–29. Springer, Heidelberg (2013). doi:10.1007/978-3-642-38527-8_4

83. Delling, D., Holzer, M., Müller, K., Schulz, F., Wagner, D., High-performance multi-level routing. In: The Shortest Path Problem: Ninth DIMACS Implementation Challenge, DIMACS Book, vol. 74, pp. 73–92. American Mathematical Society (2009)

84. Delling, D., Italiano, G.F., Pajor, T., Santaroni, F.: Better transit routing by exploiting vehicle GPS data. In: Proceedings of the 7th ACM SIGSPATIAL International Workshop on Computational Transportation Science. ACM Press, November 2014

85. Delling, D., Katz, B., Pajor, T.: Parallel computation of best connections in public transportation networks. ACM J. Exp. Algorithm. **17**(4), 4. 1–4. 26 (2012)

86. Delling, D., Kobitzsch, M., Luxen, D., Werneck, R.F.: Robust mobile route planning with limited connectivity. In: Proceedings of the 14th Meeting on Algorithm Engineering and Experiments (ALENEX 2012), pp. 150–159. SIAM (2012)

87. Delling, D., Kobitzsch, M., Werneck, R.F.: Customizing driving directions with GPUs. In: Silva, F., Dutra, I., Santos Costa, V. (eds.) Euro-Par 2014. LNCS, vol. 8632, pp. 728–739. Springer, Heidelberg (2014). doi:10.1007/978-3-319-09873-9_61

88. Delling, D., Nannicini, G.: Core routing on dynamic time-dependent road networks. Informs J. Comput. **24**(2), 187–201 (2012)

89. Delling, D., Pajor, T., Wagner, D.: Accelerating multi-modal route planning by access-nodes. In: Fiat, A., Sanders, P. (eds.) ESA 2009. LNCS, vol. 5757, pp. 587–598. Springer, Heidelberg (2009). doi:10.1007/978-3-642-04128-0_53

90. Delling, D., Pajor, T., Wagner, D.: Engineering time-expanded graphs for faster timetable information. In: Ahuja, R.K., Möhring, R.H., Zaroliagis, C.D. (eds.) Robust and Online Large-Scale Optimization. LNCS, vol. 5868, pp. 182–206. Springer, Heidelberg (2009). doi:10.1007/978-3-642-05465-5_7

91. Delling, D., Pajor, T., Wagner, D., Zaroliagis, C.: Efficient route planning in flight networks. In: Proceedings of the 9th Workshop on Algorithmic Approaches for Transportation Modeling, Optimization, and Systems (ATMOS 2009), OpenAccess Series in Informatics (OASIcs) (2009)

92. Delling, D., Pajor, T., Werneck, R.F.: Round-based public transit routing. In: Proceedings of the 14th Meeting on Algorithm Engineering and Experiments (ALENEX 2012), pp. 130–140. SIAM (2012)

93. Delling, D., Pajor, T., Werneck, R.F.: Round-based public transit routing. Transp. Sci. **49**, 591–604 (2014)

94. Delling, D., Sanders, P., Schultes, D., Wagner, D.: Engineering route planning algorithms. In: Lerner, J., Wagner, D., Zweig, K.A. (eds.) Algorithmics of Large and Complex Networks. LNCS, vol. 5515, pp. 117–139. Springer, Heidelberg (2009). doi:10.1007/978-3-642-02094-0_7

95. Delling, D., Sanders, P., Schultes, D., Wagner, D.: Highway hierarchies star. In: The Shortest Path Problem: Ninth DIMACS Implementation Challenge, DIMACS Book, vol. 74, pp. 141–174. American Mathematical Society (2009)

96. Delling, D., Wagner, D.: Landmark-based routing in dynamic graphs. In: Demetrescu, C. (ed.) WEA 2007. LNCS, vol. 4525, pp. 52–65. Springer, Heidelberg (2007). doi:10.1007/978-3-540-72845-0_5

97. Delling, D., Wagner, D.: Pareto paths with SHARC. In: Vahrenhold, J. (ed.) SEA 2009. LNCS, vol. 5526, pp. 125–136. Springer, Heidelberg (2009). doi:10.1007/978-3-642-02011-7_13

98. Delling, D., Wagner, D.: Time-dependent route planning. In: Ahuja, R.K., Möhring, R.H., Zaroliagis, C.D. (eds.) Robust and Online Large-Scale Optimization. LNCS, vol. 5868, pp. 207–230. Springer, Heidelberg (2009). doi:10.1007/978-3-642-05465-5_8

99. Delling, D., Werneck, R.F.: Customizable point-of-interest queries in road networks. In: Proceedings of the 21st ACM SIGSPATIAL International Symposium on Advances in Geographic Information Systems (GIS 2013), pp. 490–493. ACM Press (2013)

100. Delling, D., Werneck, R.F.: Faster customization of road networks. In: Bonifaci, V., Demetrescu, C., Marchetti-Spaccamela, A. (eds.) SEA 2013. LNCS, vol. 7933, pp. 30–42. Springer, Heidelberg (2013). doi:10.1007/978-3-642-38527-8_5

101. Demetrescu, C., Goldberg, A.V., Johnson, D.S. (eds.): The Shortest Path Problem: Ninth DIMACS Implementation Challenge, DIMACS Book, vol. 74. American Mathematical Society, Providence (2009)

102. Demiryurek, U., Banaei-Kashani, F., Shahabi, C.: A case for time-dependent shortest path computation in spatial networks. In: Proceedings of the 18th ACM SIGSPATIAL International Conference on Advances in Geographic Information Systems (GIS 2010), pp. 474–477 (2010)

103. Denardo, E.V., Fox, B.L.: Shortest-route methods: 1. reaching, pruning, and buckets. Oper. Res. **27**(1), 161–186 (1979)

104. Dial, R.B.: Algorithm 360: shortest-path forest with topological ordering [H]. Commun. ACM **12**(11), 632–633 (1969)

105. Dibbelt, J., Pajor, T., Strasser, B., Wagner, D.: Intriguingly simple and fast transit routing. In: Bonifaci, V., Demetrescu, C., Marchetti-Spaccamela, A. (eds.) SEA 2013. LNCS, vol. 7933, pp. 43–54. Springer, Heidelberg (2013). doi:10.1007/978-3-642-38527-8_6

106. Dibbelt, J., Pajor, T., Wagner, D.: User-constrained multi-modal route planning. In: Proceedings of the 14th Meeting on Algorithm Engineering and Experiments (ALENEX 2012), pp. 118–129. SIAM (2012)

107. Dibbelt, J., Strasser, B., Wagner, D.: Customizable contraction hierarchies. In: Gudmundsson, J., Katajainen, J. (eds.) SEA 2014. LNCS, vol. 8504, pp. 271–282. Springer, Heidelberg (2014). doi:10.1007/978-3-319-07959-2_23

108. Dijkstra, E.W.: A note on two problems in connexion with graphs. Numer. Math. **1**, 269–271 (1959)

109. Disser, Y., Müller–Hannemann, M., Schnee, M.: Multi-criteria shortest paths in time-dependent train networks. In: McGeoch, C.C. (ed.) WEA 2008. LNCS, vol. 5038, pp. 347–361. Springer, Heidelberg (2008). doi:10.1007/978-3-540-68552-4_26

110. Drews, F., Luxen, D.: Multi-hop ride sharing. In: Proceedings of the 5th International Symposium on Combinatorial Search (SoCS 2012), pp. 71–79. AAAI Press (2013)

111. Dreyfus, S.E.: An appraisal of some shortest-path algorithms. Oper. Res. **17**(3), 395–412 (1969)

112. Efentakis, A., Pfoser, D.: Optimizing landmark-based routing and preprocessing. In: Proceedings of the 6th ACM SIGSPATIAL International Workshop on Computational Transportation Science, pp. 25:25–25:30. ACM Press, November 2013

113. Efentakis, A., Pfoser, D.: GRASP. Extending graph separators for the single-source shortest-path problem. In: Schulz, A.S., Wagner, D. (eds.) ESA 2014. LNCS, vol. 8737, pp. 358–370. Springer, Heidelberg (2014). doi:10.1007/978-3-662-44777-2_30

114. Efentakis, A., Pfoser, D., Vassiliou., Y.: SALT: a unified framework for all shortest-path query variants on road networks. CoRR, abs/1411.0257 (2014)

115. Efentakis, A., Pfoser, D., Voisard, A.: Efficient data management in support of shortest-path computation. In: Proceedings of the 4th ACM SIGSPATIAL International Workshop on Computational Transportation Science, pp. 28–33. ACM Press (2011)

116. Efentakis, A., Theodorakis, D., Pfoser,D.: Crowdsourcing computing resources for shortest-path computation. In: Proceedings of the 20th ACM SIGSPATIAL International Symposium on Advances in Geographic Information Systems (GIS 2012), pp. 434–437. ACM Press (2012)

117. Ehrgott, M., Gandibleux, X.: Multiple Criteria Optimization: State of the Art Annotated Bibliographic Surveys. Kluwer Academic Publishers Group, New York (2002)

118. Eisenstat, D.: Random road networks: the quadtree model. In: Proceedings of the Eighth Workshop on Analytic Algorithmics and Combinatorics (ANALCO 2011), pp. 76–84. SIAM, January 2011

119. Eisner, J., Funke, S.: Sequenced route queries: getting things done on the way back home. In: Proceedings of the 20th ACM SIGSPATIAL International Symposium on Advances in Geographic Information Systems (GIS 2012), pp. 502–505. ACM Press (2012)

120. Eisner, J., Funke, S.: Transit nodes - lower bounds and refined construction. In: Proceedings of the 14th Meeting on Algorithm Engineering and Experiments (ALENEX 2012), pp. 141–149. SIAM (2012)

121. Eisner, J., Funke, S., Herbst, A., Spillner, A., Storandt, S.: Algorithms for matching and predicting trajectories. In: Proceedings of the 13th Workshop on Algorithm Engineering and Experiments (ALENEX 2011), pp. 84–95. SIAM (2011)

122. Eisner, J., Funke, S., Storandt, S.: Optimal route planning for electric vehicles in large network. In: Proceedings of the Twenty-Fifth AAAI Conference on Artificial Intelligence. AAAI Press, August 2011

123. Eppstein, D., Goodrich, M.T.: Studying (non-planar) road networks through an algorithmic lens. In: Proceedings of the 16th ACM SIGSPATIAL International Conference on Advances in Geographic Information Systems (GIS 2008), pp. 1–10. ACM Press (2008)

124. Erb, S., Kobitzsch, M., Sanders, P.: Parallel bi-objective shortest paths using weight-balanced B-trees with bulk updates. In: Gudmundsson, J., Katajainen, J. (eds.) SEA 2014. LNCS, vol. 8504, pp. 111–122. Springer, Heidelberg (2014). doi:10.1007/978-3-319-07959-2_10

125. Firmani, D., Italiano, G.F., Laura, L., Santaroni, F.: Is timetabling routing always reliable for public transport? In: Proceedings of the 13th Workshop on Algorithmic Approaches for Transportation Modeling, Optimization, and Systems (ATMOS 2013), OpenAccess Series in Informatics (OASIcs), pp. 15–26 (2013)

126. Floyd, R.W.: Algorithm 97: shortest path. Commun. ACM 5(6), 345 (1962)

127. Ford, Jr., L.R.: Network flow theory. Technical report P-923, Rand Corporation, Santa Monica, California (1956)

128. Foschini, L., Hershberger, J., Suri, S.: On the complexity of time-dependent shortest paths. Algorithmica **68**(4), 1075–1097 (2014)

129. Fredman, M.L., Tarjan, R.E.: Fibonacci heaps and their uses in improved network optimization algorithms. J. ACM **34**(3), 596–615 (1987)

130. Fu, L., Sun, D., Rilett, L.R.: Heuristic shortest path algorithms for transportation applications: state of the art. Comput. Oper. Res. **33**(11), 3324–3343 (2006)

131. Funke, S., Nusser, A., Storandt, S.: On k-path covers and their applications. In: Proceedings of the 40th International Conference on Very Large Databases (VLDB 2014), pp. 893–902 (2014)

132. Funke, S., Nusser, A., Storandt, S.: Placement of loading stations for electric vehicles: no detours necessary! In: Proceedings of the Twenty-Eighth AAAI Conference on Artificial Intelligence. AAAI Press (2014)

133. Funke, S., Storandt, S.: Polynomial-time construction of contraction hierarchies for multi-criteria objectives. In: Proceedings of the 15th Meeting on Algorithm Engineering and Experiments (ALENEX 2013), pp. 31–54. SIAM (2013)

134. Gavoille, C., Peleg, D.: Compact and localized distributed data structures. Distrib. Comput. **16**(2–3), 111–120 (2003)

135. Gavoille, C., Peleg, D., Pérennes, S., Raz, R.: Distance labeling in graphs. J. Algorithms **53**, 85–112 (2004)

136. Geisberger, R.: Contraction of timetable networks with realistic transfers. In: Festa, P. (ed.) SEA 2010. LNCS, vol. 6049, pp. 71–82. Springer, Heidelberg (2010). doi:10.1007/978-3-642-13193-6_7

137. Geisberger, R.: Advanced route planning in transportation networks. Ph.D. thesis, Karlsruhe Institute of Technology, February 2011

138. Geisberger, R., Kobitzsch, M., Sanders, P.: Route planning with flexible objective functions. In: Proceedings of the 12th Workshop on Algorithm Engineering and Experiments (ALENEX 2010), pp. 124–137. SIAM (2010)

139. Geisberger, R., Luxen, D., Sanders, P., Neubauer, S., Volker, L.: Fast detour computation for ride sharing. In: Proceedings of the 10th Workshop on Algorithmic Approaches for Transportation Modeling, Optimization, and Systems (ATMOS 2010), OpenAccess Series in Informatics (OASIcs), vol. 14, pp. 88–99 (2010)

140. Geisberger, R., Rice, M., Sanders, P., Tsotras, V.: Route planning with flexible edge restrictions. ACM J. Exp. Algorithm. **17**(1), 1–20 (2012)

141. Geisberger, R., Sanders, P.: Engineering time-dependent many-to-many shortest paths computation. In: Proceedings of the 10th Workshop on Algorithmic Approaches for Transportation Modeling, Optimization, and Systems (ATMOS 2010), OpenAccess Series in Informatics (OASIcs), vol. 14 (2010)

142. Geisberger, R., Sanders, P., Schultes, D., Vetter, C.: Exact routing in large road networks using contraction hierarchies. Transp. Sci. **46**(3), 388–404 (2012)

143. Geisberger, R., Schieferdecker, D.: Heuristic contraction hierarchies with approximation guarantee. In: Proceedings of the 3rd International Symposium on Combinatorial Search (SoCS 2010). AAAI Press (2010)

144. Geisberger, R., Vetter, C.: Efficient routing in road networks with turn costs. In: Pardalos, P.M., Rebennack, S. (eds.) SEA 2011. LNCS, vol. 6630, pp. 100–111. Springer, Heidelberg (2011). doi:10.1007/978-3-642-20662-7_9

145. Goerigk, M., Heße, S., Müller-Hannemann, M., Schmidt, M.: Recoverable robust timetable information. In: Proceedings of the 13th Workshop on Algorithmic Approaches for Transportation Modeling, Optimization, and Systems (ATMOS 2013), OpenAccess Series in Informatics (OASIcs), pp. 1–14 (2013)

146. Goerigk, M., Knoth, M., Müller-Hannemann, M., Schmidt, M., Schöbel, A.: The price of strict and light robustness in timetable information. Transp. Sci. **48**, 225–242 (2014)

147. Goldberg, A.V.: A practical shortest path algorithm with linear expected time. SIAM J. Comput. **37**, 1637–1655 (2008)

148. Goldberg, A.V., Harrelson, C.: Computing the shortest path: A* search meets graph theory. In: Proceedings of the 16th Annual ACM-SIAM Symposium on Discrete Algorithms (SODA 2005), pp. 156–165. SIAM (2005)

149. Goldberg, A.V., Kaplan, H., Werneck, R.F.: Reach for A*: shortest path algorithms with preprocessing. In: The Shortest Path Problem: Ninth DIMACS Implementation Challenge, DIMACS Book, vol. 74, pp. 93–139. American Mathematical Society (2009)

150. Goldberg, A.V., Werneck, R.F.: Computing point-to-point shortest paths from external memory. In: Proceedings of the 7th Workshop on Algorithm Engineering and Experiments (ALENEX 2005), pp. 26–40. SIAM (2005)

151. Goldman, R., Shivakumar, N.R., Venkatasubramanian, S., Garcia-Molina, H.: Proximity search in databases. In: Proceedings of the 24th International Conference on Very Large Databases (VLDB 1998), pp. 26–37. Morgan Kaufmann, August 1998

152. Goodrich, M.T., Pszona, P.: Two-phase bicriterion search for finding fast and efficient electric vehicle routes. In: Proceedings of the 22nd ACM SIGSPATIAL International Conference on Advances in Geographic Information Systems. ACM Press, November 2014

153. Gunkel, T., Schnee, M., Müller-Hannemann, M.: How to find good night train connections. Networks **57**(1), 19–27 (2011)

154. Gutman, R.J., Reach-based routing: a new approach to shortest path algorithms optimized for road networks. In: Proceedings of the 6th Workshop on Algorithm Engineering and Experiments (ALENEX 2004), pp. 100–111. SIAM (2004)

155. Hansen, P.: Bricriteria path problems. In: Fandel, G., Gal, T. (eds.) Multiple Criteria Decision Making - Theory and Application. LNEMS, vol. 177, pp. 109–127. Springer, Heidelberg (1979). doi:10.1007/978-3-642-48782-8_9

156. Hart, P.E., Nilsson, N., Raphael, B.: A formal basis for the heuristic determination of minimum cost paths. IEEE Trans. Syst. Sci. Cybern. **4**, 100–107 (1968)

157. Hilger, M., Köhler, E., Möhring, R.H., Schilling, H., Fast point-to-point shortest path computations with arc-flags. In: The Shortest Path Problem: Ninth DIMACS Implementation Challenge, DIMACS Book, vol. 74, pp. 41–72. American Mathematical Society (2009)

158. Hliněný, P., Moriš, O.: Scope-based route planning. In: Demetrescu, C., Halldórsson, M.M. (eds.) ESA 2011. LNCS, vol. 6942, pp. 445–456. Springer, Heidelberg (2011). doi:10.1007/978-3-642-23719-5_38

159. Holzer, M.: Engineering planar-separator and shortest-path algorithms. Ph.D. thesis, Karlsruhe Institute of Technology (KIT)'- Department of Informatics (2008)

160. Holzer, M., Schulz, F., Wagner, D.: Engineering multilevel overlay graphs for shortest-path queries. ACM J. Exp. Algorithm. **13**(2.5), 1–26 (2008)

161. Holzer, M., Schulz, F., Wagner, D., Willhalm, T.: Combining speed-up techniques for shortest-path computations. ACM J. Exp. Algorithm. **10**(2.5), 1–18 (2006)

162. Horvitz, E., Krumm, J.: Some help on the way: opportunistic routing under uncertainty. In: Proceedings of the 2012 ACM Conference on Ubiquitous Computing (Ubicomp 2012), pp. 371–380. ACM Press (2012)

163. Ikeda, T., Hsu, M.-Y., Imai, H., Nishimura, S., Shimoura, H., Hashimoto, T., Tenmoku, K., Mitoh, K.: A fast algorithm for finding better routes by AI search techniques. In: Proceedings of the Vehicle Navigation and Information Systems Conference (VNSI 1994), pp. 291–296. ACM Press (1994)

164. Jing, N., Huang, Y.-W., Rundensteiner, E.A.: Hierarchical encoded path views for path query processing: an optimal model and its performance evaluation. IEEE Trans. Knowl. Data Eng. **10**(3), 409–432 (1998)

165. Jung, S., Pramanik, S.: An efficient path computation model for hierarchically structured topographical road maps. IEEE Trans. Knowl. Data Eng. **14**(5), 1029–1046 (2002)

166. Kaindl, H., Kainz, G.: Bidirectional heuristic search reconsidered. J. Artif. Intell. Res. **7**, 283–317 (1997)

167. Kaufmann, H.: Towards mobile time-dependent route planning. Bachelor thesis, Karlsruhe Institute of Technology (2013)

168. Kieritz, T., Luxen, D., Sanders, P., Vetter, C.: Distributed time-dependent contraction hierarchies. In: Festa, P. (ed.) SEA 2010. LNCS, vol. 6049, pp. 83–93. Springer, Heidelberg (2010). doi:10.1007/978-3-642-13193-6_8

169. Kirchler, D., Liberti, L., Wolfler Calvo, R.: A label correcting algorithm for the shortest path problem on a multi-modal route network. In: Klasing, R. (ed.) SEA 2012. LNCS, vol. 7276, pp. 236–247. Springer, Heidelberg (2012). doi:10.1007/978-3-642-30850-5_21

170. Kirchler, D., Liberti, L., Pajor, T., Calvo, R.W.: UniALT for regular language constraint shortest paths on a multi-modal transportation network. In: Proceedings of the 11th Workshop on Algorithmic Approaches for Transportation Modeling, Optimization, and Systems (ATMOS 2011), OpenAccess Series in Informatics (OASIcs), vol. 20, pp. 64–75 (2011)

171. Kleinberg, J.M., Slivkins, A., Wexler, T.: Triangulation and embedding using small sets of beacons. In: Proceedings of the 45th Annual IEEE Symposium on Foundations of Computer Science (FOCS 2004), pp. 444–453. IEEE Computer Society Press (2004)

172. Knopp, S., Sanders, P., Schultes, D., Schulz, F., Wagner, D.: Computing many-to-many shortest paths using highway hierarchies. In: Proceedings of the 9th Workshop on Algorithm Engineering and Experiments (ALENEX 2007), pp. 36–45. SIAM (2007)

173. Kobitzsch, M.: HiDAR: an alternative approach to alternative routes. In: Bodlaender, H.L., Italiano, G.F. (eds.) ESA 2013. LNCS, vol. 8125, pp. 613–624. Springer, Heidelberg (2013). doi:10.1007/978-3-642-40450-4_52

174. Kobitzsch, M., Radermacher, M., Schieferdecker, D.: Evolution and evaluation of the penalty method for alternative graphs. In: Proceedings of the 13th Workshop on Algorithmic Approaches for Transportation Modeling, Optimization, and Systems (ATMOS 2013), OpenAccess Series in Informatics (OASIcs), pp. 94–107 (2013)

175. Kontogiannis, S., Zaroliagis, C.: Distance oracles for time-dependent networks. In: Esparza, J., Fraigniaud, P., Husfeldt, T., Koutsoupias, E. (eds.) ICALP 2014. LNCS, vol. 8572, pp. 713–725. Springer, Heidelberg (2014). doi:10.1007/978-3-662-43948-7_59

176. Krumm, J., Gruen, R., Delling, D.: From destination prediction to route prediction. J. Locat. Based Serv. **7**(2), 98–120 (2013)

177. Krumm, J., Horvitz, E.: Predestination: where do you want to go today? IEEE Comput. **40**(4), 105–107 (2007)

178. Lauther, U.: An experimental evaluation of point-to-point shortest path calculation on roadnetworks with precalculated edge-flags. In: The Shortest Path Problem: Ninth DIMACS Implementation Challenge, vol. 74, DIMACS Book, pp. 19–40. American Mathematical Society (2009)

179. Ken, C.K., Lee, J.L., Zheng, B., Tian, Y.: ROAD: a new spatial object search framework for road networks. IEEE Trans. Knowl. Data Eng. **24**(3), 547–560 (2012)

180. Lipton, R.J., Rose, D.J., Tarjan, R.: Generalized nested dissection. SIAM J. Numer. Anal. **16**(2), 346–358 (1979)

181. Lipton, R.J., Tarjan, R.E.: A separator theorem for planar graphs. SIAM J. Appl. Math. **36**(2), 177–189 (1979)

182. Loridan, P.: ϵ-solutions in vector minimization problems. J. Optim. Theory Appl. **43**(2), 265–276 (1984)

183. Luxen, D., Sanders, P.: Hierarchy decomposition for faster user equilibria on road networks. In: Pardalos, P.M., Rebennack, S. (eds.) SEA 2011. LNCS, vol. 6630, pp. 242–253. Springer, Heidelberg (2011). doi:10.1007/978-3-642-20662-7_21

184. Luxen, D., Schieferdecker, D.: Candidate sets for alternative routes in road networks. In: Klasing, R. (ed.) SEA 2012. LNCS, vol. 7276, pp. 260–270. Springer, Heidelberg (2012). doi:10.1007/978-3-642-30850-5_23

185. Luxen, D., Vetter, C.: Real-time routing with OpenStreetMap data. In: Proceedings of the 19th ACM SIGSPATIAL International Conference on Advances in Geographic Information Systems. ACM Press (2011)

186. Madduri, K., Bader, D.A., Berry, J.W., Crobak, J.R., Parallel shortest path algorithms for solving large-scale instances. In: The Shortest Path Problem: Ninth DIMACS Implementation Challenge, DIMACS Book, vol. 74, pp. 249–290. American Mathematical Society (2009)

187. Martins, E.Q.: On a multicriteria shortest path problem. Eur. J. Oper. Res. **26**(3), 236–245 (1984)

188. Maue, J., Sanders, P., Matijevic, D.: Goal-directed shortest-path queries using precomputed cluster distances. ACM J. Exp. Algorithm. **14**, 3.2:1–3.2:27 (2009)

189. Mehlhorn, K.: A faster approximation algorithm for the Steiner problem in graphs. Inf. Process. Lett. **27**(3), 125–128 (1988)

190. Mehlhorn, K., Sanders, P., Algorithms, D.S.: The Basic Toolbox. Springer, Heidelberg (2008)

191. Mellouli, T., Suhl, L.: Passenger online routing in dynamic networks. In: Mattfeld, D.C., Suhl, L. (eds.) Informations probleme in Transport und Verkehr, vol. 4, pp. 17–30. DS&OR Lab, Universität Paderborn (2006)

192. Meyer, U.: Single-source shortest-paths on arbitrary directed graphs in linear average-case time. In: Proceedings of the 12th Annual ACM-SIAM Symposium on Discrete Algorithms (SODA 2001), pp. 797–806 (2001)

193. Meyer, U., Sanders, P.: δ-stepping: a parallelizable shortest path algorithm. J. Algorithms **49**(1), 114–152 (2003)

194. Milosavljević, N.: On optimal preprocessing for contraction hierarchies. In: Proceedings of the 5th ACM SIGSPATIAL International Workshop on Computational Transportation Science, pp. 33–38. ACM Press (2012)

195. Modesti, P., Sciomachen, A.: A utility measure for finding multiobjective shortest paths in urban multimodal transportation networks. Eur. J. Oper. Res. **111**(3), 495–508 (1998)

196. Möhring, R.H.: Verteilte Verbindungssuche im öffentlichen Personenverkehr - Graphentheoretische Modelle und Algorithmen. In: Angewandte Mathematik insbesondere Informatik, Beispiele erfolgreicher Wege zwischen Mathematik und Informatik, pp. 192–220. Vieweg (1999)

197. Möhring, R.H., Schilling, H., Schütz, B., Wagner, D., Willhalm, T.: Partitioning graphs to speedup Dijkstra's algorithm. ACM J. Exp. Algorithm. 11(28), 1–29 (2006)

198. Moore, E.F.: The shortest path through a maze. In: Proceedings of the International Symposium on the Theory of Switching, pp. 285–292. Harvard University Press (1959)

199. Müller-Hannemann, M., Schnee, M.: Paying less for train connections with MOTIS. In: Proceedings of the 5th Workshop on Algorithmic Methods and Models for Optimization of Railways (ATMOS 2005), OpenAccess Series in Informatics (OASIcs), p. 657 (2006)

200. Müller-Hannemann, M., Schnee, M.: Finding all attractive train connections by multi-criteria pareto search. In: Geraets, F., Kroon, L., Schoebel, A., Wagner, D., Zaroliagis, C.D. (eds.) Algorithmic Methods for Railway Optimization. LNCS, vol. 4359, pp. 246–263. Springer, Heidelberg (2007). doi:10.1007/978-3-540-74247-0_13

201. Müller-Hannemann, M., Schnee, M.: Efficient timetable information in the presence of delays. In: Ahuja, R.K., Möhring, R.H., Zaroliagis, C.D. (eds.) Robust and Online Large-Scale Optimization. LNCS, vol. 5868, pp. 249–272. Springer, Heidelberg (2009). doi:10.1007/978-3-642-05465-5_10

202. Müller-Hannemann, M., Schnee, M., Weihe, K.: Getting train timetables into the main storage. Electron. Notes Theoret. Comput. Sci. 66(6), 8–17 (2002)

203. Müller-Hannemann, M., Schulz, F., Wagner, D., Zaroliagis, C.: Timetable information: models and algorithms. In: Geraets, F., Kroon, L., Schoebel, A., Wagner, D., Zaroliagis, C.D. (eds.) Algorithmic Methods for Railway Optimization. LNCS, vol. 4359, pp. 67–90. Springer, Heidelberg (2007). doi:10.1007/978-3-540-74247-0_3

204. Müller-Hannemann, M., Weihe, K.: Pareto shortest paths is often feasible in practice. In: Brodal, G.S., Frigioni, D., Marchetti-Spaccamela, A. (eds.) WAE 2001. LNCS, vol. 2141, pp. 185–197. Springer, Heidelberg (2001). doi:10.1007/3-540-44688-5_15

205. Muller, L.F., Zachariasen, M.: Fast and compact oracles for approximate distances in planar graphs. In: Arge, L., Hoffmann, M., Welzl, E. (eds.) ESA 2007. LNCS, vol. 4698, pp. 657–668. Springer, Heidelberg (2007). doi:10.1007/978-3-540-75520-3_58

206. Nachtigall, K.: Time depending shortest-path problems with applications to railway networks. Eur. J. Oper. Res. 83(1), 154–166 (1995)

207. Nannicini, G., Delling, D., Liberti, L., Schultes, D.: Bidirectional A* search on time-dependent road networks. Networks 59, 240–251 (2012). Best Paper Award

208. Orda, A., Rom, R.: Shortest-path and minimum delay algorithms in networks with time-dependent edge-length. J. ACM 37(3), 607–625 (1990)

209. Orda, A., Rom, R.: Minimum weight paths in time-dependent networks. Networks 21, 295–319 (1991)

210. Pajor, T.: Multi-modal route planning. Master's thesis, Universität Karlsruhe (TH), March 2009

211. Pallottino, S., Scutellà, M.G.: Shortest path algorithms in transportation models: Classical and innovative aspects. In: Equilibrium and Advanced Transportation Modelling, pp. 245–281. Kluwer Academic Publishers Group (1998)

212. Papadimitriou, C.H., Yannakakis, M.: On the approximability of trade-offs and optimal access of web sources. In: Proceedings of the 41st Annual IEEE Symposium on Foundations of Computer Science (FOCS 2000), pp. 86–92 (2000)

213. Paraskevopoulos, A., Zaroliagis, C.: Improved alternative route planning. In: Proceedings of the 13th Workshop on Algorithmic Approaches for Transportation Modeling, Optimization, and Systems (ATMOS 2013), OpenAccess Series in Informatics (OASIcs), pp. 108–122 (2013)

214. Parter, S.V.: The use of linear graphs in Gauss elimination. SIAM Rev. **3**(2), 119–130 (1961)

215. Peleg, D.: Proximity-preserving labeling schemes. J. Graph Theory **33**(3), 167–176 (2000)

216. Pohl, I.: Bi-directional and heuristic search in path problems. Technical report SLAC-104, Stanford Linear Accelerator Center, Stanford, California (1969)

217. Pohl, I.: Bi-directional search. In: Proceedings of the Sixth Annual Machine Intelligence Workshop, vol. 6, pp. 124–140. Edinburgh University Press (1971)

218. Pyrga, E., Schulz, F., Wagner, D., Zaroliagis, C.: Experimental comparison of shortest path approaches for timetable information. In: Proceedings of the 6th Workshop on Algorithm Engineering and Experiments (ALENEX 2004), pp. 88–99. SIAM (2004)

219. Pyrga, E., Schulz, F., Wagner, D., Zaroliagis, C.: Efficient models for timetable information in public transportation systems. ACM J. Exp. Algorithm. **12**(24), 1–39 (2008)

220. Rice, M., Tsotras, V.: Bidirectional A* search with additive approximation bounds. In: Proceedings of the 5th International Symposium on Combinatorial Search (SoCS 2012), AAAI Press (2012)

221. Rice, M.N., Tsotras, V.J.: Exact graph search algorithms for generalized traveling salesman path problems. In: Klasing, R. (ed.) SEA 2012. LNCS, vol. 7276, pp. 344–355. Springer, Heidelberg (2012). doi:10.1007/978-3-642-30850-5_30

222. Sanders, P., Mandow, L.: Parallel label-setting multi-objective shortest path search. In: 27th International Parallel and Distributed Processing Symposium (IPDPS 2013), pp. 215–224. IEEE Computer Society (2013)

223. Sanders, P., Schultes, D.: Highway hierarchies hasten exact shortest path queries. In: Brodal, G.S., Leonardi, S. (eds.) ESA 2005. LNCS, vol. 3669, pp. 568–579. Springer, Heidelberg (2005). doi:10.1007/11561071_51

224. Sanders, P., Schultes, D.: Robust, almost constant time shortest-path queries in road networks. In: The Shortest Path Problem: Ninth DIMACS Implementation Challenge, DIMACS Book, vol. 74, pp. 193–218. American Mathematical Society (2009)

225. Sanders, P., Schultes, D.: Engineering highway hierarchies. ACM J. Exp. Algorithm. **17**(1), 1–40 (2012)

226. Sanders, P., Schultes, D., Vetter, C.: Mobile route planning. In: Halperin, D., Mehlhorn, K. (eds.) ESA 2008. LNCS, vol. 5193, pp. 732–743. Springer, Heidelberg (2008). doi:10.1007/978-3-540-87744-8_61

227. Sanders, P., Schulz, C.: Distributed evolutionary graph partitioning. In: Proceedings of the 14th Meeting on Algorithm Engineering and Experiments (ALENEX 2012), pp. 16–29. SIAM (2012)

228. Sankaranarayanan, J., Alborzi, H., Samet, H.: Efficient query processing on spatial networks. In: Proceedings of the 13th Annual ACM International Workshop on Geographic Information Systems (GIS 2005), pp. 200–209 (2005)

229. Sankaranarayanan, J., Samet, H.: Query processing using distance oracles for spatial networks. IEEE Trans. Knowl. Data Eng. **22**(8), 1158–1175 (2010)

230. Sankaranarayanan, J., Samet, H.: Roads belong in databases. IEEE Data Eng. Bull. **33**(2), 4–11 (2010)
231. Schilling, H.: TomTom navigation - How mathematics help getting through traffic faster (2012). Talk given at ISMP
232. Schreiber, R.: A new implementation of sparse Gaussian elimination. ACM Trans. Math. Softw. **8**(3), 256–276 (1982)
233. Schultes, D.: Route planning in road networks. Ph.D. thesis, Universität Karlsruhe (TH), February 2008
234. Schultes, D., Sanders, P.: Dynamic highway-node routing. In: Demetrescu, C. (ed.) WEA 2007. LNCS, vol. 4525, pp. 66–79. Springer, Heidelberg (2007). doi:10.1007/978-3-540-72845-0_6
235. Schulz, F., Wagner, D., Weihe, K.: Dijkstra's algorithm on-line: an empirical case study from public railroad transport. ACM J. Exp. Algorithm. **5**(12), 1–23 (2000)
236. Schulz, F., Wagner, D., Zaroliagis, C.: Using multi-level graphs for timetable information in railway systems. In: Mount, D.M., Stein, C. (eds.) ALENEX 2002. LNCS, vol. 2409, pp. 43–59. Springer, Heidelberg (2002). doi:10.1007/3-540-45643-0_4
237. Sedgewick, R., Vitter, J.S.: Shortest paths in Euclidean graphs. Algorithmica **1**(1), 31–48 (1986)
238. Sommer, C.: Shortest-path queries in static networks. ACM Comput. Surv. **46**(4), 1–31 (2014)
239. Storandt, S.: Route planning for bicycles - exact constrained shortest paths made practical via contraction hierarchy. In: Proceedings of the Twenty-Second International Conference on Automated Planning and Scheduling, pp. 234–242 (2012)
240. Storandt, S., Funke, S.: Cruising with a battery-powered vehicle and not getting stranded. In: Proceedings of the Twenty-Sixth AAAI Conference on Artificial Intelligence. AAAI Press (2012)
241. Storandt, S., Funke, S.: Enabling e-mobility: facility location for battery loading stations. In: Proceedings of the Twenty-Seventh AAAI Conference on Artificial Intelligence. AAAI Press (2013)
242. Strasser, B., Wagner, D.: Connection scan accelerated. In: Proceedings of the 16th Meeting on Algorithm Engineering and Experiments (ALENEX 2014), pp. 125–137. SIAM (2014)
243. Theune, D.: Robuste und effiziente Methoden zur Lösung von Wegproblemen. Ph.D. thesis, Universität Paderborn (1995)
244. Thorup, M.: Integer priority queues with decrease key in constant time and the single source shortest paths problem. In: 35th ACM Symposium on Theory of Computing, pp. 149–158. ACM, New York (2003)
245. Thorup, M.: Compact oracles for reachability and approximate distances in planar digraphs. J. ACM **51**(6), 993–1024 (2004)
246. Tsaggouris, G., Zaroliagis, C.: Multiobjective optimization: improved FPTAS for shortest paths and non-linear objectives with applications. Theory Comput. Syst. **45**(1), 162–186 (2009)
247. Tulp, E., Siklóssy, L.: TRAINS, an active time-table searcher. ECAI **88**, 170–175 (1988)
248. Tulp, E., Siklóssy, L.: Searching time-table networks. Artif. Intell. Eng. Des. Anal. Manuf. **5**(3), 189–198 (1991)
249. van Vliet, D.: Improved shortest path algorithms for transport networks. Transp. Res. Part B: Methodol. **12**(1), 7–20 (1978)

250. Wagner, D., Willhalm, T.: Drawing graphs to speed up shortest-path computations. In: Proceedings of the 7th Workshop on Algorithm Engineering and Experiments (ALENEX 2005), pp. 15–24. SIAM (2005)

251. Wagner, D., Willhalm, T., Zaroliagis, C.: Geometric containers for efficient shortest-path computation. ACM J. Exp. Algorithm. **10**(1.3), 1–30 (2005)

252. Weller, M.: Optimal hub labeling is NP-complete. CoRR, abs/1407.8373 (2014)

253. White, D.J.: Epsilon efficiency. J. Optim. Theory Appl. **49**(2), 319–337 (1986)

254. Williams, J.W.J.: Algorithm 232: heapsort. J. ACM **7**(6), 347–348 (1964)

255. Winter, S.: Modeling costs of turns in route planning. GeoInformatica **6**(4), 345–361 (2002)

256. Witt, S.: Trip-based public transit routing. In: Bansal, N., Finocchi, I. (eds.) ESA 2015. LNCS, vol. 9294, pp. 1025–1036. Springer, Heidelberg (2015). doi:10.1007/978-3-662-48350-3_85

257. Lingkun, W., Xiao, X., Deng, D., Cong, G., Zhu, A.D., Zhou, S.: Shortest path and distance queries on road networks: an experimental evaluation. Proc. VLDB Endow. **5**(5), 406–417 (2012)

258. Yu, H., Lu, F.: Advanced multi-modal routing approach for pedestrians. In: 2nd International Conference on Consumer Electronics, Communications and Networks, pp. 2349–2352 (2012)

259. Zadeh, L.A.: Fuzzy logic. IEEE Comput. **21**(4), 83–93 (1988)

260. Zhong, R., Li, G., Tan, K.-L., Zhou, L.: G-tree: an efficient index for KNN search on road networks. In: Proceedings of the 22nd International Conference on Information and Knowledge Management, pp. 39–48. ACM Press (2013)

261. Zhu, A.D., Ma, H., Xiao, X., Luo, S., Tang, Y., Zhou, S.: Shortest path, distance queries on road networks: towards bridging theory and practice. In: Proceedings of the 2013 ACM SIGMOD International Conference on Management of Data (SIGMOD 2013), pp. 857–868. ACM Press (2013)

262. Zwick, U.: Exact and approximate distances in graphs — a survey. In: Heide, F.M. (ed.) ESA 2001. LNCS, vol. 2161, pp. 33–48. Springer, Heidelberg (2001). doi:10.1007/3-540-44676-1_3

Theoretical Analysis of the k-Means Algorithm – A Survey

Johannes Blömer[1], Christiane Lammersen[2], Melanie Schmidt[3(✉)], and Christian Sohler[4]

[1] Department of Computer Science, University of Paderborn, Paderborn, Germany
[2] School of Computing Science, Simon Fraser University, Burnaby, BC, Canada
[3] Computer Science Department, Carnegie Mellon University, Pittsburgh, PA, USA
melanie.schmidt@tu-dortmund.de
[4] Department of Computer Science, TU Dortmund University, Dortmund, Germany

Abstract. The k-means algorithm is one of the most widely used clustering heuristics. Despite its simplicity, analyzing its running time and quality of approximation is surprisingly difficult and can lead to deep insights that can be used to improve the algorithm. In this paper we survey the recent results in this direction as well as several extension of the basic k-means method.

1 Introduction

Clustering is a basic process in data analysis. It aims to partition a set of objects into groups called *clusters* such that, ideally, objects in the same group are similar and objects in different groups are dissimilar to each other. There are many scenarios where such a partition is useful. It may, for example, be used to structure the data to allow efficient information retrieval, to reduce the data by replacing a cluster by one or more representatives or to extract the main 'themes' in the data. There are many surveys on clustering algorithms, including well-known classics [45, 48] and more recent ones [24, 47]. Notice that the title of [47] is *Data clustering: 50 years beyond K-means* in reference to the *k-means algorithm*, the probably most widely used clustering algorithm of all time. It was proposed in 1957 by Lloyd [58] (and independently in 1956 by Steinhaus [70]) and is the topic of this survey.

The k-means algorithm solves the problem of *clustering to minimize the sum of squared errors* (SSE). In this problem, we are given a set of points $P \subset \mathbb{R}^d$ in a Euclidean space, and the goal is to find a set $C \subset \mathbb{R}^d$ of k points (not necessarily included in P) such that the sum of the squared distances of the points in P to their nearest center in C is minimized. Thus, the objective function to be minimized is

$$\text{cost}(P, C) := \sum_{p \in P} \min_{c \in C} \|p - c\|^2,$$

where $\| \cdot \|^2$ is the squared Euclidean distance. The points in C are called *centers*. The objective function may also be viewed as the attempt to minimize the variance of the Euclidean distance of the points to their nearest cluster centers.

L. Kliemann and P. Sanders (Eds.): Algorithm Engineering, LNCS 9220, pp. 81–116, 2016.
DOI: 10.1007/978-3-319-49487-6_3

Also notice that when given the centers, the partition of the data set is implicitly defined by assigning each point to its nearest center.

The above problem formulation assumes that the number of centers k is known in advance. How to choose k might be apparent from the application at hand, or from a statistical model that is assumed to be true. If it is not, then the k-means algorithm is typically embedded into a search for the correct number of clusters. It is then necessary to specify a measure that allows to compare clusterings with different k (the SSE criterion is monotonely decreasing with k and thus not a good measure). A good introduction to the topic is the overview by Venkatasubramanian [75] as well as Sect. 5 in the paper by Tibshirani et al. [71] and the summary by Gordon [39]. In this survey, we assume that k is provided with the input.

As Jain [47] also notices, the k-means algorithm is still widely used for clustering and in particular for solving the SSE problem. That is true despite a variety of alternative options that have been developed in fifty years of research, and even though the k-means algorithm has known drawbacks.

In this survey, we review the theoretical analysis that has been developed for the k-means algorithm. Our aim is to give an overview on the properties of the k-means algorithm and to understand its weaknesses, but also to point out what makes the k-means algorithm such an attractive algorithm. In this survey we mainly review theoretical aspects of the k-means algorithm, i.e. focus on the deduction part of the algorithm engineering cycle, but we also discuss some implementations with focus on scalability for big data.

1.1   The k-Means Algorithm

In order to solve the SSE problem heuristically, the k-means algorithm starts with an initial candidate solution $\{c_1, \ldots, c_k\} \subset \mathbb{R}^d$, which can be chosen arbitrarily (often, it is chosen as a random subset of P). Then, two steps are alternated until convergence: First, for each c_i, the algorithm calculates the set P_i of all points in P that are closest to c_i (where ties are broken arbitrarily). Then, for each $1 \leq i \leq k$, it replaces c_i by the mean of P_i. Because of this calculation of the 'means' of the sets P_i, the algorithm is also called *the k-means algorithm.*

THE k-MEANS ALGORITHM
Input: Point set $P \subseteq \mathbb{R}^d$
          number of centers k
1. Choose initial centers $c_1, \ldots, c_k$ of from $\mathbb{R}^d$
2. **repeat**
3. $P_1, \ldots, P_k \leftarrow \emptyset$
4.   **for each** $p \in P$ **do**
5.   Let $i = \arg\min_{i=1,\ldots,k} \|p - c_i\|^2$
6. $P_i \leftarrow P_i \cup \{p\}$
7.   **for** $i = 1$ **to** k **do**
8.   **if** $P_i \neq \emptyset$ **then** $c_i = \frac{1}{|P_i|} \sum_{p \in P_i} p$
9. **until** the centers do not change

The k-means algorithm is a local improvement heuristic, because replacing the center of a set P_i by its mean can only improve the solution (see Fact 1 below), and then reassigning the points to their closest center in C again only improves the solution. The algorithm converges, but the first important question is how many iterations are necessary until an optimal or good solution is found. The second natural question is how good the solution will be when the algorithm stops. We survey upper and lower bounds on running time and quality in Sect. 2. Since the quality of the computed solution depends significantly on the starting solution, we discuss ways to choose the starting set of centers in a clever way in Sect. 3. Then, we survey variants of the basic k-means algorithm in Sect. 4 and alternatives to the k-means algorithm in Sect. 5. In Sect. 6, we consider the complexity of the SSE problem. Finally, we describe results on the k-means problem and algorithm for Bregman divergences Sect. 7. Bregman divergences have numerous applications and constitute the largest class of dissimilarity measure for which the k-means algorithm can be applied.

2 Running Time and Quality of the Basic k-Means Algorithm

In this section, we consider the two main theoretical questions about the k-means algorithm: What is its running time, and does it provide a solution of a guaranteed quality? We start with the running time.

2.1 Analysis of Running Time

The running time of the k-means algorithm depends on the number of iterations and on the running time for one iteration. While the running time for one iteration is clearly polynomial in n, d and k, this is not obvious (and in general not true) for the number of iterations. Yet, in practice, it is often observed that the k-means algorithm does not significantly improve after a relatively small number of steps. Therefore, one often performs only a constant number of steps. It is also common to just stop the algorithm after a given maximum number of iterations, even if it has not converged. The running time analysis thus focuses on two things. First, what the asymptotic running time of one iteration is and how it can be accelerated for benign inputs. Second, whether there is a theoretical explanation on why the algorithm tends to converge fast in practice.

Running Time of One Iteration. A straightforward implementation computes $\Theta(nk)$ distances in each iteration in time $\Theta(ndk)$ and runs over the complete input point set. We denote this as the 'naive' implementation. Asymptotically, the running time for this is dominated by the number of iterations, which is in general not polynomially bounded in n in the worst case (see next subsection for details). However, in practice, the number of iterations is often manually capped, and the running time of one iteration becomes the important factor. We thus want to mention a few practical improvements.

The question is whether and how it can be avoided to always compute the distances between all points and centers, even if this does not lead to an asymptotic improvement. Imagine the following pruning rule: Let c_i be a center in the current iteration. Compute the minimum distance Δ_i between c_i and any other center in time $\Theta(kd)$. Whenever the distance between a point p and c_i is smaller than $\Delta_i/2$, then the closest center to p is c_i and computing the other $k-1$ distances is not necessary. A common observation is that points often stay with the same cluster as in the previous iteration. Thus, check first whether the point is within the safe zone of its old center. More complicated pruning rules take the movement of the points into account. If a point has not moved far compared to the center movements, it keeps its center allocation. Rules like this aim at accelerating the k-means algorithm while computing the same clustering as a naïve implementation. The example pruning rules are from [50].

Accelerating the algorithm can also be done by assigning groups of points together using *sufficient statistics*. Assume that a subset P' of points is assigned to the same center. Then finding this center and later updating it based on the new points can be done by only using three statistics on P'. These are the sum of the points (which is a point itself), the sum of the squared lengths of the points (and thus a constant) and the number of points. However, this is only useful if the statistic is already precomputed. For low-dimensional data sets, the precomputation can be done using *kd-trees*. These provide a hierarchical subdivision of a point set. The idea now is to equip each inner node with sufficient statistics on the point set represented by it. When reassigning points to centers, pruning techniques can be used to decide whether all points belonging to an inner node have the same center, or whether it is necessary to proceed to the child nodes to compute the assignment. Different algorithms based on this idea are given in [10,54,68]. Notice that sufficient statistics are used in other contexts, too, e.g. as a building block of the well-known data stream clustering algorithm BIRCH [76].

There are many ways more that help to accelerate the k-means algorithm. For an extensive overview and more pointers to the literature, see [41].

Worst-Case Analysis. Now we take a closer look at the worst-case number of iterations, starting with (large) general upper bounds and better upper bounds in special cases. Then we review results for lower bounds on the number of iterations and thus on the running time of the basic k-means algorithm. In the next section, we have a look into work on smoothed analysis for the k-means algorithm which gives indications on why the k-means algorithm often performs so well in practice.

Upper Bounds. The worst-case running time to compute a k-clustering of n points in $\mathbb{R}^d$ by applying the k-means algorithm is upper bounded by $\mathcal{O}(ndk \cdot T)$, where T is the number of iterations of the algorithm. It is known that the number of iterations of the algorithm is bounded by the number of partitionings of the input points induced by a Voronoi-diagramm of k centers. This number can be bounded by $\mathcal{O}(n^{dk^2})$ because given a set of k centers, we can move each of the $O(k^2)$ bisectors such that they coincide with d linearly independent points without changing the partition. For the special case of $d = 1$ and $k < 5$, Dasgupta [31] proved an upper bound of $\mathcal{O}(n)$ iterations. Later, for $d = 1$ and any k, Har-Peled and Sadri [44] showed an upper bound of $\mathcal{O}(n\Delta^2)$ iterations, where Δ is the ratio between the diameter and the smallest pairwise distance of the input points.

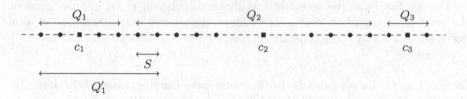

Fig. 1. Illustration of the upper bound for the k-means algorithm [44].

In the following, we will explain the idea to obtain the upper bound given in [44]. The input is a set P of n points with spread Δ from the Euclidean line $\mathbb{R}$. W.l.o.g., we can assume that the minimum pairwise distance in P is 1 and the diameter of P is Δ. For any natural number k and for any partition of P into k sets, the clustering cost of P with the means of the subsets as centers is bounded by $\mathcal{O}(n\Delta^2)$. In particular, this holds for the solution of the k-means algorithm after the first iteration. Additionally, the clustering cost of P certainly is $\omega(1)$ as we assumed that the minimum pairwise distance in P is 1. Thus, if we can show that each following iteration decreases the cost by at least some constant amount, then we are done. Let us now consider the point of time in any iteration of the k-means algorithm when the cluster centers have been moved to the means of their respective clusters and the next step is to assign each point to the new closest cluster center. In this step, there has to be a cluster that is extended or shrunk from its right end. W.l.o.g. and as illustrated in Fig. 1, let us assume that the leftmost cluster Q_1 is extended from its right end. Let S be the set of points that join cluster Q_1 to obtain cluster Q_1'. Since the minimum pairwise distance is 1, the distance of the mean of S to the leftmost point in S is at least $(|S| - 1)/2$. Similarly, the distance of the mean of Q_1 to the rightmost point in Q_1 is at least $(|Q_1| - 1)/2$. Furthermore, the distance between any point in Q_1 and any point in S is at least 1. Let $\mu(X)$ be the mean of any point set X. Then, we have $\|\mu(Q_1) - \mu(S)\| \geq (|Q_1| - 1)/2 + (|S| - 1)/2 + 1 = (|Q_1| + |S|)/2$. The movement of the mean of the leftmost cluster is at least

$$\|\mu(Q_1) - \mu(Q_1')\| = \left\|\mu(Q_1) - \frac{|Q_1|\mu(Q_1) + |S|\mu(S)}{|Q_1| + |S|}\right\|$$

$$= \frac{|S|}{|Q_1| + |S|}\|\mu(Q_1) - \mu(S)\| \geq \frac{|S|}{2} \geq \frac{1}{2}.$$

We will now need the following fact, which is proved in Sect. 6.

Fact 1. *Let*

$$\mu := \frac{1}{|P|} \sum_{p \in P} p$$

be the mean of a point set P, and let $y \in \mathbb{R}^d$ be any point. Then, we have

$$\sum_{p \in P} \|p - y\|^2 = \sum_{p \in P} \|p - \mu\|^2 + |P| \cdot \|y - \mu\|^2.$$

Due to this fact, the result of the above calculation is an improvement of the clustering cost of at least $1/4$, which shows that in each iteration the cost decreases at least by some constant amount and hence there are at most $\mathcal{O}(n\Delta^2)$ iterations.

Lower Bounds. Lower bounds on the worst-case running time of the k-means algorithm have been studied in [13,31,72]. Dasgupta [31] proved that the k-means algorithm has a worst-case running time of $\Omega(n)$ iterations. Using a construction in some $\Omega(\sqrt{n})$-dimensional space, Arthur and Vassilvitskii [13] were able to improve this result to obtain a super-polynomial worst-case running time of $2^{\Omega(\sqrt{n})}$ iterations. This has been simplified and further improved by Vattani [72] who proved an exponential lower bound on the worst-case running time of the k-means algorithm showing that k-means requires $2^{\Omega(n)}$ iterations even in the plane. A modification of the construction shows that the k-means algorithm has a worst-case running time that, besides being exponential in n, is also exponential in the spread Δ of the d-dimensional input points for any $d \geq 3$.

In the following, we will give a high-level view on the construction presented in [72]. Vattani uses a special set of n input points in $\mathbb{R}^2$ and a set of $k = \Theta(n)$ cluster centers adversarially chosen among the input points. The points are arranged in a sequence of $t = \Theta(n)$ gadgets $G_0, G_1, \ldots, G_{t-1}$. Except from some scaling, the gadgets are identical. Each gadget contains a constant number of points, has two clusters and hence two cluster centers, and can perform two stages reflected by the positions of the two centers. In one stage, gadget G_i, $0 \leq i < t$, has one center in a certain position c_i^*, and, in the other stage, the same center has left the position c_i^* and has moved a little bit towards gadget G_{i+1}. Once triggered by gadget G_{i+1}, G_i performs both of these stages twice in a row. Performing these two stages happens as follows. The two centers of gadget G_{i+1} are assigned to the center of gravity of their clusters, which results in some points of G_{i+1} are temporarily assigned to the center c_i^* of G_i. Now, the center of G_i located at c_i^* and the centers of G_{i+1} move, so that the points temporarily assigned to a center of G_i are again assigned to the centers of G_{i+1}. Then, again triggered by G_{i+1}, gadget G_i performs the same two stages once more. There is only some small modification in the arrangement of the two clusters of G_{i+1}. Now, assume that all gadgets except G_{t-1} are stable and the centers of G_{t-1} are moved to the centers of gravity of their clusters. This triggers a chain reaction, in which the gadgets perform $2^{\Omega(t)}$ stages in total. Since, each stage of a gadget corresponds to one iteration of the k-means algorithm, the algorithm needs $2^{\Omega(n)}$ iterations on the set of points contained in the gadgets.

Smoothed Analysis. Concerning the above facts, one might wonder why k-means works so well in practice. To close this gap between theory and practice, the algorithm has also been studied in the model of smoothed analysis [12,15,62]. This model is especially useful when both worst-case and average-case analysis are not realistic and reflects the fact that real-world datasets are likely to contain measurement errors or imprecise data. In case an algorithm has a low time complexity in the smoothed setting, it is likely to have a small running time on real-world datasets as well.

Next, we explain the model in more detail. For given parameters n and σ, an adversary chooses an input instance of size n. Then, each input point is perturbed by adding some small amount of random noise using a Gaussian distribution with mean 0 and standard deviation σ. The maximum expected running time of the algorithm executed on the perturbed input points is measured.

Arthur and Vassilvitskii [15] showed that, in the smoothed setting, the number of iterations of the k-means algorithm is at most $\mathrm{poly}(n^k, \sigma^{-1})$. This was improved by Manthey and Röglin [62] who proved the upper bounds $\mathrm{poly}(n^{\sqrt{k}}, 1/\sigma)$ and $k^{kd} \cdot \mathrm{poly}(n, 1/\sigma)$ on the number of iterations. Finally, Arthur et al. [12] showed that k-means has a polynomial-time smoothed complexity of $\mathrm{poly}(n, 1/\sigma)$.

In the following, we will give a high-level view on the intricate analysis presented in [12]. Arthur et al. show that after the first iteration of k-means, the cost of the current clustering is bounded by some polynomial in n, k and d. In each further iteration, either some cluster centers move to the center of gravity of their clusters or some points are assigned to a closer cluster center or even both events happen. Obviously, the clustering cost is decreased after each iteration, but how big is this improvement? Arthur et al. prove that, in expectation, an iteration of k-means decreases the clustering cost by some amount polynomial in $1/n$ and σ. This results in a polynomial-time smoothed complexity.

The key idea to obtain the above lower bound on the minimum improvement per iteration is as follows. Let us call a configuration of an iteration, defined by a partition into clusters and a set of cluster centers, *good* if in the successive iteration either a cluster center moves significantly or reassigning a point decreases the clustering cost of the point significantly. Otherwise, the configuration is called *bad*. Arthur et al. show an upper bound on the probability that a configuration is bad. The problem is now that there are many possible configurations. So we cannot take the union bound over all of these possible configurations to show that the probability of the occurrence of any bad configuration during a run of k-means is small. To avoid this problem, Arthur et al. group all configurations into a small number of subsets and show that each subset contains either only good configurations or only bad configurations. Finally, taking the union bound over all subsets of configurations leads to the desired result, i.e., proving that the occurrence of a bad configuration during a run of k-means is small.

2.2   Analysis of Quality

As mentioned above, the k-means algorithm is a local improvement heuristic. It is known that the k-means algorithm converges to a local optimum [69] and that no approximation ratio can be guaranteed [55]. Kanungo et al. [55] illustrate the latter fact by the simple example given in Fig. 2. In this example, we are given four input points on the Euclidean line depicted by the first dashed line in Fig. 2. The distances between the first and second, second and third and third and fourth point are named x, y and z, respectively. We assume that $x < y < z$, so x is the smallest distance and placing two centers in the first two points and one between the third and fourth costs $2 \cdot x^2/4 = x^2/2$, and this is the (unique) optimal solution depicted on the second dashed line.

On the third dashed line, we see a solution that is clearly not optimal because it costs $y^2/2$ and $y > x$. The approximation ratio of this solution is y^2/x^2, which can be made arbitrarily bad by moving the first point to the left and thus increasing y.

If we choose the initial centers randomly, it can happen that the k-means algorithm encounters this solution (for example when we pick the first, third and fourth point as initial centers and keep $y < z$ while increasing y). When finding the solution, the k-means algorithm will terminate because the assignment of points to the three centers is unique and every center is the mean of the points assigned to it.

Thus, the worst-case approximation guarantee of the k-means algorithm is unbounded.

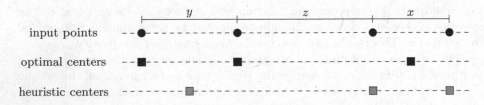

Fig. 2. Example illustrating the fact that no approximation guarantee can be given for the k-means algorithm [55].

3   Seeding Methods for the k-Means Algorithm

The k-means algorithm starts with computing an initial solution, which can be done in a number of different ways. Since the k-means algorithm is a local improvement strategy we can, in principle, start with an arbitrary solution and then the algorithms runs until it converges to a local optimum. However, it is also known that the algorithm is rather sensible to the choice of the starting centers. For example, in the situation in Fig. 2, no problem occurs if we choose the first, second and third point as the starting centers.

Often one simply chooses the starting centers uniformly at random, but this can lead to problems, for example, when there is a cluster that is far away from the remaining points and that is so small that it is likely that no point of it is randomly drawn as one of the initial centers. In such a case one must hope to eventually converge to a solution that has a center in this cluster as otherwise we would end up with a bad solution. Unfortunately, it is not clear that this happens (in fact, one can assume that it will not).

Therefore, a better idea is to start with a solution that already satisfies some approximation guarantees and let the k-means algorithm refine the solution. In this section we will present methods that efficiently pick a relatively good initial solution. As discussed later in Sect. 6 there are better approximation algorithms, but they are relatively slow and the algorithms presented in this section present a better trade-off between running time and quality of the initial solution.

3.1 Adaptive Sampling

Arthur and Vassilvitskii [14] proposed a seeding method for the k-means algorithm which applies *adaptive sampling*. They construct an initial set C of k centers in the following way: The first center is sampled uniformly at random. For the ith center, each input point p is sampled with probability $D^2(p)/\sum_{q \in P} D^2(q)$, where P is the input point set, $D^2(p) = \min_{c_1,\ldots,c_{i-1}} ||p - c_i||^2$ is the cost of p in the current solution and $c_1, \ldots c_{i-1}$ are the centers chosen so far. The sampling process is referred to as D^2-*sampling*, and the algorithm consisting of D^2-sampling followed by the k-means algorithm is called k-*means++*.

We study the progress of D^2-sampling in comparison to a fixed optimal solution. An optimal set of centers partitions P into k *optimal clusters*. If we could sample a center from each cluster uniformly at random, we would in expectation obtain a constant approximation. Since taking a point uniformly at random can also be described as first choosing the cluster and then picking the point uniformly at random, we know that the first point will be uniformly from one (unknown) cluster, which is fine. We want to make sure that this will also approximately be the case for the remaining clusters. The main problem is that there is a significant probability to sample points from a cluster which we already hit (especially, if these clusters contain a lot of points). In order to avoid this, we now sample points with probability proportional to the squared distance from the previously chosen cluster centers. In this way, it is much more likely to sample points from the remaining clusters since the reason that these points belong to a different cluster is that otherwise they would incur a high cost. One can show that in a typical situation, when one of the remaining clusters is far away from the clusters we already hit, then conditioned on the fact that we hit this cluster, the new center will be approximately uniformly distributed within the cluster. In the end, this process leads to a set of k centers that is an expected $O(\log k)$-approximation [14].

Thus, D^2-sampling is actually an approximation algorithm by itself (albeit one with a worse approximation guarantee than other approximations). It has a running time of $O(kdn)$ and is easy to implement. In addition, it serves well as a seeding method. Arthur and Vassilvitskii obtain experimental results indicating that k-means++ outperforms the k-means algorithm in practice, both in quality and running time. It also leads to better results than just using D^2-sampling as an independent algorithm.

In follow-up work, Aggarwal et al. [7] show that when sampling $\mathcal{O}(k)$ centers instead of k centers, one obtains a constant-factor approximation algorithm for SSE. This is a bicriteria approximation because in addition to the fact that the clustering cost might not be optimal, the number of clusters is larger than k.

Adaptive Sampling Under Separation Conditions. Clustering under separation conditions is an interesting research topic on its own. The idea is that the input to a clustering problem should have some structure, otherwise, clustering it would not be meaningful. Separation conditions assume that the optimal clusters cannot have arbitrary close centers or a huge overlap.

We focus on initialization strategies for the k-means algorithm. In this paragraph, we will see a result on adaptive sampling that uses a separation condition. In Sect. 3.2, we will see another example for the use of separation conditions. Other related work includes the paper by Balcan et al. [18], who proposed the idea to recover a 'true' (but not necessarily optimal) clustering and introduced assumptions under which this is possible. Their model is stronger than the model by Ostrovsky et al. [67] that we will describe next and triggered a lot of follow-up work on other clustering variants.

Ostrovsky et al. [67] analyze adaptive sampling under the following ε-separability: The input is ε-separated if clustering it (optimally) with $k-1$ instead of the desired k clusters increases the cost by a factor of at least $1/\varepsilon^2$. Ostrovsky et al. show that under this separation condition, an approach very similar to the above k-means++ seedings performs well[1]. In their seeding method, the first center is not chosen uniformly at random, but two centers are chosen simultaneously, and the probability for each pair of centers is proportional to their distance. Thus, the seeding starts by picking two centers with rather high distance instead of choosing one center uniformly at random and then picking a center with rather high distance to the first center. Ostrovsky et al. show that if the input is ε-separated, this seeding achieves a $(1+f(\varepsilon))$-approximation for SSE where $f(\varepsilon)$ is a function that goes to zero if ε does so. The success probability of this algorithm decreases exponentially in k (because there is a constant chance to miss the next cluster in every step), so Ostrovsky et al. enhance their algorithm by sampling $\mathcal{O}(k)$ clusters and using a greedy deletion process to reduce the number back to k. Thereby, they gain a linear-time constant-factor approximation algorithm (under their separation condition) that can be used as a seeding method.

Later, Awasthi et al. [16] improved this result by giving an algorithm where the approximation guarantee and the separation condition are decoupled, i.e., parameterized by different parameters. Braverman et al. [25] developed a streaming algorithm.

---

[1] Notice that though we present these results after [14] and [7] for reasons of presentation, the work of Ostrovsky et al. [67] appeared first.

Note that ε-separability scales with the number of clusters. Imagine k optimal clusters with the same clustering cost $\mathcal{C}$, i.e., the total clustering cost is $k \cdot \mathcal{C}$. Then, ε-separability requires that clustering with $k - 1$ clusters instead of k clusters costs at least $k \cdot \mathcal{C}/\varepsilon^2$. Thus, for more clusters, the pairwise separation has to be higher.

3.2 Singular Value Decomposition and Best-Fit Subspaces

In the remainder of this section, we will review a result from a different line of research because it gives an interesting result for the SSE problem when we make certain input assumptions.

Learning Mixtures of Gaussians. In machine learning, clustering is often done from a different perspective, namely as a problem of learning parameters of mixture models. In this setting, a set of observations $\mathcal{X}$ is given (in our case, points) together with a statistical model, i.e., a family of density functions over a set of parameters $\Theta = \{\Theta^1, \ldots, \Theta^\ell\}$. It is assumed that $\mathcal{X}$ was generated by the parameterized density function for one specific parameter set and the goal is to recover these parameters. Thus, the desired output are parameters which explain $\mathcal{X}$ best, e.g., because they lead to the highest likelihood that $\mathcal{X}$ was drawn.

For us, the special case that the density function is a mixture of Gaussian distributions on $\mathbb{R}^d$ is of special interest because it is very related to SSE. Here, the set of observations $\mathcal{X}$ is a point set which we denote by P. On this topic, there has been a lot of research lately, which started by Dasgupta [30] who analyzed the problem under separation conditions. Several improvements were made with separation conditions [1,11,26,28,33,53,74] and without separation conditions [21–23,36,51,65]. The main reason why this work cannot be directly applied to SSE is the assumption that the input data $\mathcal{X}$ is actually drawn from the parameterized density function so that properties of these distributions can be used and certain extreme examples become unlikely and can be ignored. However, in [56], the authors prove a result which can be decoupled from this assumption, and the paper proposes an initialization method for the k-means algorithm. So, we take a closer look at this work.

Kumar and Kannan [56] assume a given *target clustering* which is to be recovered and then show the following. If $(1-\varepsilon) \cdot |P|$ points in P satisfy a special condition which they call *proximity condition* (which depends on the target clustering), then applying a certain initialization method and afterwards running the k-means algorithm leads to a partitioning of the points that *misclassifies* at most $\mathcal{O}(k^2 \varepsilon n)$ points. Kumar and Kannan also show that in many scenarios like learning of Gaussian mixtures, points satisfy their proximity condition with high probability.

Notice that for $\varepsilon = 0$ their result implies that all points are correctly classified, i.e., the optimal partitioning is found. This in particular implies a result for the k-means algorithm which is the second step of the algorithm by Kumar and Kannan: It converges to the 'true' centers provided that the condition holds for all points. We take a closer look at the separation condition.

Separation Condition. To define the proximity condition, consider the $|P| \times d$ matrix A which has the points of P in its rows. Also define the matrix C by writing the optimal center of the point in row i of A in row i of C (this implies that there are only k different rows vectors in C). Now, let $T_1, \ldots, T_k$ be the target clustering, let μ_i be the mean of T_i, and let n_i be the number of points in T_i. Then, define

$$\Delta_{rs} := \left(\frac{ck}{\sqrt{n_r}} + \frac{ck}{\sqrt{n_s}} \right) \|A - C\|_S$$

for each $r \neq s$ with $r, s \in \{1, \ldots, k\}$, where c is some constant. The term $\|A - C\|_S$ is the *spectral norm* of the matrix $A - C$, defined by

$$\|A - C\|_S := \max_{v \in \mathbb{R}^d, \|v\| = 1} \|(A - C) \cdot v\|^2.$$

A point p from cluster T_r satisfies the *proximity condition* if, for any $s \neq r$, the projection of p onto the line between μ_r and μ_s is at least Δ_{rs} closer to μ_r than to μ_s.

We have a closer look at the definition. The term $A - C$ is the matrix consisting of the difference vectors, i. e., it gives the deviations of the points to their centers. The term $\|(A - C) \cdot v\|^2$ is the projection of these distance vectors into direction v, i. e., a measure on how much the data is scattered in this direction. Thus, $\|A - C\|_S / n$ is the largest average distance to the mean in any direction. It is an upper bound on the variance of the optimal clusters. Assume that $n_i = n/k$ for all i. Then, $\Delta_{rs}^2 = (2c)^2 k^2 \|A - C\|_S^2 / n_i$ is close to being the maximal average variance of the two clusters in any direction. It is actually larger, because $\|A - C\|_S$ includes all clusters, so Δ_{rs} and thus the separation of the points in T_r and T_s depends on all clusters even though it differs for different r, s.

Seeding Method. Given an input that is assumed to satisfy the above separation condition, Kumar and Kanan compute an initial solution by projecting the points onto a lower-dimensional subspace and approximately solving the low-dimensional instance. The computed centers form the seed to the k-means method.

The lower-dimensional subspace is the best-fit subspace V_k, i. e., it minimizes the expression $\sum_{p \in P} \min_{v \in V} \|p - v\|^2$ among all k-dimensional subspaces V. It is known that V_k is the subspace spanned by the first k eigenvectors of A, which can be calculated by *singular value decomposition (SVD)*[2], and that projecting points to V_k and solving the SSE optimally on the projected points yields a 2-approximation. Any constant-factor approximation thus gives a constant approximation for the original input.

In addition to these known facts, the result by Kumar and Kannan shows that initializing the k-means algorithm with this solution even yields an optimal solution as long as the optimal partition satisfies the proximity condition.

---

[2] The computation of the SVD is a well-studied field of research. For an in-depth introduction to spectral algorithms and singular value decompositions, see [52].

4   Variants and Extensions of the k-Means Algorithm

The k-means algorithm is a widely used algorithm, but not always in the form given above. Naming all possible variations of the algorithm is beyond the scope of this survey and may be impossible to do. We look at two theoretically analyzed modifications.

Single Point Assignment Step. We call a point in a given clustering *misclassified* if the distance to the cluster center it is currently assigned to is longer than the distance to at least one of the other cluster centers. Hence, in each iteration of the k-means algorithm, all misclassified points are assigned to their closest cluster center and then all cluster centers are moved to the means of the updated clusters. Har-Peled and Sadri [44] study a variant of the k-means algorithm in which the assignment step assigns only one misclassified point to the closest cluster center instead of all misclassified points at once as done in the original algorithm. After such an assignment step, the centers of the two updated clusters are moved to the means of the clusters. The algorithm repeats this until no misclassified points exist. Har-Peled and Sadri call their variant *SinglePnt*. Given a number of clusters k and a set P of n points with spread Δ from a Euclidean space $\mathbb{R}^d$, they show that the number of iterations of *SinglePnt* is upper bounded by some polynomial in n, Δ, and k.

In the following, we will describe the proof given in [44]. W.l.o.g., we can assume that the minimum pairwise distance in P is 1 and the diameter of P is Δ. As we have seen for the classical k-means algorithm, the cost of P is $\mathcal{O}(n\Delta^2)$ after the first iteration of *SinglePnt*. The main idea is now to show that, in each following iteration of *SinglePnt*, the improvement of the clustering cost is lower bounded by some value dependent on the ratio between the distance of the reassigned point to the two involved cluster centers and the size of the two clusters. Based on this fact, we will prove that $\mathcal{O}(kn)$ iterations are sufficient to decrease the clustering cost by some constant amount, which results in $\mathcal{O}(kn^2\Delta^2)$ iterations in total.

Let Q_i and Q_j be any two clusters such that, in an assignment step, a point $q \in Q_j$ moves from cluster Q_j to cluster Q_i, i.e., after this step we obtain the two clusters $Q_i' = Q_i \cup \{q\}$ and $Q_j' = Q_j \backslash \{q\}$. Let $\mu(X)$ be the mean of any point set $X \subset \mathbb{R}^d$. Then, the movement of the first cluster center is

$$\|\mu(Q_i) - \mu(Q_i')\| = \left\|\mu(Q_i) - \left(\frac{|Q_i|}{|Q_i|+1}\mu(Q_i) + \frac{1}{|Q_i|+1}q\right)\right\| = \frac{\|\mu(Q_i)-q\|}{|Q_i|+1}.$$

Similarly, we have $\|\mu(Q_j) - \mu(Q_j')\| = \|\mu(Q_j) - q\|/(|Q_j| - 1)$. Due to Fact 1, the movement of the first cluster center decreases the clustering cost of Q_i' by $(|Q_i| + 1)\|\mu(Q_i) - \mu(Q_i')\|^2 = \|\mu(Q_i) - q\|/(|Q_i| + 1)$, and the movement of the second cluster center decreases the clustering cost of Q_j' by $(|Q_j| - 1)\|\mu(Q_j) - \mu(Q_j')\|^2 = \|\mu(Q_j) - q\|/(|Q_j| - 1)$. It follows that the total decrease in the clustering cost is at least $(\|\mu(Q_i) - q\| + \|\mu(Q_j) - q\|)^2/(2(|Q_i| + |Q_j|))$.

The reassignment of a point $q \in P$ is called *good* if the distance of q to at least one of the two centers of the involved clusters is bigger than $1/8$. Otherwise, the reassignment is called *bad*. If a reassignment is good, then it follows from the above that the improvement of the clustering cost is at least $(1/8)^2/(2n) = 1/(128n)$. Thus, $\mathcal{O}(n)$ good reassignments are sufficient to improve the clustering cost by some constant amount. Next, we show that one out of $k+1$ reassignments must be good, which then completes the proof.

For each $i \in \{1, \ldots, k\}$, let B_i be the ball with radius $1/8$ whose center is the i-th center in the current clustering. Since the minimum pairwise distance in P is 1, each ball can contain at most one point of P. Observe that a point $q \in P$ can only be involved in a bad reassignment if it is contained in more than one ball. Let us consider the case that, due to a bad reassignment, a ball B_i loses its point $q \in P$ and so has been moved a distance of at most $1/8$ away from q. Since the minimum pairwise distance in P is 1, B_i needs a good reassignment, so that it can again contain a point from P. Next, observe that, while performing only bad reassignments, a cluster Q_i is changed by gaining or losing the point q contained in B_i. Hence, if a cluster B_i loses q, it cannot gain it back. Otherwise, the clustering cost would be increased. It follows that the total number of consecutive bad reassignments is at most k.

Generalization of Misclassification. Har-Peled and Sadri [44] study another variant of the k-means algorithm, which they call *Lazy-k-Means*. This variant works exactly like the original algorithm except that each iteration reassigns only those points which are significantly misclassified. More precisely, given a k-clustering of a set P of n points from a Euclidean space $\mathbb{R}^d$ and a precision parameter ε, $0 \leq \varepsilon \leq 1$, we call a point $q \in P$ $(1+\varepsilon)$-misclassified if q belongs to some cluster Q_j and there is some other cluster Q_i with $\|q - \mu(Q_j)\| > (1 + \varepsilon)\|q - \mu(Q_i)\|$, where $\mu(X)$ is the mean of some set $X \subset \mathbb{R}^d$. Each iteration of *Lazy-k-Means* reassigns all $(1 + \varepsilon)$-misclassified points to their closest cluster center and then moves each cluster center to the mean of its updated cluster. This process is repeated until there are no $(1 + \varepsilon)$-misclassified points. Note that, for $\varepsilon = 0$, *Lazy-k-Means* is equal to the k-means algorithm. For $0 < \varepsilon \leq 1$, Har-Peled and Sadri prove that the number of iteration of *Lazy-k-Means* is upper bounded by some polynomial in n, Δ, and ε^{-1}, where Δ is the spread of the point set P.

In the following, we will sketch the proof given in [44]. W.l.o.g., we can assume that the minimum pairwise distance in P is 1 and the diameter of P is Δ, so the clustering cost is $\mathcal{O}(n\Delta^2)$ after the first iteration of *Lazy-k-Means*. The idea is now to show that every two consecutive iterations lead to a cost improvement of $\Omega(\varepsilon^3)$, which results in $\mathcal{O}(n\Delta^2\varepsilon^{-3})$ iterations in total. The proof of the lower bound on the cost improvement is based on the following known fact (see also Fig. 3).

Fact 2. *Given two points $c, c' \in \mathbb{R}^d$ with $\|c - c'\| = \ell$, all points $q \in \mathbb{R}^d$ with $\|q - c\| > (1 + \varepsilon)\|q - c'\|$ are contained in the open ball whose radius is $R = \ell(1 + \varepsilon)/(\varepsilon(2 + \varepsilon))$ and whose center is on the line containing the segment cc' at distance $R + \ell\varepsilon/(2(2 + \varepsilon))$ from the bisector of cc' and on the same side of the bisector as c'. The ball is called ε-Apollonius ball for c' with respect to c.*

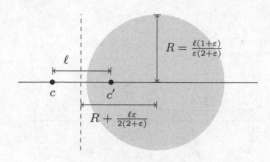

Fig. 3. Illustration of the ε-Apollonius ball for a point c' with respect to a point c.

Let $q \in\!\!\!/\, P$ be any $(1 + \varepsilon)$-misclassified point that switches its assignment from a center c to another center c' with $\ell = \|c - c'\|$. We also say that c and c' are the *switch centers* of q. Then, based on the fact that the distance of q to the bisector of cc' is at least $\ell\varepsilon/(2(2 + \varepsilon))$ (see Fact 2 and Fig. 3) and by using Pythagorean equality, one can show that the improvement of the clustering cost for q is at least

$$\|q - c\|^2 - \|q - c'\|^2 \geq \frac{\ell^2\varepsilon}{2 + \varepsilon}.$$

We call any $(1+\varepsilon)$-misclassified point $q \in P$ *strongly misclassified* if the distance between its switch centers is at least $\ell_0 := \varepsilon(2 + \varepsilon)/(16(1 + \varepsilon))$. Otherwise, a $(1+\varepsilon)$-misclassified point is called *weakly misclassified*. It follows from the above inequality that the improvement of the clustering cost caused by reassigning a strongly misclassified point is at least $\ell_0^2\varepsilon/(2 + \varepsilon) = \Omega(\varepsilon^3)$ for $0 < \varepsilon \leq 1$. Thus, if we can show that at least every second iteration of *Lazy-k-Means* reassigns some strongly misclassified point, then we are done.

Let us assume that there are only weakly misclassified points, and q is one of these points with switch centers c and c'. We know that the distance ℓ between c and c' is less than ℓ_0, which is less than $1/8$ for $0 < \varepsilon \leq 1$. Furthermore, it follows from $\ell < \ell_0$ that the radius of the ε-Apollonius ball for c' with respect to c is less than $1/16$ (see also Fig. 4). Since q is contained in this ε-Apollonius ball, the distance between c' and q is less than $1/8$. Hence, both switch centers have a distance of less than $1/4$ from q. Since the minimum pairwise distance in P is 1, every center can serve as a switch center for at most one weakly misclassified point.

Let us consider any weakly misclassified point q with switch centers c and c', where c belongs to the cluster that loses q and c' belongs to the cluster that gains q. As explained above, both centers have a distance of less than $1/4$ from q. Hence, due to reassigning q, center c is moved by a distance of less than $1/4$. It follows that, after the considered iteration, the distance between c and q is less than $1/2$. Since the minimum pairwise distance in P is 1, every other point in P has a distance of more than $1/2$ to c. Thus, c can only be a switch center for strongly misclassified points in the next iteration. Furthermore, due to reassigning q, the gaining center c' is moved towards q. Since the distance of

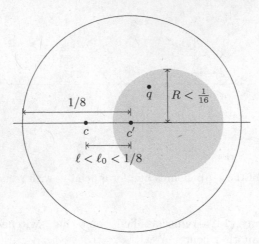

Fig. 4. Illustration of the fact that each center can serve as a switch center for at most one weakly misclassified point.

q to all the other points in P is at least 1, no other center can move closer to q than c' due to a reassignment of a weakly misclassified point. This means in the next iteration c' will still be the closest cluster center to q and q will not be $(1 + \varepsilon)$-misclassified. As a result, either there are no $(1 + \varepsilon)$-misclassified points left and the algorithm terminates or there are some strongly misclassified points. Thus, at least every second iteration reassigns some strongly misclassified points, which completes the proof.

5   Alternatives to the k-Means Algorithm for Big Data

Again, naming all alternative clustering algorithms that have been proposed is beyond the scope of this survey. However, we will take a short look at algorithms, that are developed starting from a theoretical analysis (with respect to the SSE problem), but that are also implemented and shown to be viable in practice. We have already discussed one prime example for this type of algorithm, the k-means++ algorithm by Arthur and Vassilvitskii [14]. The running time of the seeding is comparable to one iteration of the k-means algorithm (when assuming that drawing random numbers is possible in constant time), so using it as a seeding method does not have a significant influence on the running time asymptotically or in practice. However, it turns the k-means algorithm into an expected $\mathcal{O}(\log k)$-approximation algorithm. A similar example is the local search algorithm by Kanungo et al. [54] that we describe in more detail in Sect. 6. It has a polynomial worst case running time and provides a constant approximation. Additionally, it was implemented and showed very good practical behavior when combined with the k-means algorithm.

However, the research we have discussed in Sect. 2.1 aiming at accelerating the iterations of the k-means algorithm shows that there is interest in being faster than the k-means algorithm (and the constant approximation algorithms), and this interest increases with the availability of larger and larger amounts of data. The problem of solving the SSE problem for big data has been researched from a practical as well as from a theoretical side and in this section, we are interested in the intersection.

The theoretical model of choice is *streaming*. The data stream model assumes that the data can only be read once and in a given order, and that the algorithm is restricted to small space, e.g. polylogarithmic in the input it processes, but still computes an approximation. One-pass algorithms and low memory usage are certainly also desirable from a practical point of view, since random access to the data is a major slowdown for algorithms, and small memory usage might mean that all stored information actually fits into the main memory. The k-means algorithm reads the complete data set in each iteration, and a straightforward implementation of the k-means++ reads the data about k times for the seeding alone, and these are reasons why the algorithms do not scale so well for large inputs.

An old variant of the k-means algorithm, proposed independently of Lloyd's work by MacQueen [59], gives a very fast alternative to the k-means algorithm. It processes the data once, assigns each new data point to its closest center and updates this center to be the centroid of the points assigned to it. Thus, it never reassigns points. MacQueen's k-means algorithm clearly satisfies the first two requirements for a streaming algorithm, but not the third. Indeed, it is not surprising that MacQueen's algorithm does not necessarily converge to a good solution, and that the solution depends heavily on the start centers and the order of the input points. The famous streaming algorithm BIRCH [76] is also very fast and is perceived as producing better clusterings, yet, it still shares the property that there is no approximation guarantee [37].

Various data stream algorithms for the SSE problem have been proposed, see for example [29,34,35,38,42,43], achieving $(1 + \varepsilon)$-approximations in one pass over the data for constant k (and constant d, for some of the algorithms). We now look at algorithms which lie in between practical and theoretical results.

Local Search and the *Stream* Framework. Guha et al. [40] develop a framework for clustering algorithms in the data stream setting that they call *Stream*. They combine it with a constant factor approximation based on local search. The resulting algorithm is named StreamLS[3]. It computes a constant approximation in the data stream setting. StreamLS has originally been designed for the variant of the SSE problem where the distances are not squared (also called the k-median problem), but it is stated to work for the SSE problem as well with worse constants.

The Stream framework reads data in blocks of size m. For each block, it computes a set of $c \cdot k$ centers that are a constant factor approximation for the SSE problem with k centers (c is a constant) by using an approximation

---

[3] http://infolab.stanford.edu/~loc/.

algorithm A. It thus reduces m points to $c \cdot k$ points, where m is at least n^ε for some $\varepsilon > 0$. This is repeated until the number of computed centers reaches m, i.e. it is repeated for $m/(ck)$ blocks. Then, $m^2/(ck)$ points have been be reduced to m points. These are then again reduced to ck points, i.e. the computed centers are treated like as input to the same procedure, one level higher in a computation tree. On the ith level of this tree, ck points represent $(m/ck)^i$ input blocks. Thus, the height of the computation tree is at most $\mathcal{O}(\log_{m/(ck)} n/m) \in \mathcal{O}(\log_{n^\varepsilon} n)$. This is actually a constant, since

$$\log_{n^\varepsilon/(ck)} n = \frac{\log n}{\log n^\varepsilon} = \frac{1}{\varepsilon}.$$

Thus, the computation tree has constant height. It stores at most m points on each level, so the storage requirement of the algorithm is $\Theta(m) = \mathcal{O}(n^\varepsilon)$ under the assumption that A requires space that is linear in its input size. The running time of the algorithm is $\mathcal{O}(ndk)$ under the assumption that A has linear running time. Whenever an actual solution to the SSE problem is queried, it can be produced from the $\mathcal{O}(m)$ stored centers by computing a constant factor approximation by a different algorithm A'. Guha et al. show that the result is a constant factor approximation for the original input data.

Guha et al. also develop the algorithm LSEARCH which they use as the algorithm A within their framework. The algorithm StreamLS is the combination of the Stream framework with the algorithm LSEARCH. LSEARCH is a local search based algorithm that is based on algorithms for a related problem, the facility location problem. It is allowed to computed more than k centers, but additional centers are penalized. The main purpose of LSEARCH is an expected speed-up compared to other local search based methods with $\mathcal{O}(n^2)$ running time.

The experiments included in [40] actually use the SSE criterion to evaluate their results, since the intention is to compare with the k-means algorithm, which is optimized for SSE. The data sets are around fifty thousand points and forty dimensions. First, LSEARCH is compared to the k-means algorithm and found to be about three times slower than the k-means algorithm while producing results that are much better. Then, StreamLS is compared to BIRCH and to StreamKM, the algorithm resulting from embedding the k-means algorithm into the Stream framework. StreamLS and StreamKM compute solutions of much higher quality than BIRCH, with StreamLS computing the best solutions. BIRCH on the other hand is significantly faster, in particular, its running time per input point increases much less with increasing stream length.

Adaptions of k-Means++. Ailon et al. [8] use the Stream framework and combine it with different approximation algorithms. The main idea is to extend the seeding part of the k-means++ algorithm to an algorithm called k-means# and to use this algorithm within the above Stream framework description. Recall that the seeding in k-means++ is done by D^2-sampling. This method iteratively samples k centers. The first one is sampled uniformly at random. For the ith center, each input point p is sampled with probability $D^2(p)/\sum_{q \in P} D^2(q)$, where P is the input point set, $D^2(p) = \min_{c_1,...,c_{i-1}} \|p - c_i\|^2$ is the cost of p in the current solution and $c_1, \ldots c_{i-1}$ are the centers chosen so far. A set of k centers chosen in this way is an expected $\mathcal{O}(\log k)$-approximation.

The algorithm k-means# starts with choosing $3 \log k$ centers uniformly at random and then performs $k - 1$ iterations, each of which samples $3 \log k$ centers according to the above given probability distribution. This is done to ensure that for an arbitrary optimal clustering of the points, each of the clusters is 'hit' with constant probability by at least one center. Ailon et al. show that the $\mathcal{O}(k \log k)$ centers computed by k-means# are a constant factor approximation for the SSE criterion with high probability[4].

To obtain the final algorithm, the Stream framework is used. Recall that the framework uses two approximation algorithms A and A'. While A can be a bicriteria approximation that computes a constant factor approximation with $c \cdot k$ centers, A' has to compute an approximative solution with k centers. The approximation guarantee of the final algorithm is the guarantee provided by A'.

Ailon et al. sample k centers by D^2-sampling for A', thus, the overall result is an expected $\mathcal{O}(\log k)$ approximation. For A, k-means# is ran $3 \log n$ times to reduce the error probability sufficiently and then the best clustering is reported. The overall algorithm needs n^ε memory for a constant $\varepsilon > 0$.

The overall algorithm is compared to the k-means algorithm and to MacQueen's k-means algorithm on data sets with up to ten thousand points in up to sixty dimensions. While it produces solutions of better quality than the two k-means versions, it is slower than both.

Ackermann et al. [6] develop a streaming algorithm based on k-means++ motivated from a different line of work[5]. The ingredients of their algorithms look very much alike the basic building blocks of the algorithm by Ailon et al.: sampling more than k points according to the k-means++ sampling method, organizing the computations in a binary tree and computing the final clustering with k-means++. There are key differences, though.

Firstly, their work is motivated from the point of view of *coresets* for the SSE problem. A coreset S for a point set P is a smaller and weighted set of points that has approximately the same clustering cost as P *for any choice of k centers*. It thus satisfies a very strong property. Ackermann et al. show that sampling sufficiently many points according to the k-means++ sampling results in a coreset. For constant dimension d, they show that $\mathcal{O}(k \cdot (\log n)^{\mathcal{O}(1)})$ points guarantee that the clustering cost of the sampled points is within an ε-error from the true cost of P for any set of k centers[6].

Coresets can be embedded into a streaming setting very nicely by using a technique called *merge-and-reduce*. It works similar as the computation tree of

---

[4] As briefly discussed in Sect. 3.1, it is sufficient to sample $\mathcal{O}(k)$ centers to obtain a constant factor approximation as later discovered by Aggarwal et al. [7].

[5] http://www.cs.uni-paderborn.de/fachgebiete/ag-bloemer/forschung/abgeschlossene /clustering-dfg-schwerpunktprogramm-1307/streamkmpp.html.

[6] This holds with constant probability and for any constant ε.

the Stream framework: It reads blocks of data, computes a coreset for each block and merges and reduces these coresets in a binary computation tree. Now the advantage is that this tree can have superconstant height since this can be cancelled out by adjusting the error ε of each coreset computation. A maximum height of $\Theta(\log n)$ means that the block size on the lowest level can be much smaller than above (recall that in the algorithm by Ailon et al., the block size was n^ε). For the above algorithm, a height of $\Theta(\log n)$ would mean that the approximation ratio would be $\Omega(c^{\log n}) \in \Omega(n)$. By embedding their coreset construction into the merge-and-reduce technique, Ackermann et al. provide a streaming algorithm that needs $\mathcal{O}(k \cdot (\log n)^{O(1)})$ space and computes a coreset of similar size for SSE problem. They obtain a solution for the problem by running k-means++ on the coreset. Thus, the solution is an expected $\mathcal{O}(\log k)$-approximation.

Secondly, Ackermann et al. significantly speed up the k-means++ sampling approach. Since the sampling is applied again and again, this has a major impact on the running time. Notice that it is necessary for the sampling to compute $D(p)$ for all p and to update this after each center that was drawn. When computing a coreset of m points for a point of ℓ points, a vanilla implementation of this sampling needs $\Theta(dm\ell)$ time. Ackermann et al. develop a data structure called *coreset tree* which allows to perform the sampling much faster. It does, however, change the sampling procedure slightly, such that the theoretically proven bound does not necessarily hold any more.

In the actual implementation, the sample size and thus the coreset size is set to $200\,k$ and thus much smaller than it is supported by the theoretical analysis. However, experiments support that the algorithm still produces solutions of high quality, despite these two heuristic changes. The resulting algorithm is called StreamKM++.

Ackermann et al. test their algorithm on data sets with up to eleven million points in up to 68 dimensions and compare the performance to BIRCH, StreamLS, the k-means algorithm and k-means++. They find that StreamLS and StreamKM++ compute solutions of comparable quality, and much better than BIRCH. BIRCH is the fastest algorithm. However, StreamKM++ beats the running time of StreamLS by far and can e.g. compute a solution for the largest data set and $k = 30$ in 27 % of the running time of StreamLS. For small dimensions or higher k, the speed up is even larger. The k-means algorithm and k-means++ are much slower than StreamLS and thus also than StreamKM++. It is to be expected that StreamKM++ is faster than the variant by Ailon et al. as well.

Sufficient Statistics. The renown algorithm BIRCH[7] [76] computes a clustering in one pass over the data by maintaining a preclustering. It uses a data structure called *clustering feature* tree, where the term clustering feature denotes the sufficient statistics for the SSE problem. The leaves of the tree represent subsets of the input data by their sufficient statistics. At the arrival of each new

---

[7] http://pages.cs.wisc.edu/vganti/birchcode/.

point, BIRCH decides whether to add the point to an existing subset or not. If so, then it applies a rule to choose one of the subsets and to add the point to it by updating the sufficient statistics. This can be done in constant time. If not, then the tree grows and represents a partitioning with one more subset.

BIRCH has a parameter for the maximum size of the tree. If the size of the tree exceeds this threshold, then it rebuilds the tree. Notice that a subset represented by its sufficient statistics cannot be split up. Thus, rebuilding means that some subsets are merged to obtain a smaller tree. After reading the input data, BIRCH represents each subset in the partitioning by a weighted point (which is obtained from the sufficient statistics) and then runs a clustering algorithm on the weighted point set.

The algorithm is very fast since updating the sufficient statistics is highly efficient and rebuilding does not occur too often. However, the solutions computed by BIRCH are not guaranteed to have a low cost with respect to the SSE cost function.

Fichtenberger et al. [37] develop the algorithm BICO[8]. The name is a combination of the words BIRCH and coreset. BICO also maintains a tree which stores a representation of a partitioning. Each node of this tree represents a subset by its sufficient statistics.

The idea of BICO is to improve the decision if and where to add a point to a subset in order to decrease the error of the summary. For this, BICO maintains a maximum error value T. A subset is forbidden to induce more error than T. The error of a subset is measured by the squared distances of all points in the subset to the centroid because in the end of the computation, the subset will be represented by the centroid.

For a new point, BICO searches for the subset whose centroid is closest to the point. BICO first checks whether the new point lies within a certain radius of this centroid since it wants to avoid to use all the allowed error of a subset for one point. If the point lies outside of the radius, a new node is created directly beneath the root of the tree for the new point. Otherwise, the point is added to this subset if the error keeps being bounded by T. If the point does not pass this check, then it is passed on to the child node of the current node whose centroid is closest. If no child node exists or the point lies without the nodes radius, then a new child node is created based on the new point.

If the tree gets too large, then T is doubled and the tree is rebuilt by merging subsets whose error as a combined subset is below the new T.

For constant dimension d, Fichtenberger et al. show that the altered method is guaranteed to compute a summary that satisfies the coreset property for a threshold value that lies in $\Theta(k \cdot \log n)$. Combined with k-means++, BICO gives an expected $\mathcal{O}(\log k)$-approximation.

The implementation of BICO faces the same challenge as StreamKM++, k-means or k-means++, namely, it needs to again and again compute the distance between a point and its closest neighbor in a stored point set. BICO has one advantage, though, since it is only interested in this neighbor if it

---

[8] http://ls2-www.cs.uni-dortmund.de/bico.

lies within a certain radius of the new point. This helps in developing heuristics to speed up the insertion process. The method implemented in BICO has the same worst case behavior as iterating through all stored points but can be much faster.

Fichtenberger et al. compare BICO to StreamKM++, BIRCH and MacQueen's k-means algorithm on the same data sets as in [6] and one additional 128-dimensional data set. In all experiments, the summary size of BICO is set to $200\,k$, thus the summary is not necessarily a coreset. The findings are that BICO and StreamKM++ compute the best solutions, while BIRCH and MacQueen are the fastest algorithms. However, for small k, the running time of BICO is comparable to BIRCH and MacQueen. The running time of BICO is $\mathcal{O}(ndm)$, where m is the chosen summary size, thus, the increase in the running time for larger k stems from the choice $m = 200\,k$. For larger k, the running time can be decreased to lie below the running time of BIRCH by reducing m at the cost of worse solutions. In the tested instances, the quality was then still higher than for BIRCH and MacQueen.

6   Complexity of SSE

Before we consider variants of the k-means algorithm that deal with objective functions different from SSE, we conclude our SSE related study by looking at the complexity of SSE in general. We start by delivering a proof to the following fact which we already used above. We also reflect on the insights that it gives us on the structure of optimal solutions of the SSE problem.

Fact 3. *Let $\mu := \frac{1}{|P|} \sum_{p \in P} p$ be the mean of a point set P, and let $y \in \mathbb{R}^d$ be any point. Then, we have*

$$\sum_{p \in P} \|p - y\|^2 = \sum_{p \in P} \|p - \mu\|^2 + |P| \cdot \|y - \mu\|^2.$$

Proof. The result is well known and the proof is contained in many papers. We in particular follow [55]. First note that

$$\sum_{p \in P} \|p - y\|^2 = \sum_{p \in P} \|p - \mu + \mu - y\|^2$$

$$= \sum_{p \in P} \|p - \mu\|^2 + 2(\mu - y)^T \sum_{p \in P} (p - \mu) + |P| \cdot \|y - \mu\|^2.$$

Thus, the statement follows from

$$\sum_{p \in P} (p - \mu) = \sum_{p \in P} p - |P| \cdot \mu = \sum_{p \in P} p - |P| \frac{1}{|P|} \sum_{p \in P} p = 0.$$

The first consequence of Fact 3 is that the SSE problem can be solved analytically for $k = 1$: The mean μ minimizes the cost function, and the optimal cost is $\sum_{p \in P} \|p - \mu\|^2$. For $k \geq 2$, the optimal solution induces a partitioning of the input point set P into subsets of P with the same closest center. These subsets are called *clusters*. The center of a cluster is the mean of the points contained in the cluster (otherwise, exchanging the center by the mean would improve the solution). At the same time, every partitioning of the point set induces a feasible solution by computing the mean of each subset of the partitioning. This gives a new representation of an optimal solution as a partitioning of the input point set that minimizes the induced clustering cost.

Notice that we cannot easily enumerate all possible centers as there are infinitely many possibilities. By our new view on optimal solutions, we can instead iterate over all possible partitionings. However, the number of possible partitionings is exponential in n for every constant $k \geq 2$. We get the intuition that the problem is hard, probably even for small k. Next, we see a proof that this is indeed the case. Notice that there exist different proofs for the fact that SSE is NP-hard [9,32,60] and the proof presented here is the one due to Aloise et al. [9].

NP-Hardness of SSE. We reduce the following problem to SSE with $k = 2$. Given a graph $G = (V, E)$, a *cut* is a partitioning of the nodes V into subsets $X \subset V$ and $V \backslash X \subset V$. By the *density of a cut* $(X, V \backslash X)$, we mean the ratio $|E(X)|/(|X| \cdot |V \backslash X|)$, where $E(X)$ is the set of edges having one endpoint in X and the other endpoint in $V \backslash X$. Now, our version of the *densest cut problem* asks for the cut with the highest density. This problem is NP-hard because it is equivalent to finding the cut with minimal density in the complement graph, which is known to be NP-hard due to [64].

We define a type of incidence matrix M in the following way. In a $|V| \times |E|$-matrix, the entry in row i and column j is 0 if edge j is not incident to vertex i. Otherwise, let i' be the other vertex to which j is incident. Then, we arbitrarily set one of the two entries (i, j) and (i', j) to 1 and the other one to -1. For an example, see Fig. 5(a) and (b). We interpret the rows of M as points in $\mathbb{R}^{|E|}$ and name the set of these points $P(V)$. Each subset $X \subseteq V$ then corresponds to a subset $P(X) \subseteq P(V)$, and a cut $(X, V \backslash X)$ corresponds to a partitioning $(P(X), P(X \backslash V))$ of these points and thus to a 2-clustering. We take a closer look at the cost of cluster $P(X)$ which is the sum of the costs of all points in it. For each point, the cost is the squared distance to the mean of $P(X)$, and this cost can be calculated by summing up the squared differences in each coordinate. Remember that the coordinates correspond to edges in E. Thus, one way to analyze the cost is to figure out how much cost is caused by a specific edge. For each edge $e_j = (x, y)$, there are three possibilities for the clustering cost: If $x, y \in X$, then the mean of $P(X)$ has a 0 in the jth coordinate, and thus the squared distance is 0 for all coordinates except those corresponding to x and y, and it is 1 for these two. If $x, y \notin X$, then the mean of $P(X)$ also has a 0 in the jth coordinate, and as all points in $P(X)$ also have 0 at the jth coordinate, this coordinate contributes nothing to the total cost. If either $x \in X, y \notin X$ or $x \notin X, y \in X$ and thus $e_j \in E(X)$, then the mean has $\pm 1/|X|$ as its jth coordinate, which induces a squared distance of $(0 - 1/|X|)^2$ for $|X| - 1$ of the

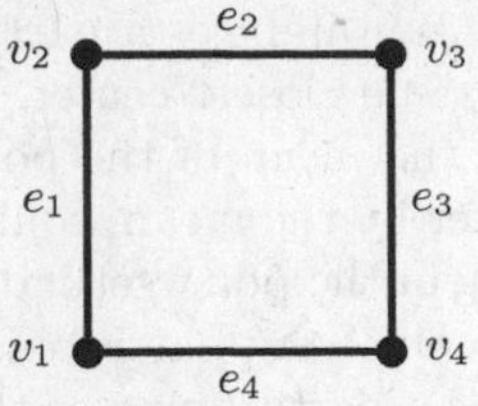

$$\begin{array}{c} \\ v_1 \\ v_2 \\ v_3 \\ v_4 \end{array} \begin{array}{c} \begin{array}{cccc} e_1 & e_2 & e_3 & e_4 \end{array} \\ \begin{pmatrix} 1 & 0 & 0 & -1 \\ -1 & 1 & 0 & 0 \\ 0 & -1 & 1 & 0 \\ 0 & 0 & -1 & 1 \end{pmatrix} \end{array}$$

(a) A simple example, (b) its corresponding matrix

$$\begin{array}{c} \\ v_1 \\ v_3 \\ v_2 \\ v_4 \end{array} \begin{array}{c} \begin{array}{cccc} e_1 & e_2 & e_3 & e_4 \end{array} \\ \begin{pmatrix} 1 & 0 & 0 & -1 \\ 0 & -1 & 1 & 0 \\ -1 & 1 & 0 & 0 \\ 0 & 0 & -1 & 1 \end{pmatrix} \end{array}$$

(c) and the cut $X = \{v_1, v_3\}$.

Fig. 5. An example for the reduction from our densest cut problem to SSE.

points, and a squared distance of $(1-1/|X|)^2$ for the one endpoint that is in X. Thus, the total cost of $P(X)$ is

$$\sum_{e_j=(x,y)\in E, x,y\in X} 2 + \sum_{e_j=(x,y)\in E(X)} \left[(|X|-1)\frac{1}{|X|^2} + (1-1/|X|)^2 \right]$$
$$= \sum_{e_j=(x,y)\in E, x,y\in X} 2 + |E(X)| \left(1 - \frac{1}{|X|}\right).$$

This analysis holds for the clustering cost of $P(V\backslash X)$ analogously. Additionally, every edge is either in $E(X)$, or it has both endpoints in either $P(X)$ or $P(V\backslash V)$. Thus, the total cost of the 2-clustering induced by X is

$$2(|E| - |E(X)|) + |E(X)| \left(2 - \frac{1}{|X|} - \frac{1}{|V\backslash X|}\right) = 2|E| - \frac{|E(X)| \cdot |V|}{|X| \cdot |V\backslash X|}.$$

Finding the optimal 2-clustering means that we minimize the above term. As $2|E|$ and $|V|$ are the same for all possible 2-clusterings, this corresponds to finding the clustering which maximizes $|E(X)|/(|X| \cdot |V\backslash X|)$. Thus, finding the best 2-clustering is equivalent to maximizing the density.

Notice that the above transformation produces clustering inputs which are $|E|$-dimensional. Thus, SSE is hard for constant k and arbitrary dimension. It is also hard for constant dimension d and arbitrary k [60]. For small dimension and a small number of clusters k, the problem can be solved in polynomial time by the algorithm of Inaba et al. [46].

Approximation Algorithms. This section is devoted to the existence of approximation algorithms for SSE. First, we convince ourselves that there is indeed hope for approximation algorithms with polynomial running time even if k or d is large. Above, we stated that we cannot solve the problem by enumerating all possible centers as there are infinitely many of them. But what if we choose centers only from the input point set? This does not lead to an optimal solution: Consider $k = 1$ and a point set lying on the boundary of a circle. Then the optimal solution is inside the circle (possibly its center) and is definitely not in the point set. However, the solution cannot be arbitrarily bad. Let $k = 1$ and let $c \in P$ be a point $p \in P$ which minimizes $\|p - \mu\|^2$, i.e., it is the point closest to the optimal center (breaking ties arbitrarily). Then,

$$\operatorname{cost}(P, \{c\}) = \sum_{p \in P} \|p - c\|^2 \stackrel{\text{Fact } 1}{=} \sum_{p \in P} \left(\|p - \mu\|^2 + \|c - \mu\|^2 \right)$$
$$\leq \sum_{p \in P} \left(\|p - \mu\|^2 + \|p - \mu\|^2 \right) = 2\operatorname{cost}(P, \{\mu\}).$$

Thus, a 2-approximated solution to the 1-means problem can be found in quadratic time by iterating through all input points. For $k > 1$, the calculation holds for each cluster in the optimal solution, and thus there exists a 2-approximate solution consisting of k input points. By iterating through all $\mathcal{O}(n^k)$ possible ways to choose k points from P, this gives a polynomial-time approximation algorithm for constant k.

For arbitrary k, we need a better way to explore the search space, i.e., the possible choices of centers out of P to gain a constant-factor approximation algorithm with polynomial running time. Kanungo et al. [55] show that a simple *swapping* algorithm suffices. Consider a candidate solution, i.e., a set $C \subseteq P$ with $|C| = k$. The swapping algorithm repeatedly searches for points $c \in C$ and $p \in P \backslash C$ with $\operatorname{cost}(P, C) > \operatorname{cost}(P, C \cup \{p\} \backslash \{c\})$, and then replaces c by p. Kanungo et al. prove that if no such swapping pair is found, then the solution is a 25-approximation of the best possible choice of centers from P. Thus, the swapping algorithm converges to a 50-approximation[9]. In addition, they show that in polynomial time by always taking swaps that significantly improve the solution, one only loses a $(1+\varepsilon)$-factor in the approximation guarantee. This gives a very simple local search algorithm with constant approximation guarantee. Kanungo et al. also refine their algorithm in two ways: First, they use a result by Matoušek [63] that says that one can find a set S of size $\mathcal{O}(n\varepsilon^{-d} \log(1/\varepsilon))$ in time $\mathcal{O}(n \log n + n\varepsilon^{-d} \log(1/\varepsilon))$ such that the best choice of centers from S is a $(1 + \varepsilon)$-approximation of the best choice of centers from $\mathbb{R}^d$. This set is used to choose the centers from instead of simply using P. Second, they use q-*swaps* instead of the 1-swaps described before. Here, $q' \leq q$ centers are simultaneously replaced by a set of q' new centers. They show that this leads to a $(9 + \varepsilon)$-approximation and also give a tight example showing that 9 is the best possible approximation ratio for swapping-based algorithms.

The work of Kanungo et al. is one step in a series of papers developing approximation algorithms for SSE. The first constant approximation algorithm

---

[9] Note that Kanungo et al. use a better candidate set and thus give a $(25 + \varepsilon)$-approximation.

was given by Jain and Vazirani [49] who developed a primal dual approximation algorithm for a related problem and extended it to the SSE setting. Inaba
et al. [46] developed the first polynomial-time $(1+\varepsilon)$-approximation algorithm for
the case of $k = 2$ clusters. Matušek [63] improved this and obtained a polynomial-
time $(1 + \varepsilon)$-approximation algorithm for constant k and d with running time
$\mathcal{O}(n \log^k n)$ if ε is also fixed. Further $(1 + \varepsilon)$-approximations were for example
given by [29, 34, 38, 43, 57, 73]. Notice that all cited $(1 + \varepsilon)$-approximation algorithms are exponential in the number of clusters k and in some cases additionally
in the dimension d.

Inapproximability Results. Algorithms with a $(1 + \varepsilon)$-guarantee are only known
for the case that k is a constant (and ε has to be a constant, too). Recently,
Awasthi et al. [17] showed that there exists an ε such that it is NP-hard to
approximate SSE within a factor of $(1 + \varepsilon)$ for arbitrary k and d. Their proof
holds for a very small value of ε, and a larger inapproximability result is not yet
known.

7 k-Means with Bregman Divergences

The k-means problem can be defined for any dissimilarity measure. An important
class of dissimilarity measures are *Bregman divergences*. Bregman divergences
have numerous applications in machine learning, data compression, speech and
image analysis, data mining, or pattern recognition. We review mainly results
known for the k-means algorithm when applied to Bregman divergences. As we
will see, for Bregman divergences the k-means method can be applied almost
without modifications to the algorithm.

To define Bregman divergences, let $\mathbb{D} \subseteq \mathbb{R}^d$, and let $\Phi : \mathbb{D} \to \mathbb{R}$ be a strictly
convex function that is differentiable on the relative interior $\mathrm{ri}(\mathbb{D})$. The *Bregman
divergence* $d_\Phi : \mathbb{D} \times \mathrm{ri}(\mathbb{D}) \to \mathbb{R}_{\geq 0} \cup \{\infty\}$ is defined as

$$d_\Phi(x, c) = \Phi(x) - \Phi(c) - (x - c)^T \nabla \Phi(c),$$

where $\nabla \Phi(c)$ is the gradient of Φ at c. The squared Euclidean distance is a
Bregman divergence. Other Bregman divergences that are used in various applications are shown on Fig. 6.

Bregman divergences have a simple geometric interpretation that is shown in
Fig. 7. For c fixed, let $f_c : \mathbb{R}^d \to \mathbb{R}$ be defined by $f_c(x) : \Phi(c) + (x - c)^T \nabla \Phi(c)$.
The function f_c is a linear approximation to Φ at point c. Then $d_\Phi(x, c)$ is
the difference between the true function value $\Phi(x)$ and the value $f_c(x)$ of the
linear approximation to Φ at c. Bregman divergences usually are asymmetric and
violate the triangle inequality. In fact, the only symmetric Bregman divergences
are the Mahalanobis divergences (see Fig. 6).

As one can see from Fig. 6, for some Bregman divergences d_Φ there exist
points x, c such that $d_\Phi(x, c) = \infty$. We call these pairs of points *singularities*. In
most results and algorithms that we describe these singularities require special
treatment or have to be defined away.

domain $\mathbb{D}$	$\Phi(x)$	$d_\Phi(p, q)$
$\mathbb{R}^d$	squared ℓ_2-norm $\|x\|_2^2$	squared Euclidean distance $\|x - c\|_2^2$
$\mathbb{R}^d$	generalized norm $x^T A x$	Mahalanobis distance $(x - c)^T A (x - c)$
$[0, 1]^d$	neg. Shannon entropy $\sum x_i \ln(x_i)$	Kullback-Leibler divergence $\sum c_i \ln(\frac{c_i}{x_i})$
$\mathbb{R}_+^d$	Burg entropy $\sum -\ln(x_i)$	Itakura-Saito divergence $\sum \frac{c_i}{x_i} - \ln(\frac{c_i}{x_i}) - 1$
$\mathbb{R}_+^d$	harmonic ($\alpha > 0$) $\sum \frac{1}{x_i^\alpha}$	harmonic divergence ($\alpha > 0$) $\sum \frac{1}{c_i^\alpha} - \frac{\alpha+1}{x_i^\alpha} + \frac{\alpha c_i}{x_i^{\alpha+1}}$
$\mathbb{R}_+^d$	norm-like ($\alpha \geq 2$) $\sum x_i^\alpha$	norm-like divergence ($\alpha \geq 2$) $\sum c_i + (\alpha - 1)x_i^\alpha + \alpha c_i x_i^{\alpha-1}$
$\mathbb{R}^d$	exponential $\sum e^{x_i}$	Exponential loss $\sum e^{c_i} - (c_i - x_i + 1)e^{x_i}$
$(-1, 1)^d$	Hellinger-like $\sum -\sqrt{1 - x_i^2}$	Hellinger-like divergence $\sum \frac{1 - c_i x_i}{\sqrt{1 - x_i^2}} - \sqrt{1 - c_i^2}$

Fig. 6. Some Bregman divergences.

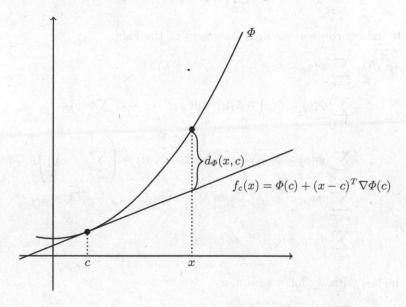

Fig. 7. Geometric interpretation of Bregman divergences

k-Means with Bregman Divergences. Similar to SSE we can define the *minimum sum-of-Bregman-errors clustering problem* (SBE). In this problem we are given a fixed Bregman divergence d_Φ with domain $\mathbb{D}$ and a set of points $P \subset \mathbb{D}$. The aim is to find a set $C \subset \mathrm{ri}(\mathbb{D})$ of k points (not necessarily included in P) such that the sum of the Bregman divergences of the points in P to their nearest center in C is minimized. Thus, the cost function to be minimized is

$$\mathrm{cost}_\Phi(P, C) := \sum_{p \in P} \min_{c \in C} d_\Phi(p, c).$$

The points in C are called *centers*. Because of the (possible) asymmetry of d_Φ the order of arguments in $d_\Phi(x, c)$ is important.

For any Bregman divergence the optimal solution for $k = 1$ is given by the mean of the points in P. More precisely, Fact 1 completely carries over to Bregman divergences (see [20]).

Fact 4. *Let $d_\Phi : \mathbb{D} \times \mathrm{ri}(\mathbb{D}) \to \mathbb{R}_{\geq 0} \cup \{\infty\}$ be a Bregman divergence and $P \subset \mathbb{D}$, $|P| < \infty$ and let*

$$\mu = \frac{1}{|P|} \sum_{p \in P}$$

be the mean of set P. For any $y \in \mathrm{ri}(\mathbb{D})$:

$$\sum_{p \in P} d_\Phi(p, y) = \sum_{p \in P} d_\Phi(p, \mu) + |P| \cdot d_\Phi(\mu, y).$$

Proof. It suffices to show the final statement of the Fact.

$$\sum_{p \in P} d_\Phi(p, y) = \sum_{p \in P} \Phi(p) - \Phi(y) - (x - s)^T \nabla \Phi(y)$$

$$= \sum_{p \in P} \Phi(p) - \Phi(\mu) + \Phi(\mu) - \Phi(y) - (x - s)^T \nabla \Phi(y)$$

$$= \sum_{p \in P} (\Phi(p) - \Phi(\mu)) + |P|(\Phi(\mu) - \Phi(y)) - \left(\sum_{p \in P} (p - y) \right)^T \nabla \Phi(y)$$

$$= \sum_{p \in P} (\Phi(p) - \Phi(\mu)) + |P|(\Phi(\mu) - \Phi(y) - (\mu - y)^T \nabla \Phi(y))$$

$$= \sum_{p \in P} d_\Phi(p, \mu) + |P| \cdot d_\Phi(\mu, y),$$

where the last equality follows from

$$\sum_{p \in P} (p - \mu)^T = 0 \quad \text{and} \quad \sum_{p \in P} (p - \mu)^T \nabla \Phi(\mu) = 0.$$

Moreover, for all Bregman divergences, any set of input points P, and any set of k centers $\{\mu_1, \ldots, \mu_k\}$, the optimal partitions for SBE induced by the centers μ_j can be separated by hyperplanes. This was first explicitly stated in [20]. More precisely, the Bregman bisector $\{x \in \mathbb{D} \mid d_\Phi(x, c_1) = d_\Phi(x, c_2)\}$ between any two points $c_1, c_2 \in \mathbb{D} \subseteq \mathbb{R}^d$ is always a hyperplane. i.e. for any pair of points c_1, c_2 there are $a \in \mathbb{R}^d, b \in \mathbb{R}$ such that

$$\{x \in \mathbb{D} \mid d_\Phi(x, c_1) = d_\Phi(x, c_2)\} = \{x \in \mathbb{D} \mid a^T x = b\}. \tag{1}$$

As a consequence, SBE can be solved for any Bregman divergence in time $O(n^{k^2 d})$. Hence for fixed k and d, SBE is solvable in polynomial time. However, in general SBE is an NP-hard problem. This was first observed in [4] and can be shown in two steps. First, let the Bregman divergence d_Φ be a Mahalanobis divergence for a symmetric, positive definite matrix A. Then there is a unique symmetric, positive definite matrix B such that $A = B^T B$, i.e. for any p, q

$$d_\Phi(p, q) = (p - q)^T A(p - q) = \|Bp - Bq\|^2. \tag{2}$$

Therefore, SBE with d_Φ is just SSE for a linearly transformed input set. This immediately implies that for Mahalanobis divergences SBE is NP-hard. Next, if Φ is sufficiently smooth, the Hessian $\nabla^2 \Phi t$ of Φ at point $t \in \text{ri}(\mathbb{D})$ is a symmetric, positive definite matrix. Therefore, d_Φ locally behaves like a Mahalanobis divergence. This can used to show that with appropriate restriction on the strictly convex function Φ SBE is NP-hard.

Approximation Algorithms and μ-Similarity. No provable approximation algorithms for general Bregman divergences are known. Approximation algorithms either work for specific Bregman divergences or for restricted classes of Bregman divergences. Chaudhuri and McGregor [27] give an $\mathcal{O}(\log(n))$ approximation algorithm for the Kullback-Leibler divergence (n is the size of the input set P). They obtain this result by exploiting relationships between the Kullback-Leibler divergence and the so-called Hellinger distortion and between the Hellinger distortion and the squared Euclidean distance.

The largest subclass of Bregman divergences for which approximation algorithms are known to exist consists of *μ-similar* Bregman divergences. A Bregman divergence d_Φ defined on domain $\mathbb{D} \times \text{ri}(\mathbb{D})$ is called μ-similar if there is a symmetric, positive definite matrix A and a constant $0 < \mu \leq 1$ such that for all $(x, y) \in \mathbb{D} \times \text{ri}(\mathbb{D})$

$$\mu \cdot d_A(x, y) \leq d_\Phi(x, y) \leq d_A(x, y). \tag{3}$$

Some Bregman divergences are (trivially) μ-similar. Others, like the Kullback-Leibler divergence or the Itakura-Saito divergence become μ-similar if one restricts the domain on which they are defined. For example, if we restrict the Kullback-Leibler divergence to $\mathbb{D} = [\lambda, \nu]^d$ for $0 < \lambda < \nu \leq 1$, then the Kullback-Leibler divergence is $\frac{\lambda}{\nu}$-similar. This can be shown by looking at the first order Taylor series expansion of the negative Shannon entropy $\Phi(x_1, \ldots, x_d) = \sum x_i \ln(x_i)$.

μ-similar Bregman divergences approximately behave like Mahalanobis divergences. Due to (2) Mahalanobis divergences behave like the squared Euclidean distance. Hence, one can hope that μ-similar Bregman divergences behave roughly like the squared Euclidean distance. In fact, it is not too difficult to show that the swapping algorithm of Kanungo et al. [55] can be generalized to μ-similar Bregman divergences to obtain approximation algorithms with approximation factor $18/\mu^2 + \epsilon$ for arbitrary $\epsilon > 0$. Whether one can combine the technique of Kanungo et al. with Matoušek's technique [63] to obtain better constant factor approximation algorithms is not known.

In the work of Ackermann et al. [5], μ-similarity has been used to obtain a probabilistic $(1 + \epsilon)$-approximation algorithm for SBE, whose running time is exponential in $k, d, 1/\epsilon$, and $1/\mu$, but linear in $|P|$. Building upon results in [57], Ackermann et al. describe and analyze an algorithm to solve the k-median problem for metric and non-metric distance measures D that satisfy the following conditions.

(1) For $k = 1$, optimal solutions to the k-median problem with respect to distance D can be computed efficiently.
(2) For every $\delta, \gamma > 0$ there is a constant $m_{\delta,\gamma}$ such that for any set P, with probability $1 - \delta$ the optimal 1-median of a random sample S of size $m_{\delta,\gamma}$ from P is a $(1 + \gamma)$-approximation to the 1-median for set P.

Together, (1) and (2) are called the $[\gamma, \delta]$-sampling property. Using the same algorithm as in [57] but a combinatorial rather than geometric analysis, Ackermann et al. show that for any distance measure D satisfying the $[\gamma, \delta]$-sampling property and any $\epsilon > 0$ there is an algorithm that with constant probability returns a $(1 + \epsilon)$-approximation to the k-median problem with distance measure D. The running time of the algorithm is linear in n, the number of input points, and exponential in $k, 1/\epsilon$, and the parameter $m_{\delta,\epsilon/3}$ from the sampling property. Finally, Ackermann et al. show that any μ-similar Bregman divergence satisfies the $[\delta, \gamma]$-sampling property with parameter $m_{\delta,\gamma} = \frac{1}{\gamma\delta\mu}$. Overall, this yields a $(1+\epsilon)$ algorithm for SBE for μ-similar Bregman divergences with running time linear in n, and exponential in $k, 1/\epsilon, 1/\mu$.

The k-Means Algorithm for Bregman Divergences. The starting point for much of the recent research on SBE for Bregman divergences is the work by Banerjee et al. [20]. They were the first to explicitly state Fact 4 and describe the k-means algorithm (see page 2) as a generic algorithm to solve SBE for arbitrary Bregman divergences. Surprisingly, the k-means algorithm cannot be generalized beyond Bregman divergences. In [19] it is shown, that under some mild smoothness conditions, any divergence that satisfies Fact 4 is a Bregman divergence. Of course, this does not imply that variants or modifications of the k-means algorithm cannot be used for distance measures other than Bregman divergences. However, in these generalizations cluster centroids cannot be used as optimizers in the second step, the re-estimation step.

Banerjee et al. already showed that for any Bregman divergence the k-means algorithm terminates after a finite number of steps. In fact, using the linear separability of intermediate solutions computed by the k-means algorithm (see Eq. 1), for any Bregman divergence the number of iterations of the k-means algorithm can be bounded by $\mathcal{O}(n^{k^2 d})$. Since the squared Euclidean distance is a Bregman divergence it is clear that no approximation guarantees can be given for the solutions the k-means algorithm finds for SBE.

1. Lower Bounds. Manthey and Röglin extended Vattani's exponential lower bound for the running time of the k-means algorithm to any Bregman divergence d_Φ defined by a sufficiently smooth function Φ. In their proof they use an approach similar to the approach used by Ackerman et al. to show that SBE is NP-hard. Using (2) Manthey and Röglin first extend Vattani's lower bound to any Mahalanobis divergence. Then, using the fact that any Bregman divergence d_Φ with sufficiently smooth Φ locally resembles some Mahalanobis divergence d_A, Manthey and Röglin show that a lower bound for the Mahalanobis divergence d_A carries over to a lower bound for the Bregman divergence d_Φ. Hence, for any smooth Bregman divergence the k-means algorithm has exponential running time. Moreover, Manthey and Röglin show that for the k-means algorithm the squared Euclidean distance, and more generally Mahalanobis divergences, are the easiest Bregman divergences.

2. Smoothed Analysis. Recall that the smoothed complexity of the k-means algorithm is polynomial in n and $1/\sigma$, when each input point is perturbed by random noise generated using a Gaussian distribution with mean 0 and standard deviation σ, a result due to Arthur, Manthey, and Röglin [12]. So far, this result has not been generalized to Bregman divergences. For almost any Bregman divergence d_Φ Manthey and Röglin [61] prove two upper bounds on the smoothed complexity of the k-means algorithm. The first bound is of the form $\mathrm{poly}(n^{\sqrt{k}}, 1/\sigma)$, the second is of the form $k^{kd} \cdot \mathrm{poly}(n, 1/\sigma)$. These bounds match bounds that Manthey and Rögin achieved for the squared Euclidean distance in [62]. Instead of reviewing their proofs, we will briefly review two technical difficulties Manthey and Röglin had to account for.

Bregman divergences $d_\Phi : \mathbb{D} \times \mathrm{ri}(\mathbb{D}) \to \mathbb{R}_{\geq 0} \cup \{\infty\}$ like the Kullback-Leibler divergence are defined on a bounded subset of some $\mathbb{R}^d$. Therefore perturbing a point in $\mathbb{D}$ may yield a point for which the Bregman divergence is not defined. Moreover, whereas the Gaussian noise is natural for the squared Euclidean distance this is by no means clear for all Bregman divergences. In fact, Banerjee et al. [20] already showed a close connection between Bregman divergences and exponential families, indicating that noise chosen according to an exponential distribution may be appropriate for some Bregman divergences. Manthey and Röglin deal with these issues by first introducing a general and abstract perturbation model parametrized by some $\sigma \in (0, 1]$. Then Manthey and Röglin give a smoothed analysis of the k-means algorithm for Bregman divergences with respect to this abstract model. It is important to note that as in the squared Euclidean case, the parameter σ measures the amount of randomness in the perturbation. Finally, for Bregman divergences like the Mahalanobis divergences, the Kullback-Leibler divergence, or the Itakura-Saito Manthey and Röglin instantiate the abstract perturbation model with some perturbations schemes using explicit distributions.

Singularities of Bregman divergences are the second technical difficulty that Manthey and Röglin have to deal with. For each Bregman divergence d_Φ they introduce two parameters $0 < \zeta \le 1$ and $\xi \ge 1$ that in some sense measures how far away d_Φ is from being a Mahalanobis divergence. This resembles the μ-similarity introduced by Ackermann et al. [5]. Whereas for many Bregman divergences the parameter μ can only be defined by restricting the domain of the divergence, this is not necessary in the approach by Manthey and Rögin. However, their upper bounds on the smoothed complexity of the k-means algorithm for Bregman divergences are not uniform, instead for any specific Bregman divergence the bound depends (polynomially) on the values ξ and $1/\zeta$.

It is still an open problem whether the polynomial bound of Arthur et al. [12] on the smoothed complexity of the k-means algorithm can be generalized to Bregman divergences. Surprisingly, even for general Mahalanobis divergences this is not known. As Manthey and Röglin mention, at this point polynomial bounds on the smoothed complexity of the k-means algorithm can only be achieved for Mahalanobis divergences d_A and input sets P, where the largest eigenvalue of A is bounded by a polynomial in $|P|$.

3. Seeding Methods. In [2] the k-means++ randomized seeding algorithm by Arthur and Vassilvitskii [14] is generalized to μ-similar Bregman divergences. Ackermann and Blömer show that for a μ-similar Bregman divergence this generalization, called Bregman++, yields a $\mathcal{O}(\mu^{-2} \log(k))$-approximation for SBE. In [3] Ackermann and Blömer generalize the result by Ostrovsky et al. [67] on adaptive sampling for ϵ-separable instances to Bregman divergences.

Nock et al. [66] generalize k-means++ to certain symmetrized versions of Bregman divergences d_Φ, called *mixed* Bregman divergences. They prove approximation factors of the form $\mathcal{O}(\rho_\psi \log k)$, where ρ_ψ is some parameter depending on d_Φ, that roughly measures how much d_Φ violates the triangle inequality. Note, however, that the mixed Bregman divergences introduced by Nock et al. are not proper Bregman divergences.

References

1. Achlioptas, D., McSherry, F.: On spectral learning of mixtures of distributions. In: Auer, P., Meir, R. (eds.) COLT 2005. LNCS (LNAI), vol. 3559, pp. 458–469. Springer, Heidelberg (2005). doi:10.1007/11503415_31
2. Ackermann, M.R., Blömer, J.: Coresets and approximate clustering for Bregman divergences. In: Proceedings of the 20th Annual ACM-SIAM Symposium on Discrete Algorithms (SODA 2009), pp. 1088–1097. Society for Industrial and Applied Mathematics (SIAM) (2009). http://www.cs.uni-paderborn.de/uploads/tx_sibibtex/CoresetsAndApproximateClusteringForBregmanDivergences.pdf
3. Ackermann, M.R., Blömer, J.: Bregman clustering for separable instances. In: Kaplan, H. (ed.) SWAT 2010. LNCS, vol. 6139, pp. 212–223. Springer, Heidelberg (2010). doi:10.1007/978-3-642-13731-0_21
4. Ackermann, M.R., Blömer, J., Scholz, C.: Hardness and non-approximability of Bregman clustering problems. In: Electronic Colloquium on Computational Complexity (ECCC), vol. 18, no. 15, pp. 1–20 (2011). http://eccc.uni-trier.de/report/2011/015/, report no. TR11-015

5. Ackermann, M.R., Blömer, J., Sohler, C.: Clustering for metric and non-metric distance measures. ACM Trans. Algorithms **6**(4), Article No. 59:1–26 (2010). Special issue on SODA 2008
6. Ackermann, M.R., Märtens, M., Raupach, C., Swierkot, K., Lammersen, C., Sohler, C.: Streamkm++: a clustering algorithm for data streams. ACM J. Exp. Algorithmics **17**, Article No. 4, 1–30 (2012)
7. Aggarwal, A., Deshpande, A., Kannan, R.: Adaptive sampling for k-means clustering. In: Dinur, I., Jansen, K., Naor, J., Rolim, J. (eds.) APPROX/RANDOM -2009. LNCS, vol. 5687, pp. 15–28. Springer, Heidelberg (2009). doi:10.1007/978-3-642-03685-9_2
8. Ailon, N., Jaiswal, R., Monteleoni, C.: Streaming k-means approximation. In: Proceedings of the 22nd Annual Conference on Neural Information Processing Systems, pp. 10–18 (2009)
9. Aloise, D., Deshpande, A., Hansen, P., Popat, P.: NP-hardness of Euclidean sum-of-squares clustering. Mach. Learn. **75**(2), 245–248 (2009)
10. Alsabti, K., Ranka, S., Singh, V.: An efficient k-means clustering algorithm. In: Proceeding of the First Workshop on High-Performance Data Mining (1998)
11. Arora, S., Kannan, R.: Learning mixtures of separated nonspherical Gaussians. Ann. Appl. Probab. **15**(1A), 69–92 (2005)
12. Arthur, D., Manthey, B., Röglin, H.: k-means has polynomial smoothed complexity. In: Proceedings of the 50th Annual IEEE Symposium on Foundations of Computer Science (FOCS 2009), pp. 405–414. IEEE Computer Society (2009)
13. Arthur, D., Vassilvitskii, S.: How slow is the k-means method? In: Proceedings of the 22nd ACM Symposium on Computational Geometry (SoCG 2006), pp. 144–153 (2006)
14. Arthur, D., Vassilvitskii, S.: **k-means++**: the advantages of careful seeding. In: Proceedings of the 18th Annual ACM-SIAM Symposium on Discrete Algorithms (SODA 2007), pp. 1027–1035. Society for Industrial and Applied Mathematics (2007)
15. Arthur, D., Vassilvitskii, S.: Worst-case and smoothed analysis of the ICP algorithm, with an application to the k-means method. SIAM J. Comput. **39**(2), 766–782 (2009)
16. Awasthi, P., Blum, A., Sheffet, O.: Stability yields a PTAS for k-median and k-means clustering. In: FOCS, pp. 309–318 (2010)
17. Awasthi, P., Charikar, M., Krishnaswamy, R., Sinop, A.K.: The hardness of approximation of Euclidean k-means. In: SoCG 2015 (2015, accepted)
18. Balcan, M.F., Blum, A., Gupta, A.: Approximate clustering without the approximation. In: SODA, pp. 1068–1077 (2009)
19. Banerjee, A., Guo, X., Wang, H.: On the optimality of conditional expectation as a Bregman predictor. IEEE Trans. Inf. Theory **51**(7), 2664–2669 (2005)
20. Banerjee, A., Merugu, S., Dhillon, I.S., Ghosh, J.: Clustering with Bregman divergences. J. Mach. Learn. Res. **6**, 1705–1749 (2005)
21. Belkin, M., Sinha, K.: Toward learning Gaussian mixtures with arbitrary separation. In: COLT, pp. 407–419 (2010)
22. Belkin, M., Sinha, K.: Learning Gaussian mixtures with arbitrary separation. CoRR abs/0907.1054 (2009)
23. Belkin, M., Sinha, K.: Polynomial learning of distribution families. In: FOCS, pp. 103–112 (2010)
24. Berkhin, P.: A survey of clustering data mining techniques. In: Kogan, J., Nicholas, C., Teboulle, M. (eds.) Grouping Multidimensional Data, pp. 25–71. Springer, Heidelberg (2006)

25. Braverman, V., Meyerson, A., Ostrovsky, R., Roytman, A., Shindler, M., Tagiku, B.: Streaming k-means on well-clusterable data. In: SODA, pp. 26–40 (2011)
26. Brubaker, S.C., Vempala, S.: Isotropic PCA and affine-invariant clustering. In: FOCS, pp. 551–560 (2008)
27. Chaudhuri, K., McGregor, A.: Finding metric structure in information theoretic clustering. In: COLT, pp. 391–402. Citeseer (2008)
28. Chaudhuri, K., Rao, S.: Learning mixtures of product distributions using correlations and independence. In: COLT, pp. 9–20 (2008)
29. Chen, K.: On coresets for k-median and k-means clustering in metric and Euclidean spaces and their applications. SIAM J. Comput. **39**(3), 923–947 (2009)
30. Dasgupta, S.: Learning mixtures of Gaussians. In: FOCS, pp. 634–644 (1999)
31. Dasgupta, S.: How fast Is k-means? In: Schölkopf, B., Warmuth, M.K. (eds.) COLT-Kernel 2003. LNCS (LNAI), vol. 2777, p. 735. Springer, Heidelberg (2003). doi:10.1007/978-3-540-45167-9_56
32. Dasgupta, S.: The hardness of k-means clustering. Technical report CS2008-0916, University of California (2008)
33. Dasgupta, S., Schulman, L.J.: A probabilistic analysis of EM for mixtures of separated, spherical Gaussians. J. Mach. Learn. Res. **8**, 203–226 (2007)
34. Feldman, D., Langberg, M.: A unified framework for approximating and clustering data. In: Proceedings of the 43th Annual ACM Symposium on Theory of Computing (STOC), pp. 569–578 (2011)
35. Feldman, D., Monemizadeh, M., Sohler, C.: A PTAS for k-means clustering based on weak coresets. In: Proceedings of the 23rd ACM Symposium on Computational Geometry (SoCG), pp. 11–18 (2007)
36. Feldman, J., O'Donnell, R., Servedio, R.A.: Learning mixtures of product distributions over discrete domains. SIAM J. Comput. **37**(5), 1536–1564 (2008)
37. Fichtenberger, H., Gillé, M., Schmidt, M., Schwiegelshohn, C., Sohler, C.: BICO: BIRCH meets coresets for k-means clustering. In: Bodlaender, H.L., Italiano, G.F. (eds.) ESA 2013. LNCS, vol. 8125, pp. 481–492. Springer, Heidelberg (2013). doi:10.1007/978-3-642-40450-4_41
38. Frahling, G., Sohler, C.: Coresets in dynamic geometric data streams. In: Proceedings of the 37th STOC, pp. 209–217 (2005)
39. Gordon, A.: Null models in cluster validation. In: Gaul, W., Pfeifer, D. (eds.) From Data to Knowledge: Theoretical and Practical Aspects of Classification, Data Analysis, and Knowledge Organization, pp. 32–44. Springer, Heidelberg (1996)
40. Guha, S., Meyerson, A., Mishra, N., Motwani, R., O'Callaghan, L.: Clustering data streams: theory and practice. IEEE Trans. Knowl. Data Eng. **15**(3), 515–528 (2003)
41. Hamerly, G., Drake, J.: Accelerating Lloyd's algorithm for k-means clustering. In: Celebi, M.E. (ed.) Partitional Clustering Algorithms, pp. 41–78. Springer, Cham (2015)
42. Har-Peled, S., Kushal, A.: Smaller coresets for k-median and k-means clustering. Discrete Comput. Geom. **37**(1), 3–19 (2007)
43. Har-Peled, S., Mazumdar, S.: On coresets for k-means and k-median clustering. In: Proceedings of the 36th Annual ACM Symposium on Theory of Computing (STOC 2004), pp. 291–300 (2004)
44. Har-Peled, S., Sadri, B.: How fast is the k-means method? In: SODA, pp. 877–885 (2005)
45. Hartigan, J.A.: Clustering Algorithms. Wiley, Hoboken (1975)

46. Inaba, M., Katoh, N., Imai, H.: Applications of weighted Voronoi diagrams and randomization to variance-based k-clustering (extended abstract). In: Symposium on Computational Geometry (SoCG 1994), pp. 332–339 (1994)
47. Jain, A.K.: Data clustering: 50 years beyond k-means. Pattern Recogn. Lett. **31**(8), 651–666 (2010)
48. Jain, A.K., Murty, M.N., Flynn, P.J.: Data clustering: a review. ACM Comput. Surv. **31**(3), 264–323 (1999)
49. Jain, K., Vazirani, V.V.: Approximation algorithms for metric facility location and k-median problems using the primal-dual schema and Lagrangian relaxation. J. ACM **48**(2), 274–296 (2001)
50. Judd, D., McKinley, P.K., Jain, A.K.: Large-scale parallel data clustering. IEEE Trans. Pattern Anal. Mach. Intell. **20**(8), 871–876 (1998)
51. Kalai, A.T., Moitra, A., Valiant, G.: Efficiently learning mixtures of two Gaussians. In: STOC, pp. 553–562 (2010)
52. Kannan, R., Vempala, S.: Spectral algorithms. Found. Trends Theoret. Comput. Sci. **4**(3–4), 157–288 (2009)
53. Kannan, R., Salmasian, H., Vempala, S.: The spectral method for general mixture models. SIAM J. Comput. **38**(3), 1141–1156 (2008)
54. Kanungo, T., Mount, D.M., Netanyahu, N.S., Piatko, C.D., Silverman, R., Wu, A.Y.: An efficient k-means clustering algorithm: analysis and implementation. IEEE Trans. Pattern Anal. Mach. Intell. **24**(7), 881–892 (2002)
55. Kanungo, T., Mount, D.M., Netanyahu, N.S., Piatko, C.D., Silverman, R., Wu, A.Y.: A local search approximation algorithm for k-means clustering. Comput. Geom. **28**(2–3), 89–112 (2004)
56. Kumar, A., Kannan, R.: Clustering with spectral norm and the k-means algorithm. In: Proceedings of the 51st Annual Symposium on Foundations of Computer Science (FOCS 2010), pp. 299–308. IEEE Computer Society (2010)
57. Kumar, A., Sabharwal, Y., Sen, S.: Linear-time approximation schemes for clustering problems in any dimensions. J. ACM **57**(2), Article No. 5 (2010)
58. Lloyd, S.P.: Least squares quantization in PCM. Bell Laboratories Technical Memorandum (1957)
59. MacQueen, J.B.: Some methods for classification and analysis of multivariate observations. In: Proceedings of the 5th Berkeley Symposium on Mathematical Statistics and Probability, vol. 1, pp. 281–297. University of California Press (1967)
60. Mahajan, M., Nimbhorkar, P., Varadarajan, K.: The planar k-means problem is NP-hard. In: Das, S., Uehara, R. (eds.) WALCOM 2009. LNCS, vol. 5431, pp. 274–285. Springer, Heidelberg (2009). doi:10.1007/978-3-642-00202-1_24
61. Manthey, B., Röglin, H.: Worst-case and smoothed analysis of k-means clustering with Bregman divergences. JoCG **4**(1), 94–132 (2013)
62. Manthey, B., Rölin, H.: Improved smoothed analysis of the k-means method. In: Proceedings of the Twentieth Annual ACM-SIAM Symposium on Discrete Algorithms, pp. 461–470. Society for Industrial and Applied Mathematics (2009)
63. Matoušek, J.: On approximate geometric k-clustering. Discrete Comput. Geom. **24**(1), 61–84 (2000)
64. Matula, D.W., Shahrokhi, F.: Sparsest cuts and bottlenecks in graphs. Discrete Appl. Math. **27**, 113–123 (1990)
65. Moitra, A., Valiant, G.: Settling the polynomial learnability of mixtures of Gaussians. In: FOCS 2010 (2010)

66. Nock, R., Luosto, P., Kivinen, J.: Mixed Bregman clustering with approximation guarantees. In: Daelemans, W., Goethals, B., Morik, K. (eds.) ECML PKDD 2008. LNCS (LNAI), vol. 5212, pp. 154–169. Springer, Heidelberg (2008). doi:10.1007/978-3-540-87481-2_11
67. Ostrovsky, R., Rabani, Y., Schulman, L.J., Swamy, C.: The effectiveness of Lloyd-type methods for the k-means problem. In: FOCS, pp. 165–176 (2006)
68. Pelleg, D., Moore, A.W.: Accelerating exact k-means algorithms with geometric reasoning. In: Proceedings of the Fifth ACM SIGKDD International Conference on Knowledge Discovery and Data Mining, pp. 277–281 (1999)
69. Selim, S.Z., Ismail, M.A.: k-means-type algorithms: a generalized convergence theorem and characterization of local optimality. IEEE Trans. Pattern Anal. Mach. Intell. (PAMI) **6**(1), 81–87 (1984)
70. Steinhaus, H.: Sur la division des corps matériels en parties. Bulletin de l'Académie Polonaise des Sciences **IV**(12), 801–804 (1956)
71. Tibshirani, R., Walther, G., Hastie, T.: Estimating the number of clusters in a dataset via the gap statistic. J. R. Stat. Soc. Ser. B (Stat. Methodol.) **63**, 411–423 (2001)
72. Vattani, A.: k-means requires exponentially many iterations even in the plane. In: Proceedings of the 25th ACM Symposium on Computational Geometry (SoCG 2009), pp. 324–332. Association for Computing Machinery (2009)
73. de la Vega, W.F., Karpinski, M., Kenyon, C., Rabani, Y.: Approximation schemes for clustering problems. In: Proceedings of the 35th Annual ACM Symposium on Theory of Computing (STOC 2003), pp. 50–58 (2003)
74. Vempala, S., Wang, G.: A spectral algorithm for learning mixture models. J. Comput. Syst. Sci. **68**(4), 841–860 (2004)
75. Venkatasubramanian, S.: Choosing the number of clusters I-III (2010). http://blog.geomblog.org/p/conceptual-view-of-clustering.html. Accessed 30 Mar 2015
76. Zhang, T., Ramakrishnan, R., Livny, M.: BIRCH: a new data clustering algorithm and its applications. Data Min. Knowl. Disc. **1**(2), 141–182 (1997)

Recent Advances in Graph Partitioning

Aydın Buluç[1], Henning Meyerhenke[2], Ilya Safro[3], Peter Sanders[2],
and Christian Schulz[2(✉)]

[1] Computational Research Division,
Lawrence Berkeley National Laboratory, Berkeley, USA
[2] Institute of Theoretical Informatics,
Karlsruhe Institute of Technology (KIT), Karlsruhe, Germany
christian.schulz@kit.edu
[3] School of Computing, Clemson University, Clemson, SC, USA

Abstract. We survey recent trends in practical algorithms for balanced
graph partitioning, point to applications and discuss future research
directions.

1 Introduction

Graphs are frequently used by computer scientists as abstractions when modeling an application problem. Cutting a graph into smaller pieces is one of the fundamental algorithmic operations. Even if the final application concerns a different problem (such as traversal, finding paths, trees, and flows), partitioning large graphs is often an important subproblem for complexity reduction or parallelization. With the advent of ever larger instances in applications such as scientific simulation, social networks, or road networks, *graph partitioning* (GP) therefore becomes more and more important, multifaceted, and challenging. The purpose of this paper is to give a structured overview of the rich literature, with a clear emphasis on explaining key ideas and discussing recent work that is missing in other overviews. For a more detailed picture on how the field has evolved previously, we refer the interested reader to a number of surveys. Bichot and Siarry [22] cover studies on GP within the area of numerical analysis. This includes techniques for GP, hypergraph partitioning and parallel methods. The book discusses studies from a combinatorial viewpoint as well as several applications of GP such as the air traffic control problem. Schloegel et al. [191] focus on fast graph partitioning techniques for scientific simulations. In their account of the state of the art in this area around the turn of the millennium, they describe geometric, combinatorial, spectral, and multilevel methods and how to combine them for static partitioning. Load balancing of dynamic simulations, parallel aspects, and problem formulations with multiple objectives or constraints are also considered. Monien et al. [156] discuss heuristics and approximation algorithms used in the multilevel GP framework. In their description they focus mostly on coarsening by matching and local search by node-swapping heuristics. Kim et al. [119] cover genetic algorithms.

L. Kliemann and P. Sanders (Eds.): Algorithm Engineering, LNCS 9220, pp. 117–158, 2016.
DOI: 10.1007/978-3-319-49487-6_4

Our survey is structured as follows. First, Sect. 2 introduces the most important variants of the problem and their basic properties such as NP-hardness. Then Sect. 3 discusses exemplary applications including parallel processing, road networks, image processing, VLSI design, social networks, and bioinformatics. The core of this overview concerns the solution methods explained in Sects. 4, 5, 6 and 7. They involve a surprising variety of techniques. We begin in Sect. 4 with basic, global methods that "directly" partition the graph. This ranges from very simple algorithms based on breadth first search to sophisticated combinatorial optimization methods that find exact solutions for small instances. Also methods from computational geometry and linear algebra are being used. Solutions obtained in this or another way can be improved using a number of heuristics described in Sect. 5. Again, this ranges from simple-minded but fast heuristics for moving individual nodes to global methods, e.g., using flow or shortest path computations. The most successful approach to partitioning large graphs – the multilevel method – is presented in Sect. 6. It successively contracts the graph to a more manageable size, solves the base instance using one of the techniques from Sect. 4, and – using techniques from Sect. 5 – improves the obtained partition when uncontracting to the original input. Metaheuristics are also important. In Sect. 7 we describe evolutionary methods that can use multiple runs of other algorithms (e.g., multilevel) to obtain high quality solutions. Thus, the best GP solvers orchestrate multiple approaches into an overall system. Since all of this is very time consuming and since the partitions are often used for parallel computing, parallel aspects of GP are very important. Their discussion in Sect. 8 includes parallel solvers, mapping onto a set of parallel processors, and migration minimization when repartitioning a dynamic graph. Section 9 describes issues of implementation, benchmarking, and experimentation. Finally, Sect. 10 points to future challenges.

2   Preliminaries

Given a number $k \in \mathbb{N}_{>1}$ and an undirected graph $G = (V, E)$ with *non-negative* edge weights, $\omega : E \to \mathbb{R}_{>0}$, the *graph partitioning problem* (GPP) asks for a partition Π of V with *blocks* of nodes $\Pi = (V_1, \ldots, V_k)$:

1. $V_1 \cup \cdots \cup V_k = V$
2. $V_i \cap V_j = \emptyset \ \forall i \neq j$.

A *balance constraint* demands that all blocks have about equal weights. More precisely, it requires that, $\forall i \in \{1, \ldots, k\} : |V_i| \leq L_{\max} := (1 + \epsilon) \lceil |V|/k \rceil$ for some imbalance parameter $\epsilon \in \mathbb{R}_{\geq 0}$. In the case of $\epsilon = 0$, one also uses the term *perfectly balanced*. Sometimes we also use weighted nodes with node weights $c : V \to \mathbb{R}_{>0}$. Weight functions on nodes and edges are extended to sets of such objects by summing their weights. A block V_i is *overloaded* if $|V_i| > L_{\max}$. A *clustering* is also a partition of the nodes. However, k is usually not given in advance, and the balance constraint is removed. Note that a partition is also a clustering of a graph. In both cases, the *goal* is to minimize or maximize a particular objective function. We recall well-known objective functions for GPP in Sect. 2.1. A node v is a *neighbor* of node u if there is an edge $\{u, v\} \in E$. If a node $v \in V_i$ has a neighbor $w \in V_j$, $i \neq j$, then it is called *boundary node*. An edge that runs between blocks is also called *cut edge*. The set $E_{ij} := \{\{u, v\} \in E : u \in V_i, v \in V_j\}$ is the set of cut edges between two blocks V_i and V_j. An abstract view of the partitioned graph is the so called *quotient graph* or *communication graph*, where nodes represent blocks, and edges are induced by connectivity between blocks. There is an edge in the quotient graph between blocks V_i and V_j if and only if there is an edge between a node in V_i and a node in V_j in the original, partitioned graph. The *degree* $d(v)$ of a node v is the number of its neighbors. An *adjacency matrix* of a graph is a $|V| \times |V|$ matrix describing node connectivity. The element $a_{u,v}$ of the matrix specifies the weight of the edge from node u to node v. It is set to zero if there is no edge between these nodes. The *Laplacian matrix* of a graph G is defined as $L = D - A$, where D is the diagonal matrix expressing node degrees, and A is the adjacency matrix. A cycle in a directed graph with negative weight is also called *negative cycle*. A *matching* $M \subseteq E$ is a set of edges that do not share any common nodes, i.e., the graph (V, M) has maximum degree one.

2.1   Objective Functions

In practice, one often seeks to find a partition that minimizes (or maximizes) an objective. Probably the most prominent objective function is to minimize the *total cut*

$$\sum_{i<j} \omega(E_{ij}). \tag{1}$$

Other formulations of GPP exist. For instance when GP is used in parallel computing to map the graph nodes to different processors, the *communication volume* is often more appropriate than the cut [100]. For a block V_i, the communication volume is defined as $\mathrm{comm}(V_i) := \sum_{v \in V_i} c(v) D(v)$, where $D(v)$ denotes the number of different blocks in which v has a neighbor node, excluding V_i. The *maximum communication volume* is then defined as $\max_i \mathrm{comm}(V_i)$, whereas the *total communication volume* is defined as $\sum_i \mathrm{comm}(V_i)$. The maximum communication volume was used in one subchallenge of the 10th DIMACS Challenge on Graph Partitioning and Graph Clustering [13]. Although some applications profit from other objective functions such as the communication volume or block shape (formalized by the block's aspect ratio [56], minimizing the cut size has been adopted as a kind of standard. One reason is that cut optimization seems to be easier in practice. Another one is that for graphs with high structural locality the cut often correlates with most other formulations but other objectives make it more difficult to use a multilevel approach.

There are also GP formulations in which balance is not directly encoded in the problem description but integrated into the objective function. For example, the *expansion* of a non-trivial cut (V_1, V_2) is defined as $\omega(E_{12})/\min(c(V_1), c(V_2))$. Similarly, the *conductance* of such a cut is defined as $\omega(E_{12})/\min(\mathrm{vol}(V_1), \mathrm{vol}(V_2))$, where $\mathrm{vol}(S) := \sum_{v \in S} d(v)$ denotes the volume of the set S.

As an extension to the problem, when the application graph changes over time, *repartitioning* becomes necessary. Due to changes in the underlying application, a graph partition may become gradually imbalanced due to the introduction of new nodes (and edges) and the deletion of others. Once the imbalance exceeds a certain threshold, the application should call the repartitioning routine. This routine is to compute a new partition Π' from the old one, Π. In many applications it is favorable to keep the changes between Π and Π' small. Minimizing these changes simultaneously to optimizing Π' with respect to the cut (or a similar objective) leads to multiobjective optimization. To avoid the complexity of the latter, a linear combination of both objectives seems feasible in practice [193].

2.2 Hypergraph Partitioning

A *hypergraph* $H = (V, E)$ is a generalization of a graph in which an edge (usually called *hyperedge* or *net*) can connect any number of nodes. As with graphs, partitioning a hypergraph also means to find an assignment of nodes to different blocks of (mostly) equal size. The objective function, however, is usually expressed differently. A straightforward generalization of the edge cut to hypergraphs is the *hyperedge cut*. It counts the number of hyperedges that connect different blocks. In widespread use for hypergraph partitioning, however, is the so-called $(\lambda - 1)$ metric, $CV(H, \Pi) = \sum_{e \in E}(\lambda(e, \Pi) - 1)$, where $\lambda(e, \Pi)$ denotes the number of distinct blocks connected by the hyperedge e and Π the partition of H's vertex set.

One drawback of hypergraph partitioning compared to GP is the necessity of more complex algorithms—in terms of implementation and running time, not necessarily in terms of worst-case complexity. Paying this price seems only worthwhile if the underlying application profits significantly from the difference between the graph and the hypergraph model.

To limit the scope, we focus in this paper on GP and forgo a more detailed treatment of hypergraph partitioning. Many of the techniques we describe, however, can be or have been transferred to hypergraph partitioning as well [33, 34, 66, 162, 208]. One important application area of hypergraph partitioning is VLSI design (see Sect. 3.5).

2.3 Hardness Results and Approximation

Partitioning a graph into k blocks of roughly equal size such that the cut metric is minimized is NP-complete (as decision problem) [79, 106]. Andreev and Räcke [4] have shown that there is no constant-factor approximation for the perfectly balanced version ($\epsilon = 0$) of this problem on general graphs. If $\epsilon \in (0, 1]$, then an $O(\log^2 n)$ factor approximation can be achieved. In case an even larger imbalance $\epsilon > 1$ is allowed, an approximation ratio of $O(\log n)$ is possible [65].

The minimum weight k-cut problem asks for a partition of the nodes into k non-empty blocks without enforcing a balance constraint. Goldschmidt et al. [88] proved that, for a fixed k, this problem can be solved optimally in $O(n^{k^2})$. The problem is NP-complete [88] if k is not part of the input.

For the unweighted minimum bisection problem, Feige and Krauthgamer [68] have shown that there is an $O(\log^{1.5} n)$ approximation algorithm and an $O(\log n)$ approximation for minimum bisection on planar graphs. The bisection problem is efficiently solvable if the balance constraint is dropped – in this case it is the minimum cut problem. Wagner et al. [211] have shown that the minimum bisection problem becomes harder the more the balance constraint is tightened towards the perfectly balanced case. More precisely, if the block weights are bounded from below by a constant, i.e., $|V_i| \geq C$, then the problem is solvable in polynomial time. The problem is NP-hard if the block weights are constrained by $|V_i| \geq \alpha n^\delta$ for some $\alpha, \delta > 0$ or if $|V_i| = \frac{n}{2}$. The case $|V_i| \geq \alpha \log n$ for some $\alpha > 0$ is open. Note that the case $|V_i| \geq \alpha n^\delta$ also implies that the general GPP with similar lower bounds on the block weights is NP-hard.

If the balance constraint of the problem is dropped and one uses a different objective function such as sparsest cut, then there are better approximation algorithms. The sparsest cut objective combines cut and balance into a single objective function. For general graphs and the sparsest cut metric, Arora et al. [7,8] achieve an approximation ratio of $O(\sqrt{\log n})$ in $\tilde{O}(n^2)$ time.

Being of high theoretical importance, most of the approximation algorithms are not implemented, and the approaches that implement approximation algorithms are too slow to be used for large graphs or are not able to compete with state-of-the-art GP solvers. Hence, mostly heuristics are used in practice.

3   Applications of Graph Partitioning

We now describe some of the applications of GP. For brevity this list is not exhaustive.

3.1   Parallel Processing

Perhaps the canonical application of GP is the distribution of work to processors of a parallel machine. Scientific computing applications such as sparse direct and iterative solvers extensively use GP to ensure load balance and minimize communication. When the problem domain does not change as the computation proceeds, GP can be applied once in the beginning of the computation. This is known as static partitioning.

Periodic repartitioning, explained in Sect. 2.1, proved to be useful for scientific computing applications with evolving computational domains such as Adaptive Mesh Refinement (AMR) or volume rendering [11]. The graph model can be augmented with additional edges and nodes to model the migration costs, as done for parallel direct volume rendering of unstructured grids [11], an important problem in scientific visualization.

Parallel Graph Computations. GP is also used to partition graphs for parallel processing, for problems such as graph eigenvalue computations [25], breadth-first search [31], triangle listing [43], PageRank and connected components [181]. In computationally intensive graph problems, such as finding the eigenvectors and eigenvalues of graphs, multilevel methods that are tailored to the characteristics of real graphs are suitable [1].

Mesh Partitioning. A *mesh* or *grid* approximates a geometric domain by dividing it into smaller subdomains. Hendrickson defines it as "the scaffolding upon which a function is decomposed into smaller pieces" [96]. Mesh partitioning involves mapping the subdomains of the mesh to processors for parallel processing, with the objective of minimizing communication and load imbalance. A partial differential equation (PDE) that is discretized over a certain grid can be solved by numerous methods such as the finite differences method or the finite elements method. The discretization also defines a system of linear equations that can be represented by a sparse matrix. While it is always possible to use that sparse matrix to do the actual computation over the mesh or grid, sometimes this can be wasteful when the matrix need not be formed explicitly. In the absence of an explicit sparse matrix, the GP solvers first define a graph from the mesh. The right mesh entity to use as the nodes of the graph can be ambiguous and application dependent. Common choices are mesh nodes, groups of mesh nodes that need to stay together, and the dual of mesh nodes. Choosing groups of mesh nodes (such as small regular meshes [74]) with appropriate weighting as graph nodes makes GP cost effective for large problem sizes when the overhead for per-node partitioned graphs would be too big. Recent work by Zhou et al. [222] gives a thorough treatment of extreme-scale mesh partitioning and dynamic repartitioning using graph models. A variety of solution methodologies described in Sect. 6, such as the multilevel and geometric methods, has been successfully applied to mesh partitioning.

3.2 Complex Networks

In addition to the previously mentioned task of network data distribution across a cluster of machines for fast parallel computations, complex networks introduced numerous further applications of GPP. A common task in these applications is to identify groups of similar entities whose similarity and connectivity is modeled by the respective networks. The quality of the localizations is quantified with different domain-relevant objectives. Many of them are based on the principle of finding groups of entities that are weakly connected to the rest of the network. In many cases such connectivity also represents similarity. In the context of optimization problems on graphs, by complex networks we mean weighted graphs with non-trivial structural properties that were created by real-life or modelling processes [159]. Often, models and real-life network generation processes are not well understood, so designing optimization algorithms for such graphs exhibit a major bottleneck in many applications.

Power Grids. Disturbances and cascading failures are among the central problems in power grid systems that can cause catastrophic blackouts. Splitting a power network area into self-sufficient islands is an approach to prevent the propagation of cascading failures [132]. Often the cut-based objectives of the partitioning are also combined with the load shedding schemes that enhance the robustness of the system and minimize the impact of cascading events [133]. Finding vulnerabilities of power systems by GPP has an additional difficulty. In some applications, one may want to find more than one (nearly) minimum partitioning because of the structural difference between the solutions. Spectral GP (see Sect. 4.2) is also used to detect contingencies in power grid vulnerability analysis by splitting the network into regions with excess generation and excess load [60].

Geographically Embedded Networks. Recent advances of location-aware devices (such as GPS) stimulated a rapid growth of streaming spatial network data that has to be analyzed by extremely fast algorithms. These networks model entities (nodes) tied to geographic places and links that represent flows such as migrations, vehicle trajectories, and activities of people [54]. In problems related to spatial data and geographical networks, the cut-based objective of GP (and clustering) is often reinforced by the spatial contiguity constraints.

Biological Networks. Many complex biological systems can be modeled by graph-theoretic representations. Examples include protein-protein interactions, and gene co-expression networks. In these networks nodes are biological entities (such as genes and proteins) and edges correspond to their common participation in some biological process. Such processes can vary from simple straightforward interactions (such as protein-protein interaction and gene-gene co-expression) to more complex relationships in which more than two entities are involved. Partitioning and clustering of such networks may have several goals. One of them is related to data reduction given an assumption that clustered nodes behave biologically similarly to each other. Another one is the detection of some biological processes by finding clusters of involved nodes. For details see [109,154].

Social Networks. Identification of community structure is among the most popular topics in social network science. In contrast to the traditional GPP, community detection problems rarely specify the number of clusters *a priori*. Notwithstanding this difference, GP methods contributed a lot of their techniques to the community detection algorithms [76]. Moreover, GP solvers are often used as first approximations for them. We refer the reader to examples of methods where GP is used for solving the community detection problem [158].

3.3   Road Networks

GP is a very useful technique to speed up route planning [48,52,118,129,138, 153]. For example, edges could be road segments and nodes intersections.[1]

---

[1] Sometimes more complex models are used to model lanes, turn costs etc.

Lauther [129] introduced the arc-flags algorithm, which uses a geometric partitioning approach as a preprocessing step to reduce the search space of Dijkstra's algorithm. Möhring et al. [153] improved this method in several ways. Using high quality graph partitions turns out to be one key improvement here since this reduces the preprocessing cost drastically. One reason is that road networks can be partitioned using surprisingly small cuts but these are not easy to find.

Schulz et al. [196] propose a multilevel algorithm for routing based on pre-computing connections between border nodes of a graph partition. This was one of the first successful speedup technique for shortest paths. It was outclassed later by other hierarchy based methods, and, somewhat surprisingly resurfaced after Delling et al. [48,52] did thorough algorithm engineering for this approach. Again, a key improvement was to use high quality graph partitions. Since the approach excels at fast recomputation of the preprocessing information when the edge weights change, the method is now usually called *customizable route planning*. Luxen and Schieferdecker [138] use GP to efficiently compute candidate sets for alternative routes in road networks and Kieritz et al. [118] parallelize shortest-path preprocessing and query algorithms. Maue et al. [141] show how to use precomputed distances between blocks of a partition to make the search goal directed. Here, block diameter seems more relevant than cut size, however.

3.4 Image Processing

Image segmentation is a fundamental task in computer vision for which GP and clustering methods have become among the most attractive solution techniques. The goal of image segmentation is to partition the pixels of an image into groups that correspond to objects. Since the computations preceding segmentation are often relatively cheap and since the computations after segmentation work on a drastically compressed representation of the image (objects rather than pixels), segmentation is often the computationally most demanding part in an image processing pipeline. The image segmentation problem is not well-posed and can usually imply more than one solution. During the last two decades, graph-based representations of an image became very popular and gave rise to many cut-based approaches for several problems including image segmentation. In this representation each image pixel (or in some cases groups of pixels) corresponds to a node in a graph. Two nodes are connected by a weighted edge if some similarity exists between them. Usually, the criteria of similarity is a small geodesic distance which can result in mesh-like graphs with four or more neighbors for each node. The edge weights represent another measure of (dis)similarity between nodes such as the difference in the intensity between the connected pixels (nodes).

GP can be formulated with different objectives that can explicitly reflect different definitions of the segmented regions depending on the applications. The classical minimum cut formulation of the GP objective (1) can lead in practice to finding too small segmented objects. One popular modification of the objective that was adopted in image segmentation, called *normalized cut*, is given by $\mathrm{ncut}(A, B) = \omega(E_{AB})/\mathrm{vol}(A) + \omega(E_{AB})/\mathrm{vol}(B)$. This objective is similar to the conductance objective described in Sect. 2.1. Many efficient algorithms were proposed for solving GPP with the normalized cut objective. Among the most successful are spectral and multilevel approaches. Another relevant formulation of the partitioning objective which is useful for image segmentation is given by optimizing the isoperimetric ratio for sets [89]. For more information on graph partitioning and image segmentation see [32,169].

3.5 VLSI Physical Design

Physical design of digital circuits for very large-scale integration (VLSI) systems has a long history of being one of the most important customers of graph and hypergraph partitioning, often reinforced by several additional domain relevant constraints. The partitioning should be accomplished in a reasonable computation time, even for circuits with millions of modules, since it is one of the bottlenecks of the design process. The goal of the partitioning is to reduce the VLSI design complexity by partitioning it into smaller components (that can range from a small set of field-programmable gate arrays to fully functional integrated circuits) as well as to keep the total length of all the wires short. The typical optimization objective (see (1)) is to minimize the total weight of connections between subcircuits (blocks), where nodes are the cells, i.e., small logical or functional units of the circuit (such as gates), and edges are the wires. Because the gates are connected with wires with more than two endpoints, hypergraphs model the circuit more accurately. Examples of additional constraints for the VLSI partitioning include information on the I/O of the circuit, sets of cells that must belong to the same blocks, and maximum cut size between two blocks. For more information about partitioning of VLSI circuits see [45,110].

4   Global Algorithms

We begin our discussion of the wide spectrum of GP algorithms with methods that work with the entire graph and compute a solution directly. These algorithms are often used for smaller graphs or are applied as subroutines in more complex methods such as local search or multilevel algorithms. Many of these methods are restricted to bipartitioning but can be generalized to k-partitioning for example by recursion.

After discussing exact methods in Sect. 4.1 we turn to heuristic algorithms. Spectral partitioning (Sect. 4.2) uses methods from linear algebra. Graph growing (Sect. 4.3) uses breadth first search or similar ways to directly add nodes to a block. Flow computations are discussed in Sect. 4.4. Section 4.5 summarizes a wide spectrum of geometric techniques. Finally, Sect. 4.5 introduces *streaming* algorithms which work with a very limited memory footprint.

4.1   Exact Algorithms

There is a large amount of literature on methods that solve GPP optimally. This includes methods dedicated to the bipartitioning case [5,6,28,49,51,69,70,93,94,

111,134,197] and some methods that solve the general GPP [71,198]. Most of the methods rely on the branch-and-bound framework [126].

Bounds are derived using various approaches: Karisch et al. [111] and Armbruster [5] use semi-definite programming, and Sellman et al. [197] and Sensen [198] employ multi-commodity flows. Linear programming is used by Brunetta et al. [28], Ferreira et al. [71], Lisser and Rendl [134] and by Armbruster et al. [6]. Hager et al. [93,94] formulate GPP in form of a continuous quadratic program on which the branch and bound technique is applied. The objective of the quadratic program is decomposed into convex and concave components. The more complicated concave component is then tackled by an SDP relaxation. Felner [70] and Delling et al. [49,51] utilize combinatorial bounds. Delling et al. [49,51] derive the bounds by computing minimum s-t cuts between partial assignments (A, B), i.e., $A, B \subseteq V$ and $A \cap B = \emptyset$. The method can partition road networks with more than a million nodes, but its running time highly depends on the bisection width of the graph.

In general, depending on the method used, two alternatives can be observed. Either the bounds derived are very good and yield small branch-and-bound trees but are hard to compute. Or the bounds are somewhat weaker and yield larger trees but are faster to compute. The latter is the case when using combinatorial bounds. On finite connected subgraphs of the two dimensional grid without holes, the bipartitioning problem can be solved optimally in $O(n^4)$ time [69]. Recent work by Bevern et al. [19] looks at the parameterized complexity for computing balanced partitions in graphs.

All of these methods can typically solve only very small problems while having very large running times, or if they can solve large bipartitioning instances using a moderate amount of time [49,51], highly depend on the bisection width of the graph. Methods that solve the general GPP [71,198] have immense running times for graphs with up to a few hundred nodes. Moreover, the experimental evaluation of these methods only considers small block numbers $k \leq 4$.

4.2   Spectral Partitioning

One of the first methods to split a graph into two blocks, spectral bisection, is still in use today. Spectral techniques were first used by Donath and Hoffman [58,59] and Fiedler [73], and have been improved subsequently by others [15,26,98,172,200]. Spectral bisection infers global information of the connectivity of a graph by computing the eigenvector corresponding to the second smallest eigenvalue of the Laplacian matrix L of the graph. This eigenvector z_2 is also known as *Fiedler vector*; it is the solution of a relaxed integer program for cut optimization. A partition is derived by determining the median value $\overline{m}$ in z_2 and assigning all nodes with an entry smaller or equal to $\overline{m}$ to V_1 and all others to V_2.

The second eigenvector can be computed using a modified Lanczos algorithm [125]. However, this method is expensive in terms of running time. Barnard and Simon [15] use a multilevel method to obtain a fast approximation of the Fiedler vector. The algorithmic structure is similar to the multilevel method explained in Sect. 6, but their method coarsens with independent node sets and performs local improvement with Rayleigh quotient iteration. Hendrickson and Leland [98] extend the spectral method to partition a graph into more than two blocks by using multiple eigenvectors; these eigenvectors are computationally inexpensive to obtain. The method produces better partitions than recursive bisection, but is only useful for the partitioning of a graph into four or eight blocks. The authors also extended the method to graphs with node and edge weights.

4.3   Graph Growing

A very simple approach for obtaining a bisection of a graph is called graph growing [81,113]. Most of its variants are based on breadth-first search. Its simplest version works as follows. Starting from a random node v, the nodes are assigned to block V_1 using a breadth-first search (BFS) starting at v. The search is stopped after half of the original node weights are assigned to this block and V_2 is set to $V \setminus V_1$. This method can be combined with a local search algorithm to improve the partition. Multiple restarts of the algorithm are important to get a good solution. One can also try to find a good starting node by looking at a node that has maximal distance from a random seed node [81]. Variations of the algorithm always add the node to the block that results in the smallest increase in the cut [113]. An extension to $k > 2$ blocks and with iterative improvement is described in Sect. 5.5.

4.4   Flows

The well-known max-flow min-cut theorem [75] can be used to separate two node sets in a graph by computing a maximum flow and hence a minimum cut between them. This approach completely ignores balance, and it is not obvious how to apply it to the balanced GPP. However, at least for random regular graphs with small bisection width this can be done [29]. Maximum flows are also often used as a subroutine. Refer to Sect. 5.4 for applications to improve a partition and to Sect. 6.4 for coarsening in the context of the multilevel framework. There are also applications of flow computations when quality is measured by expansion or conductance [3,127].

4.5   Geometric Partitioning

Partitioning can utilize the coordinates of the graph nodes in space, if available. This is especially useful in finite element models and other geometrically-defined graphs from traditional scientific computing. Here, geometrically "compact" regions often correspond to graph blocks with small cut. Partitioning using nodal coordinates comes in many flavors, such as recursive coordinate bisection (RCB) [200] and inertial partitioning [67,221]. In each step of its recursion, RCB projects graph nodes onto the coordinate axis with the longest expansion of the domain and bisects them through the median of their projections.

The bisecting plane is orthogonal to the coordinate axis, which can create partitions with large separators in case of meshes with skewed dimensions. Inertial partitioning can be interpreted as an improvement over RCB in terms of worst case performance because its bisecting plane is orthogonal to a plane L that minimizes the moments of inertia of nodes. In other words, the projection plane L is chosen such that it minimizes the sum of squared distances to all nodes.

The random spheres algorithm of Miller et al. [83,152] generalizes the RCB algorithm by stereographically projecting the d dimensional nodes to a random $d + 1$ dimensional sphere which is bisected by a plane through its center point. This method gives performance guarantees for planar graphs, k-nearest neighbor graphs, and other "well-behaved" graphs.

Other representatives of geometry-based partitioning algorithms are space-filling curves [14,105,171,223] which reduce d-dimensional partitioning to the one-dimensional case. Space filling curves define a bijective mapping from V to $\{1, \ldots, |V|\}$. This mapping aims at the preservation of the nodes' locality in space. The partitioning itself is simpler and cheaper than RCB once the bijective mapping is constructed. A generalization of space-filling curves to general graphs can be done by so-called graph-filling curves [190].

A recent work attempts to bring information on the graph structure into the geometry by embedding arbitrary graphs into the coordinate space using a multilevel graph drawing algorithm [121]. For a more detailed, albeit not very recent, treatment of geometric methods, we refer the interested reader to Schloegel et al. [191].

4.6   Streaming Graph Partitioning (SGP)

Streaming data models are among the most popular recent trends in big data processing. In these models the input arrives in a data stream and has to be processed on the fly using much less space than the overall input size. SGP algorithms are very fast. They are even faster than multilevel algorithms but give lower solution quality. Nevertheless, many applications that require extremely fast repartitioning methods (such as those that deal with dynamic networks) can still greatly benefit from the SGP algorithms when an initial solution obtained by a stronger (static data) algorithm is supplied as an initial ordering. For details on SGP we refer the reader to [160,203,209].

5   Iterative Improvement Heuristics

Most high quality GP solvers iteratively improve starting solutions. We outline a variety of methods for this purpose, moving from very fine-grained localized approaches to more global techniques.

5.1   Node-Swapping Local Search

Local search is a simple and widely used metaheuristic for optimization that iteratively changes a solution by choosing a new one from a neighborhood. Defining the neighborhood and the selection strategy allows a wide variety of techniques. Having the improvement of paging properties of computer programs in mind, Kernighan and Lin [117] were probably the first to define GPP and to provide a local search method for this problem. The selection strategy finds the swap of node assignments that yields the largest decrease in the total cut size. Note that this "decrease" is also allowed to be negative. A round ends when all nodes have been moved in this way. The solution is then reset to the best solution encountered in this round. The algorithm terminates when a round has not found an improvement.

A major drawback of the KL method is that it is expensive in terms of asymptotic running time. The implementation assumed in [117] takes time $O(n^2 \log n)$ and can be improved to $O(m \max(\log n, \Delta))$ where Δ denotes the maximum degree [64]. A major breakthrough is the modification by Fiduccia and Mattheyses [72]. Their carefully designed data structures and adaptations yield the KL/FM local search algorithm, whose asymptotic running time is $O(m)$. Bob Darrow was the first who implemented the KL/FM algorithm [72].

Karypis and Kumar [114] further accelerated KL/FM by only allowing boundary nodes to move and by stopping a round when the edge cut does not decrease after x node moves. They improve quality by random tie breaking and by allowing additional rounds even when no improvements have been found.

A highly localized version of KL/FM is considered in [161]. Here, the search spreads from a single boundary node. The search stops when a stochastic model of the search predicts that a further improvement has become unlikely. This strategy has a better chance to climb out of local minima and yields improved cuts for the GP solvers KaSPar [161] and KaHIP [183].

Rather than swapping nodes, Holtgrewe et al. move a single node at a time allowing more flexible tradeoffs between reducing the cut or improving balance [102].

Helpful Sets by Diekmann et al. [55, 155] introduce a more general neighborhood relation in the bipartitioning case. These algorithms are inspired by a proof technique of Hromkovič and Monien [103] for proving upper bounds on the bisection width of a graph. Instead of migrating single nodes, whole sets of nodes are exchanged between the blocks to improve the cut. The running time of the algorithm is comparable to the KL/FM algorithm, while solution quality is often better than other methods [155].

5.2 Extension to k-way Local Search

It has been shown by Simon and Teng [201] that, due to the lack of global knowledge, recursive bisection can create partitions that are very far away from the optimal partition so that there is a need for k-way local search algorithms. There are multiple ways of extending the KL/FM algorithm to get a local search algorithm that can improve a k-partition.

One early extension of the KL/FM algorithm to k-way local search uses $k(k-1)$ priority queues, one for each type of move (source block, target block)

[97,182]. For a single movement one chooses the node that maximizes the gain, breaking ties by the improvement in balance.

Karypis and Kumar [114] present a k-way version of the KL/FM algorithm that runs in linear time $O(m)$. They use a single global priority queue for all types of moves. The priority used is the maximum local gain, i.e., the maximum reduction in the cut when the node is moved to one of its neighboring blocks. The node that is selected for movement yields the maximum improvement for the objective and maintains or improves upon the balance constraint.

Most current local search algorithms exchange nodes between blocks of the partition trying to decrease the cut size while also maintaining balance. This highly restricts the set of possible improvements. Sanders and Schulz [186,195] relax the balance constraint for node movements but globally maintain (or improve) balance by combining multiple local searches. This is done by reducing the combination problem to finding negative cycles in a graph, exploiting the existence of efficient algorithms for this problem.

5.3   Tabu Search

A more expensive k-way local search algorithm is based on tabu search [86,87], which has been applied to GP by [16–18,78,175]. We briefly outline the method reported by Galinier et al. [78]. Instead of moving a node exactly once per round, as in the traditional versions of the KL/FM algorithms, specific types of moves are excluded only for a number of iterations. The number of iterations that a move (v, block) is excluded depends on an aperiodic function f and the current iteration i. The algorithm always moves a non-excluded node with the highest gain. If the node is in block A, then the move (v, A) is excluded for $f(i)$ iterations after the node is moved to the block yielding the highest gain, i.e., the node cannot be put back to block A for $f(i)$ iterations.

5.4   Flow Based Improvement

Sanders and Schulz [183,185] introduce a max-flow min-cut based technique to improve the edge cut of a given bipartition (and generalize this to k-partitioning by successively looking at pairs of blocks that are adjacent in the quotient graph). The algorithm constructs an s-t flow problem by growing an area around the given boundary nodes/cut edges. The area is chosen such that each s-t cut in this area corresponds to a feasible bipartition of the original graph, i.e., a bipartition that fulfills the balance constraint. One can then apply a max-flow min-cut algorithm to obtain a min-cut in this area and hence a nondecreased cut between the blocks. There are multiple improvements to extend this method, for example, by iteratively applying the method, searching in larger areas for feasible cuts, or applying a heuristic to output better balanced minimum cuts by using the given max-flow.

5.5   Bubble Framework

Diekmann et al. [57] extend graph growing and previous ideas [216] to obtain an iterative procedure called *Bubble framework*, which is capable of partitioning into $k > 2$ *well-shaped* blocks. Some applications profit from good geometric block shapes, e. g., the convergence rate of certain iterative linear solvers.

Graph growing is extended first by carefully selecting k seed nodes that are evenly distributed over the graph. The key property for obtaining a good quality, however, is an iterative improvement within the second and the third step – analogous to Lloyd's k-means algorithm [135]. Starting from the k seed nodes, k breadth-first searches grow the blocks analogous to Sect. 4.3, only that the breadth-first searches are scheduled such that the smallest block receives the next node. Local search algorithms are further used within this step to balance the load of the blocks and to improve the cut of the resulting partition, which may result in unconnected blocks. The final step of one iteration computes new seed nodes for the next round. The new center of a block is defined as the node that minimizes the sum of the distances to all other nodes within its block. To avoid their expensive computation, approximations are used. The second and the third step of the algorithm are iterated until either the seed nodes stop changing or no improved partition was found for more than 10 iterations. Figure 1 illustrates the three steps of the algorithm. A drawback of the algorithm is its computational complexity $O(km)$.

Subsequently, this approach has been improved by using distance measures that better reflect the graph structure [144, 151, 189]. For example, Schamberger [189] introduced the usage of diffusion as a growing mechanism around the initial seeds and extended the method to weighted graphs. More sophisticated diffusion schemes, some of which have been employed within the Bubble framework, are discussed in Sect. 5.6.

5.6   Random Walks and Diffusion

A *random walk* on a graph starts on a node v and then chooses randomly the next node to visit from the set of neighbors (possibly including v itself) based on transition probabilities. The latter can for instance reflect the importance of an edge. This iterative process can be repeated an arbitrary number of times.

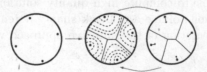

Fig. 1. The three steps of the Bubble framework. Black nodes indicate the seed nodes. On the left hand side, seed nodes are found. In the middle, a partition is found by performing breadth-first searches around the seed nodes and on the right hand side new seed nodes are found.

It is governed by the so-called *transition matrix* $\mathbf{P}$, whose entries denote the edges' transition probabilities. More details can be found in Lovasz's random walk survey [136].

Diffusion, in turn, is a natural process describing a substance's desire to distribute evenly in space. In a discrete setting on graphs, diffusion is an iterative process which exchanges splittable entities between neighboring nodes, usually until all nodes have the same amount. Diffusion is a special random walk; thus, both can be used to identify dense graph regions: Once a random walk reaches a dense region, it is likely to stay there for a long time, before leaving it via one of the relatively few outgoing edges. The relative size of $\mathbf{P}^t_{u,v}$, the probability of a random walk that starts in u to be located on v after t steps, can be exploited for assigning u and v to the same or different clusters. This fact is used by many authors for graph clustering, cf. Schaeffer's survey [188].

Due to the difficulty of enforcing balance constraints, works employing these approaches for partitioning are less numerous. Meyerhenke et al. [148] present a similarity measure based on diffusion that is employed within the Bubble framework. This diffusive approach bears some conceptual resemblance to spectral partitioning, but with advantages in quality [150]. Balancing is enforced by two different procedures that are only loosely coupled to the actual partitioning process. The first one is an iterative procedure that tries to adapt the amount of diffusion load in each block by multiplying it with a suitable scalar. Underloaded blocks receive more load, overloaded ones less. It is then easier for underloaded blocks to "flood" other graph areas as well. In case the search for suitable scalars is unsuccessful, the authors employ a second approach that extends previous work [219]. It computes a migrating flow on the quotient graph of the partition. The flow value f_{ij} between blocks i and j specifies how many nodes have to be migrated from i to j in order to balance the partition. As a key and novel property for obtaining good solutions, to determine *which* nodes should be migrated in which order, the diffusive similarity values computed before within the Bubble framework are used [146,148].

Diffusion-based partitioning has been subsequently improved by Pellegrini [165], who combines KL/FM and diffusion for bipartitioning in the tool Scotch. He speeds up previous approaches by using *band graphs* that replace unimportant graph areas by a single node. An extension of these results to k-way partitioning with further adaptations has been realized within the tools DibaP [143] and PDibaP for repartitioning [147]. Integrated into a multilevel method, diffusive partitioning is able to compute high-quality solutions, in particular with respect to communication volume and block shape. It remains further work to devise a faster implementation of the diffusive approach without running time dependence on k.

6   Multilevel Graph Partitioning

Clearly the most successful heuristic for partitioning large graphs is the *multilevel graph partitioning* approach. It consists of the three main phases outlined in Fig. 2: coarsening, initial partitioning, and uncoarsening. The main goal of the coarsening (in many multilevel approaches implemented as *contraction*) phase is to gradually approximate the original problem and the input graph with fewer degrees of freedom. In multilevel GP solvers this is achieved by creating a hierarchy of successively coarsened graphs with decreasing sizes in such a way that cuts in the coarse graphs reflect cuts in the fine graph. There are multiple possibilities to create graph hierarchies. Most methods used today *contract* sets of nodes on the fine level. Contracting $U \subset V$ amounts to replacing it with a single node u with $c(u) := \sum_{w \in U} c(w)$. Contraction (and other types of coarsening) might produce parallel edges which are replaced by a single edge whose weight accumulates the weights of the parallel edges (see Fig. 3). This implies that balanced partitions on the coarse level represent balanced partitions on the fine level with the same cut value.

Coarsening is usually stopped when the graph is sufficiently small to be *initially partitioned* using some (possibly expensive) algorithm. Any of the basic algorithms from Sect. 4 can be used for initial partitioning as long as they are able to handle general node and edge weights. The high quality of more expensive methods that can be applied at the coarsest level does not necessarily translate into quality at the finest level, and some GP multilevel solvers rather run several faster but diverse methods repeatedly with different random tie breaking instead of applying expensive global optimization techniques.

Uncoarsening consists of two stages. First, the solution obtained on the coarse level graph is mapped to the fine level graph. Then the partition is improved, typically by using some variants of the improvement methods described in Sect. 5. This process of uncoarsening and local improvement is carried on until the finest hierarchy level has been processed. One run of this simple coarsening-uncoarsening scheme is also called a *V-cycle* (see Fig. 2).

There are at least three intuitive reasons why the multilevel approach works so well: First, at the coarse levels we can afford to perform a lot of work per node without increasing the overall execution time by a lot. Furthermore, a single node move at a coarse level corresponds to a big change in the final solution. Hence, we might be able to find improvements easily that would be difficult to find on the finest level. Finally, fine level local improvements are expected to run fast since they already start from a good solution inherited from the coarse level. Also

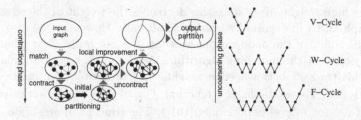

Fig. 2. The multilevel approach to GP. The left figure shows a two-level contraction-based scheme. The right figure shows different chains of coarsening-uncoarsening in the multilevel frameworks.

multilevel methods can benefit from their iterative application (such as chains of V-cycles) when the previous iteration's solution is used to improve the quality of coarsening. Moreover, (following the analogy to multigrid schemes) the inter-hierarchical coarsening-uncoarsening iteration can also be reconstructed in such way that more work will be done at the coarser levels (see F-, and W-cycles in Fig. 2, and [183,212]). An important technical advantage of multilevel approaches is related to parallelization. Because multilevel approaches achieve a global solution by local processing only (though applied at different levels of coarseness) they are naturally parallelization-schemes friendly.

6.1   Contracting a Single Edge

A minimalistic approach to coarsening is to contract only two nodes connected by a single edge in the graph. Since this leads to a hierarchy with (almost) n levels, this method is called n-level GP [161]. Together with a k-way variant of the highly localized local search from Sect. 5.1, this leads to a very simple way to achieve high quality parti-

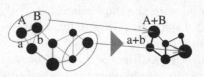

Fig. 3. An example matching and contraction of the matched edges.

tions. Compared to other techniques, n-level partitioning has some overhead for coarsening, mainly because it needs a priority queue and a dynamic graph data structure. On the other hand, for graphs with enough locality (e.g. from scientific computing), the n-level method empirically needs only sublinear work for local improvement.

6.2   Contracting a Matching

The most widely used contraction strategy contracts (large) matchings, i.e., the contracted sets are pairs of nodes connected by edges and these edges are not allowed to be incident to each other. The idea is that this leads to a geometrically decreasing size of the graph and hence a logarithmic number of levels, while subsequent levels are "similar" so that local improvement can quickly find good solutions. Assuming linear-time algorithms on all levels, one then gets linear overall execution time. Conventional wisdom is that a good matching contains many high weight edges since this decreases the weight of the edges in the coarse graph and will eventually lead to small cuts. However, one also wants a certain uniformity in the node weights so that it is not quite clear what should be the objective of the matching algorithm. A successful recent approach is to delegate this tradeoff between edge weights and uniformity to an *edge rating* function [1,102]. For example, the function $f(u,v) = \frac{\omega(\{u,v\})}{c(v)c(u)}$ works very well [102,183] (also for n-level partitioning [161]). The concept of algebraic distance yields further improved edge ratings [179].

The weighted matching problem itself has attracted a lot of interest motivated to a large extent by its application for coarsening. Although the maximum

close to optimal. This is particularly striking for bipartitioning, where recent exact results suggest that heuristics often find the optimal solution. In contrast, theoretical results state that we cannot even find constant-factor approximations in polynomial time. On the other hand, the sophisticated theoretical methods developed to obtain approximation guarantees are currently not used in the most successful solvers. It would be interesting to see to what extent these techniques can yield a practical contribution. There is a similar problem for exact solvers, which have made rapid progress for the case $k = 2$. However, it remains unclear how to use them productively for larger graphs or in case $k > 2$, for example as initial partitioners in a multilevel system or for pair-wise local improvement of subgraphs. What *is* surprisingly successful, is the use of solvers with performance guarantees for subproblems that are easier than partitioning. For example, KaHIP [187] uses weighted matching, spanning trees, edge coloring, BFS, shortest paths, diffusion, maximum flows, and strongly connected components. Further research into this direction looks promising.

Difficult Instances. The new "complex network" applications described in Sect. 3.2 result in graphs that are not only very large but also difficult to handle for current graph partitioners. This difficulty results from an uneven degree distribution and much less locality than observed in traditional inputs. Here, improved techniques within known frameworks (e.g., better coarsening schemes) and even entirely different approaches can give substantial improvements in speed or quality.

Another area where large significant quality improvements are possible are for large k. Already for the largest value of k considered in the Walshaw benchmark (64), the spread between different approaches is considerable. Considering graphs with billions of nodes and parallel machines reaching millions of processors, $k \leq 64$ increasingly appears like a special case. The multilevel method loses some of its attractiveness for large k since even initial partitioning must solve quite large instances. Hence new ideas are required.

Multilevel Approach. While the multilevel paradigm has been extremely successful for GP, there are still many algorithmic challenges ahead. The variety of continuous systems multilevel algorithms (such as various types of multigrid) turned into a separate field of applied mathematics, and optimization. Yet, multilevel algorithms for GPP still consist in practice of a very limited number of multilevel techniques. The situation with other combinatorial optimization problems is not significantly different. One very promising direction is bridging the gaps between the theory and practice of multiscale computing and multilevel GP such as introducing nonlinear coarsening schemes. For example, a novel multilevel approach for the minimum vertex separator problem was recently proposed using the continuous bilinear quadratic program formulation [92], and a *hybrid of the geometric multigrid, and full approximation scheme* for continuous problem was used for graph drawing, and VLSI placement problems [45,177]. Development of more sophisticated coarsening schemes, edge ratings, and metrics of nodes' similarity that can be propagated throughout the hierarchies are among the future challenges for graph partitioning as well as any attempt of their rigorous analysis.

Parallelism and Other Hardware Issues. Scalable high quality GP (with quality comparable to sequential partitioners) remains an open problem. With the advent of exascale machines with millions of processors and possibly billions of threads, the situation is further aggravated. Traditional "flat" partitions of graphs for processing on such machines implies a huge number of blocks. It is unclear how even sequential partitioners perform for such instances. Resorting to recursive partitioning brings down k and also addresses the hierarchical nature of such machines. However, this means that we need parallel partitioners where the number of available processors is much bigger than k. It is unclear how to do this with high quality. Approaches like the band graphs from PT-Scotch are interesting but likely to fail for complex networks.

Efficient implementation is also a big issue since complex memory hierarchies and heterogeneity (e.g., GPUs or FPGAs) make the implementation complicated. In particular, there is a mismatch between the fine-grained discrete computations predominant in the best sequential graph partitioners and the massive data parallelism (SIMD-instructions, GPUs,...) in high performance computing which better fits highly regular numeric computations. It is therefore likely that high quality GP will only be used for the higher levels of the machine hierarchy, e.g., down to cluster nodes or CPU sockets. At lower levels of the architectural hierarchy, we may use geometric partitioning or even regular grids with dummy values for non-existing cells (e.g. [74]).

While exascale computing is a challenge for high-end applications, many more applications can profit from GP in cloud computing and using tools for high productivity such as Map/Reduce [47], Pregel [139], GraphLab [137], Combinatorial BLAS [30], or Parallel Boost Graph Library [90]. Currently, none of these systems uses sophisticated GP software.

These changes in architecture also imply that we are no longer interested in algorithms with little computations but rather in data access with high locality and good energy efficiency.

Beyond Balanced k-partitioning with Cut Minimization. We have intentionally fixed our basic model assumptions above to demonstrate that even the classical setting has a lot of open problems. However, these assumption become less and less warranted in the context of modern massively parallel hardware and huge graphs with complex structure. For example, it looks like the assumptions that low total cut is highly correlated with low bottleneck cut or communication volume (see Sect. 2.1) is less warranted for complex network [31]. Eventually, we would like a dynamic partition that adapts to the communication requirements of a computation such as PageRank or BFS with changing sets of active nodes and edges. Also, the fixed value for k becomes questionable when we want to tolerate processor failures or achieve "malleable" computations that adapt their resource usage to the overall situation, e.g., to the arrival or departure of high priority jobs. Techniques like overpartitioning, repartitioning (with changed k), and (re)mapping will therefore become more important. Even running time as the bottom-line performance goal might be replaced by energy consumption [199].

Acknowledgements. We express our gratitude to Bruce Hendrickson, Dominique LaSalle, and George Karypis for many valuable comments on a preliminary draft of the manuscript.

References

1. Abou-Rjeili, A., Karypis, G.: Multilevel algorithms for partitioning power-law graphs. In: 20th International Parallel and Distributed Processing Symposium (IPDPS). IEEE (2006)
2. Akhremtsev, Y., Sanders, P., Schulz, C.: (Semi-)external algorithms for graph partitioning and clustering. In: 15th Workshop on Algorithm Engineering and Experimentation (ALENEX), pp. 33–43 (2015)
3. Andersen, R., Lang, K.J.: An algorithm for improving graph partitions. In: 19th ACM-SIAM Symposium on Discrete Algorithms, pp. 651–660 (2008)
4. Andreev, K., Räcke, H.: Balanced graph partitioning. Theory Comput. Syst. **39**(6), 929–939 (2006)
5. Armbruster, M.: Branch-and-cut for a semidefinite relaxation of large-scale minimum bisection problems. Ph.D. thesis, U. Chemnitz (2007)
6. Armbruster, M., Fügenschuh, M., Helmberg, C., Martin, A.: A comparative study of linear and semidefinite branch-and-cut methods for solving the minimum graph bisection problem. In: Lodi, A., Panconesi, A., Rinaldi, G. (eds.) IPCO 2008. LNCS, vol. 5035, pp. 112–124. Springer, Heidelberg (2008). doi:10.1007/978-3-540-68891-4_8
7. Arora, S., Hazan, E., Kale, S.: $O(\sqrt{\log n})$ approximation to sparsest cut in $\tilde{O}(n^2)$ time. SIAM J. Comput. **39**(5), 1748–1771 (2010)
8. Arora, S., Rao, S., Vazirani, U.: Expander flows, geometric embeddings and graph partitioning. In: 36th ACM Symposium on the Theory of Computing (STOC), pp. 222–231 (2004)
9. Aubanel, E.: Resource-aware load balancing of parallel applications. In: Udoh, E., Wang, F.Z. (eds.) Handbook of Research on Grid Technologies and Utility Computing: Concepts for Managing Large-Scale Applications, pp. 12–21. Information Science Reference - Imprint of: IGI Publishing, May 2009
10. Auer, B.F., Bisseling, R.H.: Graph coarsening and clustering on the GPU. In: Bader et al. [13], pp. 19–36
11. Aykanat, C., Cambazoglu, B.B., Findik, F., Kurc, T.: Adaptive decomposition and remapping algorithms for object-space-parallel direct volume rendering of unstructured grids. J. Parallel Distrib. Comput. **67**(1), 77–99 (2007). http://dx.doi.org/10.1016/j.jpdc.2006.05.005
12. Bader, D.A., Meyerhenke, H., Sanders, P., Schulz, C., Kappes, A., Wagner, D.: Benchmarking for graph clustering and graph partitioning. In: Encyclopedia of Social Network Analysis and Mining (to appear)
13. Bader, D.A., Meyerhenke, H., Sanders, P., Wagner, D. (eds.): Graph Partitioning and Graph Clustering – 10th DIMACS Impl. Challenge, Contemporary Mathematics, vol. 588. AMS, Boston (2013)
14. Bader, M.: Space-Filling Curves. Springer, Heidelberg (2013)

15. Barnard, S.T., Simon, H.D.: A fast multilevel implementation of recursive spectral bisection for partitioning unstructured problems. In: 6th SIAM Conference on Parallel Processing for Scientific Computing, pp. 711–718 (1993)
16. Benlic, U., Hao, J.K.: An effective multilevel memetic algorithm for balanced graph partitioning. In: 22nd IEEE International Conference on Tools with Artificial Intelligence (ICTAI), pp. 121–128 (2010)
17. Benlic, U., Hao, J.K.: A multilevel memetic approach for improving graph k-partitions. IEEE Trans. Evol. Comput. **15**(5), 624–642 (2011)
18. Benlic, U., Hao, J.K.: An effective multilevel tabu search approach for balanced graph partitioning. Comput. Oper. Res. **38**(7), 1066–1075 (2011)
19. van Bevern, R., Feldmann, A.E., Sorge, M., Suchý, O.: On the parameterized complexity of computing balanced partitions in graphs. CoRR abs/1312.7014 (2013). http://arxiv.org/abs/1312.7014
20. Bhatele, A., Kale, L.: Heuristic-based techniques for mapping irregular communication graphs to mesh topologies. In: 13th Conference on High Performance Computing and Communications (HPCC), pp. 765–771 (2011)
21. Bhatele, A., Jain, N., Gropp, W.D., Kale, L.V.: Avoiding hot-spots on two-level Direct networks. In: ACM/IEEE Conference for High Performance Computing, Networking, Storage and Analysis (SC), pp. 76:1–76:11. ACM (2011)
22. Bichot, C., Siarry, P. (eds.): Graph Partitioning. Wiley, Hoboken (2011)
23. Bichot, C.E.: A new method, the fusion fission, for the relaxed k-way graph partitioning problem, and comparisons with some multilevel algorithms. J. Math. Model. Algorithms **6**(3), 319–344 (2007)
24. Birn, M., Osipov, V., Sanders, P., Schulz, C., Sitchinava, N.: Efficient parallel and external matching. In: Wolf, F., Mohr, B., Mey, D. (eds.) Euro-Par 2013. LNCS, vol. 8097, pp. 659–670. Springer, Heidelberg (2013). doi:10.1007/978-3-642-40047-6_66
25. Boman, E.G., Devine, K.D., Rajamanickam, S.: Scalable matrix computations on large scale-free graphs using 2D graph partitioning. In: ACM/IEEE Conference for High Performance Computing, Networking, Storage and Analysis (SC) (2013)
26. Boppana, R.B.: Eigenvalues and graph bisection: an average-case analysis. In: 28th Symposium on Foundations of Computer Science (FOCS), pp. 280–285 (1987)
27. Brandfass, B., Alrutz, T., Gerhold, T.: Rank reordering for MPI communication optimization. Comput. Fluids **80**, 372–380 (2013). http://www.sciencedirect.com/science/article/pii/S004579301200028X
28. Brunetta, L., Conforti, M., Rinaldi, G.: A branch-and-cut algorithm for the equicut problem. Math. Program. **78**(2), 243–263 (1997)
29. Bui, T., Chaudhuri, S., Leighton, F., Sipser, M.: Graph bisection algorithms with good average case behavior. Combinatorica **7**, 171–191 (1987)
30. Buluç, A., Gilbert, J.R.: The combinatorial BLAS: design, implementation, and applications. Int. J. High Perform. Comput. Appl. **25**(4), 496–509 (2011)
31. Buluç, A., Madduri, K.: Graph partitioning for scalable distributed graph computations. In: Bader et al. [13], pp. 83–102
32. Camilus, K.S., Govindan, V.K.: A review on graph based segmentation. IJIGSP **4**, 1–13 (2012)
33. Catalyurek, U., Aykanat, C.: A hypergraph-partitioning approach for coarse-grain decomposition. In: ACM/IEEE Conference on Supercomputing (SC). ACM (2001)

34. Catalyurek, U., Boman, E., et al.: Hypergraph-based dynamic load balancing for adaptive scientific computations. In: 21st International Parallel and Distributed Processing Symposium (IPDPS). IEEE (2007)
35. Çatalyürek, Ü., Aykanat, C.: PaToH: partitioning tool for hypergraphs. In: Padua, D. (ed.) Encyclopedia of Parallel Computing. Springer, Heidelberg (2011)
36. Chan, S.Y., Ling, T.C., Aubanel, E.: The impact of heterogeneous multi-core clusters on graph partitioning: an empirical study. Cluster Comput. **15**(3), 281–302 (2012)
37. Chardaire, P., Barake, M., McKeown, G.P.: A PROBE-based heuristic for graph partitioning. IEEE Trans. Comput. **56**(12), 1707–1720 (2007)
38. Chen, J., Safro, I.: Algebraic distance on graphs. SIAM J. Sci. Comput. **33**(6), 3468–3490 (2011)
39. Chevalier, C., Pellegrini, F.: Improvement of the efficiency of genetic algorithms for scalable parallel graph partitioning in a multi-level framework. In: Nagel, W.E., Walter, W.V., Lehner, W. (eds.) Euro-Par 2006. LNCS, vol. 4128, pp. 243–252. Springer, Heidelberg (2006). doi:10.1007/11823285_25
40. Chevalier, C., Pellegrini, F.: PT-Scotch: a tool for efficient parallel graph ordering. Parallel Comput. **34**(6), 318–331 (2008)
41. Chevalier, C., Safro, I.: Comparison of coarsening schemes for multi-level graph partitioning. In: Proceedings Learning and Intelligent Optimization (2009)
42. Chierichetti, F., Kumar, R., Lattanzi, S., Mitzenmacher, M., Panconesi, A., Raghavan, P.: On compressing social networks. In: 15th ACM SIGKDD International Conference on Knowledge Discovery and Data Mining, pp. 219–228 (2009)
43. Chu, S., Cheng, J.: Triangle listing in massive networks and its applications. In: 17th ACM SIGKDD Conference on Knowledge Discovery and Data Mining, pp. 672–680 (2011)
44. Comellas, F., Sapena, E.: A multiagent algorithm for graph partitioning. In: Rothlauf, F., Branke, J., Cagnoni, S., Costa, E., Cotta, C., Drechsler, R., Lutton, E., Machado, P., Moore, J.H., Romero, J., Smith, G.D., Squillero, G., Takagi, H. (eds.) EvoWorkshops 2006. LNCS, vol. 3907, pp. 279–285. Springer, Heidelberg (2006). doi:10.1007/11732242_25
45. Cong, J., Shinnerl, J.: Multilevel Optimization in VLSICAD. Springer, Heidelberg (2003)
46. Davis, T.: The University of Florida Sparse Matrix Collection (2008). http://www.cise.ufl.edu/research/sparse/matrices/
47. Dean, J., Ghemawat, S.: MapReduce: simplified data processing on large clusters. In: 6th Symposium on Operating System Design and Implementation (OSDI), pp. 137–150. USENIX (2004)
48. Delling, D., Goldberg, A.V., Pajor, T., Werneck, R.F.: Customizable route planning. In: Pardalos, P.M., Rebennack, S. (eds.) SEA 2011. LNCS, vol. 6630, pp. 376–387. Springer, Heidelberg (2011). doi:10.1007/978-3-642-20662-7_32
49. Delling, D., Goldberg, A.V., Razenshteyn, I., Werneck, R.F.: Exact combinatorial branch-and-bound for graph bisection. In: 12th Workshop on Algorithm Engineering and Experimentation (ALENEX), pp. 30–44 (2012)
50. Delling, D., Goldberg, A.V., et al.: Graph partitioning with natural cuts. In: 25th International Parallel and Distributed Processing Symposium (IPDPS), pp. 1135–1146 (2011)
51. Delling, D., Werneck, R.F.: Better bounds for graph bisection. In: Epstein, L., Ferragina, P. (eds.) ESA 2012. LNCS, vol. 7501, pp. 407–418. Springer, Heidelberg (2012). doi:10.1007/978-3-642-33090-2_36

52. Delling, D., Werneck, R.F.: Faster customization of road networks. In: Bonifaci, V., Demetrescu, C., Marchetti-Spaccamela, A. (eds.) SEA 2013. LNCS, vol. 7933, pp. 30–42. Springer, Heidelberg (2013). doi:10.1007/978-3-642-38527-8_5

53. Devine, K.D., Boman, E.G., Heaphy, R.T., Bisseling, R.H., Catalyurek, U.V.: Parallel hypergraph partitioning for scientific computing. In: Proceedings of the IEEE International Parallel and Distributed Processing Symposium, p. 124. IPDPS 2006 (2006). http://dl.acm.org/citation.cfm?id=1898953.1899056

54. Guo, D., Ke Liao, H.J.: Power system reconfiguration based on multi-level graph partitioning. In: 7th International Conference, GIScience 2012 (2012)

55. Diekmann, R., Monien, B., Preis, R.: Using helpful sets to improve graph bisections. In: Interconnection Networks and Mapping and Scheduling Parallel Computations, vol. 21, pp. 57–73 (1995)

56. Diekmann, R., Preis, R., Schlimbach, F., Walshaw, C.: Shape-optimized mesh partitioning and load balancing for parallel adaptive FEM. Parallel Comput. 26, 1555–1581 (2000)

57. Diekmann, R., Preis, R., Schlimbach, F., Walshaw, C.: Shape-optimized mesh partitioning and load balancing for parallel adaptive FEM. Parallel Comput. 26(12), 1555–1581 (2000)

58. Donath, W.E., Hoffman, A.J.: Algorithms for partitioning of graphs and computer logic based on eigenvectors of connection matrices. IBM Tech. Discl. Bull. 15(3), 938–944 (1972)

59. Donath, W.E., Hoffman, A.J.: Lower bounds for the partitioning of graphs. IBM J. Res. Dev. 17(5), 420–425 (1973)

60. Donde, V., Lopez, V., Lesieutre, B., Pinar, A., Yang, C., Meza, J.: Identification of severe multiple contingencies in electric power networks. In: 37th N. A. Power Symposium, pp. 59–66. IEEE (2005)

61. Drake, D., Hougardy, S.: A simple approximation algorithm for the weighted matching problem. Inf. Process. Lett. 85, 211–213 (2003)

62. Drake Vinkemeier, D.E., Hougardy, S.: A linear-time approximation algorithm for weighted matchings in graphs. ACM Trans. Algorithms 1(1), 107–122 (2005)

63. Duan, R., Pettie, S., Su, H.H.: Scaling Algorithms for Approximate and Exact Maximum Weight Matching. CoRR abs/1112.0790 (2011)

64. Dutt, S.: New faster Kernighan-Lin-type graph-partitioning algorithms. In: 4th IEEE/ACM Conference on Computer-Aided Design, pp. 370–377 (1993)

65. Even, G., Naor, J.S., Rao, S., Schieber, B.: Fast approximate graph partitioning algorithms. SIAM J. Comput. 28(6), 2187–2214 (1999)

66. Fagginger Auer, B.O., Bisseling, R.H.: Abusing a hypergraph partitioner for unweighted graph partitioning. In: Bader et al. [13], pp. 19–35

67. Farhat, C., Lesoinne, M.: Automatic partitioning of unstructured meshes for the parallel solution of problems in computational mechanics. J. Numer. Methods Eng. 36(5), 745–764 (1993). http://dx.doi.org/10.1002/nme.1620360503

68. Feige, U., Krauthgamer, R.: A polylogarithmic approximation of the minimum bisection. SIAM J. Comput. 31(4), 1090–1118 (2002)

69. Feldmann, A.E., Widmayer, P.: An $\mathcal{O}(n^4)$ time algorithm to compute the bisection width of solid grid graphs. In: Demetrescu, C., Halldórsson, M.M. (eds.) ESA 2011. LNCS, vol. 6942, pp. 143–154. Springer, Heidelberg (2011). doi:10.1007/978-3-642-23719-5_13

70. Felner, A.: Finding optimal solutions to the graph partitioning problem with heuristic search. Ann. Math. Artif. Intell. 45, 293–322 (2005)

71. Ferreira, C.E., Martin, A., De Souza, C.C., Weismantel, R., Wolsey, L.A.: The node capacitated graph partitioning problem: a computational study. Math. Program. **81**(2), 229–256 (1998)
72. Fiduccia, C.M., Mattheyses, R.M.: A linear-time heuristic for improving network partitions. In: 19th Conference on Design Automation, pp. 175–181 (1982)
73. Fiedler, M.: A property of eigenvectors of nonnegative symmetric matrices and its application to graph theory. Czech. Math. J. **25**(4), 619–633 (1975)
74. Fietz, J., Krause, M.J., Schulz, C., Sanders, P., Heuveline, V.: Optimized hybrid parallel lattice Boltzmann fluid flow simulations on complex geometries. In: Kaklamanis, C., Papatheodorou, T., Spirakis, P.G. (eds.) Euro-Par 2012. LNCS, vol. 7484, pp. 818–829. Springer, Heidelberg (2012). doi:10.1007/978-3-642-32820-6_81
75. Ford, L.R., Fulkerson, D.R.: Maximal flow through a network. Can. J. Math. **8**(3), 399–404 (1956)
76. Fortunato, S.: Community Detection in Graphs. CoRR abs/0906.0612 (2009)
77. Fourestier, S., Pellegrini, F.: Adaptation au repartitionnement de graphes d'une méthode d'optimisation globale par diffusion. In: RenPar'20 (2011)
78. Galinier, P., Boujbel, Z., Fernandes, M.C.: An efficient memetic algorithm for the graph partitioning problem. Ann. Oper. Res. **191**(1), 1–22 (2011)
79. Garey, M.R., Johnson, D.S., Stockmeyer, L.: Some simplified NP-complete problems. In: 6th ACM Symposium on Theory of Computing, pp. 47–63. STOC, ACM (1974)
80. Garey, M.R., Johnson, D.S.: Computers and Intractability: A Guide to the Theory of NP-Completeness. W. H. Freeman & Co., New York (1979)
81. George, A., Liu, J.W.H.: Computer Solution of Large Sparse Positive Definite Systems. Prentice-Hall, Upper Saddle River (1981)
82. Ghazinour, K., Shaw, R.E., Aubanel, E.E., Garey, L.E.: A linear solver for benchmarking partitioners. In: 22nd IEEE International Symposium on Parallel and Distributed Processing (IPDPS), pp. 1–8 (2008)
83. Gilbert, J.R., Miller, G.L., Teng, S.H.: Geometric mesh partitioning: implementation and experiments. SIAM J. Sci. Comput. **19**(6), 2091–2110 (1998)
84. Glantz, R., Meyerhenke, H., Noe, A.: Algorithms for mapping parallel processes onto grid and torus architectures. In: Proceedings of the 23rd Euromicro International Conference on Parallel, Distributed and Network-Based Processing (2015, to appear). Preliminary version: http://arxiv.org/abs/1411.0921
85. Glantz, R., Meyerhenke, H., Schulz, C.: Tree-based coarsening and partitioning of complex networks. In: Gudmundsson, J., Katajainen, J. (eds.) SEA 2014. LNCS, vol. 8504, pp. 364–375. Springer, Heidelberg (2014). doi:10.1007/978-3-319-07959-2_31
86. Glover, F.: Tabu search – part I. ORSA J. Comput. **1**(3), 190–206 (1989)
87. Glover, F.: Tabu search – part II. ORSA J. Comput. **2**(1), 4–32 (1990)
88. Goldschmidt, O., Hochbaum, D.S.: A polynomial algorithm for the k-cut problem for fixed k. Math. Oper. Res. **19**(1), 24–37 (1994)
89. Grady, L., Schwartz, E.L.: Isoperimetric graph partitioning for image segmentation. IEEE Trans. Pattern Anal. Mach. Intell. **28**, 469–475 (2006)
90. Gregor, D., Lumsdaine, A.: The parallel BGL: a generic library for distributed graph computations. In: Parallel Object-Oriented Scientific Computing (POOSC) (2005)
91. Gutfraind, A., Meyers, L.A., Safro, I.: Multiscale Network Generation. CoRR abs/1207.4266 (2012)

92. Hager, W.W., Hungerford, J.T., Safro, I.: A multilevel bilinear programming algorithm for the vertex separator problem. CoRR abs/1410.4885 (2014). arXiv:1410.4885

93. Hager, W.W., Krylyuk, Y.: Graph partitioning and continuous quadratic programming. SIAM J. Discrete Math. **12**(4), 500–523 (1999)

94. Hager, W.W., Phan, D.T., Zhang, H.: An exact algorithm for graph partitioning. Math. Program. **137**(1–2), 531–556 (2013)

95. Hendrickson, B.: Chaco: Software for Partitioning Graphs. http://www.cs.sandia.gov/bahendr/chaco.html

96. Hendrickson, B.: Graph partitioning and parallel solvers: has the emperor no clothes? In: Ferreira, A., Rolim, J., Simon, H., Teng, S.-H. (eds.) IRREGULAR 1998. LNCS, vol. 1457, pp. 218–225. Springer, Heidelberg (1998). doi:10.1007/BFb0018541

97. Hendrickson, B., Leland, R.: A multilevel algorithm for partitioning graphs. In: ACM/IEEE Conference on Supercomputing 1995 (1995)

98. Hendrickson, B., Leland, R.: An improved spectral graph partitioning algorithm for mapping parallel computations. SIAM J. Sci. Comput. **16**(2), 452–469 (1995)

99. Hendrickson, B., Leland, R., Driessche, R.V.: Enhancing data locality by using terminal propagation. In: 29th Hawaii International Conference on System Sciences (HICSS 2009), vol. 1, p. 565. Software Technology and Architecture (1996)

100. Hendrickson, B., Kolda, T.G.: Graph partitioning models for parallel computing. Parallel Comput. **26**(12), 1519–1534 (2000)

101. Hoefler, T., Snir, M.: Generic topology mapping strategies for large-scale parallel architectures. In: ACM International Conference on Supercomputing (ICS 2011), pp. 75–85. ACM (2011)

102. Holtgrewe, M., Sanders, P., Schulz, C.: Engineering a scalable high quality graph partitioner. In: 24th IEEE International Parallel and Distributed Processing Symposium (IPDPS), pp. 1–12 (2010)

103. Hromkovič, J., Monien, B.: The bisection problem for graphs of degree 4 (configuring transputer systems). In: Tarlecki, A. (ed.) MFCS 1991. LNCS, vol. 520, pp. 211–220. Springer, Heidelberg (1991). doi:10.1007/3-540-54345-7_64

104. Huang, S., Aubanel, E., Bhavsar, V.C.: PaGrid: a mesh partitioner for computational grids. J. Grid Comput. **4**(1), 71–88 (2006)

105. Hungershöfer, J., Wierum, J.-M.: On the quality of partitions based on space-filling curves. In: Sloot, P.M.A., Hoekstra, A.G., Tan, C.J.K., Dongarra, J.J. (eds.) ICCS 2002. LNCS, vol. 2331, pp. 36–45. Springer, Heidelberg (2002). doi:10.1007/3-540-47789-6_4

106. Hyafil, L., Rivest, R.: Graph partitioning and constructing optimal decision trees are polynomial complete problems. Technical report 33, IRIA - Laboratoire de Recherche en Informatique et Automatique (1973)

107. Jeannot, E., Mercier, G., Tessier, F.: Process placement in multicore clusters: algorithmic issues and practical techniques. IEEE Trans. Parallel Distrib. Syst. **PP**(99), 1–1 (2013)

108. Jerrum, M., Sorkin, G.B.: The metropolis algorithm for graph bisection. Discret. Appl. Math. **82**(1–3), 155–175 (1998)

109. Junker, B., Schreiber, F.: Analysis of Biological Networks. Wiley, Hoboken (2008)

110. Kahng, A.B., Lienig, J., Markov, I.L., Hu, J.: VLSI Physical Design - From Graph Partitioning to Timing Closure. Springer, Heidelberg (2011)

111. Karisch, S.E., Rendl, F., Clausen, J.: Solving graph bisection problems with semidefinite programming. INFORMS J. Comput. **12**(3), 177–191 (2000)

112. Karypis, G., Kumar, V.: Parallel multilevel k-way partitioning scheme for irregular graphs. In: ACM/IEEE Supercomputing 1996 (1996)
113. Karypis, G., Kumar, V.: A fast and high quality multilevel scheme for partitioning irregular graphs. SIAM J. Sci. Comput. **20**(1), 359–392 (1998)
114. Karypis, G., Kumar, V.: Multilevel k-way partitioning scheme for irregular graphs. J. Parallel Distrib. Comput. **48**(1), 96–129 (1998)
115. Karypis, G., Kumar, V.: Multilevel k-way hypergraph partitioning. In: 36th ACM/IEEE Design Automation Conference, pp. 343–348. ACM (1999)
116. Karypis, G., Kumar, V.: Parallel multilevel series k-way partitioning scheme for irregular graphs. SIAM Rev. **41**(2), 278–300 (1999)
117. Kernighan, B.W., Lin, S.: An efficient heuristic procedure for partitioning graphs. Bell Syst. Tech. J. **49**(1), 291–307 (1970)
118. Kieritz, T., Luxen, D., Sanders, P., Vetter, C.: Distributed time-dependent contraction hierarchies. In: Festa, P. (ed.) SEA 2010. LNCS, vol. 6049, pp. 83–93. Springer, Heidelberg (2010). doi:10.1007/978-3-642-13193-6_8
119. Kim, J., Hwang, I., Kim, Y.H., Moon, B.R.: Genetic approaches for graph partitioning: a survey. In: 13th Genetic and Evolutionary Computation (GECCO), pp. 473–480. ACM (2011). http://doi.acm.org/10.1145/2001576.2001642
120. Kim, Y.M., Lai, T.H.: The complexity of congestion-1 embedding in a hypercube. J. Algorithms **12**(2), 246–280 (1991). http://www.sciencedirect.com/science/article/pii/019667749190004I
121. Kirmani, S., Raghavan, P.: Scalable parallel graph partitioning. In: High Performance Computing, Networking, Storage and Analysis, SC 2013. ACM (2013)
122. Korosec, P., Silc, J., Robic, B.: Solving the mesh-partitioning problem with an ant-colony algorithm. Parallel Comput. **30**(5–6), 785–801 (2004)
123. Kunegis, J.: KONECT - the Koblenz network collection. In: Web Observatory Workshop, pp. 1343–1350 (2013) ·
124. Lafon, S., Lee, A.B.: Diffusion maps and coarse-graining: a unified framework for dimensionality reduction, graph partioning and data set parametrization. IEEE Trans. Pattern Anal. Mach. Intell. **28**(9), 1393–1403 (2006)
125. Lanczos, C.: An iteration method for the solution of the eigenvalue problem of linear differential and integral operators. J. Res. Natl Bur. Stand. **45**(4), 255–282 (1950)
126. Land, A.H., Doig, A.G.: An automatic method of solving discrete programming problems. Econometrica **28**(3), 497–520 (1960)
127. Lang, K., Rao, S.: A flow-based method for improving the expansion or conductance of graph cuts. In: Bienstock, D., Nemhauser, G. (eds.) IPCO 2004. LNCS, vol. 3064, pp. 325–337. Springer, Heidelberg (2004). doi:10.1007/978-3-540-25960-2_25
128. Lasalle, D., Karypis, G.: Multi-threaded graph partitioning. In: 27th International Parallel and Distributed Processing Symposium (IPDPS), pp. 225–236 (2013)
129. Lauther, U.: An extremely fast, exact algorithm for finding shortest paths in static networks with geographical background. In: Münster GI-Days (2004)
130. Leighton, F.T.: Introduction to Parallel Algorithms and Architectures: Arrays, Trees, Hypercubes. Morgan Kaufmann Publishers, Burlington (1992)
131. Lescovec, J.: Stanford network analysis package (SNAP). http://snap.stanford.edu/index.html
132. Li, H., Rosenwald, G., Jung, J., Liu, C.C.: Strategic power infrastructure defense. Proc. IEEE **93**(5), 918–933 (2005)
133. Li, J., Liu, C.C.: Power system reconfiguration based on multilevel graph partitioning. In: PowerTech, pp. 1–5 (2009)

134. Lisser, A., Rendl, F.: Graph partitioning using linear and semidefinite programming. Math. Program. **95**(1), 91–101 (2003)
135. Lloyd, S.: Least squares quantization in PCM. IEEE Trans. Inf. Theory **28**(2), 129–137 (1982)
136. Lovász, L.: Random walks on graphs: a survey. Comb. Paul Erdös is Eighty **2**, 1–46 (1993)
137. Low, Y., Gonzalez, J., Kyrola, A., Bickson, D., Guestrin, C., Hellerstein, J.M.: Distributed GraphLab: a framework for machine learning in the cloud. PVLDB **5**(8), 716–727 (2012)
138. Luxen, D., Schieferdecker, D.: Candidate sets for alternative routes in road networks. In: Klasing, R. (ed.) SEA 2012. LNCS, vol. 7276, pp. 260–270. Springer, Heidelberg (2012). doi:10.1007/978-3-642-30850-5_23
139. Malewicz, G., Austern, M.H., Bik, A.J.C., Dehnert, J.C., Horn, I., Leiser, N., Czajkowski, G.: Pregel: a system for large-scale graph processing. In: ACM SIGMOD International Conference on Management of Data (SIGMOD), pp. 135–146. ACM (2010)
140. Maue, J., Sanders, P.: Engineering algorithms for approximate weighted matching. In: Demetrescu, C. (ed.) WEA 2007. LNCS, vol. 4525, pp. 242–255. Springer, Heidelberg (2007). doi:10.1007/978-3-540-72845-0_19
141. Maue, J., Sanders, P., Matijevic, D.: Goal directed shortest path queries using precomputed cluster distances. ACM J. Exp. Algorithmics **14**, 3.2:1–3.2:27 (2009)
142. Meuer, H., Strohmaier, E., Simon, H., Dongarra, J.: June 2013 — TOP500 supercomputer sites. http://top.500.org/lists/2013/06/
143. Meyerhenke, H., Monien, B., Sauerwald, T.: A new diffusion-based multilevel algorithm for computing graph partitions. J. Parallel Distrib. Comput. **69**(9), 750–761 (2009)
144. Meyerhenke, H., Monien, B., Schamberger, S.: Accelerating shape optimizing load balancing for parallel FEM simulations by algebraic multigrid. In: 20th IEEE International Parallel and Distributed Processing Symposium (IPDPS), p. 57 (CD) (2006)
145. Meyerhenke, H., Sanders, P., Schulz, C.: Partitioning complex networks via size-constrained clustering. In: Gudmundsson, J., Katajainen, J. (eds.) SEA 2014. LNCS, vol. 8504, pp. 351–363. Springer, Heidelberg (2014). doi:10.1007/978-3-319-07959-2_30
146. Meyerhenke, H.: Disturbed diffusive processes for solving partitioning problems on graphs. Ph.D. thesis, Universität Paderborn (2008)
147. Meyerhenke, H.: Shape optimizing load balancing for MPI-parallel adaptive numerical simulations. In: Bader et al. [13], pp. 67–82
148. Meyerhenke, H., Monien, B., Schamberger, S.: Graph partitioning and disturbed diffusion. Parallel Comput. **35**(10–11), 544–569 (2009)
149. Meyerhenke, H., Sanders, P., Schulz, C.: Parallel graph partitioning for complex networks. In: Proceeding of the 29th IEEE International Parallel & Distributed Processing Symposium, (IPDPS 2015) (2015 to appear). Preliminary version: http://arxiv.org/abs/1404.4797
150. Meyerhenke, H., Sauerwald, T.: Beyond good partition shapes: an analysis of diffusive graph partitioning. Algorithmica **64**(3), 329–361 (2012)
151. Meyerhenke, H., Schamberger, S.: Balancing parallel adaptive FEM computations by solving systems of linear equations. In: Cunha, J.C., Medeiros, P.D. (eds.) Euro-Par 2005. LNCS, vol. 3648, pp. 209–219. Springer, Heidelberg (2005). doi:10.1007/11549468_26

152. Miller, G., Teng, S.H., Vavasis, S.: A unified geometric approach to graph separators. In: 32nd Symposium on Foundations of Computer Science (FOCS), pp. 538–547 (1991)
153. Möhring, R.H., Schilling, H., Schütz, B., Wagner, D., Willhalm, T.: Partitioning graphs to speedup Dijkstra's algorithm. ACM J. Exp. Algorithmics **11**, 1–29 (2006, 2007)
154. Mondaini, R.: Biomat 2009: International Symposium on Mathematical and Computational Biology, Brasilia, Brazil, 1–6. World Scientific (2010). http://books.google.es/books?id=3tiLMKtXiZwC
155. Monien, B., Schamberger, S.: Graph partitioning with the party library: helpfulsets in practice. In: 16th Symposium on Computer Architecture and High Performance Computing, pp. 198–205 (2004)
156. Monien, B., Preis, R., Schamberger, S.: Approximation algorithms for multilevel graph partitioning. In: Gonzalez, T.F. (ed.) Handbook of Approximation Algorithms and Metaheuristics, chap. 60, pp. 60-1–60-15. Taylor & Francis, Abingdon (2007)
157. Moulitsas, I., Karypis, G.: Architecture aware partitioning algorithms. In: Bourgeois, A.G., Zheng, S.Q. (eds.) ICA3PP 2008. LNCS, vol. 5022, pp. 42–53. Springer, Heidelberg (2008). doi:10.1007/978-3-540-69501-1_6
158. Newman, M.E.J.: Community detection and graph partitioning. CoRR abs/1305.4974 (2013)
159. Newman, M.: Networks: An Introduction. Oxford University Press Inc., New York (2010)
160. Nishimura, J., Ugander, J.: Restreaming graph partitioning: simple versatile algorithms for advanced balancing. In: 19th ACM SIGKDD International Conference on Knowledge Discovery and Data Mining (KDD) (2013)
161. Osipov, V., Sanders, P.: n-level graph partitioning. In: Berg, M., Meyer, U. (eds⟩) ESA 2010. LNCS, vol. 6346, pp. 278–289. Springer, Heidelberg (2010). doi:10.1007/978-3-642-15775-2_24
162. Papa, D.A., Markov, I.L.: Hypergraph partitioning and clustering. In: Gonzalez, T.F. (ed.) Handbook of Approximation Algorithms and Metaheuristics, chap. 61, pp. 61-1–61-19. CRC Press, Boca Raton (2007)
163. Pellegrini, F.: Scotch home page. http://www.labri.fr/pelegrin/scotch
164. Pellegrini, F.: Static mapping by dual recursive bipartitioning of process and architecture graphs. In: Scalable High-Performance Computing Conference (SHPCC), pp. 486–493. IEEE, May 1994
165. Pellegrini, F.: A parallelisable multi-level banded diffusion scheme for computing balanced partitions with smooth boundaries. In: Kermarrec, A.-M., Bougé, L., Priol, T. (eds.) Euro-Par 2007. LNCS, vol. 4641, pp. 195–204. Springer, Heidelberg (2007). doi:10.1007/978-3-540-74466-5_22
166. Pellegrini, F.: Scotch and libScotch 5.0 user's guide. Technical report, LaBRI, Université Bordeaux I, December 2007
167. Pellegrini, F.: Static mapping of process graphs. In: Bichot, C.E., Siarry, P. (eds.) Graph Partitioning, chap. 5, pp. 115–136. Wiley, Hoboken (2011)
168. Pellegrini, F.: Scotch and PT-Scotch graph partitioning software: an overview. In: Naumann, U., Schenk, O. (eds.) Combinatorial Scientific Computing, pp. 373–406. CRC Press, Boca Raton (2012)
169. Peng, B., Zhang, L., Zhang, D.: A survey of graph theoretical approaches to image segmentation. Pattern Recognit. **46**(3), 1020–1038 (2013)
170. Pettie, S., Sanders, P.: A simpler linear time $2/3 - \epsilon$ approximation for maximum weight matching. Inf. Process. Lett. **91**(6), 271–276 (2004)

171. Pilkington, J.R., Baden, S.B.: Partitioning with space-filling curves. Technical report CS94-349, UC San Diego, Department of Computer Science and Engineering (1994)
172. Pothen, A., Simon, H.D., Liou, K.P.: Partitioning sparse matrices with eigenvectors of graphs. SIAM J. Matrix Anal. Appl. **11**(3), 430–452 (1990)
173. Preis, R.: Linear time 1/2-approximation algorithm for maximum weighted matching in general graphs. In: Meinel, C., Tison, S. (eds.) STACS 1999. LNCS, vol. 1563, pp. 259–269. Springer, Heidelberg (1999). doi:10.1007/3-540-49116-3_24
174. Raghavan, U.N., Albert, R., Kumara, S.: Near linear time algorithm to detect community structures in large-scale networks. Phys. Rev. E **76**(3) (2007)
175. Rolland, E., Pirkul, H., Glover, F.: Tabu search for graph partitioning. Ann. Oper. Res. **63**(2), 209–232 (1996)
176. Ron, D., Wishko-Stern, S., Brandt, A.: An algebraic multigrid based algorithm for bisectioning general graphs. Technical report MCS05-01, Department of Computer Science and Applied Mathematics, The Weizmann Institute of Science (2005)
177. Ron, D., Safro, I., Brandt, A.: A fast multigrid algorithm for energy minimization under planar density constraints. Multiscale Model. Simul. **8**(5), 1599–1620 (2010)
178. Ron, D., Safro, I., Brandt, A.: Relaxation-based coarsening and multiscale graph organization. Multiscale Model. Simul. **9**(1), 407–423 (2011)
179. Safro, I., Sanders, P., Schulz, C.: Advanced coarsening schemes for graph partitioning. In: Klasing, R. (ed.) SEA 2012. LNCS, vol. 7276, pp. 369–380. Springer, Heidelberg (2012)
180. Safro, I., Temkin, B.: Multiscale approach for the network compression-friendly ordering. J. Discret. Algorithms **9**(2), 190–202 (2011)
181. Salihoglu, S., Widom, J.: GPS: a graph processing system. In: Proceedings of the 25th International Conference on Scientific and Statistical Database Management, SSDBM, pp. 22:1–22:12. ACM (2013). http://doi.acm.org/10.1145/2484838.2484843
182. Sanchis, L.A.: Multiple-way network partitioning. IEEE Trans. Comput. **38**(1), 62–81 (1989)
183. Sanders, P., Schulz, C.: Engineering multilevel graph partitioning algorithms. In: Demetrescu, C., Halldórsson, M.M. (eds.) ESA 2011. LNCS, vol. 6942, pp. 469–480. Springer, Heidelberg (2011). doi:10.1007/978-3-642-23719-5_40
184. Sanders, P., Schulz, C.: Distributed evolutionary graph partitioning. In: 12th Workshop on Algorithm Engineering and Experimentation (ALENEX), pp. 16–29 (2012)
185. Sanders, P., Schulz, C.: High quality graph partitioning. In: Bader et al. [13], pp. 19–36
186. Sanders, P., Schulz, C.: Think locally, act globally: highly balanced graph partitioning. In: Bonifaci, V., Demetrescu, C., Marchetti-Spaccamela, A. (eds.) SEA 2013. LNCS, vol. 7933, pp. 164–175. Springer, Heidelberg (2013). doi:10.1007/978-3-642-38527-8_16
187. Sanders, P., Schulz, C.: KaHIP - Karlsruhe High Quality Partitioning Homepage. http://algo2.iti.kit.edu/documents/kahip/index.html
188. Schaeffer, S.E.: Graph clustering. Comput. Sci. Rev. **1**(1), 27–64. http://dx.doi.org/10.1016/j.cosrev.2007.05.001
189. Schamberger, S.: On partitioning FEM graphs using diffusion. In: HPGC Workshop of the 18th International Parallel and Distributed Processing Symposium (IPDPS 2004). IEEE Computer Society (2004)

190. Schamberger, S., Wierum, J.M.: A locality preserving graph ordering approach for implicit partitioning: graph-filling curves. In: 17th International Conference on Parallel and Distributed Computing Systems (PDCS), ISCA, pp. 51–57 (2004)

191. Schloegel, K., Karypis, G., Kumar, V.: Graph partitioning for high-performance scientific simulations. In: Dongarra, J., Foster, I., Fox, G., Gropp, W., Kennedy, K., Torczon, L., White, A. (eds.) Sourcebook of parallel computing, pp. 491–541. Morgan Kaufmann Publishers, Burlington (2003)

192. Schloegel, K., Karypis, G., Kumar, V.: Multilevel diffusion schemes for repartitioning of adaptive meshes. J. Parallel Distrib. Comput. **47**(2), 109–124 (1997)

193. Schloegel, K., Karypis, G., Kumar, V.: A unified algorithm for load-balancing adaptive scientific simulations. In: Supercomputing 2000, p. 59 (CD). IEEE Computer Society (2000)

194. Schloegel, K., Karypis, G., Kumar, V.: Parallel static and dynamic multi-constraint graph partitioning. Concurr. Comput.: Pract. Exp. **14**(3), 219–240 (2002)

195. Schulz, C.: High quality graph partititioning. Ph.D. thesis. epubli GmbH (2013)

196. Schulz, F., Wagner, D., Zaroliagis, C.: Using multi-level graphs for timetable information in railway systems. In: Mount, D.M., Stein, C. (eds.) ALENEX 2002. LNCS, vol. 2409, pp. 43–59. Springer, Heidelberg (2002). doi:10.1007/3-540-45643-0_4

197. Sellmann, M., Sensen, N., Timajev, L.: Multicommodity flow approximation used for exact graph partitioning. In: Battista, G., Zwick, U. (eds.) ESA 2003. LNCS, vol. 2832, pp. 752–764. Springer, Heidelberg (2003). doi:10.1007/978-3-540-39658-1_67

198. Sensen, N.: Lower bounds and exact algorithms for the graph partitioning problem using multicommodity flows. In: Heide, F.M. (ed.) ESA 2001. LNCS, vol. 2161, pp. 391–403. Springer, Heidelberg (2001). doi:10.1007/3-540-44676-1_33

199. Shalf, J., Dosanjh, S., Morrison, J.: Exascale computing technology challenges. In: Palma, J.M.L.M., Daydé, M., Marques, O., Lopes, J.C. (eds.) VECPAR 2010. LNCS, vol. 6449, pp. 1–25. Springer, Heidelberg (2011). doi:10.1007/978-3-642-19328-6_1

200. Simon, H.D.: Partitioning of unstructured problems for parallel processing. Comput. Syst. Eng. **2**(2), 135–148 (1991)

201. Simon, H.D., Teng, S.H.: How good is recursive bisection? SIAM J. Sci. Comput. **18**(5), 1436–1445 (1997)

202. Soper, A.J., Walshaw, C., Cross, M.: A combined evolutionary search and multilevel optimisation approach to graph-partitioning. J. Glob. Optim. **29**(2), 225–241 (2004)

203. Stanton, I., Kliot, G.: Streaming graph partitioning for large distributed graphs. In: 18th ACM SIGKDD International Conference on Knowledge discovery and data mining (KDD), pp. 1222–1230. ACM (2012)

204. Stock, L.: Strategic logistics management. Cram101 Textbook Outlines, Lightning Source Inc. (2006). http://books.google.com/books?id=1LyCAQAACAAJ

205. Sui, X., Nguyen, D., Burtscher, M., Pingali, K.: Parallel graph partitioning on multicore architectures. In: Cooper, K., Mellor-Crummey, J., Sarkar, V. (eds.) LCPC 2010. LNCS, vol. 6548, pp. 246–260. Springer, Heidelberg (2011). doi:10.1007/978-3-642-19595-2_17

206. Tang, L., Liu, H., Zhang, J., Nazeri, Z.: Community evolution in dynamic multimode networks. In: 14th ACM SIGKDD International Conference on Knowledge discovery and data mining (KDD), pp. 677–685. ACM (2008)

207. Teresco, J., Beall, M., Flaherty, J., Shephard, M.: A hierarchical partition model for adaptive finite element computation. Comput. Method. Appl. Mech. Eng. **184**(2–4), 269–285 (2000). http://www.sciencedirect.com/science/article/pii/S0045782599002315

208. Trifunović, A., Knottenbelt, W.J.: Parallel multilevel algorithms for hypergraph partitioning. J. Parallel Distrib. Comput. **68**(5), 563–581 (2008)

209. Tsourakakis, C.E., Gkantsidis, C., Radunovic, B., Vojnovic, M.: Fennel: streaming graph partitioning for massive scale graphs. Technical report MSR-TR-2012-113, Microsoft Research (2000)

210. Ucar, B., Aykanat, C., Kaya, K., Ikinci, M.: Task assignment in heterogeneous computing systems. J. Parallel Distrib. Comput. **66**(1), 32–46 (2006). http://www.sciencedirect.com/science/article/pii/S0743731505001577

211. Wagner, D., Wagner, F.: Between min cut and graph bisection. In: Borzyszkowski, A.M., Sokołowski, S. (eds.) MFCS 1993. LNCS, vol. 711, pp. 744–750. Springer, Heidelberg (1993). doi:10.1007/3-540-57182-5_65

212. Walshaw, C.: Multilevel refinement for combinatorial optimisation problems. Ann. Oper. Res. **131**(1), 325–372 (2004)

213. Walshaw, C., Cross, M.: Mesh partitioning: a multilevel balancing and refinement algorithm. SIAM J. Sci. Comput. **22**(1), 63–80 (2000)

214. Walshaw, C., Cross, M.: Parallel mesh partitioning on distributed memory systems. In: Topping, B. (ed.) Computational Mechanics Using High Performance Computing, pp. 59–78. Saxe-Coburg Publications, Stirling (2002). Invited chapter

215. Walshaw, C., Cross, M.: JOSTLE: parallel multilevel graph-partitioning software - an overview. In: Mesh Partitioning Techniques and Domain Decomposition Techniques, pp. 27–58. Civil-Comp Ltd. (2007)

216. Walshaw, C., Cross, M., Everett, M.G.: A localized algorithm for optimizing unstructured mesh partitions. J. High Perform. Comput. Appl. **9**(4), 280–295 (1995)

217. Walshaw, C.: Variable partition inertia: graph repartitioning and load balancing for adaptive meshes. In: Parashar, M., Li, X. (eds.) Advanced Computational Infrastructures for Parallel and Distributed Adaptive Applications, pp. 357–380. Wiley Online Library, Hoboken (2010)

218. Walshaw, C., Cross, M.: Multilevel mesh partitioning for heterogeneous communication networks. Future Gener. Comp. Syst. **17**(5), 601–623 (2001)

219. Walshaw, C., Cross, M., Everett, M.G.: Dynamic load-balancing for parallel adaptive unstructured meshes. In: Proceedings of the 8th SIAM Conference on Parallel Processing for Scientific Computing (PPSC 1997) (1997)

220. Laboratory of Web Algorithms, University of Macedonia: Datasets. http://law.dsi.unimi.it/datasets.php, http://law.dsi.unimi.it/datasets.php

221. Williams, R.D.: Performance of dynamic load balancing algorithms for unstructured mesh calculations. Concurr.: Pract. Exp. **3**(5), 457–481 (1991)

222. Zhou, M., Sahni, O., et al.: Controlling unstructured mesh partitions for massively parallel simulations. SIAM J. Sci. Comput. **32**(6), 3201–3227 (2010)

223. Zumbusch, G.: Parallel Multilevel Methods: Adaptive Mesh Refinement and Load-balancing. Teubner, Stuttgart (2003)

How to Generate Randomized Roundings with Dependencies and How to Derandomize Them

Benjamin Doerr[1] and Magnus Wahlström[2(✉)]

[1] Ecole Polytechnique de Paris, Palaiseau, France
[2] Royal Holloway, University of London, London, UK
Magnus.Wahlstrom@rhul.ac.uk

Abstract. We give a brief survey on how to generate randomized roundings that satisfy certain constraints with probability one and how to
compute roundings of comparable quality deterministically (derandomized randomized roundings). The focus of this treatment of this broad
topic is on how to actually compute these randomized and derandomized roundings and how the different algorithms with similar proven
performance guarantees compare in experiments and the applications
of computing low-discrepancy point sets, low-congestion routing, the
max-coverage problem in hypergraphs, and broadcast scheduling. While
mostly surveying results of the last 5 years, we also give a simple, unified
proof for the correctness of the different dependent randomized rounding
approaches.

1 Introduction

Randomized rounding is a core primitive of randomized algorithmics (see, e.g.,
the corresponding chapter in the textbook [40]). One central application going
back to Raghavan and Thompson [47,48] is to round non-integral solutions of
linear systems to integer ones. By rounding the variables independently, large
deviations bounds of Chernoff-Hoeffding type can be exploited, leading to good
performance guarantees and low rounding errors. This has been successfully
applied to a broad set of algorithmic problems.

More recently, a need for roundings that also satisfy certain hard constraints
was observed. Here, independent randomized rounding performs not so well—
the chance that a single such constraint is satisfied can easily be as low as
$O(1/\sqrt{n})$, where n is the number of variables. Repeatedly generating independent randomized roundings, even for a single constraint and when one is willing to pay an $O(\sqrt{n})$ runtime loss, is surprisingly not admissible as noted by
Srinivasan [56]. Consequently, the better solution is to generate the randomized roundings not independently, but in a way that they immediately satisfy

Work while both authors were affiliated with the Max Planck Institute for
Informatics, Saarbrücken, Germany. Supported by the German Science Foundation
(DFG) through grants DO 749/4-1, DO 749/4-2, and DO 749/4-3 in the priority
programme SPP 1307 "Algorithm Engineering".

© Springer International Publishing AG 2016
L. Kliemann and P. Sanders (Eds.): Algorithm Engineering, LNCS 9220, pp. 159–184, 2016.
DOI: 10.1007/978-3-319-49487-6_5

the desired constraints. This was most successfully done by Srinivasan in his seminal paper [56], who showed a way to generate randomized roundings that satisfy the constraint that the sum of all variables is not changed in the rounding process (provided, of course, that the sum of the original variables is integral).[1] These roundings provably satisfy the same large deviation bounds that were known to hold for independent randomized rounding. This work extended to hard constraints of the bipartite edge weight rounding type in [35,36], however for restricted applications of large deviation bounds. A completely different approach to generating randomized roundings respecting hard constraints was proposed in [16]. It satisfies the same large deviation bounds, hence yields the same guarantees on rounding errors and approximation ratios as the previous approach, but had the additional feature that it could be derandomized easily. Further extensions followed, see, e.g., Chekuri et al. [9,10]. Throughout these works, several applications of the roundings were given, in particular to LP-rounding based approximation algorithms.

The existence of two very different algorithms for this important problem that from the proven performance guarantees look very similar spurred a sequence of algorithm engineering works. While mostly experimental in nature, both concerning test problems and classic algorithmic problems, these works also led to a derandomization of the approach of [36] and to the invention of a hybrid approach (both for the randomized and derandomized setting) combining features of both previous ones. The aim of this work is to survey these results, which currently are spread mostly over several conference papers. By presenting them in a concise and coherent manner, we hope to make these methods easily accessible also to the non-expert. To complete the picture, we also review some applications of the tools to concrete problems (as opposed to studying the roundings in isolation). Furthermore, we also give an elementary and unified proof that all three approaches to generate randomized roundings with cardinality constraints are actually correct. For this, only separate proofs, all quite technical, existed so far.

The field of non-independent randomized rounding and related topics has seen several other breakthrough results in the last years. We mention them here, but for reasons of brevity have to point the reader to the relevant literature. These include the algorithmic breakthroughs for the Lovász local lemma by Moser and Tardos [42,58] and for Spencer's combinatorial discrepancy result [54] by Bansal [5,6], both of which represent efficient algorithms for computing objects whose existence was previously only guaranteed by non-constructive probabilistic methods. There are also several variants of rounding procedures which are out of scope for the present chapter, including the *entropy rounding* method [51], iterative rounding [33,41] and the problem-specific polytope rounding used by Saha and Srinivasan for resource allocation problems [52].

---

[1] Note that some earlier solutions for special cases exist, e.g., for sums of variables adding up to one [47] or the hypergraph discrepancy problem [14,15], which is the rounding problem with all variables being 1/2 and the rounding errors defined by a binary matrix.

Finally, approximation algorithms based on semi-definite programming frequently contain rounding steps which are quite different in nature from the above; see, e.g., [3, 39, 49, 50].

2 Classical Independent Randomized Rounding and Its Derandomization

Randomized rounding as a method to transform non-integral solutions of linear systems into integer ones, was introduced by Raghavan and Thompson [47, 48] already in the late eighties. The key idea is to round the variables to adjacent integers with the probability of rounding upward being equal to the fractional part of the number. By this, the expectation of the new random variable equals the original value. By linearity of expectation, this extends to linear combinations of variables. When each variable is rounded independently, then, in addition, Chernoff-type large deviation inequalities allow to bound deviations from the mean for such linear combinations.

To make things precise, for a number $x \in \mathbb{R}$, we denote by $\lfloor x \rfloor$ its integral part and by $\{x\} = x - \lfloor x \rfloor$ its fractional part. We say that a random variable y is a *randomized rounding* of x when

$$\Pr[y = \lfloor x \rfloor] = 1 - \{x\},$$
$$\Pr[y = \lfloor x \rfloor + 1] = \{x\}.$$

Often, we can assume without loss of generality that $x \in [0, 1]$. In this case, a randomized rounding y of x is one with probability x and zero otherwise. As said already, we have $E[y] = x$ in any case.[2]

For a family $x = (x_1, \ldots, x_n)$ of numbers, we say that $y = (y_1, \ldots, y_n)$ is a randomized rounding of x when each y_j is a randomized rounding of x_j. By linearity of expectation, this implies $E[\sum_{j \in [n]} a_j y_j] = \sum_{j \in [n]} a_j x_j$ for all coefficients $a_j \in \mathbb{R}$.

When thinking of x as a solution of a linear system $Ax = b$, then our aim is to keep the *rounding errors* $(Ay)_i - (Ax)_i$ small. Note first that these rounding errors are independent of the integral part of x, which is why we often assume $x \in [0, 1]^n$. When y is an *independent randomized rounding* of x, that is, the random variables $y_1, \ldots, y_n$ are mutually independent, then the usual Chernoff-Hoeffding large deviation bounds can be used to bound the rounding errors. For example, when $A \in [0, 1]^{m \times n}$ and $\delta \in [0, 1]$, we have

$$\Pr[|(Ay)_i - (Ax)_i| \geq \delta(Ax)_i] \leq 2 \exp(-\delta^2 (Ax)_i / 3).$$

By the union bound, this implies that with constant probability the rounding errors are bounded by $O(\max\{\sqrt{(Ax)_i \log m}, \log m\})$ for all rows i simultaneously.

---

[2] Note that, in fact, $E[y] = x$ and $y \in \{\lfloor x \rfloor, \lceil x \rceil\}$ is equivalent to saying that y is a randomized rounding of x.

Randomized rounding can be *derandomized*, that is, there is a way to deterministically compute roundings that satisfy large deviations bounds essentially
with the same deviations that the randomized version would satisfy with positive probability. This comes, however, at the price of an increased computational
complexity of at least $\Omega(mn)$. Note that Raghavan's derandomization works for
arbitrary $A \in (\mathbb{Q} \cap [0,1])^{m \times n}$ only when we assume that exponentials of rational
numbers can be computed with arbitrary precision, otherwise, e.g., in the RAM
model of computation, it only works for binary A. In [57], a method was given to
obtain the same large deviations also for general A, however in $O(mn^2 \log(mn))$
time. In [17], a method was presented that gives large deviations larger by a
constant factor, but computationally simpler and with a complexity of $O(mn)$
when the entries of A have a finite length binary representation.

3   Randomized Rounding Under Constraints

Before we go into the engineering aspects, let us review the problem of randomized rounding under hard constraints from the perspective of what can theoretically be achieved.

In the general situation, we have a family fractional variables $x = (x_1, \ldots, x_n)$
(where we assume that $x_i \in [0,1]$; see Sect. 2) and (optionally) sets of *hard*
and *soft* constraints that all hold for x, and we wish to generate a randomized
rounding y of x such that all hard constraints hold for y with certainty, while
the violation of the soft constraints is as small as possible (usually achieved via
Chernoff-Hoeffding-style concentration bounds on the latter). Additionally, one
can also consider the derandomization problem, where the goal is to *deterministically* generate a point y satisfying all hard constraints, with bounded violation
of the soft constraints. Depending on the type of hard constraints, this task can
be either impossible (e.g., if the hard constraints are unsatisfiable by an integral
solution), or possible with various restrictions on the supported types of soft
constraints.

Two frameworks for describing hard constraints (in this context) have been
considered in the literature. In the more general, the hard constraints are
described as a *polytope* $P \subseteq [0,1]^n$, where the constraint is $x \in P$. Naturally,
P should be an *integral polytope*, i.e., with all vertices in $\{0,1\}^n$, to guarantee
that integral roundings y exist. The second framework is a special case of this,
where the hard constraints take the form of a collection of *equality* or *cardinality*
constraints $\sum_{i \in S} x_i = \sum_{i \in S} y_i$, $S \subseteq [n] := \{1, \ldots, n\}$ (optionally, we can accept
rounding errors less than 1, if the left hand side is not integral). This perspective
will mostly suffice for the rest of this chapter, but in this section we will need
the polytope perspective.

Let us first observe that there is a certain inevitable tradeoff between the
types of hard and soft constraints. On the one hand, if there are no hard constraints, then one may simply apply independent randomized rounding (or one
of its derandomizations) and get Chernoff-Hoeffding concentration bounds for
any collection of linear soft constraints as in Sect. 2.

On the other hand, we can easily create systems of hard constraints where randomized rounding is possible, but which only allow for extremely limited concentration bounds. Consider an integral polytope P and a fractional point $x \in P$, and let x be expressed as a convex combination $\sum_i \alpha_i p_i$ over vertices p_i of P (i.e., $0 \le \alpha_i \le 1$ for each i, and $\sum_i \alpha_i = 1$); note that such an expression always exists. If we can compute such an expression for any $x \in P$, then we can produce a randomized rounding by simply letting $x = p_i$ with probability α_i for each i. Such a blunt rounding algorithm would in general not allow for any interesting concentration bounds. (See [23] for the corresponding statement in the setting of hard cardinality constraints.) Concretely, we may consider a polytope with only two integral points $(0, 1, 0, 1, \ldots), (1, 0, 1, 0, \ldots) \in [0, 1]^n$; this may also be described via cardinality constraints $(x_i + x_{i+1} = 1)$ for $1 \le i < n$. Given a fractional point $x = (\xi, 1 - \xi, \xi, \ldots)$, we can create a randomized rounding y of x by letting $y = (1, 0, 1, 0, \ldots)$ with probability ξ, and $y = (0, 1, 0, 1, \ldots)$ otherwise. It is clear that this produces a randomized rounding of x, but the procedure only allows for very specific and restricted concentration bounds. To get useful concentration bounds, we must consider weaker classes of hard constraints.

Cases with Complete Negative Correlation. Having seen that when no hard constraints are present, independent randomized rounding allows large deviation inequalities on all variables, a natural question is for which systems of hard constraints in general we can obtain unrestricted large deviation bounds. The standard approach to this is via *negative correlation*. A set of variables $y = \{y_1, \ldots, y_n\} \in \{0, 1\}^n$ are negatively correlated (over all subsets, also referred to as *complete* negative correlation) if, for each $S \subseteq [n]$ and each $b = 0, 1$ it holds that

$$\Pr[\bigwedge_{i \in S} y_i = b] \le \prod_{i \in S} \Pr[y_i = b].$$

Since negative correlation suffices for the classic large deviation bounds to hold [46], the question is which hard constraints allow randomized rounding in a way that the rounded variables are negatively correlated. Chekuri et al. [9] showed that this is possible for every point in a polytope P exactly when P is a kind of *matroid* polytope. Specifically, they show the following.

Theorem 1 ([9]). *Let P be a polytope with vertices in $\{0, 1\}^V$. Then the following two properties are equivalent:*

1. *For any $x \in P$, there exists a probability distribution over vertices of P such that a random vertex y drawn from this distribution satisfies $E[y] = x$ and the coordinates $\{y_i\}_{i \in V}$ are negatively correlated.*
2. *P is a projection of a matroid base polytope, in the sense that there is a matroid $M = (V', \mathcal{I})$ such that $V \subseteq V'$ and p is a vertex of P iff $p = 1_{B \cap V}$ for some base B of M.*

Additionally, their algorithm provides a guarantee of sharp concentration bounds for any *submodular* function; see [9]. While this is a very powerful result, it requires the use of algorithms of forbidding complexity (both in terms of implementation and running time); see Sect. 4.4. Interesting special cases include spanning trees (e.g., [4]) and cardinality constraints. The latter is covered in detail in Sect. 4.

Partial Negative Correlation. To gain more expressive power in the hard constraints, we have to give up some generality for the soft constraints. The first result in this direction was by Gandhi et al. [35,36], who covered the case of edge-rounding in bipartite graphs, with hard cardinality constraints (and negative correlation) over sets of edges incident on a common vertex; see Sect. 5 for details. This was generalized by Chekuri et al. to *matroid intersection* constraints, with negative correlation over subsets of variables corresponding to *equivalence classes* of the matroids; see [9]. Unlike for complete negative correlation, we have no complete characterization for this case (i.e., no "only if" statement corresponding to the second part of Theorem 1).

Further Extensions. Chekuri et al. [10] showed that by relaxing the conditions slightly, one can achieve roundings that are in a sense *almost* randomized roundings (up to a factor $(1 - \epsilon)$ for a given $\epsilon > 0$), which satisfy a set of hard constraints generalizing all cases above, and such that Chernoff-Hoeffding concentration bounds apply for *any* linear function over the variables (i.e., not restricted to certain variable subsets). In particular, they show the following.

Theorem 2. *Let P be either a matroid intersection polytope or a (not necessarily bipartite) graph matching polytope. For every fixed $0 < \varepsilon \leq \frac{1}{2}$, there is an efficient randomized rounding procedure, such that given a point $x \in P$ it outputs a random feasible solution R corresponding to an (integer) vertex of P, such that $E[\sum R] = (1 - \varepsilon) \sum_i x_i$, and such that for any linear function $a(R) = \sum_{i \in R} a_i$, with $a_i \in [0,1]$ and $E[a(R)] = \mu$, exponential concentration bounds apply to $E[a(R)]$. The bounds are dimensionless, i.e., the coefficients in the concentration bounds depend on ε and μ, but not on $n = |x|$.*

See [10] for details. They mention that the results can be further generalized to non-bipartite b-matching. Since cardinality constraints are special cases of bipartite b-matchings, and since both matroid intersection and non-bipartite b-matching are covered by the above result, this result properly generalizes all the above-given results (except for the factor $(1 - \varepsilon)$ and the exact factors involved in the concentration bounds).

4 Disjoint Cardinality Constraints

In this section, we describe methods for generating dependent randomized roundings subject to disjoint hard cardinality constraints and admitting the usual large deviation bounds on arbitrary subsets of the variables (complete negative correlation). This includes a simple proof uniformly showing the correctness of all approaches proposed so far.

To ease the presentation, we shall assume that there is only a single global cardinality constraint, that is, that we do not tolerate any rounding error in the sum of all variables (this implies that we assume that $\sum_{i=1}^{n} x_i$ is integral). It will be immediately clear how to extend all of the below to disjoint cardinality constraints (that is, more than one and possibly not covering all variables), and even to cardinality constraints forming a laminar system (for each two constraints, the two sets of variables concerned are disjoint or one is a subset of the other).

4.1   Algorithms

We now describe three algorithms that have been proposed for generating randomized roundings with a global cardinality constraint [16, 30, 56]. All three approaches (as well as the preliminary works [14, 15]) use the same basic idea of breaking down the rounding process to suitably rounding pairs of variables. We thus first describe this common core, then fill in the details of how each algorithm works. (The rounding algorithm for matroid constraints of Chekuri et al. [9] can also be phrased in this framework, though it is not the perspective taken in [9].)

Pair Roundings. Let (x_i, x_j) be a pair of fractional variables in x. A *pair-rounding step* is to take such a pair (x_i, x_j) and modify their values as follows. Let $\delta^+, \delta^- > 0$ be two values chosen by the respective algorithm, and adjust (x_i, x_j) to $(x_i + \delta, x_j - \delta)$, with $\delta \in \{\delta^+, -\delta^-\}$ chosen randomly so that $E[\delta] = 0$. The values δ^+, δ^- and the choice of the pair (x_i, x_j) vary according to the algorithm; see below. Clearly, each pair-rounding step preserves the sum of all values, keeps x_i in $[0, 1]$ and does not change the expectation of x_i (hence the final y_i is a randomized rounding of x_i). Negative correlation also follows, as shown next.

Theorem 3. *Let $x \in [0, 1]^n$ be rounded by a sequence of pair-rounding steps to some $y \in \{0, 1\}^n$ (with δ^+, δ^- chosen throughout so that all coordinates remain in the range $[0, 1]$). Then $|\sum_i x_i - \sum_i y_i| < 1$ and y is a randomized rounding of x with negative correlation over all sets of coordinates.*

Proof. The first two claims are clear; we need to show that for any $S \subseteq [n]$ we have $\Pr[\bigwedge_{i \in S} y_i = b] \leq \prod_{i \in S} \Pr[y_i = b]$ for $b = 0, 1$. We give a proof via induction over the number of pair-rounding steps. As a base case, assume that no pair-rounding steps are taken. In that case x is integral, $y = x$, and for each choice of S and b, $\Pr[\bigwedge_{i \in S} y_i = b] = \prod_{i \in S}[x_i = b]$; thus the statements hold. For the inductive case, let $S \subseteq [n]$ be an arbitrary set and consider $P := \Pr[\bigwedge_{t \in S} x_t = 1]$. Let (x_i, x_j) be the pair of variables in the first pair-rounding step, and observe $\Pr[\delta = \delta^+] = \frac{\delta^-}{\delta^+ + \delta^-}$. Also let $S' = S \setminus \{x_i, x_j\}$ and $P' = \Pr[\bigwedge_{t \in S'} x_t = 1] \leq \prod_{t \in S'} x_t$, by the inductive hypothesis. Now the statement follows by simple manipulations. If $|S \cap \{x_i, x_j\}| = 1$, say $x_i \in S$, then

$$P = (\frac{\delta^-}{\delta^+ + \delta^-}(x_i + \delta^+) + \frac{\delta^+}{\delta^+ + \delta^-}(x_i - \delta^-))P' = x_i P';$$

if $S = S'$, then $P = P'$ and we are done. Otherwise, we have

$$P = (\frac{\delta^-}{\delta^+ + \delta^-}(x_i + \delta^+)(x_j - \delta^+) + \frac{\delta^+}{\delta^+ + \delta^-}(x_i - \delta^-)(x_j + \delta^-))P'$$
$$= (x_i x_j - \frac{\delta^+(\delta^-)^2}{\delta^+ + \delta^-} - \frac{\delta^-(\delta^+)^2}{\delta^+ + \delta^-})P' \le x_i x_j P'.$$

The case of $\bigwedge_{t \in S} x_t = 0$ is analogous, replacing each x_t by $1 - x_t$.

Srinivasan's Method. In [56], the details of the above scheme are filled in as follows. Let (x_i, x_j) be a pair of fractional variables (chosen arbitrarily), and let $\delta^+ = \min(1 - x_i, x_j)$ and $\delta^- = \min(x_i, 1 - x_j)$. Working through the above description of the choice of δ, we find that $\delta = \delta^+$ with probability $\delta^-/(\delta^+ + \delta^-)$, and $\delta = -\delta^-$ with complementary probability. Observe that in each case, at least one of the new values $x_i + \delta, x_j - \delta$ is integral, and will hence not be chosen for further rounding steps. In particular, this implies that there are only $O(n)$ rounding steps. While the choice of pairs (x_i, x_j) has no impact on the theoretical behavior of the algorithm, in practice it had some importance; see below.

Bitwise Roundings. A different approach to dependent rounding was taken by Doerr [16]. For this method, we must assume that the variables $\{x_1, \ldots, x_n\}$ have finite *bit-depth* ℓ, i.e., that each variable x_i can be written as $c_i \cdot 2^{-\ell}$ for some integers c_i and ℓ. In this case, we round variables as follows. Let (x_i, x_j) be a pair of variables with the least significant bit (LSB) set to 1 (i.e., $c_i \bmod 2 = c_j \bmod 2 = 1$). If no such variables exist, we may rewrite x with a shorter bit-depth $\ell' < \ell$; also note that under the assumption that the total cardinality $\sum_{i=1}^n x_i$ is integral, there cannot be only a single variable x_i with non-zero LSB. We round (x_i, x_j) as described above by letting $\delta = \pm 2^{-\ell}$ with equal probability; note that after this step, both variables will have a LSB of 0, regardless of choice of δ. Hence, after $O(n)$ rounding steps there will be no further variables with non-zero LSB, and we may consider our variables to have a smaller bit-depth $\ell - 1$. After ℓ such phases, and consequently $O(n\ell)$ rounding steps, all variables will be integral.

The advantage and original motivation of this scheme is that the ℓ rounding phases are (arguably) simpler than the previous case, both to implement and to analyze. In [16], each individual rounding phase was performed in a plug-in fashion by the independent randomized rounding method of Raghavan and Thompson, allowing for the first announced derandomized algorithm for this problem. However, later it was observed (in [31]) that the standard method of pessimistic estimators can be applied directly to all these schemes (see below). The complexity of $O(n\ell)$ is noticeably worse than the $O(n)$ of Srinivasan's method, except for variables with small, constant bit-depth, but the approach of bit-wise rounding turned out useful for the more general case of bipartite graphs; see Sect. 5.

A Hybrid Scheme. Motivated by differences observed in running time and solution quality for the case of bipartite graphs (see Sect. 5), a third variant of rounding scheme was considered in [30]. In brief, this variant consists of picking pairs of variables (x_i, x_j) as in bitwise rounding, but picking the adjustment δ as in Srinivasan's method (i.e., so that one of x_i and x_j becomes integral). Observe that this adjustment δ will inevitably have a LSB of 1, implying that in each rounding step, we both fix one variable as integral and decrease the bit-depth of the other. The correctness of this method (in particular the negative correlation) follows directly from Srinivasan [56], as the only technical difference between the two is the choice of variable pairs, which is left unspecified in [56].

4.2 Derandomization

All of the above rounding schemes can be derandomized using the methods developed by Raghavan for classical randomized rounding [47]. Let us outline how these methods work for the independent rounding case before we review how they can be adapted to cases of dependent rounding.

The first ingredient is known as *method of conditional probabilities* [32, 47, 55]. Let x be as above, and let $P(x)$ be the probability of some undesirable event, e.g., the probability that an independent randomized rounding y of x has a rounding error larger than some bound μ. Assume that P can be efficiently computed, and that $P(x) < 1$. We can then produce a rounding y of x by iteratively rounding each variable x_i in turn, at each step picking the value $y_i \in \{0, 1\}$ that minimizes $P(\cdot)$. Let x' resp. x'' be x modified as $x_i \leftarrow 1$ resp. $x_i \leftarrow 0$. Then $P(x) = x_i P(x') + (1 - x_i) P(x'')$, as $P(x')$ and $P(x'')$ are simply the conditional probabilities of failure given x_i. Hence $\min(P(x'), P(x'')) \leq P(x) < 1$, and we maintain the invariant that $P(x^*) < 1$ for every generated point x^*. By induction we have $P(y) < 1$, where y is the final rounding of x generated this way, and since y is integral we conclude that $P(y) = 0$, e.g., y produces a rounding error less than our bound μ, and we are done.

To extend this to cases where $P(x)$ is unknown or too expensive to compute (as is the case for rounding errors in a linear system $Ax = b$), we may use a *pessimistic estimator* $F(x)$ in place of $P(x)$. Such an estimator is an efficiently computable function $F(x)$ such that $F(x) \geq P(x)$ for all x, $F(x) < 1$ for the initial point x, and for every two modifications x', x'' of a point x as above, $\min(F(x'), F(x'')) \leq F(x)$. By identical arguments as above, using a pessimistic estimator $F(x)$ in place of the probability $P(x)$, we may deterministically produce a rounding y of x which satisfies our condition. The art, or course, is finding such pessimistic estimators. Raghavan [47] showed that certain technical expressions occurring in the proof of Chernoff-Hoeffding bounds are pessimistic estimators. This has the advantage that they can applied to systems $Ax = b$ of soft linear constraints whenever the corresponding Chernoff-Hoeffding bound shows that with positive probability a solution with a certain rounding error exists; see Sect. 2, and [47] for details.

To adapt the above to the dependent cases, we proceed as follows. Let x be a point, and consider a pair-rounding step on variables x_i, x_j. Recall that here we adjust $x \leftarrow x + \delta(e_i - e_j)$ for some $\delta \in \{\delta^+, \delta^-\}$. Let $F(x)$ be the above pessimistic estimator, and define $f(\delta) = F(x + \delta(e_i - e_j))$. It was shown in [31] that $f(\delta)$ is a concave function, meaning that for any pair $\delta^+, \delta^- \geq 0$, at least one of the values $f(\delta^+), f(-\delta^-)$ is at most $F(x)$. We may now proceed greedily, as above, at every pair-rounding step selecting that value of δ which minimizes $F(x)$. As before, this can be done in $O(mn)$ time, for n variables and m soft constraints. (Similarly to Theorem 3, this can be used to derandomize any pair-rounding-based algorithm with the same guarantee for the rounding errors.)

Historically, the derandomization of the bit-wise method progressed through several generations, from the initial derandomization in [16] with significantly worse constant factors in the rounding error guarantees, via a partial improvement given in [31], until the general form of the above method was realized [30].

In practice, though the pessimistic estimators are far from perfect (e.g., due to the use of a union bound), the greedy aspect of the derandomization process makes for a powerful heuristic, as points with smaller value $F(x)$ also tend to lead to smaller rounding errors. Although the theoretical guarantees for the resulting rounding error are comparable to the expected outcome of a randomized process, in applications and experiments we repeatedly find that derandomized, greedy methods significantly outperform randomized ones. (See the experiments in this section for more.)

Implementation Notes. A few potential issues suggest themselves with respect to implementation of the above. The first is the source of randomness for the randomized methods. While we did not have access to a "real" (hardware) randomness source, we found no indication in our experiments that the choice of pseudo-random number generator would have a very powerful impact on the results. The second potential issue lies in the use of floating-point arithmetics. As noted in Sect. 2, exact computation of pessimistic estimators is only possible in the Real RAM model, and alternatives for the standard model are very costly. Instead, our implementation (as is usual) uses CPU-native floating point arithmetics. While this "usually" works "reasonably" well, there are potential issues of accumulated imprecision (in particular since the pessimistic estimators become repeatedly adjusted throughout the process). However, in experiments we found no indication of such problems within the scope of this and the next section.

4.3   Experimental Evaluations

Seemingly, not much experimental work on randomized rounding, even without constraints, is published. Consequently, we include independent randomized rounding in the following summary of experimental comparisons of the above algorithms. They are mainly from [31], augmented by [29,30], which relate to later developments (the hybrid rounding scheme and improvements in derandomization). For all details, we refer to these papers. All conclusions below are supported for the latest versions of the respective programs (unless explicitly stated otherwise). All experiments reported below use inputs of full bit-depth.

First, regarding the running time and general program feasibility (code complexity and numerical stability), the conclusions are generally positive; the randomized versions of independent rounding and Srinivasan's method rounded

1,000,000 variables in 0.05–0.14 s, with the bit-wise method being slower at approximately one second. For the derandomized versions, rounding 10,000 variables subject to 10,000 soft constraints took 52 s for independent rounding, 75 s for Srinivasan's method, and in excess of ten minutes with bit-wise rounding. Later engineering of the code base reduced these times, eventually allowing derandomization instances for a geometric discrepancy problem with $2^{15} = 32,768$ variables and an equal number of soft constraints to be rounded in 37 s (with special-purpose code) [28]; see Sect. 6.1. No issues of numerical stability were encountered. The hybrid method was not tested here, but based on the results in [30] there is no reason to expect that the outcome would be noticeably different from the other applications of Srinivasan's method tested here.

Next, we consider solution quality (i.e., rounding errors). All considered methods have identical theoretical concentration bounds (in the randomized case) respectively identical theoretical upper bounds (in the derandomized case), including the classical independent, non-constraint-preserving roundings. For the bit-wise method, as noted above, the derandomization used in [31] had a worse constant factor than the latest versions, thus we focus first on the other methods. Taking the performance of independent randomized rounding as a reference (100%), the experiments of [31] showed that adding a cardinality constraint led to no worse rounding errors, and in some cases to a reduction of rounding errors if the soft constraints have large cardinality (e.g., on instances with a dense random matrix of soft constraints, a hard cardinality constraint reduced rounding errors by 15%). No clear difference between the dependent randomized methods was found. Using a derandomization reduced the rounding error by approximately 50% on random instances; more on structured instances stemming from experiments reported in Sect. 6.1. Comparing the independent derandomized rounding with Srinivasan's method revealed no clear difference, though perhaps an advantage for Srinivasan's method of a few percent. In particular, there seemed to be no significant "price of hard constraints" in terms of solution quality. All algorithms outperformed their theoretical bounds on rounding error by a factor of 2–3 (presumably due to the latter's use of union bounds).

This data supports the general expectation that derandomized methods produce significantly smaller rounding errors, a conclusion that was consistently arrived at in all our experiments. This advantage persisted when compared to generating a large number of random solutions and keeping the best one (note that computing the rounding error requires $O(nm)$ time).

Finally, regarding the derandomized bit-wise method, the version used in [31] performed worse than the other two (with rounding errors at 55–65% of those of randomized rounding). Experiments in [29,30] (see later) using newer versions tend to confirm a (modest) advantage of the derandomization of Srinivasan's method over that of the bit-wise method, though we have no good explanation for this. However, we did find that particular combinations of soft constraints and order of variable comparison led to very poor quality solutions for Srinivasan's method; see [31] regarding tree shape (but note that in later investigations, the effect has been found to be less general than originally implied). In this respect, the hybrid method (viewed as a tweak on Srinivasan's method) always creates the balanced tree shape recommended in [31].

4.4 Extension: Matroid Constraints

As noted in Sect. 3, the above methods can be extended to the setting of *matroid* constraints [9]. Matroids are powerful objects, whose usage unifies many results from combinatorics (see, e.g., [45,53]); hence this extension is a powerful result. However, it comes with a significant impact to practicality. While the algorithm of [9] is reasonable (being combinatorial in nature), it works on the basis of a *decomposition* of the input point $x \in \mathbb{R}^n$ into a convex combination of matroid bases, and such a decomposition is difficult to obtain, both in terms of computational complexity (the best bound for the general case being $O(n^6)$ time [11]) and in terms of implementation difficulty. Using a more traditional pair-rounding approach (in line with the algorithms of Sect. 4.1; see [8]) would save us from having to provide an explicit decomposition, but instead requires the ability to test the *fractional membership* of a point x in the corresponding polytope; for the general case, this is again as difficult as obtaining a decomposition [11]. (Chekuri et al. [9] note that in some applications, such a decomposition is provided along with the point x, in which case these objections do not apply.)

One particularly interesting special case of this result are *spanning tree* constraints, i.e., creating a random spanning tree for a graph according to some given edge probabilities. This was used in the breakthrough $O(\log n/ \log \log n)$-approximation result for Asymmetric TSP of Asadpour et al. [4] (although [4] used the heavy machinery of maximum entropy sampling). However, the cost of the above-noted primitives for the spanning tree polytope is still non-trivial, e.g., decomposition requires $O(n^2)$ calls to a max-flow algorithm [34]. The best present bound for max-flow is $O(nm)$ time due to Orlin [44].

5   Pipage Rounding: Rounding in Bipartite Graphs

In this section, we move on to a more general constraint type, which can be described either as rounding fractional edges in a bipartite graph (e.g., *bipartite b-matching*) or as rounding a fractional point x subject to membership in the *assignment polytope*. We will employ the bipartite graph perspective. Dependent randomized roundings for this setting were provided by Gandhi et al. [35,36] and by Doerr [16]. In this setting, the hard constraints are represented by a bipartite graph $G = (U \cup V, E)$, with edge set $E = \{e_1, \ldots, e_m\}$, and with one variable $x_i \in [0,1]$ for every edge e_i. Let $x = (x_1, \ldots, x_m)$ denote the family of variables, and for a vertex $w \in U \cup V$ let $\delta(w)$ denote the edges incident on w. The hard constraints that we must observe are then to preserve the values of $\sum_{e_i \in \delta(w)} x_i$, for all vertices $w \in U \cup V$; in other words, we are rounding the values of the fractional edges e_i, subject to preserving the *fractional degree* in every vertex. We refer to these constraints as *vertex constraints* $\delta(w)$. For ease of presentation, we assume that each vertex constraint $\delta(w)$ is integral, by adding two dummy vertices u_0, v_0 and up to $|U| + |V| + 1$ dummy edges [30].

We will not be able to guarantee complete negative correlation, as may be realized by considering, e.g., an even cycle C_{2n} of $\frac{1}{2}$-edges. There are exactly two rounded solutions for this instance, and if the edges are numbered $e_1, e_2, \ldots, e_{2n}$ in order, in each solution we have $x_1 = x_3 = \ldots = x_{2n-1}$ and $x_2 = x_4 = \ldots = x_{2n}$. However, the above results show that one can generate roundings subject to the above, with negative correlation within all subsets $\delta(w)$ for $w \in U \cup V$. We will review the above algorithms, and recall the conclusions of some experiments, both in general terms and for particular applications. We also briefly report on theoretical results that extend the above situation, again to a matroid setting.

5.1 Algorithms

As in Sect. 4, three different rounding schemes are available for solving the problem, corresponding to the three schemes of Sect. 4.1, with a common algorithmic core referred to as *pipage rounding*. Thus we first describe this common core, then review how each rounding scheme can be applied to it.

Pipage Rounding. The common principle behind these algorithms is the idea of *pipage rounding*, due to Ageev and Sviridenko [1]. Let $C \subseteq E$ be a set of edges that induce a simple cycle in E; w.l.o.g. assume that $C = \{e_1, \ldots, e_{2t}\}$, numbered in order along the cycle C, and let $x_C = \{x_1, \ldots, x_{2t}\}$ be the corresponding set of variables. We will perform a pair-rounding step at every vertex incident to C, similarly as in Sect. 4, but this time, in order to maintain all cardinality constraints $\delta(w)$ the adjustments need to *cascade*. Concretely, for some $\delta \in (-1, 1)$ we will adjust the values x_i for all edges $e_i \in C$ so that $x_{2i-1} \leftarrow x_{2i-1} + \delta$ and $x_{2i} \leftarrow x_{2i} - \delta$; the adjustment δ is chosen randomly as $\delta \in \{\delta^+, -\delta^-\}$ with $E[\delta] = 0$. The choice of δ^+, δ^- and the cycle C is algorithm-specific. We refer to such an adjustment as a *pipage rounding step*. We review the effects of applying the various rounding schemes to this outline. Note that when considering $\delta(w)$ in isolation, $w \in U \cup V$, the above scheme acts exactly like a pair-rounding algorithm, implying both negative correlation and derandomization as in Sect. 4.

Gandhi et al. In [35,36], the details are chosen much as in Srinivasan's method for pair-rounding. That is, δ^+ and δ^- are defined as the largest values such that using an adjustment of $\delta = \delta^+$ (resp. $\delta = -\delta^-$) leaves all variables $x_i \in [0,1]$; necessarily, at least one variable x_i must become integral in such an adjustment. Each time, the cycle C is chosen arbitrarily among the edges that still have fractional values. (By the integrality of each vertex constraint, no vertex is incident to exactly one fractional edge, hence such a cycle always exists.) We get an upper bound of $O(m)$ pipage rounding steps; as each step may involve a cycle C of $O(n)$ edges, the total running time (pessimistically) becomes $O(nm)$. (A better bound may be $O(mp)$, where p is the average cycle length, but this is hard to estimate theoretically.)

Bit-Wise. In [16], the bit-wise rounding scheme is applied to the above. Concretely, we assume that each variable x_i has a finite bit-depth of ℓ. Let E_ℓ be the set of all edges e_i whose corresponding variables x_i have an LSB of 1. By the integrality of the vertex constraints, these edges E_ℓ form an even graph, which hence decomposes into cycles. Let C be an arbitrary cycle in the graph formed by E_ℓ, and pick $\delta = \pm 2^{-\ell}$ uniformly at random. Now note that *every* edge $e_i \in C$ gets an LSB of 0 after such an adjustment, hence for each bit level, each edge is adjusted at most once. Consequently, the total work over all pipage rounding steps is simply $O(m\ell)$, which compares favorably to the bound $O(mn)$ for the previous method.

Hybrid. Finally, we consider the hybrid method, developed in [30] for this case. Here, the cycle C is chosen as in the bit-wise method, while the values δ^+, δ^- are chosen as in Gandhi et al. Again we find that both δ^+ and δ^- have an LSB of 1, hence after each pipage rounding step, we both decrease the bit-depth of all adjusted edges and make at least one edge variable integral. In practice, this method proves to be faster than both the above methods (see below).

Derandomization and Implementation Notes. All these algorithms can be derandomized, in the same sense as in Sect. 4, assuming that each soft constraint has its support contained in $\delta(w)$ for some vertex w; as noted above, we may simply use the same pessimistic estimators as in Sect. 4.2. As for implementation notes, we found that there is some undesirable interaction between the derandomization and the hard constraints, if one uses inexact (e.g., floating-point) arithmetics. In many applications we may have hard cardinality constraints which make the solutions to the problem more costly (e.g., the limits on broadcasting in Sect. 6.3). If a cardinality constraint of, say, 2 is misinterpreted as a fractional constraint of $2 + \varepsilon$ (where $\varepsilon > 0$ is some small number stemming from floating point imprecision), then a greedy derandomized algorithm may opt to "round" this constraint up to 3. (Note that this would be extremely unlikely for a randomized algorithm.) Based on our experiences in [30], rather than trying to "plug" such holes in an ad-hoc manner, we recommend to transform the input to fixed precision (say, 64-bit integers) before commencing any rounding. (This 64-bit data will need to be cleaned up for integrality after the transformation, but this only has to be done once.)

5.2   Experimental Evaluations

We now report briefly on conclusions from [30] on random bipartite graphs, regarding running time and solution quality; an application to broadcast scheduling is covered in Sect. 6.3.

As noted above, the method of Gandhi et al. needs $O(mp)$ time where p is the average cycle length, while the bit-wise method needs $O(m\ell)$ time where ℓ is the bit-depth (and the hybrid method needs the smaller of these). The derandomized versions incur an additional cost due to the need to update pessimistic estimators; with the implementation used in [30], this cost is roughly proportional to the total number of pessimistic estimators for all methods. (With a more advanced implementation, it should be possible to reduce this factor to the number of *affected* pessimistic estimators, which should further boost the differences in running times; however, this was not attempted.) Naturally, the value of p depends on the graph structure; in experiments with random 5-regular bipartite graphs with n vertices, we found that the total number of edge visits for the method of Gandhi et al. scaled as $O(n^{1.37})$, while it remained linear for the two other methods. (The difference in running time scaled proportionally to this.) In concrete numbers, for 5-regular graphs on 1000 vertices, the average running times for the derandomized versions were 14.6 s, 10.2 s, resp. 8 s for the method of Gandhi et al., bit-wise, resp. the hybrid method; for 20-regular graphs on 1000 vertices the times were 109 s, 67 s, resp. 65 s; and for random graphs with $m = 20,000$ and $n = 400$, the times were 56 s, 46 s, resp. 33 s. In other words, for this range of m, the order of Gandhi - bit-wise - hybrid is stable, with a total gap of roughly factor of two. The randomized methods were roughly two orders of magnitude faster on these instances, which though noteworthy is a less drastic difference than in Sect. 4. In terms of rounding error, the general order was that the method of Gandhi et al. produced smaller rounding errors, and the bit-wise method larger errors, e.g., for the random graphs with $m = 20,000$ and $n = 400$, the average rounding errors were 4.38 for Gandhi et al., 6.09 for the bit-wise method, and 5.43 for the hybrid method. However, in the application experiments (reported in Sect. 6.3), this order was not preserved (there, instead, the hybrid was both fastest and produced the best-quality solutions). The randomized methods again produced rounding errors similar to each other, up to twice as large as the derandomized methods.

5.3 Extension: Matroid Intersection

As noted in Sect. 3, there is a far-reaching generalization of the above into so-called *matroid intersection* constraints [9]. The algorithm, as in Sect. 4.4, is based on a convex decomposition of the input point x, though the individual steps are more complicated (being pipage rounding steps rather than simple pair-rounding steps). However, the extent of negative correlation is more limited (covering vertex constraints as above, but perhaps not much beyond this). In a further extension, the authors also produce "approximate roundings" for settings including matroid intersection; see Sect. 3 and [10]. (An interesting future question is how the concentration bounds of these "approximate" roundings play out in practice against other methods.) Naturally, the complexity drawbacks reported in Sect. 4.4 apply equally strongly here.

6   Some Applications

To get a feeling for the behavior of the algorithms "in practice," we now review some work on applying the above methods to (real or artificial) instances of concrete optimization problems. We cover three topics: Low-discrepancy pointsets (in Sect. 6.1), routing and covering problems (in Sect. 6.2), and problems of broadcast scheduling (in Sect. 6.3). These represent various areas where methods of dependent randomized rounding have been proposed for approximation algorithms. Additionally, we report on advances in randomized rumor spreading (Sect. 6.4), which not only has a natural interpretation as rounding problem, but moreover turned out to be a very useful test-case to investigate the influence of additional dependencies in the random experiment.

6.1 Low-Discrepancy Point Sets

The first application is from the area of geometric discrepancy. Specifically, we consider the following discrepancy problem. Let P be a set on n points in $[0, 1]^d$. The L^∞-*star discrepancy* of P (*star discrepancy* for short) is defined as $d_\infty^*(P) = \sup_{x \in [0,1]^d} |\frac{1}{n}|T \cap [0, x[| - \text{vol}[0, x[|$, where $[0, x[$ is the d-dimensional half-open box $[0, x_1[\times \ldots \times [0, x_d[$; that is, the star discrepancy is defined with respect to a range space of boxes anchored in 0. Our task is to create a point set P for given values of n and d, with $d_\infty^*(P)$ as small as possible. Such point sets have important applications in numerical integration, in particular in financial mathematics; see, e.g., [43]. In [22], a randomized method is proposed for this problem. They define a subdivision of $[0, 1]^d$ into a (non-regular) *grid* of k^d grid *boxes*, for some $k = k(d, n)$ chosen by the algorithm. The point sets P are then created in a two-stage process, where in the first stage it is decided how many points each grid box should contain, and in the second stage the placement of these points inside the boxes are chosen. The first stage naturally corresponds to a dependent rounding problem, with one variable x_i for each box B_i of the grid, with initial value $x_i = n \cdot \text{vol}(B_i)$, and with a hard constraint $\sum_i x_i = n$. Note that this can be done in a randomized or derandomized manner, as the range space above reduces to the set of all k^d corner-anchored boxes one can form from the grid, each of which can be treated as a soft constraint. This leaves the design of the grid, for which see [22], and the choice of the parameter $k = k(d, n)$. In [22], k was chosen so as to balance the contribution to the discrepancy of the two stages, based on theoretical bounds on the rounding and placement errors. The resulting bounds were superior to previous work for a domain of intermediate dimension d and relatively few points n. (For a related method, see [26,27].)

In [28], this was tried out experimentally. In line with the experiences reported in Sect. 4.3, the immediate experiences were that contrary to theory, there was a very big difference in rounding error between the derandomized and randomized methods, and even the randomized methods produced rounding errors significantly smaller than the union bound-based theoretical bound. On tests with $d = 7$, $n \approx 150$, the median rounding error (using Srinivasan's method) was 0.026, while the median star discrepancy was 0.139, i.e., five times larger; tests with $d = 9$, $n \approx 90$ revealed even larger differences. Therefore, the recommendation would be to pick a larger k than $k(n, d)$, perhaps up to the limits formed by the growth of the number of variables k^d.

To enable larger tests, the derandomization code was carefully engineered, and special-purpose code was written for the case $k = 2$, enabling derandomization of instances with $k = 2$, $d = 20$ (with over a million variables and soft constraints), taking 10.5 h to compute. However, the investigation was hindered by the difficulty of computing discrepancies. Note that the formula for $d_\infty^*(P)$ discretizes into $O(n^d)$ tests, which is highly impractical. Unfortunately, though improvements exist [13,38], no practical method for upper-bounding $d_\infty^*(P)$ for larger d is known, and it is now known that computing $d_\infty^*(P)$ in time $O(n^{o(d)})$ would contradict certain complexity-theoretical assumptions [37]. Therefore, the final conclusions of the experiments remain tentative.

6.2 Routing and Covering Problems

We now move on to more typical approximation applications of randomized rounding. We briefly review two applications, namely a low-congestion routing problem and max coverage in hypergraphs; the results are taken from [29]. Both problems are classical examples of the randomized rounding approach to approximation algorithms, involving solving an LP-relaxation of the problem, rounding it randomly to an integral solution, and using probabilistic arguments to show that this integral solution is with good probability a good approximation. For both problems, we find that the proposed algorithms work well in practice, but furthermore we find that various heuristic improvements can be used to improve the result quite significantly. We cover each problem in turn.

Low-Congestion Routing. The routing problem we considered is the following variant. We are given a (directed) network $G = (V, E)$ and a set of k routing requests consisting of a source vertex s_i, a sink vertex t_i, and a demand r_i. The objective is to find a set of paths (i.e., an integral multi-commodity flow) such that for each $i \in [k]$, r_i units of commodity i are routed from s_i to t_i. The optimization goal is to minimize the congestion, i.e., the maximum total amount of flow over any single edge. This variant, where every r_i is an integer, is also called *integer-splittable flow*. Raghavan and Thompson [48] gave an approximation algorithm for the case where all demands are 1, using randomized rounding; Srinivasan [56] extended this to integer demands using dependent rounding. The algorithm works as follows. First, we solve an LP relaxation of the problem; this is simply the usual multi-commodity flow LP, with km variables $x_{i,e}$ for $i \in [k]$, $e \in E$, and with a flow-conservation constraint for each commodity i and each vertex v (properly adjusted for the vertices s_i and t_i). The LP is solved with an optimization goal of minimizing the congestion $C = \max_{e \in E} \sum_{i \in [k]} x_{i,e}$ (note that this is easily implemented in an LP, using C as an extra variable). This gives a set of fractional variables $x_{i,e}$ which will be the basis for the solution. Next, we apply *path stripping* to these variables, creating for each $i \in [k]$ a finite set $\mathcal{P}_i$ of fractional paths from s_i to t_i, of total weight r_i. For each such fractional path P, we let $x(P)$ denote the weight that P was given in the path stripping; hence $\sum_{P \in \mathcal{P}_i} x(P) = r_i$ for each $i \in [k]$. Finally, we *round* these fractional weights $x(P)$ to integers $y(P)$, using $\sum_{P \in \mathcal{P}_i} x(P)$, $i \in [k]$ as a collection of (disjoint) hard cardinality constraints. Effectively, this means that we select for each $i \in [k]$ a collection of integral paths from s_i to t_i, of total weight r_i, creating a solution for the problem. As for the congestion, observe that for every $e \in E$ the congestion over e is simply $\sum_{P : e \in P} x(P)$, i.e., the congestion can be treated as a set of m soft linear constraints, and as long as the congestion C of the LP solution is not too low, large deviation bounds imply that the congestion of the integral solution y is on expectation close to C (e.g., the expected congestion is $C + O(\sqrt{C \log m})$ if $C > \log m$). The derandomized version of this works out of the box with pessimistic estimators, giving the same approximation guarantee.

So far the theory; let us now focus on the algorithm engineering perspective. In [29], the above was tested with the following setup. As instances, we used $n \times n$-size bidirectional grids with randomly generated requests, with demands around $r_i = 3$ (the details of placements of requests (s_i, t_i) and choice of r_i were not found to affect the outcome too much). In the experiments, we found that the proposed algorithms generally work well and tend to perform better than the theoretical upper bound, and that (once again) the derandomized versions clearly outperform the randomized versions by a factor of two or more (in terms of rounding error relative to the fractional solution). Although the derandomized versions were slower than the randomized ones, both running times were completely dwarfed by the time required to solve the LP, making a strong recommendation towards using the derandomizations. Inspired by inspection of the structure of the fractional and integral solutions, we also proposed a heuristic modification of the LP, intended to produce fractional solutions that are easier to round (e.g., with fewer edges at maximum initial congestion C); see [29] for details. This modification, while making the LP still more expensive to solve, paid off in terms of still further reduced rounding errors for all methods. Concretely, for a 15×15 grid with 75 requests, with the basic LP the randomized methods produced over 40% overhead and the derandomizations produced 25.8% (bitwise) resp. 18.2% (Srinivasan) overhead; with the improvement, the numbers were 28.5% (randomized) resp. 12.9% (derandomized Srinivasan).

Max Coverage. We next consider a different problem. Let $\mathcal{S} = \{S_1, \ldots, S_n\}$ be a collection of subsets of some ground set U; without loss of generality, let $U = [m]$. In the basic version of the problem, we are given a budget L, and asked to select L sets of $\mathcal{S}$ to maximize the size of the union; in the weighted version, we additionally have a cost c_i associated with every set S_i, and a profit w_i associated with every element $i \in U$ of the ground set. Two algorithms have been proposed for this problem, both with approximation guarantees of $(1 - 1/e)$; unless $P = NP$, this is also the best possible approximation guarantee. The first algorithm is the *greedy* algorithm, which (for the basic variant) simply repeatedly selects the set S_i which would cover the largest number of so-far uncovered elements. To adapt this for the weighted case, we may instead select the set S_i to maximise the profit/cost ratio. (To guarantee a good approximation ratio for the weighted case, special care must be taken with sets of very large cost; however, the instances in our experiments did not include any such sets.) The second algorithm is an *LP-rounding* algorithm, which solves the natural LP-relaxation of the problem and produces a solution by dependent rounding, developed in [1,56]. Concretely, the LP has variables x_i, $i \in [n]$, signifying that set S_i is taken, and e_i, $i \in U$, signifying that element i is covered; the constraints are the budget constraint $\sum_i c_i x_i \leq L$, and coverage constraints $e_i \leq \sum_{j : i \in S_j} x_j$

for each $i \in U$; the optimization goal is $\max \sum_i w_i e_i$. For the basic variant, for the rounding case we can simply treat the values x_i as values to be rounded, subject to a hard constraint $\sum_i x_i = L$. Let $y \in \{0, 1\}^n$ be the rounded version of x. For a single element $i \in U$, the probability that i is covered by y equals

$$1 - \Pr[\bigwedge_{j : i \in S_j} (y_j = 0)] \geq 1 - \prod_{j : i \in S_j} \Pr[y_j = 0] = 1 - \prod_{j : i \in S_j} (1 - y_j);$$

the inequality is due to the negative correlation of the rounding. By standard methods, it follows that $\Pr[e_i \text{ is covered}] \geq (1 - 1/e)y_i$, and the approximation ratio follows by linearity of expectation. To adapt this for the weighted case, we must replace the simple cardinality constraint by a *weighted constraint* $\sum_i c_i x_i \leq L + \delta$, where δ is a noise term bounded by the cost of the most expensive set. Such budget-preserving roundings are given in [29, 35]. For derandomization, we simply use as guide the function

$$F(x) = \sum_{i \in U} w_i (1 - \prod_{j : i \in S_j} (1 - y_j)),$$

representing the expected solution value for an independent randomized rounding of x. With similar arguments as in Sect. 4.1 regarding concavity, we can show that it is possible to keep the value of $F(x)$ non-decreasing during the rounding process, meaning that we end up with a final profit of $F(y) \geq F(x) \geq (1 - 1/e) \sum_i w_i e_i$. For details, see [29]. Again, some further care must be taken if there are sets of very large cost, but we will ignore this aspect.

In [29], we implemented the above, and tested it on instances adapted from facility location problems. As before, the results were generally positive; all algorithms listed above performed well, and produced good-quality outputs (at least compared to the theoretical guarantees), derandomized rounding significantly outperformed randomized rounding, and the running time requirement for solving the LP generally dominated all other computation steps. (Observe that the expected outcome for a randomized rounding is simply $F(x)$. An alternate randomized strategy of producing 1000 random outputs and keeping the best solution was tried, but found to be both slower than derandomized rounding and producing worse-quality solutions.) However, it was also found that the greedy algorithm frequently produced very high-quality solutions, and was partially complementary to the LP-rounding algorithm. For this reason, we considered several ways of incorporating further greedy aspects into the LP-rounding algorithm. In particular, we complemented the pair-rounding step with a greedy selection of the particular pair (x_i, x_j) of variables to round (i.e., we make that adjustment $(x_i, x_j) \leftarrow (x_i + \delta, x_j - \delta)$ which locally best improves the value of $F(x)$). Thanks to the nature of the function $F(x)$, this decision may be taken based on the *gradients* $\partial F(x)/\partial x_i$, which in turn can be computed and updated efficiently; see [29] for details. In experiments, this gradient-based rounding performed impressively well, outperforming both the greedy and basic LP-rounding algorithms, at no significant practical cost to the running time. Concretely, for the instance br818-400 (see [29]), the initial $F(x) = 22, 157,$

greedy achieves 28054, derandomized Srinivasan 27, 397, and the gradient derandomization 28, 448. Optimum was found via exhaustive search to be 28, 709. We also considered an alternate way of combining greedy and LP-rounding, of *seeding* the LP-algorithm by using some fraction εL of the budget for greedy pre-selection before solving the remaining instance as above; this was found in some cases to further improve solution quality.

6.3   Broadcast Scheduling

Finally, we briefly cover some problems in broadcast scheduling, where the bipartite edge roundings of Sect. 5 have found approximation applications; this material is from [30,36]. In this problem, a server has a set of pages P and a set of requests for pages; each request has an issue date, a weight, and optionally a deadline after which it cannot be satisfied. Time is divided into discrete slots, and the server is able to transmit (broadcast) one page (or a bounded number of pages) per time slot; each broadcast satisfies all requests that are live at the time (i.e., between issue date and deadline). Different optimization goals are possible; we will focus mostly on the max throughput case, where the goal is to maximize the weight of the satisfied requests. We are considering the offline setting, i.e., all requests are known in advance.

The reduction of the max throughput problem into bipartite edge rounding goes as follows. We first solve an LP-relaxation of the problem, creating a fractional schedule x_t^p; here a variable x_t^p, for a time slot t and page $p \in P$, represents the decision that page p is to be broadcast at time slot t. These are constrained so that $\sum_{p \in P} x_t^p = 1$ for each time slot t. (The LP formulation also contains further variables to keep track of the optimization goal. Concretely, for each request i we have a variable r_i signifying that the request is satisfied; if request i is for page p_i to be transmitted between time slots t_i and d_i, then we have a constraint $r_i \leq \sum_{t=t_i}^{d_i} x_t^{p_i}$. The goal is $\max \sum_i w_i r_i$ where w_i is the weight of request i.) This fractional schedule is then converted into a bipartite graph as follows. First, for every page p, the transmissions $\{t : x_t^p > 0\}$ of p are split into consecutive *windows* W_j^p, such that in each window except the first and the last, the page p is transmitted exactly once by the schedule. (Note that to achieve this, we may have to split a single transmission x_t^p between two windows W_j^p, W_{j+1}^p.) Call those windows which are not the first or the last window *internal* windows. The amount by which p is broadcast during its first window W_1^p is decided by a random variable z. Next, we form a graph $G = (U \cup V, E)$, where U contains one vertex u_t for each time slot t, and V contains one vertex $v_{j,p}$ for each window W_j^p formed above; the edges E connect the windows and the time slots in the natural way, with values based on x_t^p. In summary, this gives us a weighted bipartite graph where every vertex representing a time slot or an internal window is incident to edges of weight exactly 1. The first window of each page is incident to edges of weight exactly z, where $z \in (0, 1]$ is a *page shift* value chosen uniformly at random. This graph G is then put through the rounding procedure, and in the end this produces a schedule where each time slot is incident to exactly one integral edge, and each page is transmitted once for each internal window (and possibly up to twice more). It is shown in [36] that each request satisfied in the LP-solution has a chance of at least 3/4 to be satisfied in the integral solution, leading in particular to a 3/4-approximation assuming that there is a fractional solution which satisfies all requests.

In [30], this algorithm was implemented and tested, on broadcast scheduling instances derived from Wikipedia access logs (see the paper for details). We complemented the randomized algorithm above with a simple greedy algorithm and a derandomization of the above. The greedy algorithm simply proceeds time slot by time slot, and in each time slot broadcasts that page which would satisfy the greatest number of remaining requests. For the derandomization, observe that there are two randomized aspects to the above algorithm, namely the choice of shifts and the decisions in the pipage rounding. The latter can be derandomized via ad-hoc pessimistic estimators; the former can be derandomized by selecting for each page p that value z_p which maximizes the sum of its associated pessimistic estimators. In the experiments, we found that the greedy algorithm performed the worst (unlike in Sect. 6.2, where it was quite competitive), and that the two aspects of the derandomization (choice of z and using derandomized pipage rounding) both strongly improved the solution quality. Concretely, for the larger instance tested in [30], the greedy algorithm and the randomized algorithms all achieve a value of 24.6, while derandomizing both aspects gives value 26.6 (bitwise), 27 (Gandhi et al.), resp. 27.3 (hybrid). The LP has a value of 27.5. (However, it should be noted that in all derandomized versions, the original "fairness" condition that each request has a 3/4 chance of being satisfied naturally no longer holds.)

We also tested the goal of minimum average delay, which was also covered in [36]. However, for this goal, the LP-rounding approach does not seem to be warranted, as the greedy algorithm was found to be both much faster and to produce better solutions.

6.4   Randomized Rumor Spreading

While the main focus of this survey is how to efficiently generate randomized roundings with dependencies, there is the equally important question what dependencies to use. For many applications it is simple the desired solution that makes it obvious which dependencies to add. However, our experiments sketched in Sect. 4.3 also suggest that adding dependencies even where not demanded by the structure of the solution can improve the performance. To gain more insight into this phenomenon, we exemplarily regarded the classic push-protocol to disseminate a piece of information in a network ("*randomized rumor spreading*"). This has a natural interpretation as randomized rounding problem. Due to its simple structure, several dependent approaches suggest itself. Interestingly, the most astonishing improvement was achieved on preferential attachment graph, hence on networks trying to imitate the structure of real-world networks.

Randomized rumor spreading is a basic process to disseminate a piece of information in a network. Initiated by a node wanting to spread a piece of information ("rumor"), this round-based protocol works as follows. In each round of the process, each node knowing the rumor already calls a random neighbor and gossips the rumor to it. This process has been observed to be a very robust and scalable method to disseminate information, consequently if found many applications both in replicated databases [12] and wireless sensor networks [2].

Randomized rumor spreading has a natural interpretation as randomized rounding problem. Note that for a node u of degree d at each time step for each neighbor v the probability that u calls v is $x_{tv} = 1/d$. An actual run of the protocol leads to a rounding y_{tv} defined by $y_{tv} = 1$ if and only if u actually called v in round t. This rounding problem comes with the natural dependency $\sum_v y_{tv} = \sum_v x_{tv} = 1$, but as we shall see, adding further dependencies can be useful.

In [24,25], it was suggested that nodes should not take independent actions over time, but rather it should be avoided, e.g., that a node calls the same other node twice in a row. To keep the bookkeeping effort low, it was proposed that each node has a cyclic permutation of his neighbors. When first informed, it chooses a random starting point in this cyclic order, but from then on deterministically follows the order of the list. Note that this also massively reduced the number of random bits needed by the process. Despite using much less randomness, this process was proven to have a mostly similar or slightly better performance than the classic independent rumor spreading. In [21], an experimental investigation was undertaken that confirmed speed-ups for several settings where the theoretical works could not prove a difference of the protocols. Also, it was observed that the particular choice of the lists can make a difference, e.g., for 2D grids with diagonal adjacencies a low-discrepancy order to serve the directions was shown to be much better than a clock-wise order.

Interestingly, the most significant improvement stemming from dependencies (and in fact very low dependencies) was found on preferential attachment graphs [18,20]. These graphs were introduced by Barabási and Albert [7] as a model for real-world networks. For these graphs, surprisingly, a very minor fine-tuning turned out to change the asymptotic runtime [18,20]. While classic protocol with high probability needs $\Omega(\log n)$ rounds to inform all vertices, this changes to $O(\log n/\log \log n)$ when the independent choice is replaced by talking to a neighbor chosen uniformly at random from all neighbors except the one called in the very previous round. That this asymptotic improvement is visible also for realistic network sizes was shown in [19]. We are not aware of previous results showing that such a minor fine-tuning of a randomized algorithm can lead to such gains for real-world network structures.

7   Conclusions

All results presented in the article indicate that randomized rounding and its derandomization can be adapted to respect additional hard cardinality constraints without incurring significant losses compared to classical independent randomized rounding as introduced by Raghavan and Thomspon [47,48]. For disjoint cardinality constraints, when using Srinivasan's approach or the hybrid approach, we did not observe that generating the roundings or the derandomizations took more time or was significantly more complicated. Also, we generally did not observe larger rounding errors when additional hard constraints were present (rather the opposite, in particular, adding a global cardinality constraint may in fact slightly descrease the rounding errors). For the choice of the rounding method to be used, the experimental results clearly indicate that for disjoint cardinality constraints, Srinivasan's or the hybrid approach should be preferred, where as for the bipartite edge weight setting, the bit-wise or the hybrid approach are more efficient.

Acknowledgements. The authors are grateful to the German Science Foundation for generously supporting this research through their priority programme *Algorithm Engineering*, both financially and by providing scientific infrastructure. We are thankful to our colleagues in the priority programme for many stimulation discussions. A particular thank goes to our collaborators and associated members of the project, namely Carola Doerr née Winzen (University of Kiel, then MPI Saarbrücken, now Université Pierre et Marie Curie—Paris 6), Tobias Friedrich (MPI Saarbrücken, now University of Jena), Michael Gnewuch (University of Kiel, now University of Kaiserslautern), Peter Kritzer (University of Linz), Marvin Künnemann (MPI Saarbrücken), Friedrich Pillichshammer (University of Linz), and Thomas Sauerwald (MPI Saarbrücken, now University of Cambridge).

References

1. Ageev, A.A., Sviridenko, M.: Pipage rounding: a new method of constructing algorithms with proven performance guarantee. J. Comb. Optim. **8**(3), 307–328 (2004)
2. Al-Karaki, J.N., Kamal, A.E.: Routing techniques in wireless sensor networks: a survey. Wirel. Commun. IEEE **11**(6), 6–28 (2004)
3. Arora, S., Rao, S., Vazirani, U.V.: Expander flows, geometric embeddings and graph partitioning. J. ACM **56**(2) (2009)
4. Asadpour, A., Goemans, M.X., Madry, A., Gharan, S.O., Saberi, A.: An O(log n/ log log n)-approximation algorithm for the asymmetric traveling salesman problem. In: SODA, pp. 379–389 (2010)
5. Bansal, N.: Constructive algorithms for discrepancy minimization. In: FOCS, pp. 3–10 (2010)
6. Bansal, N., Spencer, J.: Deterministic discrepancy minimization. Algorithmica **67**(4), 451–471 (2013)
7. Barabási, A.L., Albert, R.: Emergence of scaling in random networks. Science **286**, 509–512 (1999)
8. Chekuri, C., Vondrák, J., Zenklusen, R.: Dependent randomized rounding for matroid polytopes and applications (2009). http://arxiv.org/pdf/0909.4348v2.pdf
9. Chekuri, C., Vondrák, J., Zenklusen, R.: Dependent randomized rounding via exchange properties of combinatorial structures. In: FOCS, pp. 575–584 (2010)
10. Chekuri, C., Vondrák, J., Zenklusen, R.: Multi-budgeted matchings and matroid intersection via dependent rounding. In: SODA, pp. 1080–1097 (2011)
11. Cunningham, W.H.: Testing membership in matroid polyhedra. J. Comb. Theory Ser. B **36**(2), 161–188 (1984)

12. Demers, A.J., Greene, D.H., Hauser, C., Irish, W., Larson, J., Shenker, S., Sturgis, H.E., Swinehart, D.C., Terry, D.B.: Epidemic algorithms for replicated database maintenance. Oper. Syst. Rev. **22**, 8–32 (1988)
13. Dobkin, D.P., Eppstein, D., Mitchell, D.P.: Computing the discrepancy with applications to supersampling patterns. ACM Trans. Graph. **15**(4), 354–376 (1996)
14. Doerr, B.: Multi-color discrepancies. dissertation, Christian-Albrechts-Universität zu Kiel (2000)
15. Doerr, B.: Structured randomized rounding and coloring. In: Freivalds, R. (ed.) FCT 2001. LNCS, vol. 2138, pp. 461–471. Springer, Heidelberg (2001). doi:10. 1007/3-540-44669-9_53
16. Doerr, B.: Generating randomized roundings with cardinality constraints and derandomizations. In: Durand, B., Thomas, W. (eds.) STACS 2006. LNCS, vol. 3884, pp. 571–583. Springer, Heidelberg (2006). doi:10.1007/11672142_47
17. Doerr, B.: Randomly rounding rationals with cardinality constraints and derandomizations. In: Thomas, W., Weil, P. (eds.) STACS 2007. LNCS, vol. 4393, pp. 441–452. Springer, Heidelberg (2007). doi:10.1007/978-3-540-70918-3_38
18. Doerr, B., Fouz, M., Friedrich, T.: Social networks spread rumors in sublogarithmic time. In: STOC, pp. 21–30. ACM (2011)
19. Doerr, B., Fouz, M., Friedrich, T.: Experimental analysis of rumor spreading in social networks. In: MedAlg, pp. 159–173 (2012)
20. Doerr, B., Fouz, M., Friedrich, T.: Why rumors spread so quickly in social networks. Communun. ACM **55**, 70–75 (2012)
21. Doerr, B., Friedrich, T., Künnemann, M., Sauerwald, T.: Quasirandom rumor spreading: an experimental analysis. JEA **16**. Article 3.3 (2011)
22. Doerr, B., Gnewuch, M.: Construction of low-discrepancy point sets of small size by bracketing covers and dependent randomized rounding. In: Keller, A., Heinrich, S., Niederreiter, H. (eds.) Monte Carlo and Quasi-Monte Carlo Methods 2006, pp. 299–312. Springer, Heidelberg (2008)
23. Doerr, B.: Non-independent randomized rounding. In: SODA, pp. 506–507 (2003)
24. Doerr, B., Friedrich, T., Sauerwald, T.: Quasirandom rumor spreading. In: SODA, pp. 773–781 (2008)
25. Doerr, B., Friedrich, T., Sauerwald, T.: Quasirandom rumor spreading: expanders, push vs. pull, and robustness. In: ICALP, pp. 366–377 (2009)
26. Doerr, B., Gnewuch, M., Kritzer, P., Pillichshammer, F.: Component-by-component construction of low-discrepancy point sets of small size. Monte Carlo Meth. Appl. **14**(2), 129–149 (2008)
27. Doerr, B., Gnewuch, M., Wahlström, M.: Implementation of a component-by-component algorithm to generate small low-discrepancy samples. In: L'Ecuyer, P., Owen, A.B. (eds.) Monte Carlo and Quasi-Monte Carlo Methods 2008, pp. 323–338. Springer, Heidelberg (2009)
28. Doerr, B., Gnewuch, M., Wahlström, M.: Algorithmic construction of low-discrepancy point sets via dependent randomized rounding. J. Complex. **26**(5), 490–507 (2010)
29. Doerr, B., Künnemann, M., Wahlström, M.: Randomized rounding for routing and covering problems: experiments and improvements. In: Festa, P. (ed.) SEA 2010. LNCS, vol. 6049, pp. 190–201. Springer, Heidelberg (2010). doi:10.1007/ 978-3-642-13193-6_17
30. Doerr, B., Künnemann, M., Wahlström, M.: Dependent randomized rounding: the bipartite case. In: ALENEX, pp. 96–106 (2011)
31. Doerr, B., Wahlström, M.: Randomized rounding in the presence of a cardinality constraint. In: ALENEX, pp. 162–174 (2009)

32. Erdős, P., Selfridge, J.L.: On a combinatorial game. J. Combinatorial Theory Ser. A **14**, 298–301 (1973)
33. Fleischer, L., Jain, K., Williamson, D.P.: Iterative rounding 2-approximation algorithms for minimum-cost vertex connectivity problems. J. Comput. Syst. Sci. **72**(5), 838–867 (2006)
34. Gabow, H.N., Manu, K.S.: Packing algorithms for arborescences (and spanning trees) in capacitated graphs. Math. Program. **82**, 83–109 (1998)
35. Gandhi, R., Khuller, S., Parthasarathy, S., Srinivasan, A.: Dependent rounding in bipartite graphs. In: FOCS, pp. 323–332 (2002)
36. Gandhi, R., Khuller, S., Parthasarathy, S., Srinivasan, A.: Dependent rounding and its applications to approximation algorithms. J. ACM **53**, 324–360 (2006)
37. Giannopoulos, P., Knauer, C., Wahlström, M., Werner, D.: Hardness of discrepancy computation and epsilon-net verification in high dimension. J. Complexity **28**(2), 162–176 (2012)
38. Gnewuch, M., Wahlström, M., Winzen, C.: A new randomized algorithm to approximate the star discrepancy based on threshold accepting. SIAM J. Numerical Anal. **50**(2), 781–807 (2012)
39. Goemans, M.X., Williamson, D.P.: Improved approximation algorithms for maximum cut and satisfiability problems using semidefinite programming. J. ACM **42**(6), 1115–1145 (1995)
40. Hromkovič, J.: Design and Analysis of Randomized Algorithms. Introduction to Design Paradigms. Texts in Theoretical Computer Science An EATCS Series. Springer, Berlin (2005)
41. Jain, K.: A factor 2 approximation algorithm for the generalized Steiner network problem. Combinatorica **21**(1), 39–60 (2001)
42. Moser, R.A., Tardos, G.: A constructive proof of the general Lovász local lemma. J. ACM **57**(2) (2010)
43. Niederreiter, H.: Random number generation and Quasi-Monte Carlo methods. In: CBMS-NSF Regional Conference Series in Applied Mathematics, vol. 63. Society for Industrial and Applied Mathematics (SIAM), Philadelphia, PA (1992)
44. Orlin, J.B.: Max flows in O(nm) time, or better. In: STOC, pp. 765–774 (2013)
45. Oxley, J.: Matroid Theory. Oxford Graduate Texts in Mathematics. OUP Oxford, Oxford (2011)
46. Panconesi, A., Srinivasan, A.: Randomized distributed edge coloring via an extension of the Chernoff-Hoeffding bounds. SIAM J. Comput. **26**, 350–368 (1997)
47. Raghavan, P.: Probabilistic construction of deterministic algorithms: approximating packing integer programs. J. Comput. Syst. Sci. **37**, 130–143 (1988)
48. Raghavan, P., Thompson, C.D.: Randomized rounding: a technique for provably good algorithms and algorithmic proofs. Combinatorica **7**, 365–374 (1987)
49. Raghavendra, P.: Optimal algorithms and inapproximability results for every CSP? In: STOC, pp. 245–254 (2008)
50. Raghavendra, P., Steurer, D.: How to round any CSP. In: FOCS, pp. 586–594 (2009)
51. Rothvoß, T.: The entropy rounding method in approximation algorithms. In: SODA, pp. 356–372 (2012)
52. Saha, B., Srinivasan, A.: A new approximation technique for resource-allocation problems. In: ICS, pp. 342–357 (2010)
53. Schrijver, A.: Combinatorial Optimization: Polyhedra and Efficiency. Algorithms and Combinatorics, vol. 24. Springer, Heidelberg (2003)
54. Spencer, J.: Six standard deviations suffice. Trans. Amer. Math. Soc. **289**, 679–706 (1985)

55. Spencer, J.: Ten Lectures on the Probabilistic Method. SIAM, Philadelphia (1987)
56. Srinivasan, A.: Distributions on level-sets with applications to approximations algorithms. In: FOCS, pp. 588–597 (2001)
57. Srivastav, A., Stangier, P.: Algorithmic Chernoff-Hoeffding inequalities in integer programming. Random Struct. Algorithms **8**, 27–58 (1996)
58. Szegedy, M.: The Lovász local lemma - a survey. In: CSR, pp. 1–11 (2013)

External-Memory State Space Search

Stefan Edelkamp[✉]

Fakultät Mathematik Und Informatik, Universität Bremen,
Am Fallturm 1, 28359 Bremen, Germany
edelkamp@tzi.de

Abstract. Many state spaces are so big that even in compressed form
they fail to fit into main memory. As a result, during the execution of a
search algorithm, only a part of the state space can be processed in main
memory at a time; the remainder is stored on a disk.

In this paper we survey research efforts in external-memory search
for solving state space problems, where the state space is generated by
applying rules. We study different form of expressiveness and the effect of
guiding the search into the direction of the goal. We consider outsourcing
the search to disk as well as its additional parallelization to many-core
processing units. We take the sliding-tile puzzle as a running example.

1 Introduction

A multitude of algorithmic tasks in a variety of application domains can be
formalized as a *state space problem*. A typical example is the sliding-tile puzzle –
in square arrangement called the $(n^2 - 1)$-puzzle (see Fig. 1). Numbered tiles in a
rectangular grid have to be moved to a designated goal location by successively
sliding tiles into the only empty square. The state space grows rapidly: the
8-puzzle has 181,440, the 15-puzzle $20,922,789,888,000/2 \approx 10$ trillion, and the
24-puzzle $15,511,210,043,330,985,984,000,000/2 \approx 7.75 \times 10^{25}$ states.

More generally, a *state space problem* $P = (S, A, s, T)$ consists of a set of
states S, an initial state $s \in S$, a set of goal states $T \subseteq S$, and a finite set
of actions A where each $a \in A$ transforms a state into another one. Usually,
a subset of actions $A(u) \subseteq A$ is applicable in each state u. A *solution* π is an
ordered sequence of actions $a_i \in A$, $i \in \{1, \ldots, k\}$ that transforms the initial
state s into one of the goal states $t \in T$, i.e., there exists a sequence of states
$u_i \in S$, $i \in \{0, \ldots, k\}$, with $u_0 = s$, $u_k = t$ and u_i is the outcome of applying a_i
to u_{i-1}, $i \in \{1, \ldots, k\}$. A *cost (or weight) function* $w : A \to I\!R_{\geq 0}$ induces the
cost of a solution consisting of actions $a_1, \ldots, a_k$ as $\sum_{i=1}^{k} w(a_i)$. In the usual
case of unit-cost domains, for all $a \in A$ we have $w(a) = 1$. A solution is *optimal*
if it has minimum cost among all solutions.

A *state space problem graph* $G = (V, E, s, T)$ for the state space problem
$P = (S, A, s, T)$ is defined by $V = S$ as the set of nodes, $s \in S$ as the initial
node, T as the set of goal nodes, and $E \subseteq V \times V$ as the set of edges that connect
nodes to nodes with $(u, v) \in E$ if and only if there exists an $a \in A$ with $a(u) = v$.
Solving state space problems, however, is best characterized as a search in an

© Springer International Publishing AG 2016
L. Kliemann and P. Sanders (Eds.): Algorithm Engineering, LNCS 9220, pp. 185–225, 2016.
DOI: 10.1007/978-3-319-49487-6_6

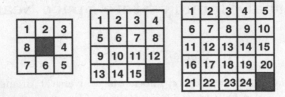

Fig. 1. $(n^2 - 1)$-puzzle instances: 8-puzzle, 15-puzzle, 24-puzzle.

implicit graph. The difference is that not all edges have to be *explicitly* stored, but are generated by a set of rules (such as in games). We have an initial node $s \in V$, a set of goal nodes determined by a predicate *Goal*: $V \rightarrow I\!B = \{false, true\}$. The basic operation is called *node expansion* (a.k.a., *node exploration*), which means generation of all neighbors of a node u. The resulting nodes are called *successors* (a.k.a., *children*) of u, and u is called a *parent* or *predecessor*. We will write $Succ(u) = \{v \in S \mid \exists a \in A(u) \mid a(u) = v\}$ for the successor set.

In more general state-space search models, by applying one action the successor of a state is no longer unique. For the non-deterministic case, we have $Succ(u, a) = \{v \in S \mid a \in A(u)\}$. For a *Markov decision problem* (MDPs) with probabilities $p(v \mid u, a)$ we additionally impose $\sum_{v \in Succ(u,a)} p(v \mid u, a) = 1$.

All nodes have to be reached at least once on a path from the initial node through successor generation. Consequently, we can divide the set of *reached nodes* into the set of *expanded nodes* and the set of *generated nodes* that are not yet expanded. In AI literature the former set is often referred to as the *Closed list* or the *search frontier*, and the latter set as the *Open list*. The denotation as a *list* refers to the legacy of the first implementation, namely as a simple linked list. However, realizing them using the right data structures (e.g., a hash table for duplicate elimination and a priority queue for best-first exploration) is crucial for the search algorithm's characteristics and performance.

Refined algorithms have led to first optimal solutions for challenging combinatorial games. Besides computation time, space is a crucial computational resource. For the Rubik's Cube with 43,252,003,274,489,856,000 states the exact diameter is 20 moves [73]. The computation for the lower bound took 35 CPU years on several computers using (pattern) databases [62]. Rokicki et al. [83] partitioned the states into 2,217,093,120 sets of 19,508,428,800 states each, reduced the count of sets needed to solve to 55,882,296 using symmetry and set covering. Only solutions of length of at most 20 were found with a program that solved a single set in about 20 s. The Towers-of-Hanoi problem (with 4 pegs and 30 disks) spawns a space of 1,152,921,504,606,846,976 states and was solved in 17 days using 400 GBs of disk space [67]. To show that Checkers is draw (assuming optimal play) [86,87], endgame databases of up to 10 pieces were built, for any combination of kings and checkers. The database size amounts to 39 trillion states. The number of states in the proof for a particular opening took about one month on an average of 7 processors, with a longest line of 67 moves. The standard problem for Connect 4 has 4,531,985,219,092 reachable states [31].

It is won for the first player [4,5]. Most other states have been classified via an external-memory hybrid of explicit-state and symbolic retrograde analysis [33].

Current domain-independent action planning systems solve Blocksworld problems with 50 blocks and more, and produce close-to cost-optimal plans in Logistics with hundreds of steps [24,49,50,52,82]. For planning with numbers, potentially infinite search spaces have to be explored [48,52]. With the external-memory search, in some cases, optimal plans can be obtained [54].

External-memory search algorithms have also helped finding bugs in software [8,26,34,35,54,74,92]. Different model checkers have been externalized and enhanced by directing the search toward system errors. Search heuristics accelerate symbolic model checkers for analyzing hardware, on-the-fly verifiers for analyzing compiled software units, and industrial tools for exploring real-time domains and finding resource-optimal schedules. Given a large and dynamically changing state vector, external-memory and parallel exploration scaled best. A *sweep-line* scans the search space according to a given partial order [71], while [55] implements a model-checking algorithm on top of external-memory A*, [56] provides a distributed implementation of the algorithm of [55] for model checking safety properties, while [27] extends the approach to general (LTL) properties. Iterative broadening has been suggested in the context of model checking real-time domains by [26], and some recent algorithms include perfect hash functions [12,13] in what has been denoted as semi-external-memory search [35,36].

External-memory search is also among the best-known methods for optimally solving multiple sequence alignment problems [69,88,96]. The graphs for the some challenging problems required days of CPU time to be explored [30]. Monte-Carlo tree search [17,60,84] is effective especially for post-hoc optimization [42].

The text kicks off with introducing external-memory search algorithms (Sect. 2) and continues with engineering the delayed detection (and elimination) of duplicates (Sect. 3). It then turns to pattern databases (Sect. 4), before addressing more general state space formalisms (Sect. 5), as well as parallelization options on CPUs (Sect. 6) and GPUs (Sect. 7). The work refers to prior publications of the author. E.g., Sects. 2.1 and 2.5 contain content from [29], Sect. 5 refers to [28], Sects. 6.1 and 6.3 are based on [56], Sect. 6.4 is based on [32], and Sect. 7 contains content from [37,40].

2   External-Memory Search

The commonly used model for comparing the performance of external algorithms consists of a single processor, a small internal memory that can hold up to M data items, and an unlimited secondary memory. The size of the input problem (in terms of the number of records) is abbreviated by N. Moreover, the *block size* B governs the bandwidth of memory transfers. It is usually assumed that at the beginning of the algorithm, the input data is stored in contiguous blocks on external memory, and the same must hold for the output. Only the number of block read and writes are counted, computations in internal memory do not incur any cost (see Fig. 2).

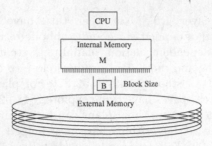

Fig. 2. The external memory model.

It is convenient to express the complexity of external-memory algorithms using two frequently occurring primitive operations [1,85]. The simplest operation is *external scanning*, reading a stream of N records stored consecutively on secondary memory. In this case, it is trivial to exploit disk (D) and block (B) parallelism. The number of I/Os is $scan(N) = N/DB$. *External sorting* is essential to arrange similar states together, for example, in order to find duplicates. Although we will mainly be concerned only with the case of a single disk ($D = 1$), it is possible to make optimal use of multiple disks with $sort(N) = O((N/DB)\log_{M/B}(N/B))$ I/Os. In practice, one pass in a multi-way merging process suffices so that we have $sort(N) = O(scan(N))$.

The advantage of state space search is that the (implicit) problem graph $G = (V, E)$ is generated on-the-fly by a set of rules, and, hence, no disk accesses for the adjacency lists are required. Moreover, considering the I/O complexities, bounds like those that include $|V|$ are rather misleading, since we often avoid generating the entire state space.

Many external-memory algorithms arrange the data flow in a directed acyclic graph, with nodes representing physical sources. Every node writes or reads *streams* of elements. *Pipelining*, a technique inherited from the database community, improves algorithms that reads data from and writes data to buffered files [2]. It enables a procedure to feed the output as a data stream directly to the algorithm that consumes the output, rather than writing it to the disk first.

2.1 External-Memory Breadth-First Search

There is no fundamental difference in the external-memory BFS algorithm by Munagala and Ranade [77] for explicit and implicit unweighted graphs. However, the access efforts are by far larger for the explicit graphs, even though the extension by Mehlhorn and Meyer [76] was successful in breaking the $O(|V|)$ I/O barrier for explicit graphs. The variant of Munagala and Ranade's algorithm in implicit graphs has been coined with the term *delayed duplicate detection* (DDD) [63,70]. The algorithm maintains BFS layers on disk. Layer $Open(i-1)$ is scanned and the set of successors are put into a buffer of size close to the main memory capacity. If the buffer becomes full, internal sorting followed by a duplicate elimination phase generates a sorted duplicate-free node sequence in the buffer that is flushed to disk. The outcome of this phase are k pre-sorted files. Note that duplicate elimination can be improved by using hash tables for the blocks before flushed to disk. Since the node set in the hash table has to be stored anyway, the savings by early duplicate detection are considerably small.

In the next step, *external-memory (multi-way) merging* is applied to unify the files into $Open(i)$ by a simultaneous scan. The size of the output files is chosen such that a single pass suffices. Duplicates are eliminated (even though semantically more insightful for the ease of notation not renaming the files into *Closed*). Since the files were pre-sorted, the complexity is given by the scanning time of all files. One also has to eliminate $Open(i-1)$ and $Open(i-2)$ from $Open(i)$ to avoid re-computations; that is, nodes extracted from the disk-based queue are not immediately deleted, but kept until the layer has been completely generated and sorted, at which point duplicates can be eliminated using a parallel scan. The process is repeated until $Open(i-1)$ becomes empty, or the goal has been found. The algorithm applies $O(sort(|Succ(Open(i-1))|)+scan(|Open(i-1)|+|Open(i-2)|))$ I/Os. By $\sum_i |Succ(Open(i))| = O(|E|)$ and $\sum_i |Open(i)| = O(|V|)$, the total execution time is $O(sort(|E|) + scan(|V|))$ I/Os.

In search problems with bounded branching factor we have $|E| = O(|V|)$, and thus the complexity for external-memory BFS reduces to $O(sort(|V|))$ I/Os. If we keep each $Open(i)$ in a separate file for sparse problem graphs (e.g. simple chains) file opening and closing would accumulate to $O(|V|)$ I/Os. The solution for this case is to store the nodes in $Open(i)$, $Open(i+1)$, and so forth consecutively in internal memory. Therefore, I/O is needed, only if a level has at most B nodes.

Let s be the initial node, and $Succ$ be the successor generation function. The algorithm extends to integer weighted graphs $G = (V, E, w)$ with bounded locality $loc_G = \max\{\delta(s, u) - \delta(s, v) + w(u, v) \mid u \in S, v \in Succ(u)\}$, where $\delta(s, u)$ is the shortest path distance from s to u. The locality determines the *thickness* of the search frontier needed to prevent duplicates in the search.

In external-memory search the exploration fully resides on disk. As pointers are not available solutions are reconstructed by saving the predecessor together with every state, by scanning with decreasing depth the stored files, and by looking for matching predecessors. Any reached node that is a predecessor of the current node is its predecessor on an optimal solution path. This results in a I/O complexity of $O(scan(|V|))$. Even if conceptually simpler, there is no need to store the the the search frontier $Open(i)$, $i \in \{0, 1, \ldots, k\}$, in different files.

By completely enumerating the state space the external-memory BFS exploration results showed that an instance of the 15-puzzle requires at most 80 steps [68]. The result has been validated in [80] on a distributed-memory system with 32 nodes (128 CPUs) in 66h.

2.2 External-Memory Breadth-First Branch-and-Bound

With general cost functions we hardly can omit states in the search. However, if $f = g+h$ with current path cost g and a *consistent heuristic* h, with $h(u)-h(v) \leq w(u, v)$ for all successors v of u, we may prune the exploration. For the domains where cost $f = g + h$ is monotonically increasing, *external-memory breadth-first branch-and-bound* (external-memory BFBnB) (with DDD) does not prune any node that is on the optimal solution path and ultimately finds the best solution. External BFBnB simulates memory-limited breadth-first heuristic search [93]. Let U be an upper bound on the solution cost. States with $f(v) > U$ are pruned and the expansion of states with $f(v) \leq U$ induce an updated bound.

In external-memory BFBnB with cost function $f = g + h$, where g is the depth of the search and h a consistent search heuristic, every duplicate with a smaller depth has been explored with a smaller f-value. This is simple to see as the h-values of the query node and the duplicate node match, and BFS generates duplicate with smaller g-value first. Moreover, u is safely pruned if $f(u)$ exceeds the current threshold, as an extension of the path to u to a solution will have a larger f-value. External BFBnB is optimal, since expands all nodes u with $f(u)$ smaller than the optimal solution cost f^*.

2.3 External-Memory Enforced Hill Climbing

Enforced hill climbing (EHC) [53] is a conservative variant of hill-climbing search. Given that the estimated goal distance value, often called *heuristic* in the area AI search, is minimized, the term *enforced downhill* would be a better fit. EHC has been adapted to both propositional and numerical planning by [52]. Starting from the initial state, a (breadth-first) search for a successor with a better heuristic value is started. As soon as such a successor is found, the hash tables are cleared and a fresh search is started. The process continues until the goal with distance value zero is reached (see Fig. 3, left). Since the algorithm performs a complete search on every seed state and will end up with a strictly better heuristic value, it is guaranteed to find a solution in directed graphs without dead-ends. In directed search spaces it can be trapped without finding a solution. Moreover, while often good, its results are not provably optimal.

Having external-memory BFS in hand, an external algorithm for EHC can easily be derived by utilizing the heuristic estimates. Figure 3 considers parts of an exploration for solving a planning problem in a histogram showing the number of nodes in BFS layers for external EHC in a typical search problem.

Let $h(s)$ be the heuristic estimate of the initial state s then the I/O complexity is bounded by the number of calls to BFS times the I/O complexity of each run, i.e., by $O(h(s) \cdot (scan(|V|) + sort(|E|)))$ I/Os.

2.4 External-Memory A*

In the following, we study how to extend external-memory BFS to A* [47]. The main advantage of A* with respect to BFS is that, due to the use of a lower bound on the goal distance, it often traverses a much smaller part of the search space to establish an optimal solution. Since A* only changes the traversal ordering, it is advantageous to BFS only if both algorithms terminate at a goal node.

In A*, the cost for node u is $f(u) = g(u) + h(u)$, with g being the cost of the path from the initial node to u and $h(u)$ being the estimate of the remaining costs from u to the goal. In each step, a node u with minimum f-value is removed

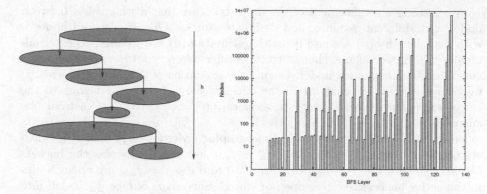

Fig. 3. Schematic view of enforced hill climbing, incrementally queuing down to better goal distance values, restarting each time the exit of a plateau is reached (left). Typical memory profile in external-memory enforced hill climbing of a particular benchmark planning problem (right): the x-axis provides an index for the concatenatenation of all the BFS-layers encountered during the search, while the y-axis denotes the number of states stored and expanded (height of bars), for the according index (on log scale).

from *Open*, and the new value $f(v)$ of a successor v of u is updated to the minimum of its current value and $f(v) = g(v) + h(v) = g(u) + w(u,v) + h(v) = f(u) + w(u,v) - h(u) + h(v)$; in this case, it is inserted into *Open* itself.

In our algorithm, we first assume a consistent heuristic, where for all u and v we have $w(u,v) \geq h(u) - h(v)$, and a uniformly weighted undirected problem graph. These conditions are often met in practice, since many problem graphs in single-agent search, e.g., in Rubik's cube and sliding-tile puzzles are uniformly weighted and undirected and many heuristics, e.g., pattern database estimates [66] are consistent. Under these assumptions, we have $h(u) \leq h(v) + 1$ for every node u and every successor v of u. Since the problem graph is undirected this implies $|h(u) - h(v)| \leq 1$ and $h(v) - h(u) \in \{-1, 0, 1\}$. If the heuristic is consistent, then on each search path, the evaluation function f is non-decreasing. No successor will have a smaller f-value than the current one. Therefore, A*, which traverses the node set in f-order, expands each node at most once.

In the $(n^2 - 1)$-puzzle, for example, the *Manhattan distance* is defined as the sum of the horizontal and vertical differences between actual and goal configurations, for all tiles. The heuristic is consistent, since for two successive nodes u and v the difference of the according estimate evaluations $h(v) - h(u)$ is either -1 or 1. The f-values of nodes u and successor nodes v of are either the same or $f(v) = f(u) + 2$.

As above, *external-memory A** [29] maintains the search frontier on disk, possibly partitioned into main-memory-sized sequences. In fact, the disk files correspond to a bucket implementation of a priority queue data structure. In the course of the algorithm, each bucket addressed with index i contains all nodes u in the set *Open* that have priority $f(u) = i$. A disk-based representation of this data structure will store each bucket in a different file [64].

We introduce a refinement of the data structure that distinguishes between nodes with different g-values, and designates bucket $Open(i, j)$ to all nodes u with path length $g(u) = i$ and heuristic estimate $h(u) = j$. Similar to external-memory BFS, we do not change the identifier $Open$ (to $Closed$) to separate generated from expanded nodes. During the execution of A*, bucket $Open(i, j)$ may refer to elements that are in the current search frontier or belong to the set of expanded nodes. During the exploration process, only nodes from one currently *active bucket* $Open(i, j)$ with $i + j = f_{\min}$ are expanded, up to its exhaustion. Buckets are selected in lexicographic order for (i, j); then, the buckets $Open(i', j')$ with $i' < i$ and $i' + j' = f_{\min}$ are closed, whereas the buckets $Open(i', j')$ with $i' + j' > f_{\min}$ or with $i' > i$ and $i' + j' = f_{\min}$ are open. Nodes in the active bucket are either open or closed. Successors of $Open(g, h)$ fall into $Open(g + 1, h - 1)$, $Open(g + 1, h)$, or $Open(g + 1, h + 1)$, so that the number of buckets needed is bounded by $O((f^*)^2)$ with f^* being the optimal solution cost.

By the restriction of f-values in the $(n^2 - 1)$-puzzle only about half the number of buckets have to be allocated. Figure 4 shows the memory profile of external-memory A* on a 35-puzzle puzzle instance (with 14 tiles permuted). The exploration starts in bucket $(50, 0)$ and terminated while expanding bucket $(77, 1)$. Similar to external-memory BFS but in difference to ordinary A*, external-memory A* terminates while generating the goal, since all frontier states with smaller g-value have already been expanded.

We can restrict the removal of duplicates to buckets of the same h-value. since for all i, i', j, j' with $j \neq j'$ we have $Open(i, j) \cap Open(i', j') = \emptyset$. In undirected problem graphs duplicates of a node with BFS-level i can at most occur in levels $i, i - 1$ and $i - 2$. In addition, if $u = v$ we have $h(u) = h(v)$.

For ease of describing the algorithm, we consider each bucket for the *Open* list as a different file. Very sparse graphs can lead to bad I/O performance, as they may lead to buckets that contain by far less than B elements and dominate I/O access. Hence, we generally assume graphs with $(f^*)^2 = O(scan(|V|))$.

Algorithm 1.1 depicts the pseudo-code of the external-memory A* algorithm for consistent estimates and uniform graphs. The algorithm maintains the two values $g_{\min}$ and $f_{\min}$ to address the currently considered buckets. The buckets of $f_{\min}$ are traversed for increasing $g_{\min}$ up to $f_{\min}$. According to their different h-values, successors are arranged into three different frontier lists $A(f_{\min})$, $A(f_{\min} + 1)$, and $A(f_{\min} + 2)$; hence, at each instance only four buckets have to be accessed by I/O operations. For each of them, we keep a separate buffer of size $B/4$; this will reduce the internal memory requirements to B. If a buffer becomes full then it is flushed to disk. As in BFS it is practical to pre-sort buffers in one bucket immediately by an efficient internal algorithm to ease merging, but we could equivalently sort the unsorted buffers for one bucket externally.

There can be two cases that can give rise to duplicates within an active bucket (see Fig. 5, black bucket): two different nodes of the *same* predecessor bucket generating a common successor, and two nodes belonging to *different* predecessor buckets generating a duplicate. These two cases can be dealt with by merging all the pre-sorted buffers corresponding to the same bucket, resulting in

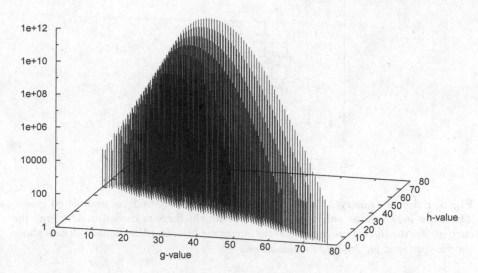

Fig. 4. Typical memory profile of external-memory A* in a selected sliding-tile bench-mark problem instance. The (g, h)-value surface is a grid of buckets, and each bucket corresponds to a file. The search starts with g-value zero, ends with h-value zero, and progresses is made in expanding all states within a bucket with increasing $g + h$-value, tie breaking on a lower g-value. The height of the bars (log scale) denotes the file sizes (each state correspond to a fixed number of bytes).

one sorted file. This file can then be scanned to remove the duplicate nodes from it. In fact, both the merging and duplicates removal can be done simultaneously.

Another special case of the duplicate nodes exists when the nodes that have already been evaluated in the upper layers are generated again (see Fig. 5). These duplicate nodes have to be removed by a file subtraction process for the next active bucket $Open(g_{min} + 1, h_{max} - 1)$ by removing any node that has appeared in $Open(g_{min}, h_{max} - 1)$ and $Open(g_{min} - 1, h_{max} - 1)$ (Buckets shaded in light gray). This file subtraction can be done by a mere parallel scan of the pre-sorted files and by using a temporary file in which the intermediate result is stored. It suffices to remove duplicates only in the bucket that is expanded next, i.e., $Open(g_{min} + 1, h_{max} - 1)$.

When merging the pre-sorted sets with the previously existing $Open$ buckets (both residing on disk), duplicates are eliminated, leaving the sets $Open(g_{min} + 1, h_{max} - 1)$, $Open(g_{min} + 1, h_{max})$ and $Open(g_{min} + 1, h_{max} + 1)$ duplicate-free. Then the next active bucket $Open(g_{min} + 1, h_{max} - 1)$ is refined not to contain any node in $Open(g_{min} - 1, h_{max} - 1)$ or $Open(g_{min}, h_{max} - 1)$. This can be achieved through a parallel scan of the pre-sorted files and by using a temporary file in which the intermediate result is stored, before $Open(g_{min} + 1, h_{max} - 1)$ is updated. It suffices to perform file subtraction lazily only for the bucket that is

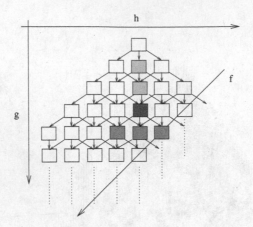

Fig. 5. External-memory A* with consistent heuristic in a uniform undirected graph. The arrow indicates the order of matrix traversal, the buckets shaded in dark gray the current file, the shaded buckets below the successor files, and the shaded buckets above the files are used for duplicate elimination.

expanded next. Since external-memory A* only modifies the order of states with the same f-value, completeness and optimality are inherited from internal A*.

By simulating internal A*, DDD ensures that each edge in the problem graph is looked at most once, so that $O(sort(|Succ(Open(g_{\min}+1, h_{\max}-1))|))$ I/Os are needed to eliminate duplicates in the successor lists. Since each node is expanded at most once, this adds $O(sort(|E|))$ I/Os to the overall run time. Filtering, evaluating nodes, and merging lists is available in scanning time of all buckets in consideration. During the exploration, each bucket $Open$ will be referred to at most six times, once for expansion, at most three times as a successor bucket and at most two times for duplicate elimination as a predecessor of the same h-value as the currently active bucket. Therefore, evaluating, merging and file subtraction add $O(scan(|V|) + scan(|E|))$ I/Os to the overall run time.

If $|E| = O(|V|)$ the complexity reduces to $O(sort(|V|))$ I/Os. It is not difficult to generalize the result to directed graphs with bounded locality, since in this case subtraction amounts to $O(loc_G \cdot scan(|V|)) = O(scan(|V|))$ I/Os.

By setting the weight of all edges (u, v) to $h(u) - h(v) + 1$ for a consistent heuristic h, A* can be cast as a variant of Dijkstra's algorithm. To reconstruct a solution path, we store predecessor information with each node on disk (thus doubling the state vector size), and apply backward chaining, starting with the target node. However, this is not strictly necessary: For a node in depth g, we intersect the set of possible predecessors with the buckets of depth $g - 1$. Any node that is in the intersection is reachable on an optimal solution path, so that we can iterate the construction process. Time is bounded by $O(scan(|V|))$ I/Os.

Let us consider how to externally solve 15-puzzle problem instances that cannot be solved internally with A* and the Manhattan distance estimate. Internal sorting is implemented by applying Quicksort [51]. Multi-way external-memory

Procedure External-Memory A*
Input: Problem graph with start node s
Output: Optimal solution path

$Open(0, h(s)) \leftarrow \{s\}$	;; Initialize frontier bucket
$f_{min} \leftarrow h(s)$	;; Initialize merit
while $(f_{min} \neq \infty)$	;; Termination criterion for full exploration
$\quad g_{min} \leftarrow \min\{i \mid Open(i, f_{min} - i) \neq \emptyset\}$	;; Determine minimal depth
$\quad$ **while** $(g_{min} \leq f_{min})$	;; As far as merit not exceeded
$\quad\quad h_{max} \leftarrow f_{min} - g_{min}$	;; Determine corresponding h-value
$\quad\quad A(f_{min}), A(f_{min} + 1), A(f_{min} + 2) \leftarrow Succ(Open(g_{min}, h_{max}))$	;; Successors
$\quad\quad Open(g_{min} + 1, h_{max} + 1) \leftarrow A(f_{min} + 2)$	;; New bucket
$\quad\quad Open(g_{min} + 1, h_{max}) \leftarrow A(f_{min} + 1) \cup Open(g_{min} + 1, h_{max})$	;; Merge
$\quad\quad Open(g_{min} + 1, h_{max} - 1) \leftarrow A(f_{min}) \cup Open(g_{min} + 1, h_{max} - 1)$	;; Merge
$\quad\quad$ **if** $(Goal(Open(g_{min} + 1, h_{max} - 1)))$	;; Terminal state in set
$\quad\quad\quad$ **return** $Construct(Open(g_{min} + 1, h_{max} - 1))$	;; Generate solution path
$\quad\quad Open(g_{min} + 1, h_{max} - 1) \leftarrow$	;; Simplify list
$\quad\quad\quad RemoveDuplicates(Open(g_{min} + 1, h_{max} - 1))$	;; Sort/scan
$\quad\quad Open(g_{min} + 1, h_{max} - 1) \leftarrow Open(g_{min} + 1, h_{max} - 1) \backslash$	;; Omit duplicates from
$\quad\quad\quad (Open(g_{min}, h_{max} - 1) \cup Open(g_{min} - 1, h_{max} - 1))$	;; ... previous levels
$\quad\quad g_{min} \leftarrow g_{min} + 1$	;; Increase depth
$\quad f_{min} \leftarrow \min\{i + j > f_{min} \mid Open(i, j) \neq \emptyset\} \cup \{\infty\}$	;; Find minimal f-value

Algorithm 1.1. External-memory A* for consistent and integer heuristics.

merging maintains file pointers for every flushed buffer and joins them into a single sorted file. Internally, a heap is used (its engineered implementation is crucial for the efficiency of the sorting). Duplicate removal and bucket subtraction are performed on single passes through the bucket file. Table 1 illustrates the impact of duplicate removal (dr) and bucket subtraction (sub) on the number of generated states for problem instances of increasing complexity. In some cases, the experiment is terminated because of the limited hard disk capacity.

One interesting feature of our approach from a practical point of view is the ability to pause and resume the program execution in large problem instances. This is desirable, e.g. in the case when the limits of secondary storage are reached, as one can resume the execution with more disk space. External sorting can be avoided to some extent, by a single or a selection of hash functions that splits larger files into smaller pieces until they fit into main memory. As with the h-value in the above case a node and its duplicate will have the same hash address.

While external-memory A* requires a constant amount of memory for the internal read and write buffers, iterative-deepening A* (IDA*) [61] that applies depth-first bounded searches with an increasing optimal solution cost threshold, requires very little memory that scales linear with the search depth. External-memory A* removes all duplicates from the search, but require slow disk to succeed. Moreover, in search practice disk space is limited, too. Therefore, one

Table 1. Impact of duplicate removal and bucket subtraction.

Instance	States	States$_{dr}$	States$_{dr+sub}$
1	530,401	2,800	1,654
2	71,751,166	611,116	493,990
3	<out of disk space>	7,532,113	5,180,710
4	<out of disk space>	<out of disk space>	297,583,236
5	<out of disk space>	<out of disk space>	2,269,240,000
6	<out of disk space>	<out of disk space>	2,956,384,330

Table 2. Combining IDA* with external-memory A* in a 24-puzzle problem.

Split at f-value	Solution length	Nodes generated
68 (IDA*)	82	94,769,462
72 (Hybrid)	82	127,777,529
76 (Hybrid)	82	63,733,384
80 (Hybrid)	82	96,612,234
84 (External A*)	82	171,814,208

option is to combine the advantages of IDA* and external-memory A*. Starting with external-memory A*, the buckets up to a predefined f-value f_{sp} (the *split value*) are generated. Then, with increasing depth, all buckets on the f_{sp} diagonal are read, and all states contained in the buckets are fed into IDA* as initial states, which is initialized to an anticipated solution length $U = f_{sp}$. As a side effect of all such runs being pairwise independent they can be easily distributed.

Table 2 shows results of solving a 24-puzzle instance according to different f-value splits to document the potential of such hybrid algorithm. By its breadth-first ordering, external A* expands the entire f^*-diagonal, while IDA* stops at the first goal generated. Another instance (with an optimal plan of 100 moves) and a split value of 94 generated 367,243,074,706 nodes using 4.9 GB disk, while split value of 98 resulted in 451,034,974,741 generated nodes and 169 GB disk.

2.5 Non-uniformly Weighted Graphs

For integer weights in $\{1, \ldots, C\}$, due to consistency of the heuristic, it holds for every node u and every successor v of u that $h(v) \geq h(u) - w(u,v)$. Moreover, since the graph is undirected, we equally have $h(u) \geq h(v) - w(u,v)$, or $h(v) \leq h(u) + w(u,v)$; hence, $|h(u) - h(v)| \leq w(u,v)$, so that The I/O complexity for external A* in an implicit unweighted and undirected graph, where the weights are in $\{1, \ldots, C\}$, with a consistent estimate, is bounded by $O(sort(|E|) + C \cdot scan(|V|))$. The difference to the uniform case is that each bucket is referred to at most $2C + 1$ times for bucket subtraction and expansion, so that each edge in the problem graph is considered at most once. If we do not impose a bound C on the maximum integer weight, or if we allow directed graphs, the run time increases to $O(sort(|E|) + f^* \cdot scan(|V|))$ I/Os. For larger edge weights, small-sized buckets have to be handled with care.

3   Duplicate Detection

We have seen that sorting-based duplicate detection is essential for disk-based search, as ordinary hash functions are not access-locality preserving. This effect is known as *thrashing*, the computer's virtual memory subsystem is in a constant state of paging, rapidly exchanging data in memory for data on disk.

3.1   Hash-Based Duplicate Detection

Hash-based duplicate detection is designed to avoid the complexity of sorting. It is based on either one or two orthogonal hash functions. The primary hash function distributes the nodes to different files. Once a file of successors has been generated, duplicates are eliminated. The assumption is that all nodes with the same primary hash address fit into main memory. The secondary hash function (if available) maps all duplicates to the same hash address. This approach can be illustrated by sorting a card deck of 52 cards. For 13 internal memory places the best strategy is to hash cards to different files based on their suit in one scan. Next, we individually read each of the files to main memory to sort the cards or search for duplicates.

By iterating this Bucket Sort process we obtain an external-memory version of Radix Sort that scans the files more than once according to a radix representation of the key values. For the 15-puzzle problem in ordinary vector representation with a number for each board position, we have 16 phases for radix sort using 16 buckets.

3.2   Structured Duplicate Detection

Structured duplicate detection (SDD) [94] incorporates a hash function that maps nodes into an *abstract* problem graph; this reduces the successor scope of nodes that have to be kept in main memory. Such hash projections are state space homomorphisms, such that for each pair of consecutive abstract nodes the pair of original nodes is also connected. A bucket now corresponds to the set of original states, which all map to the same abstract state. In difference to DDD, SDD detects duplicates *early*; as soon as they are generated. Before expanding a bucket, not only the bucket itself, but all buckets that are potentially affected by successor generation have to be loaded and, consequently, fit into main memory. This gives rise to a different definition of locality, which determines a handle for the duplicate-detection scope. In difference to the *locality for DDD* the *locality for SDD* is defined as the the maximum node branching factor

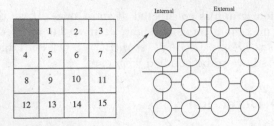

Fig. 6. Example for structured duplicate detection; problem instance (left) is mapped to one node in the abstract graph (right). For expanding all states mapped to an abstract node, for the elimination of duplicates only states stored in the abstract successors nodes need to be loaded in main memory.

$b_{\max} = \max_{v \in \phi(S)} |Succ(v)|$ in the abstract state space $\phi(S)$. If there are different abstractions to choose from, we take those that have the smallest ratio of maximum node branching factor $b_{\max}$ and abstract state space size $|\phi(S)|$. The idea is that smaller abstract state space sizes should be preferred but usually lead to larger branching factors.

In the example of the 15-puzzle (see Fig. 6), the projection is based on nodes that have the same blank position. This state-space abstraction also preserves the additional property that the successor set and the expansion sets are disjoint, yielding no self-loops in the abstract problem graph. The duplicate scope defines the successor buckets that have to be read into main memory.

The method is crucially dependent on the availability and selection of suitable abstraction functions ϕ that adapt to the internal memory constraints. In contrast, DDD does not rely on any partitioning beside the heuristic function and it does not require the duplicate scope to fit in main memory. A time-space trade-off refinement called *edge partitioning* [97] generates successors only along one edge at a time.

SDD is compatible with ordinary and hash-based duplicate detection, as in case the files that have to be loaded into main memory do no longer fit, we have to delay. However, the structured partitioning may have truncated the file sizes for duplicate detection to a manageable number. Each heuristic or hash function defines a partitioning of the search space but not all partitions provide a good locality with respect to the successor or predecessor states.

4   External-Memory Pattern Databases

While earlier versions of heuristic search via abstraction generate heuristics estimates on demand, *pattern databases* precompute and store the goal distances for the entire abstract search space in a lookup table [20]. Successful approaches additionally combine the heuristics of multiple smaller pattern databases, either by maximizing, or by cumulating the values, which is admissible under certain disjointness conditions [66]. To save space, the computation of the database can

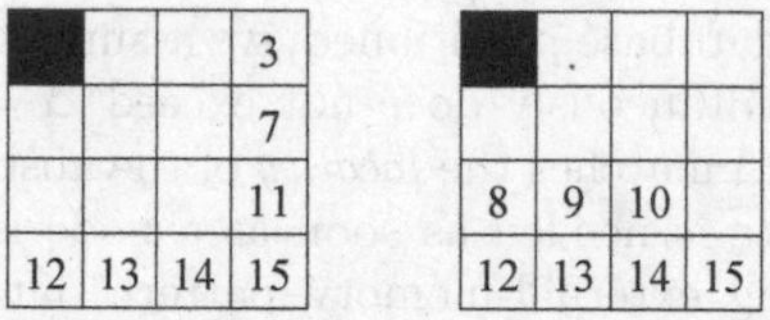

Fig. 7. Fringe and corner target pattern for the 15-puzzle.

Table 3. Effect of pattern databases in the 15-puzzle.

Heuristic	Nodes	Mean heuristic value
Manhattan distance	401,189,630	36.942
Linear conflict heuristic	40,224,625	38.788
5-tile pattern database	5,722,922	41.562
6-tile pattern database	3,788,680	42.924

be restricted using an upper bound on the length of an optimal solution path; and by exploiting specialized data compression schemes.

For the $(n^2 - 1)$-puzzle problem abstraction consists of ignoring a selected subset of tiles on the board. Their labels are replaced by a special *don't care* symbol; the remaining set of tiles is referred to as the *pattern* (see Fig. 7).

In experiments it has been showed that taking the maximum of the Manhattan distance and a singleton pattern database reduces the number of expanded nodes by two orders of magnitude of the algorithm using only the Manhattan distance. Using both databases together even leads to an improvement according to three orders of magnitude. Table 3 shows some exploration results for the 15-puzzle in reducing the number of search nodes and in increasing the mean of the heuristic value.

Larger pattern databases correspond to complete explorations of abstract state spaces that don't fit into main memory. Most frequently they correspond to external-memory BFS with DDD. The construction of external-memory pattern databases is especially suited to *frontier search*, as no solution path has to be reconstructed. They have been used together with SDD [95] and in different representations [25].

During the construction each BFS-layer i has been assigned to an individual file B_i. All states in B_i have the same goal distance, and all states that map to a state in i share the heuristic estimate i. For determining the h-value for some given state u in algorithms we first have to scan the files to find u. As this is a cost-intensive operation, whenever possible, pattern database lookup should be delayed, so that the heuristic estimates for a larger set of states can be retrieved in one scan. For example, external-memory A* distributes the set of successor states of each bucket according to their heuristic estimates. Hence, it can be adapted to delayed lookup, intersecting the set of successor states with (the state set represented by) the abstract states in the file of a given h-value.

To keep the pattern database partitioned, we assume that the number of files that can be opened simultaneously does not exceed $\Delta = \max\{h(v) - h(u)\} + 1 \mid u, v \in Succ(u)\}$, i.e., Δ matches the *locality* of the abstract state space graph.

If a heuristic estimate is needed as soon as a node is generated, an appropriate choice for creating external-memory pattern databases is a backwards BFS with SDD, as SDD already provides locality with respect to a state space abstraction function. After the construction patterns are arranged according to pattern blocks, one for each abstract state. When a concrete heuristic search algorithm expands nodes, it must check if the pattern form the pattern-lookup scope are in main memory, and, if not, it reads them from disk. Pattern blocks that do not belong to the current pattern-lookup scope are removed. When the part of internal memory is full, the search algorithm must decide, which pattern block to remove, e.g., by adopting the least-recently used strategy.

Larger pattern databases provide better bounds and thus allow more guidance in the search. For the 15-puzzle puzzle, a 10-tile 28 GB pattern database has been built [81], while [23] computed 9-9-6, 9-8-7, and 8-8-8 pattern database sets for the 24-puzzle that are up to three orders of magnitude larger (up to 1.4 TB) than the standard 6-6-6-6 pattern database set. This was possible by performing a parallel breadth-first search in the compressed pattern space. Experiments indicate an average 8-fold improvement of the 9-9-6 set over the 6-6-6-6 set. Combining several large pattern databases yielded on average a 13-fold improvement. A massive parallel search based on the *map-and-reduce* paradigm [21] using these databases was proposed by [89].

If we consider the example of the 35-puzzle with x tiles in the pattern, the abstract state space consists of $36!/(36 - x)!$ states. A perfect hash-table for the 35-puzzle has space requirements of 43.14 MB ($x = 5$), 1.3 GB ($x = 6$), and 39.1 GB ($x = 7$). The latter has successfully been constructed on disk by [32].

5   External-Memory Value Iteration

We now discuss an approach for extending the search model to cover uncertainty. More precisely, we extend the *value iteration* procedure to work on large state spaces that cannot fit into the RAM. There is a tight relation to the shortest-paths algorithm of Bellman and Ford (see [19]). Different guidance heuristics for improving the update have been proposed [7,10,45].

A *Markov decision process problem* (MDP) is a tuple (S, A, w, p), where S is the underlying state space, A is the set of actions, $w : S \times A \rightarrow I\!R$ is the cost or immediate reward function and $p(v \mid u, a)$ is the probability that action a in state u will lead to state v. The goal is to minimize the (expected or discounted) accumulated costs or, equivalently, to maximize the (expected or discounted) accumulated rewards.

A common way of calculating an optimal policy is by means of dynamic programming using *value iteration* based on the Bellman equation

$$f^*(u) = \min_{a \in A} \left\{ w(u, a) + \sum_{v \in S} p(v \mid u, a) \cdot f^*(v) \right\}.$$

In some cases, we apply a *discount* δ to allow assigning values to infinite paths. Roughly speaking, we can define the value of a state as the total reward/cost an agent can expect to accumulate when traversing the graph according to its policy, starting from that state. The discount factor defines how much more we should value immediate costs/rewards, compared to costs/rewards that are only attainable after two or more steps. Formally, the corresponding equation according to the principle of optimality is

$$f^*(u) = \min_{a \in A} \left\{ w(u,a) + \delta \cdot \sum_{v \in S} p(v \mid u,a) \cdot f^*(v) \right\}.$$

Value iteration improves the estimated *cost-to-go* function f by successively performing the following operation for each state u:

$$f(u) \leftarrow \min_{a \in A} \left\{ w(u,a) + \sum_{v \in S} p(v \mid u,a) \cdot f(v) \right\}.$$

The algorithm exits if an error bound on the policy evaluation falls below a user-supplied threshold ϵ, or a maximum number of iterations have been executed. If the optimal cost f^* is known for each state, the optimal policy can be easily extracted by choosing an operation according to a single application of the Bellman equation. The procedure takes a heuristic h for initializing the value function as an additional parameter.

The error bound on the value function is also called the *residual*, and can for example be computed in form $\max_{u \in S} |f_t(u) - f_{t-1}(u)|$. A residual of zero denotes that the process has converged. An advantage of other methods like policy iteration is that it converges to the exact optimum, while value iteration usually only reaches an approximation. On the other hand, the latter technique is usually more efficient on large state spaces.

For implicit search graphs, value iteration proceed in two phases. In the first phase, the whole state space is generated from the initial state s. In this process, an entry in a hash table (or vector) is allocated in order to store the f-value for each state u; this value is initialized to the cost of u if $u \in T$, or to a given (non-necessarily admissible) heuristic estimate (or zero if no estimate is available) if u is non-terminal. In the second phase, iterative scans of the state space are performed updating the values of non-terminal states u as:

$$f(u) = \min_{a \in A(u)} q(u,a), \tag{1}$$

where $q(u,a)$, which depends on the search model.

Value iteration converges to the solution optimal value function provided that its values are finite for all $u \in S$. In the case of MDPs, which may have cyclic solutions, the number of iterations is not bounded and value iteration typically only converges in the limit. For this reason, for MDPs, value iteration is often terminated after a predefined bound of $t_{\max}$ iterations are performed, or when the residual falls below a given $\epsilon > 0$.

For *external-memory value iteration* [28] instead of working on states, we work on edges for reasons that shall become clear soon. In our case, an edge is a 4-tuple $(u, v, a, f(v))$, where u is called the predecessor state, v the stored state, a the action that transforms u into v, and $f(v)$ is the current assignment of the value function to v. Clearly, v must belong to $Succ(a, u)$. In deterministic problems, v is determined by u and a and so it can be completely dropped, but for the non-deterministic problems, it is a necessity. Similarly to the internal value iteration, the external-memory version works in two phases. A forward phase, where the state space is generated, and a backward phase, where the heuristic values are repeatedly updated until an ϵ-optimal policy is computed, or $t_{\max}$ iterations are performed.

Forward Phase: State Space Generation. Typically, a state space is generated by a depth-first or a breadth-first exploration that uses a hash table to avoid re-expansion of states. We choose an external breadth-first exploration to handle large state spaces. Since in an external setting a hash table is not affordable, we rely on DDD. It consists of two phases, first removing duplicates within the newly generated layer, and then removing duplicates with respect to previously generated layers. Note that an edge $(u, v, a, f(v))$ is a duplicate, if and only if its predecessor u, its state v, and the action a match an existing edge. Thus, in undirected graphs, there are two different edges for each undirected edge. In our case, sorting-based DDD is best suited as the sorted order is further exploited during the backward phase. For each depth value d the algorithm maintains the BFS layers $Layer(d)$ on disk. The first phase ends up by concatenating all layers into one *Open* list that contains all edges reachable from s. For bounded locality, the complexity of this phase is $O(sort(|E|))$ I/Os.

Backward Phase: Update of Values. This is the most critical part of the approach and deserves more attention. To perform the update on the value of state v, we have to bring together the value of its successor states. As they both are contained in one file, and there is no arrangement that can bring all successor states close to their predecessor states, we make a copy of the entire graph (file) and deal with the current state and its successor differently. To establish the adjacencies, the second copy, called $Temp$, is sorted with respect to the node u. Remember that *Open* is sorted with respect to the node v.

A parallel scan of files *Open* and $Temp$ gives us access to all the successors and values needed to perform the update on the value of v. This scenario is shown in Fig. 8 for the graph in the example. The contents of $Temp$ and $Open_t$, for $t = 0$, are shown along with the heuristic values computed so far for each edge (u, v). The arrows show the flow of information (alternation between dotted and dashed arrows is just for clarity). The results of the updates are written to the file $Open_{t+1}$ containing the new values for each state after $t + 1$ iterations. Once $Open_{t+1}$ is computed, the file $Open_t$ can be removed as it is no longer needed.

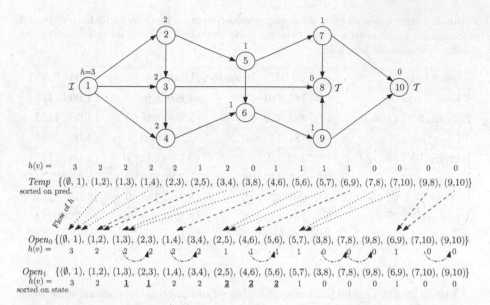

$h(v) =$ 3   2   2   2   2   1   2   0   1   1   1   1   0   0   0   0

$Temp$ {($\emptyset$, 1), (1,2), (1,3), (1,4), (2,3), (2,5), (3,4), (3,8), (4,6), (5,6), (5,7), (6,9), (7,8), (7,10), (9,8), (9,10)}
sorted on pred.

$Open_0$ {($\emptyset$, 1), (1,2), (1,3), (2,3), (1,4), (3,4), (2,5), (4,6), (5,6), (5,7), (3,8), (7,8), (9,8), (6,9), (7,10), (9,10)}
$h(v) =$ 3   2   2   2   2   2   1   1   1   1   0   0   0   1   0   0

$Open_1$ {($\emptyset$, 1), (1,2), (1,3), (2,3), (1,4), (3,4), (2,5), (4,6), (5,6), (5,7), (3,8), (7,8), (9,8), (6,9), (7,10), (9,10)}
$h(v) =$ 3   2   **1**   **1**   2   2   **2**   **2**   **2**   1   0   0   0   1   0   0
sorted on state

Fig. 8. An example graph with initial f-values and one backward phase in external-memory value iteration. A parallel scan of $Open_0$ and $Temp$ is done from left to right. The file $Open_1$ is the result of the first update; f-values that changed in the first update are shown with bold underline typeface.

The backward update algorithms first copies the $Open_t$ list in $Temp$ using buffered I/O operations, and sorts the new $Temp$ list according to the predecessor states u. The algorithm then iterates on all edges from $Open_t$ and searches for the successors in $Temp$. Since $Open_t$ is sorted with respect to states v, *the algorithm never goes back and forth in any of the* Open$_t$ *or* Temp *files.* Note that all reads and writes are buffered and thus can be carried out very efficiently by always doing I/O operations in blocks. Four different cases arise when an edge $(u, v, a, f(v))$ is read from $Open_t$. (States from Fig. 8 are referred in parentheses.)

- *Case I:* v is terminal (states 8 &10). Since no update is necessary, the edge can be written to $Open_{t+1}$.
- *Case II:* v is the same as the last updated state (state 3). Write the edge to $Open_{t+1}$ with such last value. (Case shown in Fig. 8 with curved arrows.)
- *Case III:* v has no successors. That means that v is a terminal state and so is handled by case I.
- *Case IV:* v has one or more successors (remaining states). For each action $a \in A(v)$, compute the value $q(a, v)$ by summing the products of the probabilities and the stored values. Such value is kept in the array $q(a)$.

For edges (x, y, a', f') read from $Temp$, we have

- *Case A:* y is the initial state, implying $x = ptyset$. Skip this edge since there is nothing to do. By taking $ptyset$ as the smallest element, the sorting of $Temp$ brings all such edges to the front of the file. (Case not shown.)

Table 4. Performance of external-memory value iteration on deterministic $(p = 1)$ and probabilistic variants $(p = 0.9)$ of the 8-puzzle with and without initialization to the Manhattan distance heuristic.

Algorithm	p	$|S|/|E|$	Iteration	Updates	$h(s)$	$f^*(s)$	RAM
VI $(h = 0)$	1.0	181,440	27	4,898,880	0	14.00	21M
External-VI $(h = 0)$	1.0	483,839	32	5,806,048	0	14.00	11M
VI (h_{MD})	1.0	181,440	20	3,628,800	10	14.00	21M
External-VI (h_{MD})	1.0	483,839	28	5,080,292	10	14.00	11M
VI $(h = 0)$	0.9	181,440	37	6,713,280	0	15.55	21M
External $-$ VI $(h = 0)$	0.9	967,677	45	8,164,755	0	15.55	12M
VI (h_{MD})	0.9	181,440	35	6,350,400	10	15.55	21M
Ext-VI (h_{MD})	0.9	967,677	43	7,801,877	10	15.55	12M

– *Case B:* $x = v$, i.e. the predecessor of this edge matches the current state from $Open_t$. This calls for an update in the $q(a)$-value.

The array $q : A \to I\!R$ is initialized to the edge weight $w(a, v)$, for each $a \in A(v)$. Once all the successors are processed, the new value for v is the minimum of the values stored in the q-array for all applicable actions.

The backward phase performs at most $t_{\max}$ iterations. Each iteration consists of one sorting and two scanning operations for a total of $O(t_{\max} \cdot sort(|E|))$ I/Os.

For the sliding-tile puzzles we performed two experiments: one with deterministic moves, and the other with noisy actions that achieve their intended effects with probability $p = 0.9$ and no effect with probability $1 - p$. Table 4 shows the results for random instances of the 8-puzzle for both experiments. The rectangular 3×4 sliding-tile puzzle with $p = 0.9$ cannot be solved with internal value iteration because the state space did not fit in RAM. External-memory value iteration generated a total of 1,357,171,197 edges taking 45 GBs of disk space. The backward update finished successfully after 21 days in 72 iterations using 1.4 GBs RAM. The value function for initial state converged to 28.8889 with a residual smaller than $\epsilon = 10^{-4}$.

6   Parallel External-Memory Search

Combined parallel and disk-based search executes an exploration in distributed environments like multi-processor machines and workstation clusters.

Recent parallel implementation of A* and its derivatives on multi-core machines have been proposed by [59] with a subsequent scaling analysis in [58] and by [15, 16]. Our focus is the interplay of parallel and external-memory search.

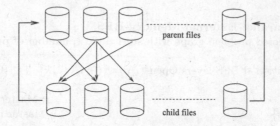

Fig. 9. Externally stored state space with parent and children files.

6.1 Parallel External-Memory Breadth-First Search

In *parallel external-memory breadth-first search* the state space is partitioned into different files using a global hash function. For example in state spaces like the 15-puzzle that are regular permutation games, each node can be perfectly hashed to a unique index, and some prefix of the state vector can be used to for partitioning. If state spaces are undirected, *frontier search* [70] can distinguish neighboring nodes that have already been explored from those that have not, in order to omit the *Closed* list. Figure 9 depicts the layered exploration on the external partition of the state space. A hash function partitions both the current *parent* layer and the *children* layer for the successors into files. If a layer is done, children files are renamed into parent files to iterate the exploration.

Even on a single processor, multiple threads maximize the performance of the disks. The reason is that a single-threaded implementation will block until the read from or write to disk has completed.

Hash-based duplicate detection generates a suitable partition for the 15-puzzle. Within one iteration, most file accesses can be performed independently. Only if one simultaneously expands two parent files have a children file in common, the two processes will be in conflict. To realize parallel processing a work queue is maintained, which contains parent files waiting to be expanded, and child files waiting to be merged. At the start of each iteration, the queue is initialized to contain all parent files. Once all parents of a child file are expanded, the child file is inserted into the queue for *early* merging.

Each process works as follows. It first locks the work queue. The algorithm checks whether the first parent file conflicts with any other file expansion. If so, it scans the queue for a parent file with no conflicts. It swaps the position of that file with the one at the head of the queue, grabs the non-conflicting file, unlocks the queue, and expands the file. For each file it generates, it checks if all of its parents have been expanded. If so, it puts the children file at the head of the queue for expansion, and then returns to the queue for more work. If there is no more work in the queue, any idle process wait for the current iteration to complete. At the end of each iteration the work queue is re-initialized to contain all parent files for the next iteration. Algorithm 1.2 shows a pseudo-code implementation.

Procedure Parallel-External-Memory-BFS
Input: Undirected problem graph with start node s, number of processes N
hash function ψ
Output: Partitioned BFS layers $Open_j(i)$, $i \in \{0, 1, \ldots, k\}$, $j \in \{0, 1, \ldots, N\}$

$g \leftarrow 0$	;; Master initializes layer
$Open_0(g) \leftarrow \{s\}$	;; Master initializes search
while $(\cup_{i=1}^{N} Open_i(g) = \emptyset)$	;; Search not terminated
for each $j \in \{1, \ldots, N\}$ **in parallel**	;; Distribute computation
if $(Goal(Open_j(g))$	;; Terminal state in set
return $Announce(GoalFound)$	;; Generate solution path
$A_j \leftarrow Succ(Open_j(g))$	;; Generated successors
$RemoveDuplicates(A_j)$	;; Sorting/Scanning current elements
for each $j \in \{1, \ldots, N\}$ **in parallel**	;; Distribute computation
$A'_j \leftarrow \{v \in \cup_{i=1}^{N} A_i \mid \psi(v) = j\}$	;; Acquire nodes to sort
$RemoveDuplicates(A'_j)$	;; Sorting/scanning
$Open_j(g+1) \leftarrow A'_j \setminus (Open_j(g) \cup Open_j(g-1))$	;; Frontier subtraction
$g \leftarrow g+1$	;; Increase depth
return $Open_j(i)$, $i \in \{0, 1, \ldots, k\}$, $j \in \{0, 1, \ldots, N\}$	

Algorithm 1.2. Parallel external-memory breadth-first search for state space enumeration.

6.2   Parallel Structured Duplicate Detection

SDD performs early duplicate detection in the RAM. Each abstract state represents a file containing every concrete states mapping to it. As all adjacent abstract states were load into main memory, duplicate detection for concrete successor states remains in the RAM. We assume breadth-first heuristic search as the underlying algorithm, which generates the search space with increasing depth, but prunes it with respect to the f-value, provided that the optimal solution length is known. If not, external-memory A* applies.

SDD extends nicely to a parallel implementation. In *parallel SDD* [98] abstract states together with their abstract neighbors are assigned to a process. We assume that the parallelization takes care of synchronization after one breadth-first search iteration has been completed, as a concurrent expansion in different depths likely affects the algorithm's optimality.

If in one BFS-layer, two abstract nodes together with their successor do not overlap, their expansion can be executed fully independently on different processors. More formally, let $\phi(u_1)$ and $\phi(u_2)$ be the two abstract nodes, then the scopes of $\phi(u_1)$ and $\phi(u_2)$ are disjoint if $Succ(\phi(u_1)) \cap Succ(\phi(u_2)) = \emptyset$. This parallelization maintains locks only for the abstract space. No locks for individual states are needed.

The approach applies to both, shared and distributed memory architectures. In the shared implementation each processor has a private memory pool. As soon as this is exhausted it asks the master process (that has spawned it as a child

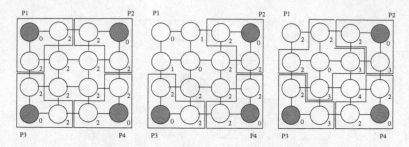

Fig. 10. Example for parallel SDD with 4 processes: before P_1 releases its work, after P_1 has released his work, after P_1 has allocated new work.

process) for more memory that might have been released using a completed exploration by some other process. For a proper (conflict-free) distribution of work, numbers $I(\phi(u))$ were assigned to each abstract node $\phi(u)$, denoting the accumulated influence that currently imposed to this node by running processes. If $I(\phi(u)) = 0$ the abstract node $\phi(u)$ can be picked for expansion from every processor that is currently idle. Function I is updated as follows. In a first step, for all $\phi(v) \neq \phi(u)$ with $\phi(u) \in Succ(\phi(v))$ value $\phi(v)$ is incremented by one: all abstract nodes that include $\phi(u)$ in their scope cannot be expanded, since $\phi(u)$ is chosen for expansion. In a second step, for all $\phi(v) \neq \phi(u)$ with $\phi(v) \in Succ(\phi(u))$ and all $\phi(w) \neq \phi(v)$ with $\phi(w) \in Succ(\phi(v))$ value $\phi(v)$ is incremented by one: all abstract nodes that include any $\phi(v)$ as a successor of $\phi(u)$ cannot be expanded, since they are also assigned to the processor.

Figure 10 illustrates the working of parallel structural duplicate detection for the 15-puzzle with the currently expanded abstract nodes shaded. The left-most part of figure shows the abstract problem graph together with 4 processes working independently at expanding abstract states. The numbers $I(\phi(u))$ are associated with each abstract node $\phi(u)$. The middle part of the figure depicts the situation after one process has finished, the right part shows the situation after process has been assigned to a new abstract state.

6.3   Parallel External-Memory A*

The distributed version of external-memory A*, called *parallel-external-memory A** is based on the observation that the internal work in each individual bucket of external-memory A* can be parallelized among different processes. More precisely each two states in a bucket $Open(g, h)$ can be expanded in different processes at the same time. An illustration is given in Fig. 11, indicating a uniform partition available for each $Open(g, h)$-bucket. We discuss disk-based message queues to distribute the load among different processes.

To organize the communication between the processes a work queue is maintained on disk. The work queue contains the requests for exploring parts of a (g, h)-bucket together with the part of the file that has to be considered (as processes may have different computational power and processes can dynamically

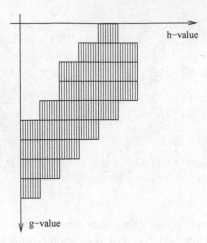

Fig. 11. Partitioning of buckets in parallel external-memory A*.

join and leave the exploration, the size of the state space partition does not necessarily have to match the number of processes. By utilizing a queue, one also may expect a process to access a bucket multiple times. However, for the ease of a first understanding, it is simpler to assume that the jobs are distributed uniformly among the processes.) For improving the efficiency, we assume a distributed environment with one master and several slave processes. In the implementation, the *master* is in fact an ordinary process defined as the one that finalized the work for a bucket. The applies to both the cases when each slave has its own hard disk or if they work together on one hard disk e.g. residing on the master. We do not expect all processes to run on one machine, but allow slaves to log-on the master machine, suitable for workstation clusters. Message passing between the master and slave processes is purely done on files, so that all processes are fully autonomous. Even if slave processes are killed, their work can be re-done by any other idle process that is available.

One file that we call the *expand-queue*, contains all current requests for exploring a node set that is contained in a file. The filename consists of the current g- and h-value. In case of larger files, file-pointers for processing parts of a file are provided, to allow for better load balancing. There are different strategies to split a file into equi-distance parts or into chunks depending on the number and performance of logged-on slaves. As we want to keep the exploration process distributed, we select the file pointer windows into equidistant parts of a fixed number of C bytes for the nodes to be expanded. For improved I/O, the number C is supposed to divide the system's block size B. As concurrent read operations are allowed for most operating systems, multiple processes reading the same file impose no concurrency conflicts.

The expand-queue is generated by the master process and is initialized with the first block to be expanded. Additionally, we maintain the total number of requests, i.e., the size of the queue, and the current number of satisfied requests.

Any logged-on slave reads a request and increases the count once it finishes. During the expansion process, in a subdirectory indexed by the slave's name it generates files that are indexed by the g- and h-value of the successor nodes.

The other queue is the *refine-queue* also generated by the master process once all processes are done. It is organized in a similar fashion as the expand queue and allows slaves to request work. The refine-queue contains filenames that have been generated above, namely the slave-name (that does not have to match with the one of the current process), the block number, and the g- and h-value. For a suitable processing the master process will move the files from subdirectories indexed by the slave's name to ones that are indexed by the block number. As this is a sequential operation executed by the master thread, changing the file locations is fast in practice. To avoid redundant work, each process eliminates the requests from the queue. Moreover, after finishing the job, it writes an acknowledge to an associated file, so that each process can access the current status of the exploration, and determine if a bucket has been completely explored or sorted.

All communication between different processes can be shared files, so that a message passing unit is not required. However, a mechanism for mutual exclusion is necessary. A rather simple but efficient method to avoid concurrent writes accesses is the following. Whenever a process has to write on a shared file, it issues an operating system command to rename the file. If the command fails, it implies that the file is currently being used by another process.

For each bucket that is under consideration, we establish four stages in the algorithm with a pseudo-code shown in Algorithm 1.3. The four phases are visualized in Fig. 12 (top to bottom). Zig-zag curves illustrate the order of the nodes in the files wrt. the comparison function used. As the states are presorted in internal memory, every peak correspond to a flushed buffer. The sorting criteria itself is defined first by the node's hash key and then by the low-level comparison based on the (compressed) state vector.

In the *exploration stage* (generating the first row in the figure), each process p flushes the successors with a particular g- and h-value to its own file (g, h, p). Each process has its own hash table and eliminates some duplicates already in main memory. The hash table is based on chaining, with chains sorted along the node comparison function. However, if the output buffer exceeds memory capacity it writes the entire hash table to disk. By the use of the sorting criteria as given above, this can be done using a mere scan of the hash table.

- In the *first sorting stage* (generating the second row in the figure), each process sorts its own file. In the distributed setting we exploit the advantage that the files can be sorted in parallel, reducing internal processing time. Moreover, the number of file pointers needed is restricted by the number of flushed buffers, illustrated by the number of peaks in the figure. Based on this restriction, we only need a merge of different sorted buffers.
- In the *distribution stage* (generating the third row in the figure), all nodes in the presorted files are distributed according to the hash value's range. As all input files are presorted this is a mere scan. No all-including file is generated,

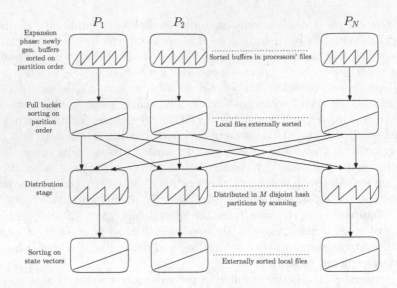

Fig. 12. Stages of bucket expansions in parallel external-memory A*.

keeping the individual file sizes small. This stage can be a bottleneck to the parallel execution, as processes have to wait until the distribution stage is completed. However, if we expect the files to reside on different hard drives, traffic for file copying can be parallelized.

- In the *second sorting stage* (generating the last row in the figure), processes resort the files (with buffers presorted wrt. the hash value's range), to find further duplicates. The number of peaks in each individual file is limited by the number of input files (=number of processes), and the number of output files is determined by the selected partitioning of the hash index range. Using the hash index as the sorting key we establish that the concatenation of files is sorted.

Figure 13 shows the distribution of a bucket among three processors.

6.4   Parallel Pattern Database Search

Disjoint pattern databases can be constructed embarrassingly parallel. The subsequent search, however, faces the problem of high memory consumption due to many large pattern databases, since loading pattern databases on demand significantly slows down the performance.

One solution is to distribute the lookup to multiple processes. For external-memory A* this works as follows. As buckets are fully expanded, the order in a bucket does not matter, so that we can distribute the work for expansion, evaluation and duplicate elimination. For the 35-puzzle we choose one master to distribute generated states to 35 client processes P_i, each one responsible for

Procedure Parallel-External-Memory-A*

Input: Undirected problem graph with start node s, predicate $Goal$, N processes
hash function ψ
Output: Optimal solution path

$g \leftarrow 0; h \leftarrow h(s)$;; Initial bucket
$Open_0(g, h) \leftarrow \{s\}$;; Master initializes search
while not $(goalFound)$;; Search not terminated
   for each $j \in \{1, \ldots, N\}$ **in parallel**   ;; Distribute computation
     if $(Goal(Open_j(g, h))$;; Terminal state in set
       return $Announce(GoalFound)$;; Generate solution path
     $A_j(h-1), A_j(h), A_j(h+1) \leftarrow Succ(Open_j(g, h))$;; Generated successors
     $Open_j(g+1, h+1) \leftarrow A_j(h+1)$;; Prepare next level
     $Open_j(g+1, h) \leftarrow A_j(h) \cup Open_j(g+1, h)$;; Prepare next level
     $RemoveDuplicates(A_j(h-1))$;; Sorting/Scanning
   for each $j \in \{1, \ldots, N\}$ **in parallel**   ;; Distribute computation
     $A'_j(h-1) \leftarrow \{u \in \cup_{i=1}^{N} A_i(h-1) \mid \psi(u) = j\}$;; Allocate work
     $Open_j(g+1, h-1) \leftarrow A'_j(h-1) \cup Open_j(g+1, h-1)$;; Prepare next level
     $RemoveDuplicates(Open_j(g+1, h-1))$;; Sorting/scanning
     $Open_j(g+1, h-1) \leftarrow Open_j(g+1, h-1) \backslash$;; Eliminate duplicates
       $(Open_j(g, h-1) \cup Open_j(g-1, h-1))$
   $f \leftarrow \min\{k + l \mid \cup_{i=1}^{N} Open_i(k, l) \neq \emptyset\}$;; Update f-value
   $g \leftarrow \min\{l \mid \cup_{i=1}^{N} Open_i(l, f-l) \neq \emptyset\} \ h \leftarrow f - g$;; Next non-empty bucket

Algorithm 1.3. Parallel external-memory A* for consistent and integral heuristics.

one tile i for $i \in \{1, \ldots, 35\}$. All client processes operate individually on different
processing nodes and communicate via shared files.

During the expansion of a bucket (see Fig. 14), the master writes a file T_i
for each client process P_i, $i \in \{1, \ldots, 35\}$. Once it has finished the expansion
of a bucket, the master P_m announces that each P_i should start evaluating T_i.
Additionally, the client is informed on the current g- and h-value. After that,
the master P_m is suspended, and waits for all P_i's to complete their task. To
relieve the master from load, no sorting takes place during distribution. Next,
the client processes start evaluating T_i, putting their results into $E_i(h-1)$ or
$E_i(h+1)$, depending on the observed difference in the h-values. All files E_i are
additionally sorted to eliminate duplicates; internally (when a buffer is flushed)
and externally (for each generated buffer). As only 3 buckets are opened at a
time (1 for reading and 2 for writing) the associated internal buffers can be large.

After the evaluation phase is completed, each process P_i is suspended. When
all clients are done, the master P_m is resumed and merges the $E_i(h-1)$ and
$E_i(h+1)$ files into $E_m(h-1)$ and $E_m(h+1)$. The merging preserves the order in
the files $E_i(h-1)$ and $E_i(h+1)$, so that the files $E_m(h-1)$ and $E_m(h+1)$ are
already sorted with all duplicates within the bucket eliminated. The subtraction

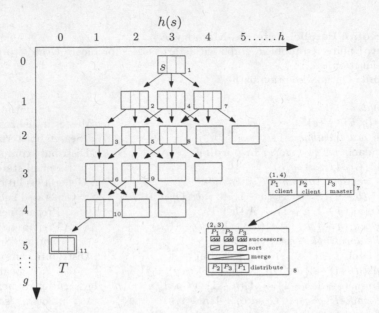

Fig. 13. Distribution of buckets in parallel external-memory A*.

of the bucket $(g-1, h-1)$ from $E_m(h-1)$ and $(g-1, h+1)$ from $E_m(h+1)$ now eliminates duplicates from the search using a parallel scan of both files.

Besides the potential for speeding up the evaluation, the chosen distribution mainly saves space. On the one hand, the master process does not need any additional memory for loading pattern databases. It can invest all its available memory for internal buffers required for the distribution, merging and subtraction of nodes. On the other hand, during the lifetime of client process P_i, it maintains only the pattern database D_j that includes tile i in its pattern.

7   Parallel Search on the GPU

In the last few years there has been a remarkable increase in the performance and capabilities of the graphics processing unit (GPU). Modern GPUs are not only powerful, but also parallel programmable processors featuring high arithmetic capabilities and memory bandwidths. High-level programming interfaces have been designed for using GPUs as ordinary computing devices. These efforts in *general purpose GPU programming* (GPGPU) has positioned the GPU as a compelling alternative to traditional microprocessors in high-performance computing. The GPU's rapid increase in both programmability and capability has inspired researchers to map computationally demanding, complex problems to it. Since the memory transfer between the card and main board on the express bus is extremely fast, GPUs have become an apparent candidate to speed-up large-scale computations. GPUs have several cores, but the programming and

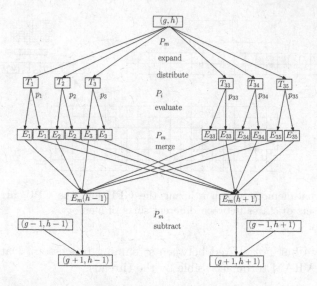

Fig. 14. Distributed expansion/evaluation of one bucket.

computational model are different from the ones on the CPU. A core is a *streaming processor* with some floating point and arithmetic logic units. Together with some special function units, streaming processor are grouped together to form streaming multiprocessors. Programming a GPU requires a special compiler, which translates the code to native GPU instructions. The GPU architecture mimics a *single instruction multiply data* computer with the same instructions running on all processors. It supports different layers for accessing memory. GPUs forbid simultaneous writes to a memory cell but support concurrent reads.

GPUs have outpaced CPUs in numerical algorithms [46,72]. Applications include studying the folding behavior of proteins by [57] and the simulation of bio-molecular systems by [79]. Since the memory transfer between the card and main board on the express bus is in the order gigabytes per second, GPUs have become an apparent candidate to speed-up large-scale computations like sorting numerical data on disk [18,44]. Its application for sorting-based delayed duplicate detection is apparent. By using perfect hash functions there is work on exploring single-agent search problems on the GPU [41], and on solving two-player games [39]. Moreover, explicit-state and probabilistic model checking problems have been ported to the GPU [11,38].

On the GPU, memory is structured hierarchically, starting with the GPU's global memory called *video RAM*, or *VRAM*. Access to this memory is slow, but can be accelerated through *coalescing*, where adjacent accesses with less than word-width number bits are combined to full word-width access. Each streaming multi-processor includes a small amount of memory called *SRAM*, which is shared between all streaming multi-processor and can be accessed at the same speed as registers. Additional registers are also located in each streaming

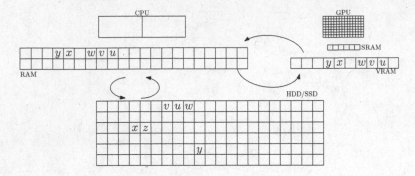

Fig. 15. External-memory search utilizing the GPU and the CPU, arrows indicate movements of sets of states between different sorts of memory.

multiprocessors but not shared between streaming processors. Data has to be copied to the VRAM to be accessible by the threads.

GPU-Based Breadth-First-Search. We assume a hierarchical GPU memory structure of SRAM (small, but fast and parallel access) and VRAM (large, but slow access). The general setting is displayed in Fig. 15. We illustrate how to perform GPU-based breadth-first search, enumerating the entire search space.

Algorithm 1.4 displays the main search algorithm running on the CPU. For each BFS-level it divides into two computational parts that are executed on the GPU: applying actions to generate the set of successors, and detecting and eliminating duplicates in a delayed fashion via GPU-based sorting. We keep the search frontier and the set of visited states distinct, as only the first one needs to be accessible in uncompressed form.

Delayed Duplicate Detection on the GPU. For delayed elimination of duplicates, we have to order a BFS level wrt. a comparison function that operates on states (sorting phase). The array is then scanned and duplicates are removed (compaction). Considering the strong set of assumptions of orthogonal, disjoint and concise hash functions, ordinary hash-based delayed duplicate detection is often infeasible. Therefore, we propose a trade-off between sorting-based and hash-based delayed duplicate detection by sorting buckets that have been filled through applying a hash function. The objective is that hashing in RAM performs more costly distant data moves, while subsequent sorting addresses local changes, and can be executed on the GPU by choosing the bucket sizes appropriately. If the buckets fit into the SRAM, they can be processed in parallel.

Disk-based sorting refers to one of the major success stories for GPU computation. Various implementations have been proposed, including variants of Bitonic Sort and GPU-based Quicksort. Applying the algorithms on larger state vectors fails as their movement within the VRAM slows down the computation significantly. Trying to sort an array of indexes also fails, as now the comparison operator exceeds the boundary of the SRAM. This leads to an alternative design of GPU sorting for state space search.

Procedure GPU-BFS
Input: State space problem with initial state s
Output: State space partitioned into layers

$g \leftarrow 0;\ Open(g) \leftarrow \{s\}$;; Initialize search
while $(Open(g) \neq \emptyset)$;; Until search levels off
  $Open(g+1) \leftarrow Closed \leftarrow OpenPart \leftarrow \emptyset$;; Initialize sets
  for each u **in** $Open(g)$;; Process BFS-level
    $OpenPart \leftarrow OpenPart \cup \{u\}$;; Add node to part
    if $(|OpenPart| = |VRAM|)$;; RAM temporary for VRAM
      $Closed \leftarrow Closed \cup GPU\text{-}Expand(OpenPart)$;; Call kernel
      $OpenPart \leftarrow \emptyset$;; Reinitialize structure
    $Closed \leftarrow Closed \cup GPU\text{-}Expand(OpenPart)$;; Call kernel function
    for each $v \in Closed$;; Consider all successors
      $H[hash(v)] \leftarrow H[hash(v)] \cup \{v\}$;; Insert in bucket
      if $H[hash(v)]$ **full**   ;; Overflow in bucket
        $Sorted \leftarrow GPU\text{-}DetectDuplicates(H)$;; Call kernel function
        $CompactedOpen \leftarrow ScanAndRemoveDuplicates(Sorted)$;; Compaction
        $DuplicateFreeOpen \leftarrow Subtract(CompactedOpen, Open(0..g))$;; Subtraction
        $Open(g+1) \leftarrow Merge(Open(g+1), DuplicateFreeOpen)$;; Combine result
        $H[0..m] \leftarrow \emptyset$;; Reset layer
    $Sorted \leftarrow GPU\text{-}DetectDuplicates(H)$;; Call kernel function
    $CompactedOpen \leftarrow ScanAndRemoveDuplicates(Sorted)$;; Compaction
    $DuplicateFreeOpen \leftarrow Subtract(CompactedOpen, Open(0..g))$;; Subtraction
    $Open(g+1) \leftarrow Merge(Open(g+1), DuplicateFreeOpen)$;; Combine result
  $g \leftarrow g+1$;; Next layer
return $Open(0..g-1)$;; Final result on disk

Algorithm 1.4. Large-scale breadth-first search on the GPU.

In *hash-based partitioning* the first phase of sorting smaller blocks in Bitonic Sort is fast, while merging the pre-sorted sequences for a total ordered slows down the performance. Therefore, we employ hash-based partitioning on the CPU to distribute the elements into buckets of adequate size (see Fig. 16). The state array to be sorted is scanned once. Using hash function h and a distribution of the VRAM into k blocks, the state s is written to the bucket with index $h'(s) = h(s)\ mod\ k$. If the distribution of the hash function is appropriate and the maximal bucket sizes are not too small, a first overflow occurs, when the entire hash table is occupied to more than a half. All remaining elements are set to a pre-defined illegal state vector that realizes the largest possible value in the ordering of states.

This hash-partitioned vector of states is copied to the graphics card and sorted by the first phase of Bitonic Sort. The crucial observation is that – due to the presorting – the array is not only partial sorted wrt. the comparison function operating on states s, but totally sorted wrt. the extended comparison function

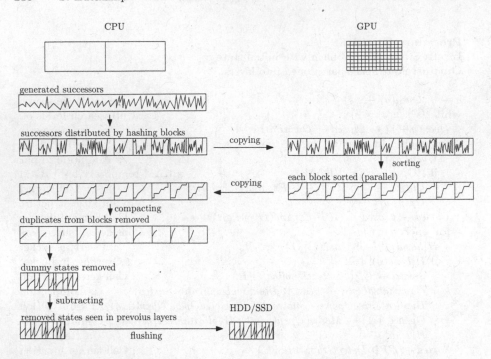

Fig. 16. Hash-based partitioning, total order for set of states is the combination of hash address (computed on the CPU) and sorting index (computed on the GPU).

operating on the pairs $(h'(s), s)$. The sorted vector is copied back from VRAM to RAM, and the array is compacted by eliminating duplicates with another scan through the elements. Subtracting visited states is made possible through scanning all previous layers residing on disk. Finally, we flush the duplicate-free file for the current BFS level to disk and iterate. To accelerate discrimination and to obey the imposed order on disk, the hash bucket value $h'(s)$ is added to the front of the state vector s.

If a BFS level becomes too large to be sorted on the GPU, we split the search frontier into parts that fit in the VRAM. This yields some additional state vector files to be subtracted to obtain a duplicate-free layer, but in practice time performance is still dominated by expansion and sorting. For the case that subtraction becomes harder, we can exploit the hash-partitioning, inserting previous states into files partitioned by the same hash value. States that have a matching hash value are mapped to the same file. Provided that the sorting order is first on the hash value then on the state, after the concatenation of files (even if sorted separately) we obtain a total order on the sets of states. This implies that we can restrict duplicate elimination to states that have matching hash values.

On the GPU, we have a fixed amount of $O(|\text{VRAM}|/|\text{SRAM}|)$ group operations, where each group is sorted by Bitonic Sort. Hence, the sorting complexity is independent from the number of elements to be sorted, as in each iteration the entire vector is processed. With a good distribution function, we assure that on the average each bucket is at least 50% filled with successor states, such that we loose less than factor 2 by not dealing with entirely filled buckets. As an example, in our case, we have $|\text{VRAM}| = 1$ GB, and $|\text{SRAM}| = (16 - c)$ KB, where c is a small constant, imposed by the internal memory requirements of the graphics card. For a state vector of 32 byte, we arrive at $k = 256$ elements in one group. Within each group Bitonic Sort is applied, known to induce $O(k \log^2 k)$ work and $O(\log k)$ iterations. In each iteration the number of comparisons that can be executed in parallel depends on the number of available threads, which in turn depends on the graphics card chosen.

Instead of sorting the buckets after they have been filled, it is possible to use chaining right away, checking each individual successor for having a duplicate against the states stored in its bucket. Keeping the list of states sorted, as in ordered hashing, accelerates the search, however, requires additional work for insertion, and does not speed up the computation, if compared to parallel sorting the buckets on the GPU. We only implemented a refinement that checks the state to be inserted in a bucket with the top element to detect some duplicates quickly.

State Compression. With a 64-bit hash address we do not encounter any collision even in very large state spaces. Henceforth, given hash function h, we compress the state vector for u to $(h(u), i(u))$, where $i(u)$ is the index of the state vector residing in RAM that is needed for expansion. We sort the pairs on the GPU with respect to the lexicographic ordering of h. The shorter the state vector, the more elements fit into one group, and the better the expected speed-up.

To estimate the probability of an error, assume a state space of $n = 2^{30}$ elements uniformly hashed to the $m = 2^{64}$ possible bit-vectors of length 64. We have $m!/(m^n(m-n)!) \geq ((m-n+1)/m)^n \geq (1 - n/m)^n$. For our case this resolves to $(1 - 2^{-34})^{2^{30}} = (.99999999994179233909)^{1073741824}$, and a confidence of at least 93.94% that no duplicate arises while hashing the entire state space to 64 bits. Recall, that missing a duplicate harms, only if the missed state is the only way to reach the error in the system. If the above confidence appears still to be too low, one may re-run the experiment with another independent hash function, showing that with $\geq 99.6\%$, no false positive has been produced during the traversal of the state space.

Expansion on the GPU. The remaining bottleneck is the CPU performance in generating the successors, which can also be reduced by applying parallel computation. For this we port the expansion for states to the GPU.

For BFS, the order of expansions within one bucket does not matter, so that no communication between threads is required. Each processor simply takes its share and starts expanding. Having fixed the set of applicable actions for each state, generating the successors in parallel on the GPU is immediate by replicating each state to be expanded by the number of applicable actions. All generated states are copied back to RAM (or GPU sorting is applied).

Procedure Rank
Input: Depth N, permutations π, π^{-1}
Output: Rank of π
if $(N = 1)$ **return** 0
$l \leftarrow \pi_{N-1}$
$Swap(\pi_{N-1}, \pi_{\pi_{N-1}^{-1}})$; $Swap(\pi_l^{-1}, \pi_{N-1}^{-1})$
return $l(N - 1)! + Rank(N - 1, \pi, \pi^{-1})$

Procedure Unrank
Input: Value N, rank r, permutation π
Side Effect: Updated global permutation
if $(N = 0)$ **return**
$l \leftarrow \lfloor r/(k - 1)! \rfloor$
$Swap(\pi_{N-1}, \pi_l)$
$Unrank(N - 1, r - l \cdot (N - 1)!, \pi)$

Algorithm 1.5. *Rank* and *Unrank* operation for permutations.

Bitvector GPU Search. Static perfect hashing has been devised in the early 70th [22,43]. Practical perfect hashing has been analyzed by [12] and an external-memory perfect hash function variant has been proposed by [13].

For the design of a minimum perfect hash function of the sliding-tile puzzles we observe that in a lexicographic ordering every two successive permutations have an alternating signature and differ by exactly one transposition. For minimal perfect hashing a $(n^2 - 1)$-puzzle state to $\{0, \ldots, n^2!/2 - 1\}$ we consequently compute the lexicographic *rank* and divide it by 2. For unranking, we now have to determine, which one of the two uncompressed permutations of puzzle is reachable. This amounts to finding the signature of the permutation, which allows to separate solvable from unsolvable states.

There is one subtle problem with the blank. Simply taking minimum perfect hash value for the alternation group in S_{n^2} does not suffice, as swapping a tile with the blank not necessarily toggles the solvability status (e.g., it may be a move). To resolve this problem, we partition state space along the position of the blank. Let $B_0, \ldots, B_{n^2-1}$ denote the sets of blank-projected states. Then each B_i contains $(n^2 - 1)!/2$ elements. Given index i and the rank inside B_i, it is simple to reconstruct the state.

Korf and Schultze [68] used lookup tables to compute lexicographic ranks, while Bonet [9] discussed different time-space trade-offs. Mares and Straka [75] proposed a linear-time algorithm for lexicographic ranking, which relies on bitvector operations in constant time. Applications of perfect hashing for bitvector state space search include Peg Solitaire [41], Nine-Men-Morris [39], and Chineese Checkers [90,91]. Bitvector-compressed pattern databases result in $\log_2 3 \approx 1.6$ bits per state [14]. Efficient permutation indices have been proposed by Myrvold and Ruskey [78]. The basic motivation is the generation of a random permutation of size N according to swapping π_i with π_r where r is a random number uniformly chosen in $0, \ldots, r$, and i decreases from $N - 1$ to 1. The (recursive) procedure *Rank* is shown in Algorithm 1.5. The permutation π and its inverse π^{-1} are initialized according with the permutation, for which a rank has to determined. The inverse π^{-1} of π can be computed by setting $\pi_{\pi_i}^{-1} = i$, for all $i \in \{0, \ldots, k - 1\}$. Take as an example permutation $\pi = \pi^{-1} = (1, 0, 3, 2)$. Then its rank is $2 \cdot 3! + Rank(102)$. This unrolls to $2 \cdot 3! + 2 \cdot 2! + 0 \cdot 1! + 0 \cdot 0! = 16$. It is also possible to compile a rank back into a permutation in linear time.

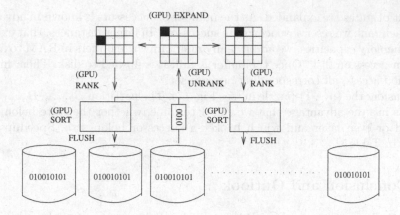

Fig. 17. GPU exploration of the 15-puzzle stored as a bitvector in RAM (GPU sorting indices is optional and was not used in the experiments).

The inverse procedure *Unrank* is initialized with the identity permutation and shown in Algorithm 1.5.

Table 5. Comparing CPU with GPU performances in Bit (vector)-BFS in various sliding-tile puzzles (o.o.m means out of memory, while o.o.t denotes out of time).

	2-Bit BFS		1-Bit BFS	
Problem	Time GPU	Time CPU	Time GPU	Time CPU
(2×6)	70 s	176 s	163 s	1517 s
(3×4)	55 s	142 s	98 s	823 s
(4×3)	64 s	142 s	104 s	773 s
(6×2)	86 s	160 s	149 s	1110 s
(7×2)	o.o.m	o.o.m	13590 s	o.o.t

In case of perfect and inversible hash functions, a bitvector exploration of the search space is fortunate. The GPU-assisted exploration will rank and unrank states during the expansion process. In constant-bit(vector) BFS search [65] the entire or the partitioned state space bitvector is kept in RAM, while copying an array of indices (ranks) to the GPU. One additional scan through the bitvector is needed to convert its bits into integer ranks, but on the GPU the work to unrank, generate the successors and rank them is identical for all threads. To avoid unnecessary memory access, the rank given to expand should be overwritten with the rank of the first child. For smaller BFS layers this means that a smaller amount of states are expanded. As the number of successors is known in advance, with each rank we reserve space for its successors. In larger instances, that exceed main memory capacities, we additionally maintain write buffers in RAM to avoid random access on disk. Once the buffer is full, it is flushed to disk. Then, in one streamed access, all corresponding bits are set.

Consider the (n^2-1)-puzzle in (see Fig. 17). The partition $B_0, \ldots, B_{n^2-1}$ into buckets has the advantage that we can determine, whether the state belongs to an odd or even layer and which bucket a successor belongs to. Speedups are shown in Table 5.

8   Conclusion and Outlook

Mechanical hard disks have provided us with reliable service over the years and, as shown in this text, lead to changes in the design of search algorithms for solving large exploration problems that exceed main memory capacity. Their dominance at least on mobile devices has changed with the advent of solid state disks (SSDs). An SSD is electrically, mechanically and software compatible with a conventional (magnetic) hard disk drive (HDD). The difference is that the storage medium is not magnetic (like a hard disk) or optical (like a CD) but solid state semiconductor (NAND flash) such as battery-backed RAM or other electrically erasable RAM-like chips. In last years, NAND flash memories outpaced RAM in terms of bit-density and the market with SSDs continues to grow. This provides faster access time than a disk, because the data can be randomly accessed and does not rely on a read/write interface head synchronizing with a rotating disk. The speed of random reads for a SSD build lies roughly at the geometric mean of the speeds of RAM and a magnetic HDD. The only factor limiting SSDs from being massively spread is their cost: the price per stored bit is still significantly higher for SSDs than for HDDs.

We observe that random read operations on SSDs are substantially faster than on mechanical disks, while other parameters are similar. For SSDs, therefore, an extension of the external-memory model is needed. Moreover, it appears natural to ask, whether it is necessary to employ *delayed duplicate detection* (DDD) known from the current I/O-efficient graph search algorithms, or if is possible to design efficient SSD algorithms using standard *immediate duplicate detection* (IDD), hashing in particular [3,6,36].

We also emphasize that not only it is true that *Disk is the New RAM* (as exploited in external-memory algorithms), but also that *RAM is the New Disk*. On large-scale main memory storage systems, e.g., on supercomputers with thousands of cores, algorithm locality is a more and more performance-critical issue. Profiling in our setting revealed that the mere retrieval of a single bit in RAM became the bottleneck in the parallel-external exploration.

References

1. Aggarwal, A., Vitter, J.S.: The input/output complexity of sorting and related problems. J. ACM **31**(9), 1116–1127 (1988)
2. Ajwani, D., Dementiev, R., Meyer, U.: A computational study of external-memory BFS algorithms. In: SODA, pp. 601–610 (2006)
3. Ajwani, D., Malinger, I., Meyer, U., Toledo, S.: Graph search on flash memory. MPI-TR (2008)
4. Allen, J.D.: The Complete Book of CONNECT 4: History, Strategy, Puzzles. Sterling Publishing, New York (2011)
5. Allis, L.V.: A knowledge-based approach to connect-four. The game is solved: white wins. Master's thesis, Vrije Univeriteit, The Netherlands (1998)
6. Barnat, J., Brim, L., Edelkamp, S., Sulewski, D., Šimeček, P.: Can flash memory help in model checking? In: Cofer, D., Fantechi, A. (eds.) FMICS 2008. LNCS, vol. 5596, pp. 150–165. Springer, Heidelberg (2009). doi:10.1007/978-3-642-03240-0_14
7. Barto, A., Bradtke, S., Singh, S.: Learning to act using real-time dynamic programming. Artif. Intell. **72**(1), 81–138 (1995)
8. Bloem, R., Ravi, K., Somenzi, F.: Symbolic guided search for CTL model checking. In: DAC, pp. 29–34 (2000)
9. Bonet, B.: Efficient algorithms to rank and unrank permutations in lexicographic order. In: AAAI-Workshop on Search in AI and Robotics (2008)
10. Bonet, B., Geffner, H.: Learning depth-first: a unified approach to heuristic search in deterministic and non-deterministic settings, and its application to MDPs. In: ICAPS, pp. 142–151 (2006)
11. Bošnački, D., Edelkamp, S., Sulewski, D.: Efficient probabilistic model checking on general purpose graphics processors. In: Păsăreanu, C.S. (ed.) SPIN 2009. LNCS, vol. 5578, pp. 32–49. Springer, Heidelberg (2009). doi:10.1007/978-3-642-02652-2_7
12. Botelho, F.C., Pagh, R., Ziviani, N.: Simple and space-efficient minimal perfect hash functions. In: WADS, pp. 139–150 (2007)
13. Botelho, F.C., Ziviani, N.: External perfect hashing for very large key sets. In: CIKM, pp. 653–662 (2007)
14. Breyer, T.M., Korf, R.E.: 1.6-bit pattern databases. In: AAAI (2010)
15. Burns, E., Lemons, S., Ruml, W., Zhou, R.: Suboptimal and anytime heuristic search on multi-core machines. In: ICAPS (2009)
16. Burns, E., Lemons, S., Zhou, R., Ruml, W.: Best-first heuristic search for multi-core machines. In: IJCAI, pp. 449–455 (2009)
17. Cazenave, T.: Nested monte-carlo search. In: IJCAI, pp. 456–461 (2009)
18. Cederman, D., Tsigas, P.: A practical quicksort algorithm for graphics processors. Technical report 2008-01, Chalmers University of Technology (2008)
19. Cormen, T., Leiserson, C., Rivest, R.: Introduction to Algorithms. MIT Press, Cambridge (1990)
20. Culberson, J.C., Schaeffer, J.: Pattern databases. Comput. Intell. **14**(4), 318–334 (1998)
21. Dean, J., Ghemawat, S.: MapReduce: simplified data processing on large clusters. In: OSDI (USENIX Association, Berkeley, CA, USA) (2004)
22. Dietzfelbinger, M., Karlin, A., Mehlhorn, K., auf der Heide, F.M., Rohnert, H., Tarjan, R.E.: Dynamic perfect hashing upper and lower bounds. SIAM J. Comput. **23**, 738–761 (1994)
23. Döbbelin, R., Schütt, T., Reinefeld, A.: Building large compressed PDBs for the sliding tile puzzle. In: Computer Games, pp. 16–27 (2013)

24. Edelkamp, S.: Planning with pattern databases. In: ECP, pp. 13–24 (2001). Reprint 2013 by AAAI Press. http://www.aaai.org/ocs/index.php/ECP/ECP01

25. Edelkamp, S.: External symbolic heuristic search with pattern databases. In: ICAPS, pp. 51–60 (2005)

26. Edelkamp, S., Jabbar, S.: Externalizing real-time model checking. In: MOCHART, pp. 67–83 (2006)

27. Edelkamp, S., Jabbar, S.: Large-scale directed model checking LTL. In: Valmari, A. (ed.) SPIN 2006. LNCS, vol. 3925, pp. 1–18. Springer, Heidelberg (2006). doi:10.1007/11691617_1

28. Edelkamp, S., Jabbar, S., Bonet, B.: External memory value iteration. In: ICAPS, pp. 414–429 (2007)

29. Edelkamp, S., Jabbar, S., Schrödl, S.: External A*. In: Biundo, S., Frühwirth, T., Palm, G. (eds.) KI 2004. LNCS (LNAI), vol. 3238, pp. 226–240. Springer, Heidelberg (2004). doi:10.1007/978-3-540-30221-6_18

30. Edelkamp, S., Kissmann, P.: Externalizing the multiple sequence alignment problem with affine gap costs. In: Hertzberg, J., Beetz, M., Englert, R. (eds.) KI 2007. LNCS (LNAI), vol. 4667, pp. 444–447. Springer, Heidelberg (2007). doi:10.1007/978-3-540-74565-5_36

31. Edelkamp, S., Kissmann, P.: Symbolic classification of general two-player games. In: Dengel, A.R., Berns, K., Breuel, T.M., Bomarius, F., Roth-Berghofer, T.R. (eds.) KI 2008. LNCS (LNAI), vol. 5243, pp. 185–192. Springer, Heidelberg (2008). doi:10.1007/978-3-540-85845-4_23

32. Edelkamp, S., Kissmann, P., Jabbar, S.: Scaling search with pattern databases. In: MOCHART, pp. 49–64 (2008)

33. Edelkamp, S., Kissmann, P., Rohte, M.: Symbolic and explicit search hybrid through perfect hash functions - a case study in connect four. In: ICAPS (2014)

34. Edelkamp, S., Leue, S., Lluch-Lafuente, A.: Directed explicit-state model checking in the validation of communication protocols. Int. J. Softw. Tools Technol. 5(2–3), 247–267 (2004)

35. Edelkamp, S., Sanders, P., Simecek, P.: Semi-external LTL model checking. In: CAV, pp. 530–542 (2008)

36. Edelkamp, S., Sulewski, D.: Flash-efficient LTL model checking with minimal counterexamples. In: SEFM, pp. 73–82 (2008)

37. Edelkamp, S., Sulewski, D.: Model checking via delayed duplicate detection on the GPU. Technical report 821, TU Dortmund (2008)

38. Edelkamp, S., Sulewski, D.: Efficient probabilistic model checking on general purpose graphics processors. In: SPIN (2010)

39. Edelkamp, S., Sulewski, D.: GPU exploration of two-player games with perfect hash functions. In: SOCS (2010)

40. Edelkamp, S., Sulewski, D.: External memory breadth-first search with delayed duplicate detection on the GPU. In: MOCHART, pp. 12–31 (2011)

41. Edelkamp, S., Sulewski, D., Yücel, C.: Perfect hashing for state space exploration on the GPU. In: ICAPS, pp. 57–64 (2010)

42. Edelkamp, S., Tang, Z.: Monte-carlo tree search for the multiple sequence alignment problem. In: SOCS, pp. 9–17 (2015)

43. Fredman, M.L., Komlós, J., Szemerédi, E.: Storing a sparse table with $o(1)$ worst case access time. J. ACM 3, 538–544 (1984)

44. Govindaraju, N.K., Gray, J., Kumar, R., Manocha, D.: GPUTeraSort: high performance graphics coprocessor sorting for large database management. In: SIGMOD, pp. 325–336 (2006)

45. Hansen, E., Zilberstein, S.: LAO*: a heuristic search algorithm that finds solutions with loops. Artif. Intell. **129**, 35–62 (2001)
46. Harris, M., Sengupta, S., Owens, J.D.: Parallel prefix sum (scan) with CUDA. In: Nguyen, H. (ed.) GPU Gems 3, pp. 851–876. Addison-Wesley, Salt Lake City (2007)
47. Hart, N., Nilsson, J., Raphael, B.: A formal basis for the heuristic determination of minimum cost paths. IEEE Trans. Syst. Sci. Cybern. **4**(2), 100–107 (1968)
48. Helmert, M.: Decidability and undecidability results for planning with numerical state variables. In: AIPS, pp. 303–312 (2002)
49. Helmert, M., Domshlak, C.: Landmarks, critical paths, abstractions: what's the difference anyway? In: ICAPS (2009)
50. Helmert, M., Haslum, P., Hoffmann, J.: Flexible abstraction heuristics for optimal sequential planning. In: ICAPS, pp. 176–183 (2007)
51. Hoare, C.A.R.: Algorithm 64: quicksort. Commun. ACM **4**(7), 321 (1961)
52. Hoffmann, J.: The metric FF planning system: translating "Ignoring the delete list" to numerical state variables. J. Artif. Intell. Res. **20**, 291–341 (2003)
53. Hoffmann, J., Nebel, B.: Fast plan generation through heuristic search. J. Artif. Intell. Res. **14**, 253–302 (2001)
54. Jabbar, S.: External memory algorithms for state space exploration in model checking and action planning. PhD thesis, TU Dortmund (2008)
55. Jabbar, S., Edelkamp, S.: I/O efficient directed model checking. In: Cousot, R. (ed.) VMCAI 2005. LNCS, vol. 3385, pp. 313–329. Springer, Heidelberg (2005). doi:10.1007/978-3-540-30579-8_21
56. Jabbar, S., Edelkamp, S.: Parallel external directed model checking with linear I/O. In: Emerson, E.A., Namjoshi, K.S. (eds.) VMCAI 2006. LNCS, vol. 3855, pp. 237–251. Springer, Heidelberg (2005). doi:10.1007/11609773_16
57. Jaychandran, G., Vishal, V., Pande, V.S.: Using massively parallel simulations, Markovian models to study protein folding: examining the Villin head-piece. J. Chem. Phys. **124**(6), 164 903–164 914 (2006)
58. Kishimoto, A., Fukunaga, A., Botea, A.: On the scaling behavior of HDA*. In: SOCS (2010)
59. Kishimoto, A., Fukunaga, A.S., Botea, A.: Scalable, parallel best-first search for optimal sequential planning. In: ICAPS (2009)
60. Kocsis, L., Szepesvári, C.: Bandit based Monte-Carlo planning. In: ICML, pp. 282–293 (2006)
61. Korf, R.E.: Linear-space best-first search. Artif. Intell. **62**(1), 41–78 (1993)
62. Korf, R.E.: Finding optimal solutions to Rubik's cube using pattern databases. In: AAAI, pp. 700–705 (1997)
63. Korf, R.E.:. Breadth-first frontier search with delayed duplicate detection. In: MOCHART, pp. 87–92 (2003)
64. Korf, R.E.: Best-first frontier search with delayed duplicate detection. In: AAAI, pp. 650–657 (2004)
65. Korf, R.E.: Minimizing disk I/O in two-bit breadth-first search. In: AAAI, pp. 317–324 (2008)
66. Korf, R.E., Felner, A.: Disjoint pattern database heuristics. In: Chips Challenging Champions: Games, Computers and Artificial Intelligence, pp. 13–26. Elsevier (2002)
67. Korf, R.E., Felner, A.: Recent progress in heuristic search: a case study of the four-peg towers of Hanoi problem. In: IJCAI, pp. 2324–2329 (2007)
68. Korf, R.E., Schultze, T.: Large-scale parallel breadth-first search. In: AAAI, pp. 1380–1385 (2005)

69. Korf, R.E., Zhang, W.: Divide-and-conquer frontier search applied to optimal sequence alignment. In: AAAI, pp. 910–916 (2000)
70. Korf, R.E., Zhang, W., Thayer, I., Hohwald, H.: Frontier search. J. ACM **52**(5), 715–748 (2005)
71. Kristensen, L., Mailund, T.: Path finding with the sweep-line method using external storage. In: ICFEM, pp. 319–337 (2003)
72. Krueger, J., Westermann, R.: Linear algebra operators for GPU implementation of numerical algorithms. ACM Trans. Graph. **22**(3), 908–916 (2003)
73. Kunkle, D., Cooperman, G.: Solving Rubik's cube: disk is the new RAM. Commun. ACM **51**(4), 31–33 (2008)
74. Kupferschmid, S., Dräger, K., Hoffmann, J., Finkbeiner, B., Dierks, H., Podelski, A., Behrmann, G.: Uppaal/DMC - abstraction-based Heuristics for directed model checking. In: TACAS, pp. 679–682 (2007)
75. Mareš, M., Straka, M.: Linear-time ranking of permutations. In: Arge, L., Hoffmann, M., Welzl, E. (eds.) ESA 2007. LNCS, vol. 4698, pp. 187–193. Springer, Heidelberg (2007). doi:10.1007/978-3-540-75520-3_18
76. Mehlhorn, K., Meyer, U.: External-memory breadth-first search with sublinear I/O. In: Möhring, R., Raman, R. (eds.) ESA 2002. LNCS, vol. 2461, pp. 723–735. Springer, Heidelberg (2002). doi:10.1007/3-540-45749-6_63
77. Munagala, K., Ranade, A.: I/O-complexity of graph algorithms. In: SODA, pp. 687–694 (1999)
78. Myrvold, W., Ruskey, F.: Ranking and unranking permutations in linear time. Inf. Process. Lett. **79**(6), 281–284 (2001)
79. Phillips, J.C., Braun, R., Wang, W., Gumbart, J., Tajkhorshid, E., Villa, E., Chipot, C., Skeel, R.D., Kale, L., Schulten, K.: Scalable molecular dynamics with NAMD. J. Comp. Chem. **26**, 1781–1802 (2005)
80. Reinefeld, A., Schütt, T.: Out-of-core parallel frontier search with MapReduce. In: Mewhort, D.J.K., Cann, N.M., Slater, G.W., Naughton, T.J. (eds.) HPCS 2009. LNCS, vol. 5976, pp. 323–336. Springer, Heidelberg (2010). doi:10.1007/978-3-642-12659-8_24
81. Reinefeld, A., Schütt, T., Döbbelin, R.: Very large pattern databases for heuristic search. In: Hariri, S., Keahey, K. (eds.), HPDC, pp. 803–809 (2010)
82. Richter, S., Helmert, M., Westphal, M.: Landmarks revisited. In: AAAI, pp. 975–982 (2008)
83. Rokicki, T., Kociemba, H., Davidson, M., Dethridge, J.: The diameter of the Rubik's cube group is twenty. SIAM J. Discrete Math. **27**(2), 1082–1105 (2013)
84. Rosin, C.D.: Nested rollout policy adaptation for Monte-Carlo tree search. In: IJCAI, pp. 649–654 (2011)
85. Meyer, U., Sanders, P., Sibeyn, J. (eds.): Algorithms for Memory Hierarchies. LNCS, vol. 2625. Springer, Heidelberg (2003)
86. Schaeffer, J., Björnsson, Y., Burch, N., Kishimoto, A., Müller, M.: Solving checkers. In: IJCAI, pp. 292–297 (2005)
87. Schaeffer, J., Burch, N., Bjrnsson, Y., Kishimoto, A., Müller, M., Lake, R., Lu, S.S.P.: Checkers is solved. Science **317**(5844), 1518–1522 (2007)
88. Schroedl, S.: An improved search algorithm for optimal multiple sequence alignment. J. Artif. Intell. Res. **23**, 587–623 (2005)
89. Schütt, T., Reinefeld, A., Maier, R.: MR-search: massively parallel heuristic search. Concurr. Comput.: Pract. Exp. **25**(1), 40–54 (2013)
90. Sturtevant, N.R.: External memory PDBs: initial results. In: SARA (2013)
91. Sturtevant, N.R., Rutherford, M.J.: Minimizing writes in parallel external memory search. In: IJCAI (2013)

92. Wijs, A.: What to do Next? Analysing and optimising system behaviour in time. PhD thesis, Vrije Universiteit Amsterdam (1999)
93. Zhou, R., Hansen, E.: Breadth-first heuristic search. In: ICAPS, pp. 92–100 (2004)
94. Zhou, R., Hansen, E.: Structured duplicate detection in external-memory graph search. In: AAAI, pp. 683–689 (2004)
95. Zhou, R., Hansen, E.: External-memory pattern databases using structured duplicate detection. In: AAAI (2005)
96. Zhou, R., Hansen, E.A.: Multiple sequence alignment using A*. In: AAAI (2002). Student abstract
97. Zhou, R., Hansen, E.A.: Edge partitioning in external-memory graph search. In: IJCAI, pp. 2410–2417 (2007)
98. Zhou, R., Hansen, E.A.: Parallel structured duplicate detection. In: AAAI, pp. 1217–1222 (2007)

Algorithm Engineering Aspects of Real-Time Rendering Algorithms

Matthias Fischer[(✉)], Claudius Jähn, Friedhelm Meyer auf der Heide,
and Ralf Petring

Heinz Nixdorf Institute, Department of Computer Science,
Paderborn University , Paderborn, Germany
mafi@upb.de

Abstract. Defining, measuring, and comparing the quality and efficiency of rendering algorithms in computer graphics is a demanding
challenge: quality measures are often application specific and efficiency
is strongly influenced by properties of the rendered scene and the used
hardware. We survey the currently employed evaluation methods for
the development process of rendering algorithms. Then, we present our
PADrend framework, which supports systematic and flexible development, evaluation, adaptation, and comparison of rendering algorithms,
and provides a comfortable and easy-to-use platform for developers of
rendering algorithms. The system includes a new evaluation method to
improve the objectivity of experimental evaluations of rendering algorithms.

1 Introduction

In the area of computer graphics, rendering describes the process of visualizing
a data set. One important aspect of rendering is, of course, how the data is presented to serve the desired application. Besides that, an algorithmic challenge
arises from the complexity of the rendered data set. Especially if the visualization has to be performed in real time, the amount of data can easily exceed
the capabilities of state of the art hardware, if only simple rendering techniques
are applied. In this paper, we focus on tools and techniques for the development of algorithms for rendering three-dimensional virtual scenes in real-time
walkthrough applications. Although the algorithmic challenges induced by complex virtual scenes traditionally play an important role in this area of computer
graphics, explicitly considering techniques supporting the developing process, or
providing a sound empirical evaluation are only considered marginally.

1.1 Context: Real-Time 3D Rendering

The input of a walkthrough application is a virtual scene usually composed of
a set of polygons, e.g., emerging from computer-aided design (CAD) data. The
user can interactively move through the scene representing a virtual observer,

© Springer International Publishing AG 2016
L. Kliemann and P. Sanders (Eds.): Algorithm Engineering, LNCS 9220, pp. 226–244, 2016.
DOI: 10.1007/978-3-319-49487-6_7

while the current view of the scene is rendered. The rendering process is normally supported by dedicated graphics hardware. Such hardware nowadays supports rendering of several million polygons at interactive frame rates (e.g., at least 10 frames per second). Considering complex virtual environments (like the CAD data of an air plane, or of a complete construction facility), the complexity of such scenes can however still exceed the capabilities of the hardware by orders of magnitudes. Thus, many real world applications require specialized rendering algorithms to reduce the amount of data processed by the graphics hardware.

The problem of rendering complex three dimensional scenes exhibits several properties that distinguishes it from many other problem areas dealing with large data sets; even influencing the process of designing, implementing, and evaluating algorithms in this area. In our opinion, the three most relevant properties are the use of dedicated graphics hardware, the influence of the input's geometric structure, and the relevance of the image quality perceived by a human observer. Dedicated graphics hardware provides a large amount of computational power, but also requires adaptations to its parallel mode of operation and the particularities of the rendering pipeline. On the one hand, the geometric structure of the virtual scene offers the opportunity to speed up the rendering process by exploiting the mutual occlusion of objects in the scene. On the other hand, the view on the scene changes for every observer position in the scene, which has to be considered in order to acquire any reasonable evaluation results on the general efficiency of a rendering algorithm. The human perception of the rendered images allows to speed up the rendering process by replacing complex parts of the scene by similar looking, but much simpler approximations. Challenges for the development of such algorithms is to actually create well looking approximations and to reasonably measure the image quality for an objective experimental evaluation.

1.2   Overview

First, we give an overview of the state of the art concerning different aspects influencing the evaluation process of rendering algorithms (Sect. 2). Then, we present the PADrend framework (Sect. 3), developed to provide a common basis for the development and evaluation and usage of rendering algorithms. The behavior of a rendering algorithm is not only depending on the visualized scene, but also on the observer's view on the scene – which is often only insufficiently considered by existing evaluation methods. We developed a special evaluation technique that tackles this issue based on globally approximated scene properties (Sect. 4). As an example, we present a meta rendering algorithm (Sect. 5.2) that uses the presented software framework and techniques to automatically assess and select other rendering algorithms for the visualization of highly complex scenes.

2   Evaluating Rendering Performance and Image Quality

The z-buffer-algorithm [6] of today's graphics hardware provides the real-time rendering of n triangles in linear time $O(n)$ [15]: The algorithm sequentially processes all triangles, whereby first the vertices of each triangle are projected into the 2-dimensional screenspace (geometric transformation), and second, each triangle is filled by coloring its pixels (rasterization).

To provide a smooth navigation through a scene about 10 images (frames) per second are necessary. Current graphics hardware can render scenes consisting of up to 15 millions triangles with 10 fps. Rendering algorithms try to overcome this limit by working like a filter [1,21] in front of the z-buffer algorithm: They reduce the number of polygons sent to the pipeline either by excluding invisible objects (visibility culling) or by approximating complex objects by less complex replacements.

In the following three sections, we discuss the challenges of objectively evaluating and comparing different rendering algorithms. An algorithm's efficiency, as well as the achieved image quality, is not only influenced by the virtual scene used as input (Sect. 2.1), but also by the user's movement through the scene (Sect. 2.2) and by the used graphics hardware (Sect. 2.3).

2.1 Influence of Input Scenes

The performance of a rendering algorithm is evaluated by experimental evaluations in which an input scene is used to perform measurements: There exist some standard objects (the Utah teapot by Martin Newell or the models from the Stanford 3D Scanning Repository[1]) that can be used for algorithms that render a single object [20,23,25]. The main difference in these scenes is mostly the number of primitives they consist of. Multiple objects are sometimes composed to form more complex scenes [3,19,35]. For walkthrough applications, there are some standard scenes [2–4,12,14,35]: the Power Plant model, the Double Eagle Tanker[2], the Boeing 777 model[3], or scenes created by software like the CityEngine[4]. These scenes differ not only in the number of primitives, but also in their geometrical structure: e.g., some consist of a large number of almost evenly distributed simple objects, while others consist of complex structures showing a large degree of mutual occlusion. Seldom, dynamic scenes are generated at runtime to compare and test rendering systems [26].

Occlusion culling algorithms, for example, work especially well on scenes having a high degree of occlusion; i.e. only a small fraction of the scene is visible at once. While some algorithms have strict requirements for the underlying structure of the scene [29], others require only some large objects that serve as occluders [35]. Some can handle almost arbitrary structured scenes by exploiting occlusion generated by any object in the scene [22]. For a suitable scene, each algorithm can increase the framerate by orders of magnitude, while an unsuitable scene can even lead to a decreased framerate compared to the simple z-buffer algorithm.

---

[1] http://graphics.stanford.edu/data/3Dscanrep/.
[2] http://www.cs.unc.edu/%7Ewalk/.
[3] http://www.boeing.com.
[4] http://www.esri.com/software/cityengine/resources/demos.

If certain properties of the input scene can be assumed (e.g., all the polygons are uniformly distributed), it can be shown, that the rendering time of several rendering algorithms is logarithmic in the complexity of the scene for any position in the scene (e.g., [7,30]).

2.2 Influence of Camera Paths

Besides the general structure of the input scene, the position and viewing direction from which the scene is rendered has a great influence on the runtime of a rendering algorithm: E.g., standing directly in front of a wall allows occlusion algorithms to cull most of the scene's occluded geometry, while from other positions huge parts of the scene might be visible.

When aiming at visualizing individual objects, only the direction from which the object is seen is an important variable for an evaluation (e.g., [23,25]). For algorithms designed for walkthrough applications it is common practice to use a representative camera paths for evaluating an algorithm. In many works (like [2–4,12,14,35]) the implementation of a rendering algorithm is evaluated by measuring a property (running time, image quality, culling results) along a camera path defined by the authors. One major drawback of this technique is, that all quantitative results only refer to the used camera path – the possible conclusions with regards to the general behavior of the algorithm for arbitrary scenes is limited. Even with a fixed scene, the algorithms are likely to behave differently depending on the chosen camera path. The documentation of the camera path used for an evaluation is important for supporting the significance of the published results. It differs largely between different works: in some, it is graphically depicted (sometimes as a video) and described [2,3,12]; in some, it is depicted or described only sparsely [4]; and in others, no description is given at all [14,35]. All considered papers state results based on statistical properties (mean, number of frames with a certain property, etc.) sampled along the used paths. Although the conclusions drawn by the authors may well relate to the general behavior of the algorithm, we think that an additional evaluation tool can help authors to generalize their results. E.g., for virtual reality applications, Yuaon et al. [34] present a framework for using actual user movements for performance analysis of rendering algorithms.

2.3 Influence of Graphics Hardware

Modern graphics cards combine a set of massively parallel processing units with dedicated memory. Normally, the high level rendering process is controlled by the CPU while the actual geometric and color calculations are performed on the graphics processing unit (GPU) in parallel. This heterogeneous design has to be reflected by the Real-time rendering systems, which reduce the number of primitives that are sent to the graphics card to the extent that a rendering with a fixed frame rate is possible. For this purpose a run-time prediction for the rendering of the primitives send to the graphics card is necessary. As today's graphics cards do not provide hard guarantees of the execution time of elementary graphics operations, the runtime predictions are imprecise and depend mainly on the chosen modeling of the hardware:

For the linear runtime estimation of the z-buffer algorithm we assume that geometric transformation and rasterization are both possible in constant time for each triangle. Practical implementations show that the runtime depends on the projected size of the triangles (counted in number of rasterized pixels). If we use an additional parameter a counting the number off all rasterized pixels, we estimate the runtime by $O(n + a)$ [15].

Funkhouser and Séquin [11] refined the model of basic graphics hardware operations in order to predict the rendering time of single polygons. The rendering time estimation take into account the number of polygons, the number of pixels and the number of pixels in the projection.

Wimmer and Wonka [33] model the graphics hardware by four major components, system tasks, and the maximum of CPU tasks and GPU tasks where the latter two are a sum of frame setup, memory management, rendering code, and idle time.

In this respect, no hard real-time rendering is possible, as in the case real-time operating systems that can provide reliably specific results within a predetermined time period. A real-time rendering system is aimed more to provide statistical guarantees for a fixed frame rate [33].

3   PADrend: Platform for Algorithm Development and Rendering

One important tool for the practical design and evaluation of algorithms is a supporting software framework. In the area of real-time rendering, there are many different software libraries and frameworks, whereas most of them focus on the use as a basis for an application and not as tool in the development process. Many game engines, for example, offer a set of powerful and well implemented rendering algorithms. For closed source engines, adding own algorithms or data structures is often not possible. Open source game engines can be extended, but, in our experience, suffer from a high internal complexity or too high level interfaces that hinder the necessary control over the underlying hardware. These problems lead to the habit of implementing a separate prototypical rendering engine for every new rendering algorithm. Besides the repeated implementation of common functionality (like scene loading, camera movement, support for runtime measurements, etc.), this especially hinders an objective comparison of different algorithmic approaches of different authors.

In this context, we developed the PADrend framework serving as one common basis for the development of rendering algorithms (first presented in [10]). For the development of PADrend, we followed several goals:

– Allow rapid development by providing a set of high level interfaces and modules for common tasks.

– Provide efficient hardware and system abstractions to allow platform independent development without dealing with too many technical details; while, at the same time, trying to provide an interface as direct to the hardware functions as possible.
– Support cooperative and concurrent development by using a modular design with only lightly coupled modules providing well defined interfaces.
– Support evaluation techniques as integral functions in the system.
– Do not restrict the usage by freely distributing all core components as open source software under the Mozilla Public License (MPL)[5].

In the following, we will discuss some of the system's aspects that directly focus on rendering algorithms and that distinguish the system from other rendering libraries and game engines.

3.1 System Design

The PADrend framework consists of several software libraries and a walkthrough application building upon these libraries (see Fig. 1). The libraries are written in C++ and among others, comprise libraries for geometrical calculations (*Geometry*), system abstraction (*Util*), rendering API abstraction (*Rendering*), and the scene graph (*MinSG*). These libraries have well defined dependencies and can also be used independently from the PADrend application. The application is based on a plugin structure written in EScript[6], an object oriented scripting language. EScript is easier to learn than C++ (in the context of a complex, existing software framework) while still offering direct access to the objects defined in C++.

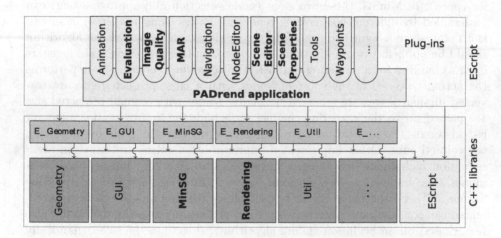

Fig. 1. Structural overview of PADrend.

---

[5] http://www.mozilla.org/MPL/.
[6] https://github.com/EScript.

3.2   Scene Graph: MinSG Library

The basis for the high level rendering processes is the Minimalistic Scene Graph (*MinSG*) library. "Minimalistic" represents the idea that the scene graph's core functionality is designed to be as compact and simplistic as possible. The library's core contains a small set of basic node types (a geometry node, a container node, and a few others) and a set of different properties (called *states*) that can be attached to nodes. The structure of the virtual scene is represented as a tree of nodes, in which the leaf nodes contain the scene's geometry as triangle meshes. Material and lighting properties are represented by states attached to the nodes. Specialized functions, data structures, and rendering algorithms can be created by building upon those core components.

A characteristic of MinSG is the way in which rendering algorithms are implemented. In most rendering engines, the rendering is defined by an external process getting the scene graph as input. In MinSG, rendering algorithms are implemented as states that are attached to inner nodes of the scene graph (like material properties). All nodes in the corresponding subtree may be influenced by this rendering state. The rendering state may use custom traversal techniques, call arbitrary rendering functions using the *Rendering* library, and dynamically store arbitrary data at any node. In this way, the rendering algorithm rather becomes a property of the scene than an external process.

Some algorithms can be seen as a combination of different high level functions. For instance, the color cubes algorithm (based on [7]) can be split up into two steps: First, identify the nodes in the scene graph whose current projected size is smaller than a given value. Then, approximate the corresponding subtrees by rendering colored cubes mimicking the original geometry (trade image quality for speed). In MinSG, these two steps (node selection and approximating) can be handled by different cooperating rendering states using *rendering channels*. Both states are assigned to a node in the scene graph for which the algorithm should be applied. Upon activation, the approximating rendering state registers itself as handler for a rendering channel. The selection rendering state performs the actual traversal of the subtree. Nodes with a large projected size are traversed further, leaves are rendered normally. Nodes with a small projected size are passed on to the rendering channel for which the approximation renderer is registered. The approximation renderer now handles the node by displaying the colored cube. Other implemented approximation states are based on mesh reduction techniques (like discrete Level of Detail [13, 21]), image-based techniques (like impostors [28]), or point-based techniques (like Progressive Blue Surfels [17]). Furthermore, the state controlling which nodes are to be approximated can consider not only a node's projected size, but its estimated visible size (e.g., based on an occlusion culling algorithm), or adjust the size dynamically to fulfill a given frame rate constraint (budget rendering). The main benefits of such a modular decomposition of algorithms are the possibility to reuse the parts for related techniques (decrease duplicated code and increase robustness) and the possibility to experiment with the recombination of different parts even at runtime. Further techniques that are implemented using rendering channels are, for example, distance sorting of transparent objects and multi-pass rendering. Techniques requiring complete control over the traversal process itself, can hoverer easily leave out rendering channels completely.

Scripted rendering states are another possibility to influence the rendering process, especially during the early development phase of an algorithm. Such states are implemented in EScript, resulting in a noticeably lower performance– but with the advantage that they can be deployed and altered at runtime. The effects of a change in the algorithm can be observed immediately even without reloading the scene. Later in the development process, such a state is normally re-implemented by translating it to C++ with little effort.

3.3   PADrend Application

The main functionality of the PADrend application is to provide a free interactive movement through the virtual scene and the modification of the scene graph. Following the overall design goals of the system, the application is split up in a set of plugins. The plugins are developed using the EScript language and can access all of the classes and functions defined in the C++ libraries (through a set of slim wrapper libraries). The basic functionality provides importing virtual scenes from different file formats and navigating through the scene using different input devices. The *Node Editor* plugin allows fine granular modification of the scene graph – including managing the rendering states (see the left window in Fig. 2). The *Scene Editor* plugin provides tools to compose new scenes out of existing 3D models to efficiently create example scenes for experimental evaluations. For instance, a brush tool can be used to place multiple trees on a surface with a single stroke to create a forest scene. Currently, there are more than thirty different optional plugins available.

3.4   Evaluating the Rendering Process

PADrend offers several evaluation tools that are described in the following:

Frame Statistics. For each rendered image (frame), different parameters are measured automatically and combined into the frame statistics. Typical parameter types include the number of rendered polygons, the time needed for rendering the frame, or the number of certain operations performed by a rendering algorithm (see the middle window in Fig. 2). Using the *Waypoints* plugin, the parameters can be measured along a predefined camera path and exported for later analysis. Similar features are common to most rendering frameworks.

Frame Events (Analyzer). A more fine granular method of observing the behavior of an algorithm is the event system. Many operations (like issuing a rendering request to the GPU) can be logged using a high resolution timer. The log can then be visualized in real time to identify bottlenecks during the

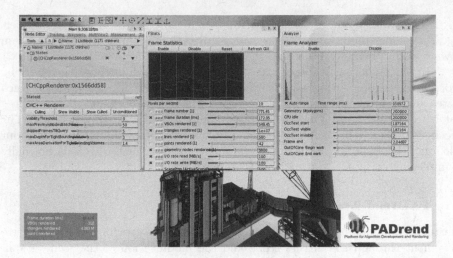

Fig. 2. Screen shot of PADrend showing the user interface of several plugins. The left window shows the configuration of a node of the scene graph with an attached rendering state. The middle window shows several frame statistics measured while moving through the scene. The right window shows a visualization of the last frame's logged rendering events: Blue lines represent rendered geometry, yellow and red lines represent occlusion queries, and empty regions show idle waiting times of the CPU – possible starting points for optimizations. (Color figure online)

rendering process (see the right window in Fig. 2). This feature is used best in combination with additional external profiler tools that are provided by major GPU manufacturers.

Image Quality Evaluation. The image compare plugin offers several functions for automatically measuring image quality. These functions require two images as input: one original image created using an exact rendering algorithm and the approximated image that is created using an approximate algorithm. Simple and widely used functions are the number of different pixels in the two images or some metrics on the color distances. We additionally support other techniques originating from the area of image compression algorithms (like JPEG compression). The structural similarity technique (SSIM) [32] detects changes in the structure of an image by comparing the pixel neighborhoods of all pixels in the two images.

To reduce the impact of small image errors compared to the dimensions of the image, like noise or aliasing artifacts, an image pyramid [5] is used additionally. SSIM is calculated for the full-size images and for multiple versions with reduced resolution. The arithmetic mean of the multiple SSIM calculations for the different resolutions is used as final image quality value. The numerical value as produced by using SSIM offers little direct meaning, but the order of different values (e.g., image A looks better than image B), aligns better with human perception than non-structure-oriented techniques like PSNR (for a discussion, see [31]).

4   Using Scene Properties for Objective Algorithm Evaluation

Common experimental evaluation based on measurements along camera paths can give a good insight into the general behavior of an algorithm. The significance of any statistical analysis is limited, as they are only valid for the chosen camera path. To improve the process of evaluating rendering algorithms' efficiency, we developed an evaluation technique to capture the global behavior of an algorithm using an adaptive sampling scheme. In the following, we give an overview of this technique (for a more detailed description, see [18]).

4.1   Position-Dependent Scene Properties

The basic observation underlying our method is that many aspects of a rendering algorithm's behavior can be expressed as position-dependent *scene properties*. Such a scene property can be expressed as function defined over $\mathbb{R}^3$ mapping to a property value from an arbitrary co-domain, like $\mathbb{R}$ or $\mathbb{N}$. In the following section, we give an overview of some property functions that proved to be useful in the evaluation of rendering algorithms.

Number of Visible Objects: One basic property is the number of the scene's objects that are visible from a position in the scene (on a pixel basis and not geometrically). This property is not bound to a specific algorithm, but can give important insight into the structure of the used scene. In our experience, almost all other properties are influenced by visibility.

A practical and efficient way of determining this property at a specific position, is to use the graphics hardware for the visibility tests. The scene is projected onto the six sides of a cube surrounding the observed position by rendering. Each object contributing at least one pixel to the final image on one of the sides is counted as visible. This can easily be measured by using hardware-assisted occlusion queries. The resolution used for rendering process should resemble the screen resolution of the walkthrough system and is an important parameter for the evaluation. If the property is evaluated in a system supporting efficient ray-casts (like a real-time ray tracing system), an omnidirectional observer can alternatively be implemented using a spherical projection, avoiding distortions in the corners of the cube.

Rendering Time: The rendering time property of an algorithm describes the time needed for an algorithm to render one frame. This value is clearly not only dependent on the position in the scene, but also on the viewing direction. To express the rendering time as meaningful position-dependent scene property, we abstract the viewing direction by taking the maximum of the values for six different directions – the six sides of a cube. The camera aperture angle is set to 120° to produce an overlap between adjacent directions. This overlap is required to reduce artifacts occurring if complex parts of the scene are not completely covered by any projection surface, although this area could be completely in view, if the camera was rotated differently.

Number of Operations: Other meaningful scene properties can be defined by the number of various operations performed by an algorithm to render a frame. This includes, for example, the number of rendered objects, the number of state changes in the graphics pipeline, or the number of issued occlusion queries. The measurement is similar to the rendering time measurement in that we can take the maximum value of the six directions.

Other possibly interesting property functions include image quality measurements or combinations of several other properties (like the difference in the performance of two different algorithms).

4.2   Global Approximation

In order to use a position dependent property for visualizing a rendering algorithm's behavior or analyzing its general performance for a given virtual scene, the next step is to approximate the property's distribution over the space of the scene. Evaluating the value of the property at some positions in the scene, we propose an adaptive sampling approach to build a data structure that yields an approximated value of the property at every position inside a predefined area (a 2D rectangle cutting through the scene, or a 3D box enclosing the scene). In order to be able to meaningfully approximate the scene property using a sampling approach, the property function has to be "well behaved". This means that on most positions, its value only changes gradually. Fortunately, this is true for many scene's visibility functions. Although there are visibility events with large changes in visibility (e.g., when the observer moves through a wall), the visibility remains almost constant if the observer only moves a small step. As the behavior of nearly all rendering algorithms is closely connected to visibility, the distribution of most relevant scene properties is coupled to the distribution of the visibility property.

The proposed adaptive sampling method aims at subdividing the considered space into regions with a mostly uniform value distribution using as few samples as possible. At places with high fluctuation in the sampled function, more samples are taken and a finer subdivision into regions is created than at places where the function is more uniform. A region is associated with one constant value calculated from all covered sample points.

Adaptive Sampling Algorithm. The input of the sampling algorithm consists of a virtual scene, a property function, a maximal number of samples to evaluate, and a region for which the property function should be approximated. This region is either a 2D rectangle cutting through the scene or a 3D bounding box.

The sampling method works as follows: Beginning with the initial region, the active region is subdivided into eight (3D case) or four (2D case) equally-sized new regions. For each of the new regions, new sample points are chosen (see description below) and the property function is evaluated at those positions. Two values are then calculated for each new region: The *property value* is the average value of all sample values that lie inside or at the border of the region. The *quality-gain* is defined by the variance of the region's sample values divided by the region's diameter – large regions with a high value fluctuation get high values, while small regions with almost uniform values get low values. The new regions are inserted into a priority queue based on their quality-gain values. The region with the highest quality-gain value is extracted from the queue and the algorithm starts over splitting this region next. The algorithm stops when the number of evaluated sample points exceeds the chosen value (or when another used defined condition is reached).

Choosing the Samples in a Region. The number of random sample points for a region is chosen by an heuristic, weighting the diameter of the region by a constant scaling-factor (typical value: 0.1 per meter). To achieve a good overall quality even for a small number of samples, the algorithm tries to spread the samples over the region while maximizing the minimum distance between two sample points. For the sampling algorithm to work in both 2D and 3D, and to support progressive refinement of the existing sampling, we chose a simple, yet flexible sampling scheme (loosely based on [8]): To choose one new sampling position, a constant number of candidate positions is generated uniformly at random inside the region (a typical number of samples is 200). The chosen candidate is the one with the largest minimum distance to any prior sample point. Additional to these random points, the corners of the region are always chosen as sample points (even if they are close to existing sample points), as these points can be shared among several neighboring regions.

4.3   Data Analysis

For analyzing globally approximated scene properties, we propose two different techniques: A graphical visualization of the evaluated values and a statistical evaluation of the value distribution.[7]

Graphical Visualization. The most intuitive way of working with an approximated scene property is its direct visualization. The simultaneous visualization of the underlying scene is optional, but usually quite helpful. Each region's associated value is mapped to a color and a transparency value for representing the region. If two-dimensional regions are used, an additional height value can be used to emphasize certain value ranges (producing a 3D plot). Figure 3 is an example for this 2D case. For three-dimensional regions, the transparency

---

[7] http://gamma.cs.unc.edu/POWERPLANT/.

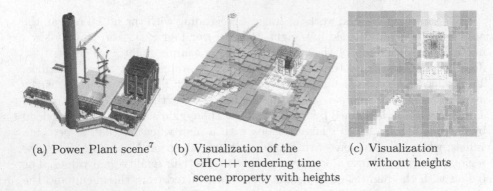

(a) Power Plant scene[7]   (b) Visualization of the   (c) Visualization
                                 CHC++ rendering time   without heights
                                 scene property with heights

Fig. 3. Visualization of the *rendering time of the CHC++ algorithm* [22] evaluated for a 2D cutting plane of the Power Plant. White regions (low height): rendering times down to 5 ms (behind the chimney, high occlusion). Red regions (large height): rendering times up to 16 ms (inside the main building, little occlusion). The approximation is based on 2 k sample points evaluated in 160 s. (image source [18]) (Color figure online)

value should be chosen in such a way that important areas can be seen from the outside. A 3D example is shown in Fig. 4(b).

The visualization makes it easy for the user to understand the algorithm's behavior and how it corresponds to the characteristics of different scene regions. Because it is intuitive and easy to use, it is a valuable tool during the development of a new algorithm. It can also be used to comprehensibly present aspects of a new algorithm in a scientific publication.

Statistical Analysis. To produce well-founded results, the approximated property function can be analyzed statistically. One has to keep in mind that the final approximation consists of differently sized regions (area in 2D, volume in 3D). In order to determine statistical properties (like the mean or median), the values have to be weighted by the corresponding region's size. To visualize the value distribution, a weighted histogram or a weighted box plot can be used. We propose using weighted violin plots [16] for summarizing the distribution of a scene property (see Fig. 4(a)). In addition to the information normally contained in a box plot (like the median and the interquartile range), a violin plot additionally adds a kernel density estimation. This is advisable, because the distribution of scene properties is likely to be multimodal.

5   Application of PADrend and Scene Properties

In Sect. 5.1 we use the well known CHC++ occlusions culling method [22] to demonstrate our proposed evaluation methods. In Sect. 5.2 we present a new rendering method that is developed with the PADrend system and show how the image quality can easily be evaluated by the system.

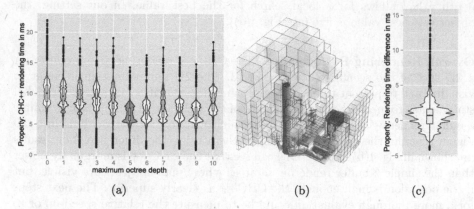

Fig. 4. (a) Violin plots for the distribution of the *CHC++ rendering time* in the Power Plant scene for different loose octree's maximum depths (4 k samples per value computed in 40 min overall). (b) Visualization of the property *rendering time: CHC++ minus simple z-buffer*. Only the negative values are shown as darker (blue) regions (two times 4 k samples). (c) Violin plot for the same property. (image source [18]) (Color figure online)

5.1   Evaluation of a Standard Occlusion Culling Algorithm

In the following, we demonstrate how our method can be used for the evaluation of different aspects of a rendering algorithm. The rendering algorithm used as subject for the analysis is the widely used CHC++ occlusion culling algorithm (Coherent-Hierarchical-Culling++ [22]). This algorithm filters out occluded geometry by applying hardware accelerated occlusion tests for bounding volumes covering parts of the scene. CHC++ applies several heuristics to minimize the number of necessary occlusion tests per frame and to reduce the average overhead that is introduced by an occlusion test. We use our method to evaluate the influence of a parameter of the chosen data structure on the algorithm to find the best parameter value for a given scene. For this value, we measure the performance gain that is achieved by using the algorithm in contrast to simple rendering. The measurements have been performed using the Power Plant scene (12.7 M triangles) on an Intel Core i7-3770 (4 × 3.40 GHz) with a NVIDIA GeForce GTX 660 (OpenGL 4.3).

Parameter Optimization: Optimal Octree Depth. The performance of the CHC++ depends on the partitioning of the scene's objects. When using a loose octree (loose factor 2.0) as a data structure, one parameter is the maximum depth of the tree. If the depth is too low, only a few nodes exist and the decisions that the CHC++ is able to make are limited. With a very high depth, the fraction of time that is used to perform occlusion tests grows too large compared to the time that is saved by the culled geometry. Creating separate global approximations (for a 3D region covering the scene) of the rendering time property for different depth values allows for a local search for the best value. In our setting, the objectively best value is five (see Fig. 4(a)).

Overall Rendering Performance. The CHC++ rendering time property for a 2D slice of the scene is shown in Fig. 3. The 2D approximation is easier to visualize and to present than a 3D approximation, but it is less suited for global statements. To measure the performance of the CHC++ in comparison to simple z-buffer rendering, we measured the running time of both algorithms in a 3D region covering the scene and combined them using the difference. The results are shown in Fig. 4(b) and (c). One can see that the CHC++ is marginally slower than the simple z-buffer rendering in areas where many objects are visible, but if the occlusion is high enough, the CHC++ is clearly superior. The next steps for a more thorough evaluation could be to measure the relative speed-up or to relate the CHC++'s overhead to the number of occlusion queries.

5.2   Multi-Algorithm-Rendering

Many large virtual 3D scenes are not structured evenly, for example, because they exhibit a high variety in their polygons' spatial distribution. For such heterogeneous data, there may be no single algorithm that constantly performs well with any type of scene and that is able to render the scene at each position fast and with high image quality. For a small set of scenes, this situation can be improved, if an experienced user is able to manually assign different rendering algorithms to particular parts of the scene. Based on the set of rendering algorithms and automatic evaluation techniques implemented in PADrend, we developed the meta rendering algorithm Multi-Algorithm-Rendering (MAR) [24], which automatically deploys different rendering algorithms simultaneously for a broad range of scene types.

In a preprocessing step, MAR first divides the scene into suitable regions. Then, at randomly chosen sample positions the expected behavior (rendering time and image quality) of all available rendering algorithms for all regions is evaluated. During runtime, this data is utilized to compute estimates for the running time and image quality for the actual observer's point of view. By solving an optimizing problem embedded in a control cycle, the frame rate can be kept almost constant, while the image quality is optimized.

Figure 5 shows an example from a highly complex and heterogeneous scene consisting of more than 100 million polygons. It is composed of models emerging from CAD, laser scans, and a procedural scene generator. The second screen shot shows the algorithm assignment MAR chose for that position from the following rendering algorithms: CHC++, normal z-buffer [6,27], Spherical Visibility Sampling [9], Color Cubes [7], discrete Level of Detail [13,21], and two variants of Progressive Blue Surfels [18]. Figure 6 shows the distribution of the achieved image quality as a scene property function. If the chosen target frame rate is increased, the faster algorithms will be preferred – resulting in a decreased image quality.

Fig. 5. Example images rendered using Multi-Algorithm-Rendering. The highlighting in the second image represents the different algorithms used by MAR.

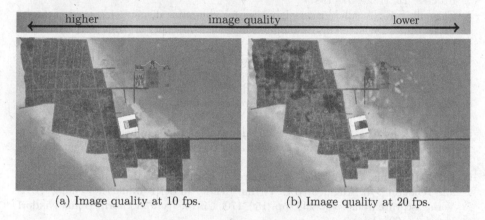

(a) Image quality at 10 fps.   (b) Image quality at 20 fps.

Fig. 6. Evaluation of MAR's image quality as scene property function. The observer positions are sampled in a slice slightly above the city.

6   Conclusion

By developing PADrend, we try to provide a common software framework for assisting the developer in each step of the practical development process of a rendering algorithm – from the initial idea to the final parameter tuning for a concrete application. New ideas can be implemented and tried out using the high level scripting interface. The algorithm's code can be altered and test scenes can be composed or altered without restarting the program. As the concept matures, the developer can build upon the optimized and tested core libraries and a large set of experimental extensions. For the experimental evaluation, standard measurement methods can be complemented by the evaluation of scene property functions. This additionally allows to adjust the algorithm's parameters for a given setting and to objectively compare the global efficiency of the new algorithm to existing state of the art algorithms.

References

1. Akenine-Möller, T., Haines, E., Hoffman, N.: Real-Time Rendering, 3rd edn. A K Peters Ltd., Wellesley (2008)
2. Baxter III., W.V., Sud, A., Govindaraju, N.K., Manocha, D.: GigaWalk: interactive walkthrough of complex environments. In: Proceedings of the 13th Eurographics Workshop on Rendering (EGWR 2002), pp. 203–214 (2002)
3. Bittner, J., Wimmer, M., Piringer, H., Purgathofer, W.: Coherent hierarchical culling: hardware occlusion queries made useful. Comput. Graph. Forum Proc. of Eurographics 2004 23(3), 615–624 (2004)
4. Brüderlin, B., Heyer, M., Pfützner, S.: Interviews3D: a platform for interactive handling of massive data sets. IEEE Comput. Graph. Appl. 27(6), 48–59 (2007)
5. Burt, P.J.: Fast filter transform for image processing. Comput. Graph. Image Process. 16(1), 20–51 (1981)
6. Catmull, E.E.: A subdivision algorithm for computer display of curved surfaces. Ph.D. thesis, Department of Computer Science, University of Utah, Salt Lake City, UT, USA (1974)
7. Chamberlain, B., DeRose, T., Lischinski, D., Salesin, D., Snyder, J.: Fast rendering of complex environments using a spatial hierarchy. In: Proceedings of Graphics Interface (GI 1996), pp. 132–141. Canadian Information Processing Society (1996)
8. Cook, R.L.: Stochastic sampling in computer graphics. ACM Trans. Graph. (TOG) 5(1), 51–72 (1986)
9. Eikel, B., Jähn, C., Fischer, M., auf der Heide, F.M.: Spherical visibility sampling. Comput. Graph. Forum Proc. of 24th Eurographics Symposium on Rendering 32(4), 49–58 (2013)
10. Eikel, B., Jähn, C., Petring, R.: PADrend: platform for algorithm development and rendering. In: Augmented & Virtual Reality in der Produktentstehung. HNI-Verlagsschriftenreihe, vol. 295, pp. 159–170. University Paderborn, Heinz Nixdorf Institute (2011)
11. Funkhouser, T.A., Séquin, C.H.: Adaptive display algorithm for interactive frame rates during visualization of complex virtual environments. In: Proceedings of the 20th Conference on Computer Graphics and Interactive Techniques (SIGGRAPH 1993), pp. 247–254 (1993)
12. Funkhouser, T.A., Séquin, C.H., Teller, S.J.: Management of large amounts of data in interactive building walkthroughs. In: Proceedings of the 1992 Symposium on Interactive 3D Graphics (I3D 1992), pp. 11–20 (1992)
13. Garland, M., Heckbert, P.S.: Simplifying surfaces with color and texture using quadric error metrics. In: Proceedings of the Conference on Visualization (VIS 1998), pp. 263–269 (1998)
14. Guthe, M., Balázs, Á., Klein, R.: Near optimal hierarchical culling: performance driven use of hardware occlusion queries. In: Akenine-Möller, T., Heidrich, W. (eds.) Proceedings of the 17th Eurographics Symposium on Rendering (EGSR 2006), pp. 207–214 (2006)
15. Heckbert, P., Garland, M.: Multiresolution modeling for fast rendering. In: Proceedings of Graphics Interface (GI 1994), pp. 43–50 (1994)
16. Hintze, J.L., Nelson, R.D.: Violin plots: a box plot-density trace synergism. Am. Stat. 52(2), 181–184 (1998)
17. Jähn, C.: Progressive blue surfels. Technical report, Heinz Nixdorf Institute, University of Paderborn (2013). CoRR. http://arxiv.org/abs/1307.0247

18. Jähn, C., Eikel, B., Fischer, M., Petring, R., Meyer auf der Heide, F.: Evaluation of rendering algorithms using position-dependent scene properties. In: Bebis, G., et al. (eds.) ISVC 2013. LNCS, vol. 8033, pp. 108–118. Springer, Heidelberg (2013). doi:10.1007/978-3-642-41914-0_12

19. Kovalčík, V., Sochor, J.: Occlusion culling with statistically optimized occlusion queries. In: The 13th International Conference in Central Europe on Computer Graphics, Visualization and Computer Vision (WSCG 2005), pp. 109–112 (2005)

20. Li, L., Yang, X., Xiao, S.: Efficient visibility projection on spherical polar coordinates for shadow rendering using geometry shader. In: IEEE International Conference on Multimedia and Expo, pp. 1005–1008 (2008)

21. Luebke, D., Reddy, M., Cohen, J.D., Varshney, A., Watson, B., Huebner, R.: Level of Detail for 3D Graphics. Computer Graphics and Geometric Modeling. Morgan Kaufman Publishers, Burlington (2003)

22. Mattausch, O., Bittner, J., Wimmer, M.: CHC++: coherent hierarchical culling revisited. Comput. Graph. Forum Proc. of Eurographics 2008 **27**(2), 221–230 (2008)

23. Meruvia-Pastor, O.E.: Visibility preprocessing using spherical sampling of polygonal patches. In: Short Presentations of Eurographics 2002 (2002)

24. Petring, R., Eikel, B., Jähn, C., Fischer, M., Meyer auf der Heide, F.: Real-time 3D rendering of heterogeneous scenes. In: Bebis, G., et al. (eds.) ISVC 2013. LNCS, vol. 8033, pp. 448–458. Springer, Heidelberg (2013). doi:10.1007/978-3-642-41914-0_44

25. Sander, P.V., Nehab, D., Barczak, J.: Fast triangle reordering for vertex locality and reduced overdraw. ACM Trans. Graph. (TOG) ACM SIGGRAPH 2007 **26**(3), 89:1–89:9 (2007)

26. Staadt, O.G., Walker, J., Nuber, C., Hamann, B.: A survey and performance analysis of software platforms for interactive cluster-based multi-screen rendering. In: Proceedings of the Workshop on Virtual Environments (EGVE 2003), pp. 261–270 (2003)

27. Straßer, W.: Schnelle Kurven- und Flächendarstellung auf graphischen Sichtgeräten. Ph.D. thesis, Technische Universität Berlin, Berlin, Germany (1974)

28. Süß, T., Jähn, C., Fischer, M.: Asynchronous parallel reliefboard computation for scene object approximation. In: Proceedings of the 10th Eurographics Symposium on Parallel Graphics and Visualization (EGPGV 2010), pp. 43–51 (2010)

29. Teller, S.J., Séquin, C.H.: Visibility preprocessing for interactive walkthroughs. In: Proceedings of the 18th Conference on Computer Graphics and Interactive Techniques (SIGGRAPH 1991), pp. 61–70 (1991)

30. Wand, M., Fischer, M., Peter, I., Meyer auf der Heide, F., Straßer, W.: The randomized z-buffer algorithm: interactive rendering of highly complex scenes. In: Proceedings of the 28th Conference on Computer Graphics and Interactive Techniques (SIGGRAPH 2001), pp. 361–370 (2001)

31. Wang, Z., Bovik, A.C.: Mean squared error: love it or leave it? A new look at signal fidelity measures. IEEE Signal Process. Mag. **26**(1), 98–117 (2009)

32. Wang, Z., Bovik, A.C., Sheikh, H.R., Simoncelli, E.P.: Image quality assessment: from error visibility to structural similarity. IEEE Trans. Image Process. **13**(4), 600–612 (2004)

33. Wimmer, M., Wonka, P.: Rendering time estimation for real-time rendering. In: Rendering Techniques 2003, Proceedings of the Eurographics Symposium on Rendering 2003, pp. 118–129 (2003)

34. Yuan, P., Green, M., Lau, R.W.H.: A framework for performance evaluation of real-time rendering algorithms in virtual reality. In: Proceedings of the ACM Symposium on Virtual Reality Software and Technology (VRST 1997), pp. 51–58 (1997)
35. Zhang, H., Manocha, D., Hudson, T., Hoff III., K.E.: Visibility culling using hierarchical occlusion maps. In: Proceedings of the 24th Conference on Computer Graphics and Interactive Techniques (SIGGRAPH 1997), pp. 77–88 (1997)

Algorithm Engineering in Robust Optimization

Marc Goerigk[1][(✉)] and Anita Schöbel[2]

[1] Management School, Lancaster University,
Lancaster LA1 4YX, United Kingdom
m.goerigk@lancaster.ac.uk
[2] Institute of Numerical and Applied Mathematics, University of Göttingen,
Lotzestraße 16-18, 37083 Göttingen, Germany
schoebel@math.uni-goettingen.de

Abstract. Robust optimization is a young and emerging field of research having received a considerable increase of interest over the last decade. In this paper, we argue that the algorithm engineering methodology fits very well to the field of robust optimization and yields a rewarding new perspective on both the current state of research and open research directions.

To this end we go through the algorithm engineering cycle of design and analysis of concepts, development and implementation of algorithms, and theoretical and experimental evaluation. We show that many ideas of algorithm engineering have already been applied in publications on robust optimization. Most work on robust optimization is devoted to analysis of the concepts and the development of algorithms, some papers deal with the evaluation of a particular concept in case studies, and work on comparison of concepts just starts. What is still a drawback in many papers on robustness is the missing link to include the results of the experiments again in the design.

1 Introduction

Similar to the approach of *stochastic* optimization, robust optimization deals with models in which the exact data is unknown, but bounded by a set of possible realizations (or scenarios). Contrary to the former approach, in robust optimization, one typically refrains from assuming a given probability distribution over the scenarios. While the first steps in robust optimization trace back to the work of Soyster [118], it has not emerged as a field of research in its own right before the late 90s with the seminal works of Ben-Tal, Nemirovski, and co-authors (see [18,19], and many more).

In this section, we first describe the general setting of robust optimization in more detail, and then discuss the algorithm engineering methodology and its application, which gives a natural structure for the remainder of the paper.

Partially supported by grant SCHO 1140/3-2 within the DFG programme *Algorithm Engineering*.

© Springer International Publishing AG 2016
L. Kliemann and P. Sanders (Eds.): Algorithm Engineering, LNCS 9220, pp. 245–279, 2016.
DOI: 10.1007/978-3-319-49487-6_8

Uncertain Optimization Problems. Nearly every optimization problem suffers from uncertainty to some degree, even if this does not seem to be the case at first sight. Generally speaking, we may distinguish two types of uncertainty: *Microscopic* uncertainty, such as numerical errors and measurement errors; and *macroscopic* uncertainty, such as forecast errors, disturbances or other conditions changing the environment where a solution is implemented.

In "classic" optimization, one would define a so-called *nominal scenario*, which describes the expected or "most typical" behavior of the uncertain data. Depending on the uncertainty type, this scenario may be, e.g., the coefficient of the given precision for numerical errors, the measured value for measurement errors, the most likely forecast for forecast errors, or an average environment for long-term solutions. Depending on the application, computing such a scenario may be a non-trivial process, see, e.g., [81].

In this paper we consider optimization problems that can be written in the form

$$(P) \qquad \min f(x)$$
$$\text{s.t. } F(x) \leq 0$$
$$x \in \mathcal{X},$$

where $F : \mathbb{R}^n \to \mathbb{R}^m$ describes the m problem constraints, $f : \mathbb{R}^n \to \mathbb{R}$ is the objective function, and $\mathcal{X} \subseteq \mathbb{R}^n$ is the variable space. In real-world applications, both the constraints and the objective may depend on parameters which are uncertain. In order to accommodate such uncertainties, instead of (P), the following parameterized *family* of problems is considered:

$$(P(\xi)) \qquad \min f(x, \xi)$$
$$\text{s.t. } F(x, \xi) \leq 0$$
$$x \in \mathcal{X},$$

where $F(\cdot, \xi) : \mathbb{R}^n \to \mathbb{R}^m$ and $f(\cdot, \xi) : \mathbb{R}^n \to \mathbb{R}$ for any fixed $\xi \in \mathbb{R}^M$. Every ξ describes a *scenario* that may occur.

Although it is in practice often not known exactly which values such a scenario ξ may take for an optimization problem $P(\xi)$, we assume that it is known that ξ lies within a given *uncertainty set* $\mathcal{U} \subseteq \mathbb{R}^M$. Such an uncertainty set represents the scenarios which are likely enough to be considered.

The *uncertain optimization problem* corresponding to $P(\xi)$ is then denoted as

$$(P(\xi), \xi \in \mathcal{U}). \tag{1}$$

Note that the uncertain optimization problem in fact consists of a whole set of parameterized problems, that is often even infinitely large. The purpose of robust optimization concepts is to transform this family of problems back into a single problem, which is called the *robust counterpart*. The choice of the uncertainty set is of major impact not only for the respective application, but also for the computational complexity of the resulting robust counterpart. It hence has to be chosen carefully by the modeler.

For a given uncertain optimization problem $(P(\xi), \xi \in \mathcal{U})$, we denote by

$$\mathcal{F}(\xi) = \{x \in \mathcal{X} : F(x, \xi) \leq 0\}$$

the feasible set of scenario $\xi \in \mathcal{U}$. Furthermore, if there exists a nominal scenario, it is denoted by $\hat{\xi} \in \mathcal{U}$. The optimal objective value for a single scenario $\xi \in \mathcal{U}$ is denoted by $f^*(\xi)$.

We say that an uncertain optimization problem $(P(\xi), \xi \in \mathcal{U})$ has convex (quasiconvex, affine, linear) uncertainty, when both functions, $F(x, \cdot) : \mathcal{U} \to \mathbb{R}^m$ and $f(x, \cdot) : \mathcal{U} \to \mathbb{R}$ are convex (quasiconvex, affine, linear) in ξ for every fixed $x \in \mathcal{X}$.

Common Uncertainty Sets. There are some types of uncertainty sets that are frequently used in current literature. These include:

1. Finite uncertainty $\mathcal{U} = \{\xi^1, \ldots, \xi^N\}$
2. Interval-based uncertainty $\mathcal{U} = [\underline{\xi}_1, \overline{\xi}_1] \times \ldots \times [\underline{\xi}_M, \overline{\xi}_M]$
3. Polytopic uncertainty $\mathcal{U} = \text{conv}\{\xi^1, \ldots, \xi^N\}$
4. Norm-based uncertainty $\mathcal{U} = \left\{\xi \in \mathbb{R}^M : \|\xi - \hat{\xi}\| \leq \alpha\right\}$ for some parameter $\alpha \geq 0$
5. Ellipsoidal uncertainty $\mathcal{U} = \left\{\xi \in \mathbb{R}^M : \sqrt{\sum_{i=1}^{M} \xi_i^2 / \sigma_i^2} \leq \Omega\right\}$ for some parameter $\Omega \geq 0$
6. Constraint-wise uncertainty $\mathcal{U} = \mathcal{U}_1 \times \ldots \times \mathcal{U}_m$, where $\mathcal{U}_i$ only affects constraint i

where $\text{conv}\{\xi^1, \ldots, \xi^N\} = \left\{\sum_{i=1}^{N} \lambda_i \xi^i : \sum_{i=1}^{N} \lambda_i = 1, \lambda \in \mathbb{R}_+^N\right\}$ denotes the convex hull of a set of points. Note that this classification is not exclusive, i.e., a given uncertainty set can belong to multiple types at the same time.

The Algorithm Engineering Methodology, and the Structure of this Paper. In the algorithm engineering approach, a feedback cycle between *design*, *analysis*, *implementations*, and *experiments* is used (see [113] for a detailed discussion). We reproduce this cycle for robust optimization in Fig. 1.

While this approach usually focuses on the design and analysis of *algorithms*, one needs to consider the important role that different *concepts* play in robust optimization. Moreover, as is also discussed later, there is a thin line between what is to be considered a robustness concept, and an algorithm – e.g., the usage of a simplified model for a robustness concept could be considered as a new concept, but also as a heuristic algorithm for the original concept. We will therefore consider the design and analysis of both, concepts and algorithms.

The algorithm engineering approach has been successfully applied to many problems and often achieved impressive speed-ups (as in routing algorithms, see, e.g. [50] and the book [102]).

Even though this aspect has not been sufficiently acknowledged in the robust optimization community, the algorithm engineering paradigm fits very well in

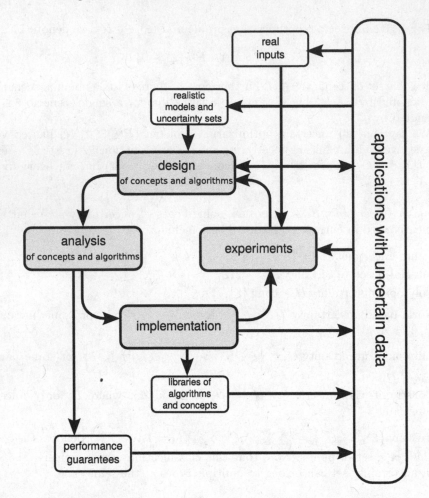

Fig. 1. The algorithm engineering cycle for robust optimization following [113].

the line of research done in this area: In algorithm engineering it is of particular importance that the single steps in the depicted cycle are not considered individually, but that special structure occurring in typical instances is identified and used in the development and analysis of concepts and algorithms. As we will show in the following sections these links to real-world applications and to the structure of the uncertain data are of special importance in particular in robust optimization. Various applications with different understandings of what defines a robust solution triggered the development of the different robustness concepts (see Sect. 2) while the particular structure of the uncertainty set led to adapted algorithms (see Sect. 3.1).

Moreover, the algorithm engineering cycle is well-suited to detect the *missing* research links to push the developed methods further into practice. A key aspect of this paper hence is to draw further attention to the potential of algorithm engineering for robust optimization.

We structure the paper along the algorithm engineering cycle, where we discuss each step separately, providing a few exemplarily papers dealing with the respective matters. Missing links to trigger further research in this areas are pointed out. Specifically, we consider

- design of robustness concepts in Sect. 2,
- analysis of robustness concepts in Sect. 3,
- design and analysis of algorithms in Sect. 4, and
- implementations and experiments in Sect. 5.

Applications of robust optimization are various, and strongly influenced the design of robustness concepts while the design of algorithms was rather driven by an analysis of the respective uncertainty sets. Some of these relations are mentioned in the respective sections. The paper is concluded in Sect. 6 where we also demonstrate on some examples how the previously mentioned results can be interpreted in the light of the algorithm engineering methodology.

2 Design of Robustness Concepts

Robust optimization started with rather conservative concepts hedging against everything that is considered as being likely enough to happen. Driven by various other situations and applications calling for "robust" solutions these concepts were further developed. In this section we give an overview on the most important older and some recent concepts. We put special emphasis on the impact applications with uncertain data have on the design of robustness concepts (as depicted in Fig. 1), and how real-world requirements influence the development of robustness models.

2.1 Strict Robustness

This approach, which is sometimes also known as classic robust optimization, one-stage robustness, min-max optimization, absolute deviation, or simply *robust optimization*, can be seen as the pivotal starting point in the field of robustness. A solution $x \in \mathcal{X}$ to the uncertain problem $(P(\xi), \xi \in \mathcal{U})$ is called *strictly robust* if it is feasible for all scenarios in $\mathcal{U}$, i.e. if $F(x, \xi) \leq 0$ for all $\xi \in \mathcal{U}$. The objective usually follows the pessimistic view of minimizing the worst-case over all scenarios. Denoting the set of strictly robust solutions with respect to the uncertainty set $\mathcal{U}$ by

$$SR(\mathcal{U}) = \bigcap_{\xi \in \mathcal{U}} \mathcal{F}(\xi),$$

the strictly robust counterpart of the uncertain optimization problem is given as

$$(SR) \qquad \min \sup_{\xi \in \mathcal{U}} \quad f(x, \xi)$$
$$\text{s.t. } x \in SR(\mathcal{U})$$
$$x \in \mathcal{X}.$$

The first to consider this type of problems from the perspective of *generalized linear programs* was Soyster [118] for uncertainty sets $\mathcal{U}$ of type

$$\mathcal{U} = K_1 \times \ldots \times K_n,$$

where the set K_i contains possible column vectors A_i of the coefficient matrix A. Subsequent works on this topic include [55,123].

However, building this approach into a strong theoretic framework is due to a series of papers by Ben-Tal, Nemirovski, El Ghaoui and co-workers [18–20,64]. A summary of their results can be found in the book [14]. Their basic underlying idea is to hedge against all scenarios that may occur. As they argue, such an approach makes sense in many settings, e.g., when constructing a bridge which must be stable, no matter which traffic scenario occurs, or for airplanes or nuclear power plants. However, this high degree of conservatism of strict robustness is not applicable to all situations which call for robust solutions. An example for this is timetabling in public transportation: being strictly robust for a timetable means that all announced arrival and departure times have to be met, no matter what happens. This may mean to add high buffer times, depending on the uncertainty set used, and thus would not result in a practically applicable timetable. Such applications triggered research in robust optimization on ways to relax the concept. We now describe some of these approaches.

2.2   Cardinality Constrained Robustness

One possibility to overcome the conservatism of strict robustness is to shrink the uncertainty set $\mathcal{U}$. This has been conceptually introduced by Bertsimas and Sim in [28] for linear programming problems. Due to this reason, this concept is sometimes also known as "the approach of Bertsimas and Sim", sometimes also under the name "Γ-robustness". Analyzing the structure of uncertainty sets in typical applications, they observed that it is unlikely that all coefficients of one constraint change simultaneously to their worst-case values. Instead they propose to hedge only against scenarios in which at most Γ uncertain parameters per constraint change to their worst-case values, i.e., they restrict the number of coefficients which are allowed to change leading to the concept of *cardinality constrained robustness*. Considering a constraint of the form

$$a_1 x_1 + \ldots + a_n x_n \leq b$$

with an uncertainty $\mathcal{U} = \{a \in \mathbb{R}^n : a_i \in [\hat{a}_i - d_i, \hat{a}_i + d_i], i = 1, \ldots, n\}$, their robustness concept requires a solution x to satisfy

$$\sum_{i=1}^{n} \hat{a}_i x_i + \max_{\substack{S \subseteq \{1,\ldots,n\}, \\ |S|=\Gamma}} \left\{ \sum_{i \in S} d_i |x_i| \right\} \leq b$$

for a given parameter $\Gamma \in \{0, \ldots, n\}$. Any solution x to this model hence hedges against all scenarios in which at most Γ many uncertain coefficients may deviate from their nominal values at the same time.

It can be shown that cardinality constrained robustness can also be considered as strict robustness using the convex hull of the *cardinality-constrained uncertainty set*

$$\mathcal{U}(\Gamma) = \{a \in \mathcal{U} : a_i \neq \hat{a}_i \text{ for at most } \Gamma \text{ indices } i\} \subseteq \mathcal{U}.$$

Since $conv(\mathcal{U}(\Gamma))$ is a polyhedral set, results on strict robustness with respect to polyhedral uncertainty can also be applied to cardinality constrained robustness.

Note that this approach also extends to fractional values of Γ. Their concept has been extended to uncertainty sets under general norms in [27]. The approach to combinatorial optimization problems has been generalized in [7,76].

2.3   Adjustable Robustness

In [16] a completely different observation of instances occurring in real-world problems with uncertain data is used: Often the variables can be decomposed into two sets. The values for the *here-and-now variables* have to be found by the robust optimization algorithm in advance, while the decision about the *wait-and-see variables* can wait until the actual scenario $\xi \in \mathcal{U}$ becomes known. Note that this is similar to two-stage programming in stochastic optimization.

We therefore assume that the variables $x = (u, v)$ are splitted into $u \in \mathcal{X}^1 \subseteq \mathbb{R}^{n_1}$ and $v \in \mathcal{X}^2 \subseteq \mathbb{R}^{n_2}$ with $n_1 + n_2 = n$, where the variables u need to be determined before the scenario $\xi \in \mathcal{U}$ becomes known, while the variables v may be determined after ξ has been realized. Thus, we may also write $x(\xi)$ to emphasize the dependence of v on the scenarios. The uncertain optimization problem $(P(\xi), \xi \in \mathcal{U})$ is rewritten as

$$P(\xi) \quad \min f(u, v, \xi)$$
$$F(u, v, \xi) \leq 0$$
$$(u, v) \in \mathcal{X}^1 \times \mathcal{X}^2.$$

When fixing the here-and-now variables, one has to make sure that for any possible scenario $\xi \in \mathcal{U}$ there exists $v \in \mathcal{X}^2$ such that (u, v) is feasible for ξ. The set of adjustable robust solutions is therefore given by

$$\text{aSR} = \{u \in \mathcal{X}^1 : \forall \xi \in \mathcal{U} \, \exists v \in \mathcal{X}^2 \text{ s.t. } (u, v) \in \mathcal{F}(\xi)\}$$
$$= \bigcap_{\xi \in \mathcal{U}} Pr_{\mathcal{X}^1}(\mathcal{F}(\xi)),$$

where $Pr_{\mathcal{X}^1}(\mathcal{F}(\xi)) = \{u \in \mathcal{X}^1 : \exists v \in \mathcal{X}^2 \text{ s.t. } (u, v) \in \mathcal{F}(\xi)\}$ denotes the projection of $\mathcal{F}(\xi)$ on $\mathcal{X}^1$.

The worst case objective for some $u \in \text{aSR}$ is given as

$$z^{\text{aSR}}(u) = \sup_{\xi \in \mathcal{U}} \inf_{v:(u,v)\in\mathcal{F}(\xi)} f(u, v, \xi).$$

The adjustable robust counterpart is then given as

$$\min\{z^{aSR}(u) : u \in aSR\}.$$

Note that this setting is also useful for another type of problem instances, namely, if auxiliary variables are used that do not represent decisions, e.g., additional variables to model the absolute value of a variable.

There are several variations of the concept of adjustable robustness. Instead of two stages, multiple stages are possible. In the approach of finitely adaptable solutions [23], instead of computing a new solution for each scenario, a set of possible static solutions is computed, such that at least one of them is feasible in each stage.

Furthermore, the development of adjustable robustness was preceded by the similar approach of Mulvey et al. [103]. They considered an uncertain linear optimization problem of the form

$$(P(B, C, e)) \qquad \min c^t u + d^t v$$
$$\text{s.t. } Au = b$$
$$\qquad Bu + Cv = e$$
$$\qquad u \in \mathbb{R}^{n_1}_+, v \in \mathbb{R}^{n_2}_+,$$

where u represents a vector of *design* variables that cannot be adjusted, and v a vector of *control* variables that can be adjusted when the realized scenario becomes known. For a finite uncertainty set $\mathcal{U} = \{(B^1, C^1, e^1), \ldots, (B^N, C^N, e^N)\}$, their robust counterpart is given as

$$(\text{Mul}) \qquad \min \sigma(u, v^1, \ldots, v^N) + \omega\rho(z^1, \ldots, z^N)$$
$$\text{s.t. } Au = b$$
$$\qquad B^i u + C^i v^i + z^i = e^i \ \forall i = 1, \ldots, N$$
$$\qquad u \in \mathbb{R}^{n_1}_+, v^i \in \mathbb{R}^{n_2}_+, z^i \in \mathbb{R}^m.$$

The variables z^i are introduced to measure the infeasibility in every scenario, i.e., the deviation from the right-hand side. The function σ represents the *solution robustness*. It can be modeled as a worst-case function of the nominal objective

$$\sigma(u, v^1, \ldots, v^N) = c^t u + \max_{i=1,\ldots,N} d^t v^i$$

or, when probabilities p^i are known, as an expected nominal objective. The function ρ on the other hand represents the *model robustness* and depends on the infeasibility of the uncertain constraints. Possible penalty functions are

$$\rho(z^1, \ldots, z^N) = \sum_{i=1}^{N} p_i \sum_{j=1}^{m} \max\{0, z^i_j\}$$

$$\text{or} = \sum_{i=1}^{N} p_i (z^i)^t z^i.$$

As (Mul) is actually a bicriteria model, ω is used as a scalarization factor to combine both objectives.

2.4   Light Robustness

The lightly robust counterpart of an uncertain optimization problem, as developed in [58] and generalized in [115] is again application driven. Originally developed for timetabling, the idea of light robustness is that a solution must not be too bad in the nominal case. For example, the printed timetable should have short travel times if everything runs smoothly and without disturbances; or a planned schedule should have a small makespan. In this sense a certain nominal quality is fixed. Among all solutions satisfying this standard, the concept asks for the most "reliable" one with respect to constraint violation. Specifically, the general lightly robust counterpart as defined in [115] is of the following form:

$$(LR) \qquad \min \sum_{i=1}^{m} w_i \gamma_i$$
$$\text{s.t. } f(x, \hat{\xi}) \leq f^*(\hat{\xi}) + \rho$$
$$F(x, \xi) \leq \gamma \qquad \forall \xi \in \mathcal{U}$$
$$x \in \mathcal{X}, \gamma \in \mathbb{R}^m,$$

where w_i models a penalty weight for the violation of constraint i and ρ determines the required nominal quality. We denote by $\hat{\xi}$ the nominal scenario, as introduced on page 3. This approach was in its first application in [58] used as a further development of the concept of cardinality constrained robustness (see Sect. 2.2).

Note that a constraint of the form $F(x, \xi) \leq 0$ is equivalent to a constraint $\lambda F(x, \xi) \leq 0$ for any $\lambda > 0$; therefore, the coefficients w_i play an important role in balancing the allowed violation of the given constraints.

2.5   Recoverable Robustness

Similar to adjustable robustness, *recoverable robustness* is again a two-stage concept. It has been developed in [43, 49, 90, 119] and has independently also been used in [54]. Its basic idea is to allow a class of *recovery algorithms* $\mathcal{A}$ that can be used in case of a disturbance. A solution x is called *recovery robust* with respect to $\mathcal{A}$ if for any possible scenario $\xi \in \mathcal{U}$ there exists an algorithm $A \in \mathcal{A}$ such that A applied to the solution x and the scenario ξ constructs a solution $A(x, \xi) \in \mathcal{F}(\xi)$, i.e., a solution which is feasible for the current scenario.

The recovery robust counterpart according to [90] is the following:

$$(RR) \qquad \min_{(x, A) \in \mathcal{F}(\hat{\xi}) \times \mathcal{A}} f(x)$$
$$\text{s.t. } A(x, \xi) \in \mathcal{F}(\xi) \; \forall \xi \in \mathcal{U}.$$

It can be extended by including the recovery costs of a solution x: Let $d(A(x,\xi))$ be a possible vector-valued function that measures the costs of the recovery, and let $\lambda \in \Lambda$ be a limit on the recovery costs, i.e., $\lambda \geq d(A(x,\xi))$ for all $\xi \in \mathcal{U}$. Assume that there is some cost function $g : \Lambda \to \mathbb{R}$ associated with λ.

Setting

$$A(x,\xi,\lambda) \in \mathcal{F}'(\xi) \iff d(A(x,\xi)) \leq \lambda \wedge A(x,\xi) \in \mathcal{F}(\xi)$$

gives the recovery robust counterpart with limited recovery costs:

$$\text{(RR-LIM)} \quad \min_{(x,A,\lambda) \in \mathcal{F}(\xi) \times \mathcal{A} \times \Lambda} f(x) + g(\lambda)$$
$$\text{s.t. } A(x,\xi,\lambda) \in \mathcal{F}'(\xi) \; \forall \xi \in \mathcal{U}.$$

Due to the generality of this robustness concept, the computational tractability heavily depends on the problem, the recovery algorithms and the uncertainty under consideration. In [65, 73–75], the concept of recoverable robustness has been considered under the usage of metrics to measure recovery costs. The aim is to minimize the costs when recovering, where they differ between recovering to a feasible solution ("recovery-to-feasibility"), and recovering to an optimal solution ("recovery-to-optimality") in the realized scenario.

2.6 Regret Robustness

The concept of regret robustness differs from the other presented robustness concepts insofar it usually only considers uncertainty in the objective function. Instead of minimizing the worst-case performance of a solution, it minimizes the difference to the objective function of the best solution that would have been possible in a scenario. In some publications, it is also called *deviation robustness*.

Let $f^*(\xi)$ denote the best objective value in scenario $\xi \in \mathcal{U}$. The min-max regret counterpart of an uncertain optimization problem with uncertainty in the objective is then given by

$$\text{(Regret)} \quad \min \sup_{\xi \in \mathcal{U}} \left(f(x,\xi) - f^*(\xi) \right)$$
$$\text{s.t. } F(x) \leq 0$$
$$x \in \mathcal{X}.$$

Regret robustness is a concept with a vast amount of applications, e.g., in location theory or in scheduling. For a survey on this concept, see [4, 89]. In a similar spirit, the concept of lexicographic α-robustness has been recently proposed [83]. Its basic idea is to evaluate a fixed solution by reordering the set of scenarios according to the performance of the solution. This performance curve is then compared to an ideal curve, where the optimization problem is solved separately for every scenario.

2.7   Some Further Robustness Concepts

Reliability. Another approach to robust optimization is to relax the constraints of strict robustness. This leads to the concept of *reliability* of Ben-Tal and Nemirovski [20], in which the constraints $F(x, \xi) \leq 0$ are replaced by $F(x, \xi) \leq \gamma$ for some $\gamma \in \mathbb{R}_{\geq 0}^m$. A solution x which satisfies

$$F(x, \xi) \leq \gamma \text{ for all } \xi \in \mathcal{U}$$

is called *reliable with respect to* γ. The goal is to find a reliable solution which minimizes the original objective function in the worst case. Similar to light robustness, one has to be careful that the representation of the constraints does not affect the reliability of the solution, otherwise one may obtain the counter-intuitive result that, although the constraints $F(x, \xi) \leq 0$ can also be written as $\Psi(F(x, \xi)) \leq 0$ for any increasing Ψ with $\Psi(0) = 0$, what is understood by a robust solution may be different if one models the constraints with F or with $\Psi(F)$.

Soft Robustness. The basic idea of *soft robustness* as introduced in [11] is to handle the conservatism of the strict robust approach by considering a nested family of uncertainty sets, and allowing more deviation in the constraints for larger uncertainties. Specifically, instead of an uncertainty set $\mathcal{U} \subseteq \mathbb{R}^M$, a family of uncertainties $\{\mathcal{U}(\varepsilon) \subseteq \mathcal{U}\}_{\varepsilon > 0}$ with $\mathcal{U}(\varepsilon_1) \subseteq \mathcal{U}(\varepsilon_2)$ for all $\varepsilon_2 \geq \varepsilon_1$ is used. The set of soft robust solutions is then given as

$$\text{softR} = \{x \in \mathcal{X} : F(x, \xi) \leq \varepsilon \; \forall \xi \in \mathcal{U}(\varepsilon), \; \varepsilon > 0\}.$$

Note that strict robustness is a special case with $\mathcal{U}(\varepsilon) = \mathcal{U}$ for all $\varepsilon > 0$.

In [11], the authors show that a solution to the softly robust counterpart – i.e., the optimization over softR with a worst-case objective – can be found by solving a sequence of strictly robust counterparts using a bisection approach over ε, and analyze the numerical performance on a bond portfolio and an asset allocation problem.

Comprehensive Robustness. While the adjustable robust approach relaxes the assumption that all decisions have to be made before the realized scenario becomes known, the approach of comprehensively robust counterparts [12] also removes the assumption that only scenarios defined in the uncertainty set $\mathcal{U}$ need to be considered. Instead, using a distance measure $dist$ in the space of scenarios, and a distance measure $\overline{dist}$ in the solution space, they assume that the further away a scenario is from the uncertainty set, the further away the corresponding solution is allowed to be from the set of feasible solutions. As in adjustable robustness, the dependence of x on the scenario ξ is allowed, and we may write $x(\xi)$. The adjustable robust counterpart is extended to the following problem:

$$(\text{CRC}) \qquad \min z$$
$$\text{s.t. } f(x(\xi), \xi) \leq z + \alpha_0 dist(\xi, \mathcal{U}) \; \forall \xi$$
$$\overline{dist}(x(\xi), \mathcal{F}(\xi)) \leq \alpha dist(\xi, \mathcal{U}) \; \forall \xi,$$

where α, α_0 denote sensitivity parameters. This formulation needs further formal specification, which is provided in [12].

Uncertainty Feature Optimization. Instead of assuming that an explicit uncertainty set is given, which may be hard to model for real-world problems, the uncertainty feature optimization (UFO) approach [53] rather assumes that the robustness of a solution is given by an explicit function. For an uncertain optimization problem $(P(\xi))$, let $\mu : \mathbb{R}^n \to \mathbb{R}^p$ be a measure for p robustness features. The UFO-counterpart of the uncertain problem is then given by

$$
\begin{aligned}
(\text{UFO}) \qquad & \text{vecmax}\,\mu(x) \\
& \text{s.t.}\, F(x) \leq 0 \\
& \qquad f(x) \leq (1 + \rho)f^*(\hat{\xi}) \\
& \qquad x \in \mathcal{X},
\end{aligned}
$$

where $f^*(\hat{\xi})$ denotes the best objective value to the nominal problem. The authors show that this approach generalizes both stochastic optimization and the concept of cardinality constrained robustness of Bertsimas and Sim.

2.8 Summary

As this section shows, we cannot actually speak of one concept or point-of-view to be "robust optimization"; instead, we should see it as a vast collection of different robustness concepts, each providing their unique advantages and disadvantages. Generally speaking, there is usually a trade-off between the degree of freedom a concept gives to react to disruptions (including what is considered as being a disruption, i.e., the choice of the uncertainty set), and its computational complexity. From an algorithm engineering point of view, the size of this "toolbox" of different concepts significantly helps with finding a suitable robustness concept for a given problem. However, as these concepts are usually application-driven, they lack a generalizing systematics: Applications tend to develop "their own approach" to robustness instead of making use of the existing body of literature, and develop their own notation and names along the way. In fact, the very same concepts are known under plenty of names. Summaries as [4,14,22,112] usually avoid this Babylonian "zoo" of robustness concepts and nomenclature by focusing only on the mainstream concepts. Thus, we suggest the following pointer to further research:

Remark 1. Robust optimization needs a unified classification scheme.

3 Analysis of Robustness Concepts

Not only the development of robustness concepts, but also their analysis is data-driven. This becomes in particular clear when looking at the structure of the underlying uncertainty set. A large amount of research in the analysis of robustness concepts is devoted to finding equivalent problem formulations that are better tractable, using the structure of the uncertainty set.

In this section we first review this line of research, and then briefly point out exemplarily which other types of structure or ideas have been used in the analysis of concepts.

3.1   Using the Structure of the Uncertainty Set

Finite Uncertainty Set. If the uncertainty set $\mathcal{U} = \{\xi^1, \ldots, \xi^N\}$ is a finite set containing not too many scenarios, most of the robustness concepts can be formulated as mathematical programs by just adding the constraints for each of the scenarios explicitly. This can straightforwardly been done for strict robustness yielding

$$(\text{SR}) \qquad \min \; z$$
$$\text{s.t. } f(x, \xi^i) \leq z \text{ for } i = 1, \ldots, N$$
$$\text{s.t. } F(x, \xi^i) \leq 0 \text{ for } i = 1, \ldots, N$$
$$x \in \mathcal{X}.$$

as the strictly robust counterpart. Reliability and light robustness can be treated analogously. In all three cases, the robust counterpart keeps many properties of the original (non-robust) problem formulation: If the original formulation was e.g., a linear program, also its robust counterpart is. The same holds for differentiability, convexity, and many other properties.

For regret robustness one needs to precompute the best objective function value for each scenario $\xi^1, i = 1, \ldots, N$ in order to receive again a straightforward reformulation. Also in adjustable and recoverable robustness mathematical programming formulations can be derived by adding a wait and see variable, or a group of recovery variables for *each* of the scenarios. This usually leads to a high number of additional variables but is (at least for linear programming) often still solvable.

Note that the concept of cardinality constrained robustness does not make much sense for a finite set of scenarios since it concerns the restriction *which* scenarios might occur. For a finite set, scenarios in which too many parameters change can be removed beforehand.

Polytopic Uncertainty. Let $f(x, \cdot)$ and $F(x, \cdot)$ be quasiconvex in ξ for any fixed $x \in \mathcal{X}$. Then there are robustness concepts in which the following *reduction result* holds: The robust counterpart w.r.t. an uncertainty set $\mathcal{U}'$ is equivalent to the robust counterpart w.r.t. $\mathcal{U} := conv(\mathcal{U}')$. In such cases the robust counterpart w.r.t. a polytopic uncertainty set $\mathcal{U} = conv\{\xi^1, \ldots, \xi^N\}$ is equivalent to the robust counterpart w.r.t. the finite uncertainty set $\{\xi^1, \ldots, \xi^N\}$, hence the formulations for finite uncertainty sets can be used to treat polytopic uncertainties.

We now review for which robustness concepts the reduction result holds. First of all, this is true for strict robustness, For affine and convex uncertainty this was mentioned in [18]; the generalization to quasiconvex uncertainty is straightforward. One of the direct consequences, namely that the robust counterpart of an uncertain linear program under these conditions is again a linear program was mentioned in [20]. The same result holds for reliability since the reliable robust counterpart can be transformed to a strictly convex counterpart by defining $\tilde{F}(x, \xi) = F(x, \xi) - \gamma$. For light robustness, the result is also true, see [115]. For the case of adjustable robustness, [16] showed that the result holds for problems with fixed recourse. Otherwise, counterexamples can be constructed. The generalization to nonlinear two-stage problems and quasiconvex uncertainty is due to [121]. For recoverable robustness there exist special cases in which the recovery robust counterpart is equivalent to an adjustable robust counterpart with fixed recourse. In these cases, the result of [16] may be applied. However, in general, recoverable robustness does not allow this property. This also holds for recovery-to-optimality.

Interval-Based Uncertainty. Interval-based uncertainty can be interpreted as a special case of polytopic uncertainty where the polytope is a box $\mathcal{U} = [\underline{\xi}_1, \overline{\xi}_1] \times \ldots \times [\underline{\xi}_M, \overline{\xi}_M]$ with 2^M extreme points $(\xi_1, \xi_2, \ldots, \xi_M)^t \in \mathbb{R}^M$, where $\xi_i \in \{\underline{\xi}_i, \overline{\xi}_i\}, i = 1, \ldots, M$. Hence, all the results for polytopic uncertainty apply. They can often be simplified by observing that not all extreme points are needed since the respective constraints often dominate each other, yielding a drastic speed-up when solving the robust counterpart.

For their concept of cardinality constrained robustness, Bertsimas and Sim [28] considered interval-based uncertainty sets for linear programs. This can be interpreted as strict robustness with a new uncertainty set $\mathcal{U}'$ only allowing scenarios in which not more than Γ uncertain parameters per constraint change their values (see also [27]). This uncertainty set $\mathcal{U}'$ is a polytope, hence the robust counterpart for cardinality constrained robustness stays a linear program for interval-based uncertainty.

Ellipsoidal Uncertainty. The case of ellipsoidal uncertainty is studied extensively for strict robustness and for adjustable robustness in [14]. It could be shown that often the constraint

$$F(x, \xi) \leq 0 \text{ for all } \xi \in \mathcal{U}$$

can be replaced by a finite number of constraints for ellipsoidal uncertainty sets. However, it has been shown in [14] that for ellipsoidal uncertainty, the structure of the strictly robust counterpart gets more complicated. For example (see [18]) the strictly robust counterpart of a linear program is a conic quadratic program, the strictly robust counterpart of a quadratic constrained quadratic program is a semidefinite program, the strictly robust counterpart of a second order cone program is a semidefinite program, and the strictly robust counterpart of a semidefinite program is NP-hard. As mentioned before, all these results can be transferred to reliability.

For light robustness, it has been shown in [115] that the lightly robust counterpart of a linear program with ellipsoidal uncertainty becomes a quadratic program. Ellipsoidal uncertainty could receive more attention also for other robustness concepts (e.g., for regret robustness, which usually only considers finite or interval-based uncertainty, see [4]), or for adjustable robustness, see [14].

3.2 Using Duality

Duality in uncertain programs has been considered as early as 1980, see [123]. In [10], it is shown that "the primal worst equals the dual best", i.e., under quite general constraints, the dual of a strictly robust counterpart (a min-max problem) amounts to optimization under the best case instead of the worst-case (a max-min problem). Since then, duality in robust optimization has been a vivid field of research, see, e.g., [82, 120]. In the following, we highlight two applications of duality for robust optimization: One for constraints, and one for objectives.

Duality in the Constraints. Duality is a useful tool for the reformulation of robust constraints. We exemplarily demonstrate this using two applications.

In [28], the authors show that the cardinality constrained robust counterpart can be linearized by using the dual of the inner maximization problem. This yields

$$\sum_{i=1}^{n} \hat{a}_i x_i + z\Gamma + \sum_{i=1}^{n} p_i \leq b$$
$$z + p_i \geq d_i y_i \qquad \forall i = 1, \ldots, n$$
$$-y_i \leq x_i \leq y_i \qquad \forall i = 1, \ldots, n$$
$$p, y, z \geq 0.$$

Note that a general, robust constraint of the form

$$f(x, \xi) \leq 0 \ \forall \xi \in \mathcal{U}$$

can be rewritten as

$$\max_{\xi \in \mathcal{U}} f(x, \xi) \leq 0.$$

This is used in [17]. With a concave function $f(x, \cdot)$ and an uncertainty set $\mathcal{U} = \{\hat{\xi} + A\zeta : \zeta \in Z\}$ with a nonempty, convex and compact set Z, applying duality yields

$$\hat{\xi}^t v + \delta^*(A^T v | Z) - f_*(v, x) \leq 0$$

where δ^* is the support function, f_* is a conjugate function, and other technical requirements are met. This gives a very general tool to compute robust counterparts; e.g., a linear constraint of the form $f(x, \xi) = \xi^t x - \beta$ and $Z = \{\zeta : \|\zeta\|_2 \leq \rho\}$ yields the counterpart $\hat{\xi}^t x + \rho \|A^t x\|_2 \leq \beta$.

Duality in the Objective. In many papers, duality is used to change the typical min-max objective of a robust counterpart into a min objective by using the dual formulation of the inner maximization problem.

This method was first applied to the spanning tree problem [126], and later extended to the general case of optimization problems with zero duality gap in [4]. Let an uncertain optimization problem of the form

$$\min c^t x$$
$$\text{s.t. } x \in \mathcal{X} = \{x \in \{0,1\}^n : Ax \geq b\}$$

with interval-based uncertainty in c be given; i.e., $c_i \in [\underline{c}_i, \overline{c}_i]$. Then we may write

$$\min_{x \in \mathcal{X}} \max_{c \in \mathcal{U}} (f(x,c) - f^*(c))$$
$$= \min_{x \in \mathcal{X}} \max_{c \in \mathcal{U}, y \in \mathcal{X}} (c^t x - c^t y)$$
$$= \min_{x \in \mathcal{X}} \left(\overline{c}x - \min_{y \in \mathcal{X}} c^{wc}(x) y \right),$$

where $c^{wc}(x)$ denotes the regret worst-case for x, given as $\overline{c}_i$ if $x_i = 1$, and $\underline{c}_i$ if $x_i = 0$. Using that the duality gap is zero, we can insert the dual to the inner optimization problem, and get the following equivalent problem:

$$\min \overline{c}x - b^t y$$
$$\text{s.t. } Ax \geq b$$
$$A^t y \leq (\overline{c} - \underline{c})x + \underline{c}$$
$$x \in \{0,1\}^n, y \in \mathbb{R}_+^n$$

This reformulation can then be solved using, e.g., a branch and bound approach.

4   Design and Analysis of Algorithms

Concerning the design and analysis of algorithms we concentrate on the most mature concept, namely on algorithms for strict robustness. Many approaches, often based on similar ideas, also exist for regret optimization – e.g., cutting plane approaches [80,95,96], or preprocessing considerations [87,126]. For the other concepts, approaches are currently still being developed.

The robust counterpart per se is a semi-infinite program; thus, all methods that apply to semi-infinite programming [93] can be used here as well. However, the special min-max structure of the robust counterpart allows improved algorithms over the general case, in particular for the reformulations based on special uncertainty sets as mentioned in Sect. 3.1.

In the following, we discuss algorithms that are generically applicable to strictly robust optimization problems.

4.1   Finite Scenarios

The case we consider here is that $\mathcal{U} = \{\xi^1, \ldots, \xi^N\}$ is a finite set; i.e., the strictly robust counterpart (SR) can be rewritten as

$$\min_{i=1,\ldots,N} \max f(x, \xi^i)$$
$$\text{s.t. } F(x, \xi^i) \leq 0 \ \forall i = 1, \ldots, N$$
$$x \in \mathcal{X}$$

Due to the lack of structure in the uncertainty set, these instances can be hard so solve, even though they have a similar structure as the nominal problem.

Branch and Bound Using Surrogate Relaxation. The following approach was introduced by [89] for discrete optimization problems with uncertainty only in the objective: For any vector $\mu \in \mathbb{R}_+^N$, the surrogate relaxation SRC(μ) of (SR) with uncertain objective function is given by

$$\text{SRC}(\mu) \quad \min \frac{1}{\sum_{\xi \in \mathcal{U}} \mu_\xi} \sum_{\xi \in \mathcal{U}} \mu_\xi f(x, \xi)$$
$$\text{s.t. } F(x) \leq 0$$
$$x \in \mathcal{X}$$

Note that the structure of the nominal problem is preserved, which allows the usage of specialized algorithms already known. Furthermore, the optimal objective value SRC*(μ) of this problem is a lower bound on the optimal objective value SR* of the strictly robust counterpart; and as the set of feasible solutions is the same, also an upper bound is provided by solving SRC(μ).

This approach is further extended by solving the problem

$$\max_{\mu \in \mathbb{R}_+^N} \text{SRC}^*(\mu),$$

i.e., by finding the multiplier μ that yields the strongest lower bound. This can be done using a sub-gradient method.

The lower and upper bounds generated by the surrogate relaxation are then used within a branch and bound framework on the x variables. The approach was further improved for the knapsack problem in [79, 122].

Local Search Heuristics. In [114], a local search-based algorithm for the knapsack problem with uncertain objective function is developed. We briefly list the main aspects. It makes use of two different search procedures: Given a feasible solution x and a list of local neighborhood moves M, let $GS(x, M)$ (the generalized search) determine the worst-case objective value of every move, and return the best move along with its objective value. Furthermore, let $RS(x, M, S)$ (the restricted search) perform a random search using the moves M with at most S steps.

The *cooperative local search* algorithm (CLS) works as follows: It first constructs a heuristic starting solution, e.g., by a greedy approach. In every iteration, a set of moves M is constructed using either the generalized search for sets with small cardinality, or the restricted search for sets with large cardinality. When a maximum number of iterations is reached, the best feasible solution found so far is returned.

Approximation Algorithms. A discussion of approximation algorithms for strict robustness with finitely many scenarios is given, e.g., in [3], where it is shown that there is an FPTAS for the shortest path, the spanning tree, and the knapsack problem when the number of scenarios is constant; but the shortest path problem is not $(2-\epsilon)$-approximable, the spanning tree problem is not $(\frac{3}{2}-\epsilon)$-approximable, and the knapsack problem is not approximable at all when the number of scenarios is considered as a non-constant input.

The basic idea for their results is to use the relationship between the strictly robust counterpart (SR) and multi-objective optimization: At least one optimal solution for (SR) is also an efficient solution in the multi-objective problem where every scenario is an objective. Thus, if the multi-objective problem has a polynomial-time α-approximation algorithm, then also (SR) has a polynomial-time α-approximation.

There exist many more approximation algorithms for specific problems. For example, in [57], robust set covering problems are considered with implicitly given, exponentially many scenarios. For a similar setting of exponentially many, implicitly given scenarios for robust network design problems (e.g., Steiner tree), [88] presents approximation results. Approximation results using finite scenario sets for two-stage robust covering problems, min-cut and shortest path can be found in [51, 78].

4.2 Infinite Scenarios

Sampling. When we cannot make use of the structure of $\mathcal{U}$ (i.e., when the reformulation approaches from Sect. 3 cannot be applied, or when we do not have a closed description of the set available), we can still solve (SR) heuristically using a finite subset of scenarios (given that we have some sampling method available). The resulting problem can be solved using the algorithms described in Sect. 4.1.

In a series of paper [37–40], the probability of a solution calculated by a sampled scenario subset being feasible for all scenarios is considered. It is shown that for a convex uncertain optimization problem, the probability of the *violation event* $V(x) = P\{\xi \in \mathcal{U} : x \notin \mathcal{F}(\xi)\}$ can be bounded by

$$P(V(x^*) > \epsilon) \le \sum_{i=0}^{n-1} \binom{N}{i} \epsilon^i (1-\epsilon)^{N-i},$$

where N is the sample size, $x^* \in \mathbb{R}^n$ is an optimal solution with respect to the sampled scenarios, and n is (as before) the dimension of the decision space. Note that the left-hand side is the probability of a probability; this is due to fact that $V(x)$ is a random variable in the sampled scenarios. In other words, if a desired probability of infeasibility ϵ is given, the accordingly required sample size can be determined. This result holds under the assumption that every subset of scenarios is feasible, and is independent of the probability distribution which is used for sampling over $\mathcal{U}$.

As the number of scenarios sampled this way may be large, the sequential optimization approach [61–63] uses sampled scenarios one by one. Using the above probability estimates, a solution generated by this method is feasible for (SR) only within a certain probability. The basic idea is the following: We consider the set $\mathcal{S}(\gamma)$ of feasible solutions with respect to a given quality level γ, i.e.,

$$\mathcal{S}(\gamma) = \{x \in \mathcal{X} : f(x) \leq \gamma, F(x, \xi) \leq 0 \ \forall \xi \in \mathcal{U}\}$$
$$= \{x \in \mathcal{X} : \nu(\gamma, x, \xi) \leq 0 \ \forall \xi \in \mathcal{U}\}$$

where
$$\nu(\gamma, x, \xi) = \left(\max\{0, f(x) - \gamma\}^2 + \max\{0, F(x, \xi)\}^2\right)^{1/2}$$

Using a subgradient on ν, the current solution is updated in every iteration using the sampled scenario ξ. Lower bounds on the number of required iterations are given to reach a desired level of solution quality and probability of feasibility.

Outer-Approximation and Cutting-Plane Methods. For this type of algorithm, the general idea is to iteratively (a) solve a robust optimization problem with a finite subset of scenarios, and (b) use a worst-case oracle that optimizes over the uncertainty set $\mathcal{U}$ for a given solution x. These steps can be done either exactly or approximately.

Algorithms of this type have often been used, see, e.g., [33, 59, 69, 99, 104, 111, 116]; sometimes even without knowledge that such an approach already exists (see also the lacking unification in robust optimization mentioned in Sect. 2.8).

The following general results should be mentioned. [104] show that this method converges under certain assumptions, and present further variations that improve the numerical performance of the algorithm. Cutting-plane methods are compared to compact formulations on general problem benchmarks in [59]. In [33], the implementation is considered in more detail: A distributed algorithm version is presented, in which each processor starts with a single uncertain constraint, and generated cutting planes are communicated.

4.3 Algorithms for Specific Problems

The goal of this section is to show how much one can benefit by using the structure a specific problem might have. To this end, we exemplarily chose three specialized algorithms: The first solves an NP-hard problem in pseudo-polynomial time, the second is a heuristic for another NP-hard problem, and the third is a polynomial-time solution approach. Note that many more such algorithms have been developed.

In [98], a dynamic programming algorithm is developed for the robust knapsack problem with cardinality constrained uncertainty in the weights. Extending the classic dynamic programming scheme to also include the number of items that are on their upper bounds, they are able to show a $\mathcal{O}(\Gamma nc)$ time complexity, where n is the number of items, and c is the knapsack budget (note that this is not a polynomial algorithm). The key idea of the dynamic program is an easy feasibility check of a solution, which is achieved by using an item sorting based on the upper weight bound $\bar{w}_i$. In computational experiments, instances with up to 5000 items can be solved in reasonable time.

The problem of min-max regret shortest paths with interval uncertainty is considered in [100]. The general idea is based on path ranking, and the conjecture that a path that ranks good on the worst-case scenario, may also rank good with respect to regret. Considering paths with respect to their worst-case performance order, they formulate a stopping criterion when the regret of a path may not improve anymore. Note that the regret of a single path can in this case easily be computed by assuming the worst-case length for all edges in the path, and the best-case length for all other edges. Experiments show a strong correlation between computation times and length of the optimal path.

While the former two problems are NP-hard (for regret shortest path, see [129]), a polynomial-time algorithm for the min-max regret 1-center on a tree with uncertain edge lengths and node weights is presented in [9]. A 1-center is a point on any edge of the tree for which the maximal weighted distance to all nodes is minimized. The algorithm runs in $\mathcal{O}(n^6)$ time, which can be reduced to $\mathcal{O}(n^2 \log(n))$ for the unweighted case. It is based on the observation that an edge that contains an optimal solution can be found in $\mathcal{O}(n^2 \log(n))$ time; however, determining its exact location for the weighted case is more complicated.

Further algorithms to be mentioned here are the polynomial algorithm for min-max regret flow-shop scheduling with two jobs from [8]; the polynomial algorithm for the min-max regret location-allocation problem from [45]; the heuristic for regret spanning arborescences from [48]; the polynomial algorithm for the min-max regret gradual covering location problem from [21]; and the PTAS for two-machine flow shop scheduling with discrete scenarios from [84].

4.4   Performance Guarantees

We now discuss performance guarantees in robust optimization. Measuring the performance of a robust solution or algorithm can be either done by developing guarantees regarding the performance of an algorithm or of a heuristic solution; but also by developing performance guarantees that compare the solutions generated by different robustness concepts.

On the algorithmic side, standard measures like the approximation ratio (i.e., the ratio between the robust objective value of the heuristic and the optimal robust solution) can be applied. There are simple, yet very general approximation algorithms presented in [4] for strict robustness and regret robustness: If the original problem is polynomially solvable, there is an N-approximation algorithm for finite uncertainty sets, where N is the number of scenarios. Furthermore, there is a 2-approximation algorithm for regret robustness with interval-based uncertainty [86] by using the mid-point scenario. These results have been extended in [47], see also the approximability survey [3] on strict and regret robustness. We do not know of approximation algorithms for other robustness concepts, which would provide interesting insight in the structural differences between the robust counterparts.

Regarding the comparison between solutions generated by different concepts, an interesting approach is to consider the quality of a strictly robust solution when used in an adjustable setting, as done in [24,25]. The authors are able to develop performance guarantees solely based on the degree of symmetry of the uncertainty set.

Concerning the evaluation of a robust solution (and not the algorithm to compute it), there is no general consent how to proceed, and surprisingly little systematic research can be found regarding this field. The so-called *robustness gap* as considered in [18] is defined as the difference between the worst-case objective of the robust solution, and the worst optimal objective value over all scenarios, i.e., as $SR^* - \sup_{\xi \in \mathcal{U}} f^*(\xi)$, where SR^* denotes the optimal value of (SR). They show that in the case of constraint-wise affine uncertainty, a compact set $\mathcal{X}$, and some technical assumptions, this gap equals zero. However, the most widely used approach is computing the so-called *price of robustness* [28], which is usually defined as the ratio between the robust solution value and the nominal solution value, i.e., as

$$\frac{\min_{x \in SR} \sup_{\xi \in \mathcal{U}} f(x, \xi)}{\min_{x \in \mathcal{F}(\hat{\xi})} f(x, \hat{\xi})}$$

As an example, [97] presents the analytical calculation of the price of robustness for knapsack problems. Using an interval-based uncertainty set on the weights (i.e., the weight of item i is in $[w_i - \underline{w}_i, w_i + \overline{w}_i]$) and a cardinality constrained robustness approach, they show that the price of robustness equals $1/(1 + \lceil \delta_{max} \rceil)$ for $\delta_{max} := \max_i \overline{w}_i / w_i$ and $\Gamma = 1$. For $\Gamma \geq 2$, the price of robustness becomes $1/(1 + \lceil 2\delta_{max} \rceil)$.

Note that this is a rather pessimistic view on robustness, as it only concentrates on the additional *costs* of a robust solution compared to the nominal objective function value of an optimal solution for the nominal case. However, if the application under consideration is affected by uncertainty, the nominal solution will not necessarily find nominal conditions, hence the robust solution may actually *save* costs compared to the nominal solution (which easily may be even infeasible). There is no general "golden rule" that would provide a fair evaluation for the performance of a robust solution.

Note that such a bound is not the kind of performance guarantee that was actually considered in [28]. Rather, they developed probability bounds for the feasibility of a solution to the cardinality constrained approach depending on Γ. Using such bounds they argue that the nominal performance of a solution can be considerably increased without decreasing the probability of being feasible too much.

Summarizing the above remarks, we claim that:

Remark 2. Performance guarantees are not sufficiently researched in robust optimization.

5 Implementation and Experiments

5.1 Libraries

In the following, we present some libraries that are designed for robust optimization. A related overview can also be found in [68].

AIMMS for Robust Optimization. AIMMS [108], which stands for "Advanced Interactive Multidimensional Modeling System", is a proprietary software that contains an algebraic modeling language (AML) for optimization problems. AIMMS supports most well-known solvers, including Cplex[1], Xpress[2] and Gurobi[3].

Since 2010, AIMMS has offered a robust optimization add-on, which was developed in a partnership with A. Ben-Tal. The extension only considers the concepts of strict and adjustable robustness as introduced in Sects. 2.1 and 2.3. As uncertainty sets, interval-based uncertainty sets, polytopic uncertainty sets, or ellipsoidal uncertainty sets are supported and transformed to mathematical programs as described in Sect. 3.1. The respective transformations are automatically done when the model is translated from the algebraic modeling language to the solver.

ROME. While AIMMS focuses on the work of Ben-Tal and co-workers, ROME [77] ("Robust Optimization Made Easy") takes its origins in the work of Bertsimas, Sim and co-workers. ROME is built in the MATLAB[4] environment, which makes it on the one hand intuitive to use for MATLAB-users, but on the other hand lacks the versatility of an AML. As a research project, ROME is free to use. It currently supports Cplex, Xpress and SDPT3[5] as solver engines.

ROME considers polytopic and ellipsoidal uncertainty sets, that can be further specified using the mean support, the covariance matrix, or directional deviations. Assuming an affine dependence of the wait-and-see variables, it then transforms the uncertain optimization problem to an adjustable robust counterpart. The strictly robust counterpart is included as a special case.

---

[1] http://www-03.ibm.com/software/products/en/ibmilogcpleoptistud.
[2] http://www.fico.com/en/products/fico-xpress-optimization-suite.
[3] http://www.gurobi.com/.
[4] http://www.mathworks.com/products/matlab/.
[5] http://www.math.nus.edu.sg/~mattohkc/sdpt3.html.

YALMIP. Similar to ROME, YALMIP [92] is a layer between MATLAB and a solver that allows the modeling of optimization problems under uncertainty. Nearly all well-known solvers are supported, including Cplex, Gurobi and Xpress.

YALMIP considers strict robustness. In order to obtain the strict robust counterpart of an uncertain optimization problems so-called *filters* are used: When presented a model with uncertainty, the software checks if one of these filters applies to generate the strictly robust counterpart. Currently, five of these automatic transformations are implemented. A duality filter (which adds dual variables according to Sect. 3.2), an enumeration filter for finite and polytopic scenario sets (which simply lists all relvant constraints), an explicit maximization filter (where a worst-case scenario is used), the Pólya filter (which is based on an inner approximation of the set of feasible solutions), and an elimination filter (which sets variables affected by uncertainty to 0 and is used as a last resort).

ROPI. The Robust Optimization Programming Interface (ROPI) [66,68] is a C++ library that provides wrapper MIP classes to support a range of solvers. Using these generic classes, a robust counterpart is automatically generated given the desired robustness concept and uncertainty set. Contrary to the previous libraries, a wider choice of robustness concepts is provided: These include strict robustness, adjustable robustness, light robustness, and different versions of recoverable robustness.

Even though a user can pick and choose between multiple robust optimization libraries, there is to the best of our knowledge no library of robust optimization *algorithms* available. All of the above implementations are based on reformulation approaches, which makes it possible to draw upon existing solvers. However, as described in Sect. 4, there are plenty of specifically designed algorithms for robust optimization available. Making them readily-implemented available to the user should be a significant concern for future work in robust optimization.

Remark 3. There is no robust optimization library available with specifically designed algorithms other than reformulation approaches.

5.2 Applications

As already stated, robust optimization has been application-driven; thus, there are abundant papers dealing with applications of some robustness approach to real-world or at least realistic problems. Presenting an exhaustive list would go far beyond the scope of this paper; examples include circuit design [94], emergency logistics [13], and load planning [32] for adjustable robustness; supply chain optimization [29] and furniture planning [5] for cardinality constrained robustness; inventory control for comprehensive robustness [15]; timetabling [58,60], and timetable information [71] for light robustness; shunting [43], timetabling [44,73], and railway rolling stock planning [36] for recoverable robustness; and airline scheduling for UFO [52].

Hence, we can state:

Remark 4. Robust optimization is application-driven.

5.3 Comparative Experiments

In this section we consider research that either compares two robustness concepts to the same problem, or two algorithms for the same problem and robustness concept. We present a list of papers on the former aspect in Table 1, and a list of papers on the latter aspect in Table 2. We do not claim completeness for these tables; rather, they should be considered as giving a general impression on recent directions of research.

Table 1. Papers presenting experiments comparing at least two different robustness concepts. "cc" abbreviates "cardinality constrained".

Year	Paper	Problem	Robustness concept
2008	[26]	Portfolio management	Strict and cc
2009	[15]	Inventory control	Adjustable and comprehensive
2009	[127]	Road improvement	Strict and scenario-based
2010	[73]	Timetabling	Strict, buffered, light, and variations of recoverable
2010	[128]	Sawmill planning	Mulvey with different recourse costs
2010	[125]	Water sensor placement	Strict and regret
2011	[74]	LP Benchmarks	Strict and recoverable
2011	[1]	Wireless network resource allocation	Finite and interval-based
2011	[91]	Newsvendor	Strict and regret
2013	[72]	Timetable information	Strict and light
2013	[70]	Timetable information	Strict and recoverable
2013	[32]	Load planning	Strict and adjustable
2013	[2]	Vehicle routing	Strict and adjustable

We conclude the following from these tables and the accompanying literature: Firstly, papers considering real-world applications that compare different robustness concepts are relatively rare. Applied studies are too often satisfied with considering only one approach of the many that are possible. Secondly, algorithmic comparisons dominantly stem from the field of min-max regret, where at the same time mostly academic problems are considered. The efficient calculation of solutions for other robustness concepts is still a relatively open and promising field of research. Summarizing, we claim that:

Remark 5. There are too few comparative studies in robust optimization.

A different aspect Table 1 reveals is that most computational studies comparing at least two robustness concepts include strict robustness as a "baseline concept"; accordingly, and unsurprisingly, the more tailor-made approaches will show an

Table 2. Papers presenting experiments comparing at least two algorithms for the same robustness concept. "cc" abbreviates "cardinality constrained".

Year	Paper	Problem	Concept	Algorithms
2005	[101]	Spanning tree	Regret	Branch and bound, MIP
2006	[99]	Spanning tree	Regret	Bender's decomp., MIP, branch and bound
2008	[106]	Spanning tree	Regret	Simulated annealing, branch and bound, Bender's decomp.
2008	[122]	Knapsack	Strict	Branch and bound with and without preprocessing
2008	[124]	Capacitated sourcing	Adjustable	Tabu search
2009	[46]	Critical path	Regret	MIP and heuristic
2010	[56]	Machine scheduling	Strict	MIP with and without cuts
2010	[30]	Wine harvesting	cc robust	MIP and scenario generation
2010	[105]	Lot allocation	Strict	Branch-and-price and heuristics
2011	[41]	Shortest path	Regret	IP with and without preprocessing
2011	[109]	Assignment	Regret	MIP, Bender's decomp., genetic algorithms
2012	[85]	Spanning tree	Regret	Tabu search and IP
2012	[59]	Diverse	cc robust	MIP and cutting planes
2012	[117]	Knapsack	Strict	Local search and branch and bound
2013	[98]	Knapsack	cc robust	Dynamic programming and IP
2013	[107]	Capacity assignment	Adjustable	Approximations

improved behavior for the application at hand. This is much similar to frequently published papers on optimization problems which compare a problem-specific method to a generic MIP solver, usually observing a better performance of the former compared to the latter.

However, while a standard MIP solver is often still competitive to problem-tailored algorithms, a robustness concept which does not capture the problem specifics at hand will nearly always be the second choice to one which uses the full problem potential.

5.4   Limits of Solvability

We show the approximate size of benchmark instances used for testing exact algorithms for a choice of robust problems in Table 3. These values should rather be considered as rough indicators on the current limits of solvability than the exact limits themselves, as problem complexities are determined by many more aspects[6].

---

[6] Number of items for finite, strict knapsack is estimated with the pegging test from [122].

Table 3. Currently considered problem sizes for exact algorithms.

Problem	Approach	Size	Source
Spanning tree	Interval regret	∼100 nodes	[110]
Knapsack	Finite strict	∼1500 items	[67]
Knapsack	Finite recoverable	∼500 items	[34]
Knapsack	cc strict	∼5000 items	[98]
Knapsack	cc recoverable	∼200 items	[35]
Shortest path	Interval regret	∼1500 nodes	[42]
Assignment	Interval regret	∼500 items	[109]

What becomes immediately obvious is that these limits are much smaller than for their nominal problem counterparts, which can go easily into the millions.

5.5 Learning from Experiments

We exemplarily show how experimental results can be used to design better algorithms for robust optimization; thus, we highlight the potential that lies in following the algorithm engineering cycle. To this end, we consider the regret shortest path problem: Given a set of scenarios consisting of arc lengths in a graph, find a path from a fixed source node to a fixed sink node which minimizes the worst-case length difference to an optimal path for each scenario.

From a theoretical perspective, the problem complexity is well-understood. For discrete uncertainty sets (and already for only two scenarios), the problem was shown to be NP-hard in the seminal monograph [89]. For interval-based uncertainty, [129] showed its NP-hardness.

Furthermore, it is known that the regret shortest path problem with a finite, but unbounded set of scenarios is not approximable within $2 - \epsilon$. For the interval-case, a very simple 2-approximation algorithm (see [86]) is known: All one needs to do is to compute the shortest path with respect to the midpoint scenario, i.e., the arc lengths which are the midpoint of the respective intervals.

To solve the interval regret problem exactly, a branch-and-bound method has been proposed [100], which branches along the worst-case path in the graph. However, computational experience shows that the midpoint solution – despite being "only" a 2-approximation – is already an optimal, or close-to-optimal solution for many of the randomly generated benchmark instances.

Examining this aspect in more detail, [42] developed an instance-dependent approximation guarantee for the midpoint solution, which is always less or equal to 2, but usually lies around $\sim 1.6 - 1.7$.

Using these two ingredients – the strong observed performance of the midpoint solution, and its instance-dependent lower bound – the branch-and-bound algorithm of [100] can be easily adapted, by using a midpoint-path-based branching strategy instead of the worst-case path, and by using the improved guarantee as a lower bound. The resulting algorithm considerably outperforms the previous version, with computation times two orders of magnitude better for some instance classes.

These modifications were possible by studying experimental results, improving thereupon the theoretical analysis, and feeding this analysis back to an algorithm. It is an example for the successful traversal of an algorithm engineering cycle, and we believe that many more such algorithmic improvements can be achieved this way.

6   Algorithm Engineering in Robust Optimization and Conclusion

In this paper we propose to use the algorithm engineering methodology to better understand the open problems and challenges in robust optimization. Doing so, we were able to point out links between algorithm engineering and robust optimization, and we presented an overview on the state-of-the-art from this perspective.

In order to further stress the usefulness of the algorithm engineering methodology, we finally present three examples. Each of them is composed of a series of papers, which together follow the algorithm engineering cycle in robust optimization.

Example 1: Development of new models based on shortcomings of previous ones.

[118] introduced the concept of strict robustness. This concept was illustrated in several examples (e.g. from linear programming, see [20] or for a cantilever arm as in [18]) and analyzed for these examples in a mathematical way. The analysis in these papers showed that the problem complexity increases then introducing robustness (e.g., the robust counterpart of an uncertain linear program with ellipsoidal uncertainty is an explicit conic quadratic program). Moreover, the authors recognized that the concept is rather conservative introducing an approximate robust counterpart with a more moderate level of conservatism. These ideas were taken up [28] to start the next run through the algorithm engineering cycle by introducing their new concept of cardinality constrained robustness, which is less conservative and computationally better tractable, but may be applied only to easier uncertainty sets. Applying this concept to train timetabling and performing experiments with it was the starting point of [58] who relaxed the constraints further and developed the concept of light robustness which was then later generalized to arbitrary uncertainty sets by [115].

Example 2: From one-stage to two-stage robustness.

Recognizing that the concept of strict robustness is too conservative, [16] proposed the first two-stage robustness approach by introducing their concept of adjustable robustness. When applying this concept to several application of railway planning within the ARRIVAL project (see [6]), [90] noted that the actions allowed to adjust a timetable do not fit the practical needs. This motivated them to integrate recovery actions in robust planning yielding the concept of recoverable robustness. Unfortunately, recovery robust solutions are hard to obtain. Research on developing practical algorithms is still ongoing. recent examples are a column-generation based approach for robust knapsack problems and shortest path problems with uncertain demand [31], an approach using Bender's decomposition for railway rolling stock planning [36], and the idea of replacing the recovery algorithm by a metric [65, 74, 75].

Example 3: Robust passenger information systems.
The following example shows the application of the algorithm engineering cycle on a specific application, namely constructing robust timetable information systems. Suppose that a passenger wants to travel from an origin to some destination by public transportation. The passenger can use a timetable information system which will provide routes with small traveling time. However, since delays are a matter of fact in public transportation, a robust route would be more valuable than just having a shortest route. In [72] this problem was considered for strictly robust routes: The model was set up, analyzed (showing that it is NP-complete), and an algorithm for its solution was designed. The experimental evaluation on real-world data showed that strictly robust routes are useless in practice: their traveling time is much too long. Based on these experiments, light robust passenger information system was considered. The light robust model was designed and analyzed; algorithms based on the strictly robust procedures could be developed. The experiments showed that this model is much better applicable in practice. However, the model was still not satisfactory, since it was assumed that a passenger stays on his/her route whatever happens. This drawback motivated to start the algorithm engineering cycle again in [70] where now recoverable robust timetables are investigated.

Considering the cycle of design, analysis, implementation, and experiments, we were also able to identify pointers for further research. We summarize our results by reproducing the most significant messages:

1. *Robust optimization is application-driven.* From the beginning, robust optimization was intended as an optimization approach which generates solutions that perform well in a realistic environment. As such, it is highly appealing to practitioners, who would rather sacrifice some nominal solution quality if the solution stability can be increased.
2. *Robust optimization needs a unified classification scheme.* While the strong connection to applications is a beneficial driver of research, it also carries problems. One striking observation is a lack of unification in robust optimization. This begins with simple nomenclature: The names for strict robustness, or the uncertainty set considered by Bertsimas and Sim are plenty. It extends to the frequent re-development of algorithmic ideas (as iterative scenario generation), and the reinvention of robustness concepts from scratch for specific applications. This lack of organization is in fact unscientific, and endangers the successful perpetuation of research. As related problems, some journals don't even offer "robust optimization" as a subject classification (even though publishing papers on robust optimization); solutions generated by some fashion that is somehow related to uncertainty call themselves "robust"; and

students that are new to the field have a hard time to identify the state-of-the-art.

3. *Performance guarantees are not sufficiently researched in robust optimization.* Also this point can be regarded as related to robust optimization being application-driven and non-unified. Performance guarantees are of special importance when comparing algorithms; hence, with a lack of comparison, there also comes a lack of performance guarantees. This includes the comparison of robust optimization concepts, of robust optimization algorithms, and even the general evaluation of a robust solution compared to a non-robust solution.

4. *There is no robust optimization library available with specifically designed algorithms other than reformulation approaches.* While libraries for robust optimization exist, they concentrate on the modeling aspects of uncertainty, and less on different algorithmic approaches. Having such a library available would prove extremely helpful not only for practitioners, but also for researches that develop new algorithms and try to compare them to the state-of-the-art.

5. *There are too few comparative studies in robust optimization.* All the above points culminate in the lack of comparative studies; however, we argue that here also lies a chance to tackle these problems. This paper is a humble step to motivate such research, and we hope for many more publications to come.

References

1. Adasme, P., Lisser, A., Soto, I.: Robust semidefinite relaxations for a quadratic OFDMA resource allocation scheme. Comput. Oper. Res. **38**(10), 1377–1399 (2011)
2. Agra, A., Christiansen, M., Figueiredo, R., Hvattum, L.M., Poss, M., Requejo, C.: The robust vehicle routing problem with time windows. Comput. Oper. Res. **40**(3), 856–866 (2013)
3. Aissi, H., Bazgan, C., Vanderpooten, D.: Approximation of min–max and min–max regret versions of some combinatorial optimization problems. Eur. J. Oper. Res. **179**(2), 281–290 (2007)
4. Aissi, H., Bazgan, C., Vanderpooten, D.: Min–max and min–max regret versions of combinatorial optimization problems: a survey. Eur. J. Oper. Res. **197**(2), 427–438 (2009)
5. Alem, D.J., Morabito, R.: Production planning in furniture settings via robust optimization. Comput. Oper. Res. **39**(2), 139–150 (2012)
6. Arrival project under contract no. FP6-021235-2. http://arrival.cti.gr/index.php/Main/HomePage
7. Atamtürk, A.: Strong formulations of robust mixed 0–1 programming. Math. Program. **108**(2), 235–250 (2006)
8. Averbakh, I.: The minmax regret permutation flow-shop problem with two jobs. Eur. J. Oper. Res. **169**(3), 761–766 (2006)
9. Averbakh, I., Berman, O.: Algorithms for the robust 1-center problem on a tree. Eur. J. Oper. Res. **123**(2), 292–302 (2000)
10. Beck, A., Ben-Tal, A.: Duality in robust optimization: primal worst equals dual best. Oper. Res. Lett. **37**(1), 1–6 (2009)

11. Ben-Tal, A., Bertsimas, D., Brown, D.B.: A soft robust model for optimization under ambiguity. Oper. Res. **58**(4–Part–2), 1220–1234 (2010)
12. Ben-Tal, A., Boyd, S., Nemirovski, A.: Extending scope of robust optimization: comprehensive robust counterparts of uncertain problems. Math. Program. **107**(1–2), 63–89 (2006)
13. Ben-Tal, A., Chung, B.D., Mandala, S.R., Yao, T.: Robust optimization for emergency logistics planning: risk mitigation in humanitarian relief supply chains. Transp. Res. Part B: Methodol. **45**(8), 1177–1189 (2011)
14. Ben-Tal, A., Ghaoui, L.E., Nemirovski, A.: Robust Optimization. Princeton University Press, Princeton (2009)
15. Ben-Tal, A., Golany, B., Shtern, S.: Robust multi-echelon multi-period inventory control. Eur. J. Oper. Res. **199**(3), 922–935 (2009)
16. Ben-Tal, A., Goryashko, A., Guslitzer, E., Nemirovski, A.: Adjustable robust solutions of uncertain linear programs. Math. Program. A **99**, 351–376 (2003)
17. Ben-Tal, A., den Hertog, D., Vial, J.P.: Deriving robust counterparts of nonlinear uncertain inequalities. Math. Program. **149**(1), 265–299 (2015)
18. Ben-Tal, A., Nemirovski, A.: Robust convex optimization. Math. Oper. Res. **23**(4), 769–805 (1998)
19. Ben-Tal, A., Nemirovski, A.: Robust solutions of uncertain linear programs. Oper. Res. Lett. **25**, 1–13 (1999)
20. Ben-Tal, A., Nemirovski, A.: Robust solutions of linear programming problems contaminated with uncertain data. Math. Program. A **88**, 411–424 (2000)
21. Berman, O., Wang, J.: The minmax regret gradual covering location problem on a network with incomplete information of demand weights. Eur. J. Oper. Res. **208**(3), 233–238 (2011)
22. Bertsimas, D., Brown, D., Caramanis, C.: Theory and applications of robust optimization. SIAM Rev. **53**(3), 464–501 (2011)
23. Bertsimas, D., Caramanis, C.: Finite adaptability in multistage linear optimization. IEEE Trans. Autom. Control **55**(12), 2751–2766 (2010)
24. Bertsimas, D., Goyal, V.: On the power of robust solutions in two-stage stochastic and adaptive optimization problems. Math. Oper. Res. **35**(2), 284–305 (2010)
25. Bertsimas, D., Goyal, V., Sun, X.A.: A geometric characterization of the power of finite adaptability in multistage stochastic and adaptive optimization. Math. Oper. Res. **36**(1), 24–54 (2011)
26. Bertsimas, D., Pachamanova, D.: Robust multiperiod portfolio management in the presence of transaction costs. Comput. Oper. Res. **35**(1), 3–17 (2008)
27. Bertsimas, D., Pachamanova, D., Sim, M.: Robust linear optimization under general norms. Oper. Res. Lett. **32**(6), 510–516 (2004)
28. Bertsimas, D., Sim, M.: The price of robustness. Oper. Res. **52**(1), 35–53 (2004)
29. Bertsimas, D., Thiele, A.: A robust optimization approach to inventory theory. Oper. Res. **54**(1), 150–168 (2006)
30. Bohle, C., Maturana, S., Vera, J.: A robust optimization approach to wine grape harvesting scheduling. Eur. J. Oper. Res. **200**(1), 245–252 (2010)
31. Bouman, P.C., Akker, J.M., Hoogeveen, J.A.: Recoverable robustness by column generation. In: Demetrescu, C., Halldórsson, M.M. (eds.) ESA 2011. LNCS, vol. 6942, pp. 215–226. Springer, Heidelberg (2011). doi:10.1007/978-3-642-23719-5_19
32. Bruns, F., Goerigk, M., Knust, S., Schöbel, A.: Robust load planning of trains in intermodal transportation. OR Spectr. **36**(3), 631–668 (2014)

33. Bürger, M., Notarstefano, G., Allgöwer, F.: A polyhedral approximation framework for convex and robust distributed optimization. IEEE Trans. Autom. Control **59**(2), 384–395 (2014)
34. Büsing, C., Koster, A.M.C.A., Kutschka, M.: Recoverable robust knapsacks: the discrete scenario case. Optim. Lett. **5**(3), 379–392 (2011)
35. Büsing, C., Koster, A., Kutschka, M.: Recoverable robust knapsacks: Γ-scenarios. In: Pahl, J., Reiners, T., Voß, S. (eds.) Network Optimization. Lecture Notes in Computer Science, vol. 6701, pp. 583–588. Springer, Heidelberg (2011)
36. Cacchiani, V., Caprara, A., Galli, L., Kroon, L., Maroti, G., Toth, P.: Railway rolling stock planning: robustness against large disruptions. Transp. Sci. **46**(2), 217–232 (2012)
37. Calafiore, G., Campi, M.: Uncertain convex programs: randomized solutions and confidence levels. Math. Program. **102**(1), 25–46 (2005)
38. Calafiore, G.C.: Random convex programs. SIAM J. Optim. **20**(6), 3427–3464 (2010)
39. Calafiore, G., Campi, M.: The scenario approach to robust control design. IEEE Trans. Autom. Control **51**(5), 742–753 (2006)
40. Campi, M.C., Garatti, S.: The exact feasibility of randomized solutions of uncertain convex programs. SIAM J. Optim. **19**(3), 1211–1230 (2008)
41. Catanzaro, D., Labbé, M., Salazar-Neumann, M.: Reduction approaches for robust shortest path problems. Comput. Oper. Res. **38**(11), 1610–1619 (2011)
42. Chassein, A.B., Goerigk, M.: A new bound for the midpoint solution in minmax regret optimization with an application to the robust shortest path problem. Eur. J. Oper. Res. **244**(3), 739–747 (2015)
43. Cicerone, S., D'Angelo, G., Stefano, G.D., Frigioni, D., Navarra, A.: Robust algorithms and price of robustness in shunting problems. In: Proceedings of the 7th Workshop on Algorithmic Approaches for Transportation Modeling, Optimization, and Systems (ATMOS 2007) (2007)
44. Cicerone, S., D'Angelo, G., Stefano, G., Frigioni, D., Navarra, A., Schachtebeck, M., Schöbel, A.: Recoverable robustness in shunting and timetabling. In: Ahuja, R.K., Möhring, R.H., Zaroliagis, C.D. (eds.) Robust and Online Large-Scale Optimization. LNCS, vol. 5868, pp. 28–60. Springer, Heidelberg (2009). doi:10.1007/978-3-642-05465-5_2
45. Conde, E.: Minmax regret location–allocation problem on a network under uncertainty. Eur. J. Oper. Res. **179**(3), 1025–1039 (2007)
46. Conde, E.: A minmax regret approach to the critical path method with task interval times. Eur. J. Oper. Res. **197**(1), 235–242 (2009)
47. Conde, E.: On a constant factor approximation for minmax regret problems using a symmetry point scenario. Eur. J. Oper. Res. **219**(2), 452–457 (2012)
48. Conde, E., Candia, A.: Minimax regret spanning arborescences under uncertain costs. Eur. J. Oper. Res. **182**(2), 561–577 (2007)
49. D'Angelo, G., Di Stefano, G., Navarra, A.: Recoverable-robust timetables for trains on single-line corridors. In: Proceedings of the 3rd International Seminar on Railway Operations Modelling and Analysis - Engineering and Optimisation Approaches (RailZurich 2009) (2009)
50. Delling, D., Sanders, P., Schultes, D., Wagner, D.: Engineering route planning algorithms. In: Lerner, J., Wagner, D., Zweig, K.A. (eds.) Algorithmics of Large and Complex Networks. Lecture Notes in Computer Science, vol. 5515, pp. 117–139. Springer, Heidelberg (2009)

51. Dhamdhere, K., Goyal, V., Ravi, R., Singh, M.: How to pay, come what may: approximation algorithms for demand-robust covering problems. In: 46th Annual IEEE Symposium on Foundations of Computer Science, FOCS 2005, pp. 367–376. IEEE (2005)
52. Eggenberg, N.: Combining robustness and recovery for airline schedules. Ph.D. thesis, EPFL (2009)
53. Eggenberg, N., Salani, M., Bierlaire, M.: Uncertainty feature optimization: an implicit paradigm for problems with noisy data. Networks 57(3), 270–284 (2011)
54. Erera, A., Morales, J., Svalesbergh, M.: Robust optimization for empty repositioning problems. Oper. Res. 57(2), 468–483 (2009)
55. Falk, J.E.: Exact solutions of inexact linear programs. Oper. Res. 24(4), 783–787 (1976)
56. de Farias, J.R., Zhao, H., Zhao, M.: A family of inequalities valid for the robust single machine scheduling polyhedron. Comput. Oper. Res. 37(9), 1610–1614 (2010)
57. Feige, U., Jain, K., Mahdian, M., Mirrokni, V.: Robust combinatorial optimization with exponential scenarios. In: Fischetti, M., Williamson, D.P. (eds.) IPCO 2007. LNCS, vol. 4513, pp. 439–453. Springer, Heidelberg (2007). doi:10.1007/978-3-540-72792-7_33
58. Fischetti, M., Monaci, M.: Light robustness. In: Ahuja, R.K., Möhring, R.H., Zaroliagis, C.D. (eds.) Robust and Online Large-Scale Optimization. LNCS, vol. 5868, pp. 61–84. Springer, Heidelberg (2009). doi:10.1007/978-3-642-05465-5_3
59. Fischetti, M., Monaci, M.: Cutting plane versus compact formulations for uncertain (integer) linear programs. Math. Program. Comput. 4(3), 239–273 (2012)
60. Fischetti, M., Salvagnin, D., Zanette, A.: Fast approaches to improve the robustness of a railway timetable. Transp. Sci. 43, 321–335 (2009)
61. Fujisaki, Y., Wada, T.: Sequential randomized algorithms for robust optimization. In: Proceedings of the 46th IEEE Conference on Decision and Control, pp. 6190–6195 (2007)
62. Fujisaki, Y., Wada, T.: Robust optimization via probabilistic cutting plane technique. In: Proceedings of the 40th ISCIE International Symposium on Stochastic Systems Theory and its Applications, pp. 137–142 (2009)
63. Fujisaki, Y., Wada, T.: Robust optimization via randomized algorithms. In: ICCAS-SICE 2009, pp. 1226–1229 (2009)
64. Ghaoui, L.E., Lebret, H.: Robust solutions to least-squares problems with uncertain data. SIAM J. Matrix Anal. Appl. 18, 1034–1064 (1997)
65. Goerigk, M.: Algorithms and concepts for robust optimization. Ph.D. thesis, Georg-August Universität Göttingen (2012)
66. Goerigk, M.: ROPI homepage (2013). http://optimierung.mathematik.uni-kl.de/~goerigk/ropi/
67. Goerigk, M.: A note on upper bounds to the robust knapsack problem with discrete scenarios. Ann. Oper. Res. 223(1), 461–469 (2014)
68. Goerigk, M.: ROPI - a robust optimization programming interface for C++. Optim. Methods Softw. 29(6), 1261–1280 (2014)
69. Goerigk, M., Deghdak, K., T'Kindt, V.: A two-stage robustness approach to evacuation planning with buses. Transp. Res. Part B: Methodol. 78, 66–82 (2015)
70. Goerigk, M., Heße, S., Müller-Hannemann, M., Schmidt, M., Schöbel, A.: Recoverable robust timetable information. In: Frigioni, D., Stiller, S. (eds.) 13th Workshop on Algorithmic Approaches for Transportation Modelling, Optimization, and Systems. OpenAccess Series in Informatics (OASIcs), vol. 33, pp. 1–14. Schloss Dagstuhl-Leibniz-Zentrum fuer Informatik, Dagstuhl (2013)

71. Goerigk, M., Knoth, M., Müller-Hannemann, M., Schmidt, M., Schöbel, A.: The price of robustness in timetable information. In: Caprara, A., Kontogiannis, S. (eds.) 11th Workshop on Algorithmic Approaches for Transportation Modelling, Optimization, and Systems. OpenAccess Series in Informatics (OASIcs), vol. 20, pp. 76–87. Schloss Dagstuhl-Leibniz-Zentrum fuer Informatik, Dagstuhl (2011)
72. Goerigk, M., Schmidt, M., Schöbel, A., Knoth, M., Müller-Hannemann, M.: The price of strict and light robustness in timetable information. Transp. Sci. **48**(2), 225–242 (2014)
73. Goerigk, M., Schöbel, A.: An empirical analysis of robustness concepts for timetabling. In: Erlebach, T., Lübbecke, M. (eds.) 10th Workshop on Algorithmic Approaches for Transportation Modelling, Optimization, and Systems (ATMOS 2010). OpenAccess Series in Informatics (OASIcs), vol. 14, pp. 100–113. Schloss Dagstuhl-Leibniz-Zentrum fuer Informatik, Dagstuhl (2010)
74. Goerigk, M., Schöbel, A.: A scenario-based approach for robust linear optimization. In: Marchetti-Spaccamela, A., Segal, M. (eds.) TAPAS 2011. LNCS, vol. 6595, pp. 139–150. Springer, Heidelberg (2011). doi:10.1007/978-3-642-19754-3_15
75. Goerigk, M., Schöbel, A.: Recovery-to-optimality: a new two-stage approach to robustness with an application to aperiodic timetabling. Comput. Oper. Res. **52**(Part A), 1–15 (2014)
76. Goetzmann, K.S., Stiller, S., Telha, C.: Optimization over integers with robustness in cost and few constraints. In: Solis-Oba, R., Persiano, G. (eds.) Approximation and Online Algorithms (WAOA 2011). Lecture Notes in Computer Science, vol. 7164, pp. 89–101. Springer, Heidelberg (2012)
77. Goh, J., Sim, M.: Robust optimization made easy with ROME. Oper. Res. **59**(4), 973–985 (2011)
78. Golovin, D., Goyal, V., Polishchuk, V., Ravi, R., Sysikaski, M.: Improved approximations for two-stage min-cut and shortest path problems under uncertainty. Math. Program. **149**(1), 167–194 (2015)
79. Iida, H.: A note on the max-min 0–1 knapsack problem. J. Comb. Optim. **3**(1), 89–94 (1999)
80. Inuiguchi, M., Sakawa, M.: Minimax regret solution to linear programming problems with an interval objective function. Eur. J. Oper. Res. **86**(3), 526–536 (1995)
81. Jenkins, L.: Selecting scenarios for environmental disaster planning. Eur. J. Oper. Res. **121**(2), 275–286 (2000)
82. Jeyakumar, V., Li, G., Srisatkunarajah, S.: Strong duality for robust minimax fractional programming problems. Eur. J. Oper. Res. **228**(2), 331–336 (2013)
83. Kalai, R., Lamboray, C., Vanderpooten, D.: Lexicographic α-robustness: an alternative to min–max criteria. Eur. J. Oper. Res. **220**(3), 722–728 (2012)
84. Kasperski, A., Kurpisz, A., Zieliński, P.: Approximating a two-machine flow shop scheduling under discrete scenario uncertainty. Eur. J. Oper. Res. **217**(1), 36–43 (2012)
85. Kasperski, A., Makuchowski, M., Zieliński, P.: A tabu search algorithm for the minmax regret minimum spanning tree problem with interval data. J. Heuristics **18**(4), 593–625 (2012)
86. Kasperski, A., Zieliński, P.: An approximation algorithm for interval data minmax regret combinatorial optimization problems. Inf. Process. Lett. **97**(5), 177–180 (2006)
87. Kasperski, A., Zieliński, P.: Minmax regret approach and optimality evaluation in combinatorial optimization problems with interval and fuzzy weights. Eur. J. Oper. Res. **200**(3), 680–687 (2010)

88. Khandekar, R., Kortsarz, G., Mirrokni, V., Salavatipour, M.R.: Two-stage robust network design with exponential scenarios. Algorithmica **65**(2), 391–408 (2013)
89. Kouvelis, P., Yu, G.: Robust Discrete Optimization and Its Applications. Kluwer Academic Publishers, Boston (1997)
90. Liebchen, C., Lübbecke, M., Möhring, R., Stiller, S.: The concept of recoverable robustness, linear programming recovery, and railway applications. In: Ahuja, R.K., Möhring, R.H., Zaroliagis, C.D. (eds.) Robust and Online Large-Scale Optimization. LNCS, vol. 5868, pp. 1–27. Springer, Heidelberg (2009). doi:10.1007/978-3-642-05465-5_1
91. Lin, J., Ng, T.S.: Robust multi-market newsvendor models with interval demand data. Eur. J. Oper. Res. **212**(2), 361–373 (2011)
92. Löfberg, J.: Automatic robust convex programming. Optim. Methods Softw. **27**(1), 115–129 (2012)
93. López, M., Still, G.: Semi-infinite programming. Eur. J. Oper. Res. **180**(2), 491–518 (2007)
94. Mani, M., Sing, A.K., Orshansky, M.: Joint design-time and post-silicon minimization of parametric yield loss using adjustable robust optimization. In: Proceedings of the 2006 IEEE/ACM International Conference on Computer-Aided Design, ICCAD 2006, pp. 19–26. ACM, New York (2006)
95. Mausser, H.E., Laguna, M.: A heuristic to minimax absolute regret for linear programs with interval objective function coefficients. Eur. J. Oper. Res. **117**(1), 157–174 (1999)
96. Mausser, H., Laguna, M.: A new mixed integer formulation for the maximum regret problem. Int. Trans. Oper. Res. **5**(5), 389–403 (1998)
97. Monaci, M., Pferschy, U.: On the robust knapsack problem. SIAM J. Optim. **23**(4), 1956–1982 (2013)
98. Monaci, M., Pferschy, U., Serafini, P.: Exact solution of the robust knapsack problem. Comput. Oper. Res. **40**(11), 2625–2631 (2013)
99. Montemanni, R.: A Benders decomposition approach for the robust spanning tree problem with interval data. Eur. J. Oper. Res. **174**(3), 1479–1490 (2006)
100. Montemanni, R., Gambardella, L.: An exact algorithm for the robust shortest path problem with interval data. Comput. Oper. Res. **31**(10), 1667–1680 (2004)
101. Montemanni, R., Gambardella, L.: A branch and bound algorithm for the robust spanning tree problem with interval data. Eur. J. Oper. Res. **161**(3), 771–779 (2005)
102. Müller-Hannemann, M., Schirra, S. (eds.): Algorithm Engineering. LNCS, vol. 5971. Springer, Heidelberg (2010)
103. Mulvey, J.M., Vanderbei, R.J., Zenios, S.A.: Robust optimization of large-scale systems. Oper. Res. **43**(2), 264–281 (1995)
104. Mutapcic, A., Boyd, S.: Cutting-set methods for robust convex optimization with pessimizing oracles. Optim. Methods Softw. **24**(3), 381–406 (2009)
105. Ng, T.S., Sun, Y., Fowler, J.: Semiconductor lot allocation using robust optimization. Eur. J. Oper. Res. **205**(3), 557–570 (2010)
106. Nikulin, Y.: Simulated annealing algorithm for the robust spanning tree problem. J. Heuristics **14**(4), 391–402 (2008)
107. Ouorou, A.: Tractable approximations to a robust capacity assignment model in telecommunications under demand uncertainty. Comput. Oper. Res. **40**(1), 318–327 (2013)
108. Paragon Decision Company: AIMMS - The Language Reference, Version 3.12, March 2012

109. Pereira, J., Averbakh, I.: Exact and heuristic algorithms for the interval data robust assignment problem. Comput. Oper. Res. **38**(8), 1153–1163 (2011)
110. Pérez-Galarce, F., Álvarez-Miranda, E., Candia-Véjar, A., Toth, P.: On exact solutions for the minmax regret spanning tree problem. Comput. Oper. Res. **47**, 114–122 (2014)
111. Reemtsen, R.: Some outer approximation methods for semi-infinite optimization problems. J. Comput. Appl. Math. **53**(1), 87–108 (1994)
112. Roy, B.: Robustness in operational research and decision aiding: a multi-faceted issue. Eur. J. Oper. Res. **200**(3), 629–638 (2010)
113. Sanders, P.: Algorithm engineering – an attempt at a definition. In: Albers, S., Alt, H., Näher, S. (eds.) Efficient Algorithms. LNCS, vol. 5760, pp. 321–340. Springer, Heidelberg (2009). doi:10.1007/978-3-642-03456-5_22
114. Sbihi, A.: A cooperative local search-based algorithm for the multiple-scenario max–min knapsack problem. Eur. J. Oper. Res. **202**(2), 339–346 (2010)
115. Schöbel, A.: Generalized light robustness and the trade-off between robustness and nominal quality. Math. Methods Oper. Res. **80**(2), 161–191 (2014)
116. Siddiqui, S., Azarm, S., Gabriel, S.: A modified Benders decomposition method for efficient robust optimization under interval uncertainty. Struct. Multidiscip. Optim. **44**(2), 259–275 (2011)
117. Song, X., Lewis, R., Thompson, J., Wu, Y.: An incomplete m-exchange algorithm for solving the large-scale multi-scenario knapsack problem. Comput. Oper. Res. **39**(9), 1988–2000 (2012)
118. Soyster, A.: Convex programming with set-inclusive constraints and applications to inexact linear programming. Oper. Res. **21**, 1154–1157 (1973)
119. Stiller, S.: Extending concepts of reliability. Network creation games, real-time scheduling, and robust optimization. Ph.D. thesis, TU Berlin (2008)
120. Suzuki, S., Kuroiwa, D., Lee, G.M.: Surrogate duality for robust optimization. Eur. J. Oper. Res. **231**(2), 257–262 (2013)
121. Takeda, A., Taguchi, S., Tütüncü, R.: Adjustable robust optimization models for a nonlinear two-period system. J. Optim. Theory Appl. **136**, 275–295 (2008)
122. Taniguchi, F., Yamada, T., Kataoka, S.: Heuristic and exact algorithms for the max–min optimization of the multi-scenario knapsack problem. Comput. Oper. Res. **35**(6), 2034–2048 (2008)
123. Thuente, D.J.: Duality theory for generalized linear programs with computational methods. Oper. Res. **28**(4), 1005–1011 (1980)
124. Velarde, J.L.G., Martí, R.: Adaptive memory programing for the robust capacitated international sourcing problem. Comput. Oper. Res. **35**(3), 797–806 (2008)
125. Xu, J., Johnson, M.P., Fischbeck, P.S., Small, M.J., VanBriesen, J.M.: Robust placement of sensors in dynamic water distribution systems. Eur. J. Oper. Res. **202**(3), 707–716 (2010)
126. Yaman, H., Karaşan, O.E., Pınar, M.Ç.: The robust spanning tree problem with interval data. Oper. Res. Lett. **29**(1), 31–40 (2001)
127. Yin, Y., Madanat, S.M., Lu, X.Y.: Robust improvement schemes for road networks under demand uncertainty. Eur. J. Oper. Res. **198**(2), 470–479 (2009)
128. Zanjani, M.K., Ait-Kadi, D., Nourelfath, M.: Robust production planning in a manufacturing environment with random yield: a case in sawmill production planning. Eur. J. Oper. Res. **201**(3), 882–891 (2010)
129. Zieliński, P.: The computational complexity of the relative robust shortest path problem with interval data. Eur. J. Oper. Res. **158**(3), 570–576 (2004)

Clustering Evolving Networks

Tanja Hartmann, Andrea Kappes, and Dorothea Wagner$^{(\boxtimes)}$

Department of Informatics, Karlsruhe Institute of Technology (KIT),
Karlsruhe, Germany
{tanja.hartmann,andrea.kappes,dorothea.wagner}@kit.edu

Abstract. Roughly speaking, clustering evolving networks aims at detecting structurally dense subgroups in networks that evolve over time. This implies that the subgroups we seek for also evolve, which results in many additional tasks compared to clustering static networks. We discuss these additional tasks and difficulties resulting thereof and present an overview on current approaches to solve these problems. We focus on clustering approaches in online information from previous time steps in order to incorporate temporal smoothness or to achieve low running time. Moreover, we describe a collection of real world networks and generators for synthetic data that are frequently used for evaluation.

1 Introduction

Clustering is a powerful tool to examine the structure of various data. Since in many fields data often entails an inherent network structure or directly derives from physical or virtual networks, clustering techniques that explicitly build on the information given by links between entities recently received great attention [58, 130]. Moreover, many real world networks are continuously evolving, which makes it even more challenging to explore their structure. Examples for evolving networks include networks based on mobile communication data, scientific publication data, and data on human interaction.

The structure that is induced by the entities of a network together with the links between is often called *graph*, the entities are called *verticals* and the links are called *edges*. However, the terms graph and network are often used interchangeably. The structural feature that is classically addressed by graph clustering algorithms are subsets of vertices that are linked significantly stronger to each other than to vertices outside the subset. In the context of mobile communication networks this could be, for example, groups of cellphone users that call each other more frequently than others. Depending on the application and the type of the underlying network, searching for this kind of subsets has many different names. Sociologists usually speak about *community detection* or *community mining* in social networks, in the context of communication networks like Twitter, people aim at *detecting emerging topics* while in citations networks the focus is on the *identification of research areas*, to name but a few. All these issues can be solved by modeling the data as an appropriate graph and applying graph clustering. The found sets (corresponding to communities, topics or

L. Kliemann and P. Sanders (Eds.): Algorithm Engineering, LNCS 9220, pp. 280–329, 2016.
DOI: 10.1007/978-3-319-49487-6_9

speaking, the idea behind this is to compare the number of edges within clusters to the expected number of edges in the same partition, if edges are randomly rewired. The most popular measure in this context is the *modularity* of a clustering as defined by Girvan and Newman [107] in 2004. Let $e(C)$ denote the number of edges connecting the vertices in cluster C. Then, the modularity $\text{mod}(\mathcal{C})$ of a (complete) clustering $\mathcal{C}$ can be defined as

$$\text{mod}(\mathcal{C}) = \sum_{C \in \mathcal{C}} \frac{e(C)}{m} - \sum_{C \in \mathcal{C}} \frac{\text{vol}(C)^2}{4m^2}.$$

Here, the first term measures the actual fraction of edges within clusters and the second the expectation of this value after random rewiring, given that the probability that a rewired edge is incident to a particular vertex is proportional to the degree of this vertex in the original graph. The larger the difference between these terms, the better the clustering is adjusted to the graph structure. The corresponding optimization problem is $\mathcal{NP}$-hard [25]. Modularity can be generalized to weighted [105] and directed [10,89] graphs, to overlapping or fuzzy clusterings [109,132], and to a local scenario, where the goal is to evaluate single clusters [34,36,94]. Part of its popularity stems from the existence of heuristic algorithms that optimize modularity and that are able to cluster very large graphs in short time [22,112,124]. In Sect. 2, we will describe some generalizations of these algorithms to the dynamic setting. Furthermore, in contrast to many other measures and definitions, modularity does not depend on any parameters. This might explain why it is still widely used, despite some recent criticism [59].

Quality Measures in Evolving Graphs. In the context of dynamic graph clustering, we aim at clusterings of high quality for each snapshot graph. Compared to the static approach, as discussed in Sect. 1, temporal smoothness becomes an additional dimension. Speaking in terms of objective functions, we would like to simultaneously optimize the two criteria quality and temporal smoothness.

As already mentioned before, one approach to obtain this has been introduced by Chakrabarti et al. [32]. The idea is to measure the *snapshot quality* sq of the current clustering $\mathcal{C}_t$ at time step t (with respect to the current snapshot $\mathcal{G}_t$) by a (static) measure for the goodness of a clustering. Similarly, the smoothness is measured by the *history cost* hc of the current clustering, which is usually defined as the distance of the current clustering to the previous clustering $\mathcal{C}_{t-1}$ at time step $t - 1$. The snapshot quality could for example be measured by modularity and the smoothness by any of the distance measures introduced in the next paragraph. The goal is then to optimize a linear combination of both measures, where α is an input parameter that determines the tradeoff between quality and smoothness:

$$\text{maximize } \alpha \cdot \text{sq}(\mathcal{C}_t, \mathcal{G}_t) - (1 - \alpha) \cdot \text{hc}(\mathcal{C}_t, \mathcal{C}_{t-1}).$$

Closely related to this approach, but not relying on an explicit distance measure, is the claim that a good clustering of the snapshot at time step t should also be a good clustering for the snapshot at time step $t - 1$. This is based on the underlying assumption that fundamental structural changes are rare. Hence, linearly combining the snapshot quality of the current clustering with respect to the current snapshot $\mathcal{G}_t$ and the previous snapshot $\mathcal{G}_{t-1}$ yields a dynamic quality measure, which can be build from any static quality measure:

$$\text{maximize } \alpha \cdot \text{sq}(\mathcal{C}_t, \mathcal{G}_t) + (1 - \alpha)\text{sq}(\mathcal{C}_t, \mathcal{G}_{t-1}).$$

This causes the clustering at time step t to also take the structure of snapshot $\mathcal{G}_{t-1}$ into account, which implicitly enforces smoothness. Takaffoli et al. [144] apply this approach in the context of modularity, and Chi et al. [35] in the context of spectral clustering; both will be discussed in Sect. 2.

Distance Measures for Clusterings. In the context of graph clustering, distance measures have three main applications. First, similar to static clustering, they can be used to measure the similarity to a given ground truth clustering. Second, they can be applied as a measure of smoothness, for example by comparing the clusterings of adjacent time steps. Third, they are useful in the context of event detection; a large distance between two consecutive clusterings may indicate an event. A plethora of different measures exist in the literature, none of which is universally accepted. For this reason, we will only introduce the measures used by the dynamic algorithms we describe in Sect. 2. This includes the probably best known index in the context of clustering, the *normalized mutual information*. If not mentioned otherwise, all clusterings considered in this section are assumed to be complete.

Mutual information has its roots in information theory and is based on the notion of the entropy of a clustering $\mathcal{C}$. For a cluster $C \in \mathcal{C}$, let $P(C) := |C|/n$. With that, the entropy $\mathcal{H}$ of $\mathcal{C}$ can be defined as

$$\mathcal{H}(\mathcal{C}) := -\sum_{C \in \mathcal{C}} P(C) \log_2 P(C)$$

Similarly, given a second clustering $\mathcal{D}$, with $P(C, D) := |C \cap D|/n$, the conditional entropy $H(\mathcal{C}|\mathcal{D})$ is defined as

$$\mathcal{H}(\mathcal{C}|\mathcal{D}) := \sum_{C \in \mathcal{C}} \sum_{D \in \mathcal{D}} P(C, D) \log_2 \frac{P(C)}{P(C, D)}$$

Now the *mutual information* $\mathcal{I}$ of $\mathcal{C}$ and $\mathcal{D}$ can be defined as

$$\mathcal{I}(\mathcal{C}, \mathcal{D}) := \mathcal{H}(\mathcal{C}) - \mathcal{H}(\mathcal{C}|\mathcal{D}) = \mathcal{H}(\mathcal{D}) - \mathcal{H}(\mathcal{D}|\mathcal{C}) = \sum_{C \in \mathcal{C}} \sum_{D \in \mathcal{D}} P(C, D) \log_2 \frac{P(C, D)}{P(C)P(D)}$$

Informally, this is a measure of the amount of information the cluster ids in $\mathcal{D}$ carry about the cluster ids in $\mathcal{C}$. Several normalizations of this measure exist; according to Fortunato [58], the most commonly used normalization is the following notion of *normalized mutual information (NMI)*, which maps the mutual information to the interval $[0, 1]$:

$$\text{NMI}(\mathcal{C}, \mathcal{D}) = \frac{2\mathcal{I}(\mathcal{C}, \mathcal{D})}{\mathcal{H}(\mathcal{C}) + \mathcal{H}(\mathcal{D})}$$

Technically, this is not a distance but a similarity measure, as high values of NMI indicate high correlation between the clustering. If need be, it can be easily turned into a distance measure by considering $1 - \text{NMI}(\mathcal{C}, \mathcal{D})$. There also exists a generalization to overlapping clusterings [86]. Among the approaches we describe in Sect. 2, Yang et al. [156], Cazabet et al. [30] and Kim and Han [81] use mutual information to compare against ground truth clusterings. In contrast to that, Lin et al. [93] apply it to compare the time step clusterings to the communities of the aggregated graph, which can be seen as both a measure of smoothness and comparison to some kind of ground truth clustering. Wang et al. [84] use NMI both to measure the similarity of a clustering to a generated ground truth clustering and to compare the results of an approximation algorithm to clusterings found by an exact algorithm (according to their definition of clusters). Lin et al. [93] refer to the *Kullback-Leibler divergence* [82], which is closely related to mutual information, both as a measure of quality and smoothness.

Aynaud and Guillaume [13] propose, as an alternative to NMI, the minimum number of vertex moves necessary to convert one clustering into the other as a measure of distance. Their main argument to consider this approach is that absolute values are far easier to interpret.

Another very intuitive measure for the distance between two partitions is the *Rand index* introduced by Rand [120] in 1971. Let $s(\mathcal{C}, \mathcal{D})$ be the number of vertex pairs that share a cluster both in $\mathcal{C}$ and $\mathcal{D}$ and $d(\mathcal{C}, \mathcal{D})$ the number of vertex pairs that are in different clusters both in $\mathcal{C}$ and $\mathcal{D}$. With that, the Rand index $\mathcal{R}$ of $\mathcal{C}$ and $\mathcal{D}$ is defined as

$$\mathcal{R}(\mathcal{C}, \mathcal{D}) := 1 - \frac{s(\mathcal{C}, \mathcal{D}) + d(\mathcal{C}, \mathcal{D})}{\binom{n}{2}}$$

This corresponds to counting the number of vertex pairs where both clusterings disagree in their classification as intracluster or intercluster pair, followed by a normalization. Delling et al. [41] argue that this measure is not appropriate in the context of graph clustering, as it does not consider the topology of the underlying graph. They propose to only consider vertex pairs connected by an edge, which leads to the *graph based* Rand index. This graph based version is used by Görke et al. [70] to measure the distance between clusterings at adjacent time steps.

Chi et al. [35] apply the *chi square statistic* to enforce and measure the similarity between adjacent clusterings. The chi square statistic was suggested by Pearson [117] in 1900 to test for independence in a bivariate distribution. In the context of comparing partitions, different variants exist [99]; the version used by Chi et al. is the following:

$$\chi^2(\mathcal{C}, \mathcal{D}) = n \cdot \left(\sum_{C \in \mathcal{C}} \sum_{D \in \mathcal{D}} \frac{|C \cap D|}{|C| \cdot |D|} - 1 \right)$$

1.3 Applications

Graph clustering has possible applications in many different disciplines, including biology and sociology. Biologists are for example interested in how diseases spread over different communities, sociologists often focus on cultural and information transmission. Many of the networks analyzed in these areas have a temporal dimension that is often neglected; taking it into account potentially increases the usefulness of clustering for the respective application and at the same time evokes new challenges like for example the involvement of temporal smoothness. In the context of social networks, the benefit of temporal smoothness becomes in particular obvious, since social relations and resulting community structures are not expected to change frequently. Giving an exhaustive list of application areas is beyond the scope of this survey; some further information can be found in the overview article of Fortunato [58]. Instead, we will give some examples where clustering approaches designed for evolving graphs have clearly motivated advantages over static approaches.

A little-known but very interesting application of graph clustering is the use in graph drawing or visualization algorithms. The general idea is to first cluster the vertices of the graph and then use this information in the layouting steps by placing vertices in the same community in proximity of each other. This has several advantages: The layout makes the community structure of the graph visible, which is desirable in many applications. Furthermore, the intracluster density versus intracluster sparsity paradigm causes many edges to be within clusters, which in turn corresponds to small edge lengths. Last but not least, layout algorithms that exploit clustering as a preprocessing step are usually quite fast. As an example, Muelder and Ma have used clustering algorithms in combination with layouts based on treemaps [103] and space filling curves [102]. A straightforward extension to these approaches is the task to visualize dynamic graphs [128]. Dynamic clustering algorithms can help in this context to reduce the running time for the preprocessing in each time step. Furthermore, if they are additionally targeted at producing smooth clusterings, this results in smoother layouts, or, in terms of layout algorithms, in a good preservation of the mental map.

Another interesting application of dynamic graph clustering is its use in routing protocols in Mobile Ad hoc Networks (MANETS). *On-Demand* forwarding schemes for this problem discover paths in the network only when receiving concrete message delivery requests. It has been shown that "routing strategies based on the discovery of modular structure have provided significant performance enhancement compared to traditional schemes" [47]. Due to the mobility of actors in the network, the resulting topology is inherently dynamic; recomputing the clustering whenever a change occurs is costly and requires global information. This motivated a number of online algorithms for modularity based dynamic clustering algorithms, with experiments showing that the use of the dynamic clustering improved the performance of forwarding schemes in this scenario [47,108]. Another interesting aspect of this application is that "consistent modular structures with minimum changes in the routing tables" [47] are desirable, again motivating temporal smoothness.

2   Online Graph Clustering Approaches

In this section we introduce current clustering algorithms and community detection approaches for evolving graphs. As discussed above, we consider only algorithms that operate in an online scenario, i.e., that do not exploit information from future time steps, and are incremental in the sense that they incorporate historic information from previous time steps to achieve temporal smoothness or better running times. We use different categories to classify the approaches presented here. Some categories are associated with particular algorithmic techniques, other categories with applications or the form of the resulting clusterings. Apart from these categories, the GRAPHSCOPE approach [140] is presented at the beginning of this section, as it is one of the first and most cited dynamic approaches. The section concludes with two further approaches, which do not fit into one of the previous categories.

GraphScope. The GRAPHSCOPE approach by Sun et al. [140] is one of the first and most cited dynamic clustering approaches so far. However, contrary to the notion of communities as densely connected subgraphs, GRAPHSCOPE follows the idea of block modeling, which is another common technique in sociology. The aim is to group actors in social networks by their role, i.e., structural equivalence. Two actors are equivalent if they interact in the same way with the same actors (not necessarily with each other). This is, the subgraph induced by such a group may be disconnected or even consisting of an independent set of vertices. The latter is the case in approaches like GRAPHSCOPE that consider bipartite graphs of source and destination vertices and seek for groups of equivalent vertices in each part, i.e., groups consisting either of source or destination vertices. Furthermore, instead of independent snapshots, GRAPHSCOPE considers whole graph segments, which are sequences of similar consecutive snapshots that (w.l.o.g.) have all the same number of sources and destinations. The main idea is the following. Given a graph segment and a partition of the vertices in each part (the same partition for all snapshots in the graph segment), the more similar the vertices are per group the cheaper are the encoding costs for the graph segment using an appropriate encoding scheme based on a form of Minimum Description Length (MDL) [123]. This is, GRAPHSCOPE seeks for two partitions, one for each part of the bipartite input graph, that minimize the encoding costs with respect to the current graph segment. It computes good partitions in that sense by an iterative greedy approach. Based on the same idea, the MDL is further used to decide whether a newly arriving snapshot belongs to the current graph segment or starts a new segment. If the new snapshot belongs to the current graph segment, the two partitions for the graph segment are updated, initialized by the previous partitions. If the new snapshot differs too much from the previous snapshots, a new segment is started. In order to find new partitions in the new segment, the iterative greedy approach is either initialized with the partitions of the previous graph segment or the iterations are done from scratch. The latter can be seen as a static version of GRAPHSCOPE. An experimental comparison on real world data proves a much better running time of the dynamic approach with respect to the static approach. Additional experiments further illustrate that the found source and destination partitions correspond to semantically meaningful clusters. Although this approach focuses on bipartite graphs, it can be easily modified to deal with unipartite graphs, by constraining the source partitions to be the same as the destination partitions [31].

Detecting Overlapping Dense Subgraphs in Microblog Streams. The approaches of Angel et al. [9] and Agarwal et al. [1] both date back to the year 2012 and aim at real-time discovery of emerging events in microblog streams, as provided for example by Twitter. To this end, they model the microblog stream as an evolving graph that represents the correlation of keywords occurring in the blogs or messages. In this keyword graph, they seek for groups of highly correlated keywords, which represent events and are updated over the time. Since a keyword may be involved in several events, these groups are allowed to overlap. The main differences between both attempts is the definition of the correlation between keywords and the definition of the desired subgraphs. Angel et al. consider two keywords as correlated if they appear together in the same message. Two keywords are the stronger correlated the more messages contain them together. The messages are considered as a stream and older messages time out. This results in atomic updates of the keyword graph. In contrast, Agarwal et al. consider multiple changes in the keyword graph resulting from a sliding time window approach. They consider two keywords as correlated if they appear in (possibly distinct) messages of the same user within the time window. Furthermore, they ignore all edges representing a correlation below a given threshold, which results in an unweighted keyword graph.

Regarding the group detection, Angel et al. introduce an algorithm called DYNDENS that considers a parameterized definition of density, which covers most standard density notions. Based on this definition, a set is dense if its density is greater than a given threshold. In order to return all dense subgraphs for each time step, a set of almost dense subgraphs is maintained over time that has the property that after a change in the keyword graph each (possibly new) dense subgraph contains one of the maintained almost dense subgraphs. Hence, the almost dense subgraphs can be iteratively grown to proper density, thus finding all new dense subgraphs after the change. With the help of an appropriate data structure the almost dense subgraphs can be maintained efficiently with respect to time and space requirements. In order to give experimental evidence of the feasibility of their approach, the authors have built a live demo for their techniques on Twitter-tweets and provide, besides a qualitative evaluation, a comparison with a simple static baseline approach that periodically recomputes all dense subgraphs. This static approach took that much time that it was able to compute the set of new events only every 48 to 96 min, compared to a real time event identification performed by the dynamic approach. Instead of dense subgraphs, Agarwal et al. seek for subgraphs that possess the property that each edge in the subgraph is part of a cycle of length at most 4. This property is highly local, and thus, can be updated efficiently. An experimental study on real-world data comparing the dynamic approach to a static algorithm that computes biconnected subgraphs confirms the efficiency of the local updates.

Other Approaches Admitting the Detection of Overlapping Clusters. While two of the following approaches indeed return overlapping clusters, the remaining approaches use a cluster definition that basically admits overlapping clusters, but the algorithmic steps for finding these clusters are designed such that they explicitly avoid overlaps. The most common reason for such an avoidance is the fact that tracking overlapping clusters over time is even more difficult than tracking disjoint clusters, which is why most of the existing tracking frameworks require disjoint clusters.

Takaffoli et al. [144] incrementally apply a method inspired by Chen et al. [34] that basically returns overlapping clusters. In order to apply, in a second step, an independent event detection framework [143] that requires disjoint clusters, they however rework this method such that it prevents overlaps. The idea is to greedily grow clusters around core sets that serve as seeds. In doing so the aim is to maximize the ratio of the average internal degree and the average external degree of the vertices in the cluster, only considering vertices with a positive internal and external degree, respectively. The reworking step then allows a vertex to also leave its initial core set, which admits the shrinking of clusters and the prevention of overlapping clusters. For the first snapshot in a dynamic scenario the initial core sets are single vertices (static version), whereas any further snapshot is clustered using the clusters of the previous snapshot as initial core sets (dynamic approach). If a previous cluster decomposes into several connected components in the current snapshot, the authors consider each of the connected components as a seed. Compared to the static version applied to each snapshot independently and also compared to the FACETNET approach [93] (which we introduce in the category of generative models), at least for the Enron network tested by the authors, the dynamic attempt results in a higher average community size and a higher *dynamic modularity* per snapshot. The latter is a linear combination of the modularity values of the current clustering with respect to the current snapshot and the previous snapshot (see also the quality measures in evolving graphs presented in Sect. 1.2).

Kim and Han [81] present an evolutionary clustering method, which incorporates temporal smoothness to SCAN [155], a popular adaption of the (static) density-based data clustering approach DBSCAN [51] to graphs. The new idea, compared to the idea of evolutionary clustering by Chakrabarti et al. [32], is that instead of minimizing a cost function that trades off the snapshot quality and the history quality at every time step, the same effect can be achieved by adapting the distance measure in SCAN. As usual for SCAN, the authors define an ε-neighborhood of a vertex with respect to a distance measure, such that the resulting ε-neighborhood consists of core vertices and border vertices. A cluster is then defined as the union of ε-neighborhoods, each of which having size at least η, that overlap in at least one core vertex. This kind of clusters can be easily found by a BFS in the graph. This initially yields clusters that may overlap in some border vertices. However, by ignoring vertices that are already assigned to a cluster during the BFS, disjoint clusters can be easily enforced. A vertex that is not found to be a member of a cluster, is classified as noise. By iteratively adapting ε the authors additionally seek for a clustering of high modularity. When a good clustering is found for the current snapshot, temporal smoothness is incorporated by adapting the distance measure that characterizes the ε-neighborhoods in the next time steps, taking the distance in the current snapshot into account. Finally, the authors also propose a method for mapping the clusters found in consecutive snapshots, based on mutual information. On synthetic networks of variable numbers of clusters the proposed approach outperformed FACETNET [93] with respect to clustering accuracy and running time.

Another approach that is also based on DBSCAN and is very similar to SCAN, is called DENGRAPH and is presented by Falkowski et al. [55]. In contrast to Kim and Han, Falkowski et al. do not focus on an evolutionary clustering approach, but introduce dynamic update techniques to construct a new clustering after the change of a vertex or an edge in the underlying graph. Again, these updates are done by locally applying an BFS (as for the static SCAN approach), but just on the vertices that are close to the change, thereby updating the cluster ids. Experiments on the Enron data set suggest that DENGRAPH is quite fast but also relatively strict as it reveals only small, very dense groups while many vertices are categorized as noise. The dynamic DENGRAPH version proposed in [55] returns disjoint clusters, while in [53] Falkowski presents a dynamic version for overlapping clusters.

A dynamic algorithm that is not based on a static clustering algorithm but also produces overlapping clusters is proposed by Cazabet et al. [30]. In each time step, clusters are updated in the following way. First, it is determined if a new seed cluster, i.e., a small clique of constant size has emerged due to edge updates. Then, existing clusters and seed clusters are extended by additional vertices. To that end, for each cluster C, two characteristics are maintained. The first of these characteristics corresponds to the average percentage of vertices a vertex in C can reach in its cluster by a path of length 2. Very similar, the second characteristic corresponds to the average percentage of vertices a vertex in C can reach in its cluster by at least two distinct path of length 2. A vertex that is not in C may be included into C if, roughly speaking, this improves both of these characteristics. In a last step, clusters that share a certain percentage of vertices with another cluster are discarded. The goal of this approach is not primarily to get good clusterings for each time step but to get a good clustering of the last time step by taking the evolution of the network into account. Nevertheless, the approach per se is capable of clustering dynamic networks in an online scenario. Other overlapping approaches are categorized according to a different focus and are thus described at another point. For an overview on all overlapping approaches presented in this survey see Table 1.

Table 1. Systematic overview on main features of the algorithms presented in Sect. 2. Checkmarks in brackets indicate a simple modification to overlapping clusters.

Reference	aims at run. time	aims at smoothn.	overlapping yes	overlapping no	based on existing static approach
Sun et al. [137] (GRAPHSCOPE)	✗	✓	✗	✓	✗
DETECTING OVERLAPPING DENSE SUBGRAPHS IN MICROBLOG STREAMS					
Angel et al. [8] (DYNDENS)	✓	✗	✓	✗	✗
Agarwal et al. [1]	✓	✗	✓	✗	✗
OTHER APPROACHES ADMITTING THE DETECTION OF OVERLAPPING CLUSTERS					
Takaffoli et al. [142]	✗	✓	(✓)	✓	Chen et al. [33]
Kim and Han [80]	✗	✓	(✓)	✓	Xu et al. [154] (SCAN)
Falkowski et al. [52, 54] (DENGRAPH)	✓	✗	✓	✓	Xu et al. [154] (SCAN)
Cazabet et al. [29]	✗	✓	✓	✗	✗
ALGORITHMS MAINTAINING AUXILIARY STRUCTURES					
Duan et al. [48]	✓	✗	✓	✗	Derényi et al. [41] (PCM)
Görke et al. [66, 67]	✓	✓	✓	✓	Flake et al. [56]
SPECTRAL GRAPH CLUSTERING METHODS					
Chi et al. [34] (EVOLSPEC)	✗	✓	✗	✓	Shi and Malik [130]
Ning et al. [107]	✓	✗	✗	✓	Shi and Malik [130]
MODULARITY BASED ALGORITHMS					
Dinh et al. [44, 46] (MIEN)	✓	✗	✗	✓	Newman and Moore [36] (CNM)
Görke et al. [69] (DGLOBAL)	✓	✓	✗	✓	Newman and Moore [36] (CNM)
Bansal et al. [15]	✓	✗	✗	✓	Newman and Moore [36] (CNM)
Aynaud and Guillaume [12]	✗	✓	✗	✓	Blondel et al. [21] (LOUVAIN)
Görke et al. [69] (TDLOCAL)	✗	✓	✗	✓	Blondel et al. [21] (LOUVAIN)
Nguyen et al. [104] (QCA)	✓	✗	✗	✓	Blondel et al. [21] (LOUVAIN)
Görke et al. [69] (DLACAL)	✓	✓	✗	✓	Blondel et al. [21] (LOUVAIN)
Nguyen et al. [43] (A³CS)	✓	✗	✗	✓	Dinh and Thai [45] and Blondel et al. [21] (LOUVAIN)
Riedy and Bader [118]	✓	✗	✗	✓	Riedy et al. [119]
LABEL PROPAGATION/DIFFUSION					
Pang et al. [111]	✓	✗	✗	✓	Raghavan et al. [116] (LABEL PROPAGATION)
Xie et al. [151] (LABELRANKT)	✓	✗	✓	✓	Xie et al. [152] (LABELRANK)
Gehweiler et al. [60] (DIDIC)	✓	✗	✗	✓	Meyerhenke et al. [95]
GENERATIVE MODELS					
Lin et al. [91] (FACETNET)	✗	✓	✓	(✓)	Yu et al. [156]
Yang et al. [155]	✗	✓	✗	✓	✗
Sun et al. [138] (EVO-NETCLUS)	✗	✓	✓	(✓)	Sun et al. [139] (NETCLUS)
FURTHER APPROACHES					
Kim et al. [79]	✗	✓	✗	✓	✗
Wang et al. [82]	✓	✗	✗	✓	✗
Xu et al. [153]	✗	✓	✗	✓	✗

Algorithms Maintaining Auxiliary Structures. The following two approaches consider atomic changes in the given graph and aim at efficiently updating structurally clearly defined clusters, which are obtained by a simple operation on an auxiliary structure. Here, atomic changes refer to the insertion or deletion of one edge or vertex. In the first approach, by Duan et al. [49], the auxiliary structure is a graph that represents the overlap of maximal cliques in the input graph, and the final clusters result from the connected components of this graph. In the second approach, by Görke et al. [68], a partial cut tree (or Gomory-Hu tree [65]) is maintained and the clusters result from the subtrees obtained by deleting a designated vertex in this tree. The latter approach further incorporates possibly given edge weights of the input graph.

The dynamic clique-clustering approach of Duan et al. [49] is a dynamic version of the clique percolation method (PCM) of Derényi et al. [42], which is again a special case of a more general clique-clustering framework proposed by Everett and Borgatti [52]. The framework by Everett and Borgatti applies an arbitrary clustering algorithm to a weighted auxiliary graph H that represents the overlap of the maximal cliques in the input graph. In the special case considered by Derényi et al. and Duan et al., the auxiliary graph H just encodes if two maximal cliques (of at least size k) overlap in at least $k - 1$ vertices, and thus, is an unweighted graph. More precisely, H is a graph where the maximal cliques in the input graph represent the vertices and two vertices are connected by an edge if and only if the corresponding cliques share at least $k - 1$ vertices. As clustering algorithm Derényi et al. and Duan et al. simply choose a DFS, which returns the connected components of H, which induce overlapping clusters in the original graph. The running time of this approach is dominated by the computation of maximal cliques, which is exponential in the number of vertices [28]. The proposed dynamic version is then straightforward. For each type of change, the authors give a procedure to update the auxiliary graph H as well as the DFS-tree T, which indicates the connected components of H. In doing so, the insertion of an edge is the only change where the computation of new maximal cliques becomes necessary in parts of the input graph. All other changes can be handled by updating the overlap of previous cliques and adapting edges in H and T. Hence, the more changes in the dynamic input graph are different from edge insertions, the better the dynamic approach outperforms the static approach, which computes all cliques from scratch after each change.

The dynamic cut-clustering approach of Görke et al. [67,68] is a dynamic version of the static cut-clustering algorithm of Flake et al. [57], based on updating a partial Gomory-Hu tree [65] of an extended input graph G_α. The graph G_α is obtained from the input graph G by inserting an artificial vertex q and artificial edges, each having weight α, between q and each vertex in G. The input parameter α determines the coarseness of the resulting clustering. A Gomory-Hu tree T for G_α then is a weighted tree on the vertices of G_α that represents a minimum s-t-cut for each vertex pair in G_α. More precisely, deleting an edge $\{s, t\}$ in T decomposes T in two subtrees inducing a minimum s-t-cut in the underlying graph. The weight assigned to the deleted edge in T further corresponds to the costs of the induced minimum s-t-cut. For two non-adjacent vertices u and v in T, the minimum u-v-cut is given by a lightest edge on the path from u to v in T. In order to obtain the final complete clustering, the artificial vertex q is deleted from T, resulting in a set of subtrees inducing the clusters. Due to the special properties of the minimum s-q-cuts that separate the resulting clusters, Flake et al. are able to prove a guarantee (depending on α) of the intercluster expansion and the intracluster expansion of the resulting clustering, which in general is NP-hard to compute (cf. the cut-based quality measures introduced in Sect. 1.2). The dynamic version of the cut-clustering algorithm determines which parts of the current Gomory-Hu tree of G_α become invalid due to an atomic change in G and describes how to update these parts depending on the type of the atomic change. The result is a cut clustering of the current graph G with respect to the same parameter value α as in the previous time step. The most difficult and also (in theory) most time consuming type of an update is the update after an edge deletion. However, in most real world instances the actual effort for this operation is still low, as shown by an experimental evaluation on real world data. We stress that there also exists another attempt [126, 127] that claims to be a dynamic version of the cut clustering algorithm of Flake et al., however, Görke et al. showed that this attempt is erroneous beyond straight-forward correction. Doll et al. [48] further propose a dynamic version of the hierarchical cut-clustering algorithm that results from varying the parameter value α, as shown by Flake et al. [57].

Spectral Graph Clustering Methods. The main idea of static spectral graph clustering is to find an r-dimensional placement of the vertices such that vertices that form a cluster in an appropriate clustering with respect to a given objective are close to each other while vertices that are assigned to different clusters are further away from each other. This can be done by considering the spectrum of a variation of the adjacency matrix, like for example the Laplacian matrix in the context of the normalized cut objective [95]. More precisely, many desirable objectives result in optimization problems that are solved by the eigenvectors associated with the top-r eigenvalues of a variation of the adjacency matrix that represents the objective. The rows of the n-by-r matrix formed by these eigenvectors then represent r-dimensional coordinates of the vertices that favor the objective. The final clustering is then obtained by applying, for example, k-means to these data points.

The EVOLSPEC algorithm by Chi et al. [35] conveys this concept to a dynamic scenario by introducing objectives that incorporate temporal smoothness. Inspired by Chakrabarti et al. [32], the authors linearly combine snapshot costs and temporal costs of a clustering at time step t, where the temporal costs either describe how well the current clustering clusters historic data in time step $t - 1$ or how different the clusterings in time step t and $t - 1$ are. For both quality measures, they give the matrices that represent the corresponding objectives, and thus, allow the use of these measures in the context of spectral graph clustering.

Ning et al. [111] show how to efficiently update the eigenvalues and the associated eigenvectors for established objectives if an edge or a vertex in the underlying graph changes. Compared to the static spectral clustering, which takes $O(n^{3/2})$ time, this linear incremental approach saves a factor of $n^{1/2}$. An experimental evaluation of the running times on Web-blog data (collected by the NEC laboratories) confirm this theoretical result. The fact that the updates yield only approximations of the desired values is not an issue, as further experiments on the approximation error and an analysis of the keywords in the found clusters show.

A concept that is closely related to spectral graph clustering is Low-rank approximations of the adjacency matrix of a graph. Tong et al. [145] do not provide a stand-alone community detection algorithm but a fast algorithm that returns a good low-rank approximation of the adjacency matrix of a graph that requires only few space. Additionally, they propose efficient updates of these matrix approximations that may enable many clustering methods that use low-rank adjacency matrix approximations to also operate on evolving graphs.

Modularity Based Algorithms. All dynamic community detection algorithms based on explicit modularity optimization are modifications of one of three static agglomerative algorithms that greedily optimize modularity.

The first of these static algorithms, commonly called CNM according to its authors Clauset, Newman and Moore [37], is similar to traditional hierarchical clustering algorithms used in data mining, such as single linkage [134]. Starting from a singleton clustering, among all clusterings that can be reached by merging two of the clusters, the one with the best modularity is chosen. This is repeated until modularity cannot be further improved by merging any of the clusters. Although modularity is an inherently global measure, the improvement of the objective function after a merge operation can be easily calculated by only considering the affected clusters. This means that the set of all possible merge operations can be maintained in a heap, which leads to a total running time of $O(n^2 \log n)$.

Dinh et al. [45,47] evaluate a straightforward dynamization of the CNM algorithm that works as follows. The graph at the first time step is clustered with the static algorithm and the resulting clustering is stored. In the next time step, we first incorporate all changes in the graph. Then, each vertex that is either newly inserted or incident to an edge that has been modified is *freed*, i.e., it is removed from its cluster and moved to a newly created singleton cluster. To arrive at the final clustering, CNM is used to determine if merging some of the clusters can again improve modularity. The authors call this framework "Modules Identification in Evolving Networks" (MIEN).

Independently, Görke et al. [70] build upon the same idea, but in a more general setting, which results in the algorithm DGLOBAL. There are two variants of this algorithm, the first one based on freeing vertices in the neighborhood of directly affected vertices and the second one based on a *backtracking* procedure. In the first variant, the subset of freed vertices can be all vertices in the same cluster, vertices within small hop distance or vertices found by a bounded breadth first search starting from the set of affected vertices. In their experiments, considering these slightly larger subsets instead of only directly affected vertices improves modularity and yields a good tradeoff between running time and quality. The second variant not only stores the clustering from the last time step but the whole sequence of merge operations in the form of a *dendrogram*. A dendrogram is a binary forest where leaves correspond to vertices in the original graph and vertices on higher levels correspond to merge operations. Additionally, if a vertex in the dendrogram is drawn in a level above another vertex, this encodes that the corresponding merge has been performed later in the algorithm. Figure 3a shows an example of a dendrogram produced by the static CNM algorithm whose resulting clustering consists of two clusters. Storing the whole dendrogram across time steps makes backtracking strategies applicable. To update the clustering for the next time step, the backtracking procedure first retracts a minimum number of merges such that certain requirements are met, which depend on the type of change. In case an intracluster edge has been inserted, the requirement is that its incident vertices are in separate clusters after the backtracking procedure. If an intercluster edge is inserted or an intracluster edge deleted, merges are retracted until both affected vertices are in singleton clusters. If an intercluster edge is deleted, the dendrogram stays unchanged. Afterwards, CNM is used to complete this preliminary clustering.

Bansal et al. [16] use a similar approach. Instead of backtracking merges in the dendrogram, their algorithm repeats all merge operations from the last time step until an affected vertex is encountered. Again, this preliminary clustering is completed with the static CNM algorithm. Figure 3 illustrates the difference between the two approaches. Both studies report a speedup in running time compared to the static algorithm, Görke et al. additionally show that their approach improves smoothness significantly. In the experiments of Bansal et al., quality in terms of modularity is comparable to the static algorithm, while Görke et al. even observe an improvement of quality on synthetic graphs and excerpts of coauthor graphs derived from arXiv. Görke et al. additionally compare the backtracking variant of DGLOBAL to the variant freeing subsets of vertices; for the test instances, backtracking was consistently faster but yielded worse smoothness values.

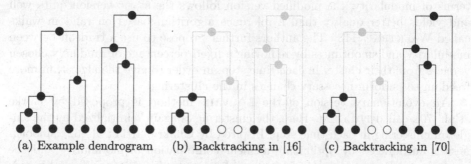

(a) Example dendrogram   (b) Backtracking in [16]   (c) Backtracking in [70]

Fig. 3. Example dendrogram and illustration of backtracking procedure by Bansal et al. [16] and Görke et al. [70] in case an intracluster edge between the white vertices is deleted.

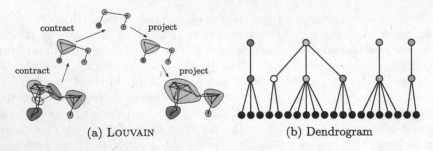

(a) LOUVAIN   (b) Dendrogram

Fig. 4. Illustration of the LOUVAIN method and the corresponding dendrogram. In the left part, the depicted edge structures show the graphs before the vertex moves, while the colored subsets depict the resulting clusters after the vertex moves on the particular level.

The second static algorithm that has been modified for the dynamic scenario is a local greedy algorithm often called LOUVAIN method [22]. Similar to CNM, the algorithm starts with a singleton clustering. Now, vertices of the graph are considered in a random order. If there is at least one cluster such that moving the current vertex v to it improves the overall modularity, v is moved to the cluster that yields the maximal gain in modularity. This process is repeated in several rounds until a local maximum is attained. Then, clusters are contracted to supernodes and edges between clusters summarized as weighted edges, whereas edges within clusters are mapped to (weighted) self loops. The local moving procedure is then repeated on the abstracted graph taking edge weights into account. Contractions and vertex moves are iterated until the graphs stays unchanged. Then, the clustering is *projected* down to the lowest level, which represents the original graph, to get the final result. Figure 4 illustrates this procedure.

Among the modifications of the LOUVAIN method to the dynamic scenario, the one by Aynaud and Guillaume [13] is the most direct. In their study, instead of the singleton clustering, the clustering from the last time step is used to initialize the clustering on the lowest level. Using a dynamic network of webblogs, they demonstrate that this modification improves smoothness significantly. In terms of modularity, the modified version follows the static version quite well and yields better quality than a reference algorithm based on random walks called WALKTRAP [118]. The authors further propose to use a tradeoff between modularity and smoothness by removing a fixed percentage of randomly chosen vertices from their cluster in each time step, in order to give the algorithm more freedom to perform necessary changes in the clustering.

An evolutionary version of the LOUVAIN method is proposed by Görke et al. [70], called TDLOCAL. Here, the clustering is again reinitialized by the singleton clustering in each time step. Inspired by Chakrabarti et al. [32], smoothness is encouraged by optimizing a linear combination of modularity and the graph based Rand index [41]. It is possible to optimize this modified objective with the LOUVAIN algorithm without increasing the asymptotic running time of one round.

In contrast to that, the algorithm QCA proposed by Nguyen et al. [108] is a modification of the LOUVAIN method based on local updates, i.e., a dynamic clustering algorithm. Depending on the kind of change, the following case distinction is used. If a new vertex v is inserted, v is greedily assigned to a cluster such that modularity is optimized, in the spirit of the original algorithm. In case an intercluster edge between u and v is inserted, the algorithm first checks if u or v can be moved to the other cluster such that modularity increases. If yes, it checks if neighbors of the moved vertex can be moved as well. In case a vertex is deleted, its cluster is potentially split by using a method similar to clique percolation [42] restricted to the affected cluster. If an intracluster edge between u and v is deleted in cluster C, where u and v have degree at least 2, the set of maximal quasi-cliques within C is determined and the clustering completed similar to static CNM. In all other cases, the clustering from the last time step is left unchanged.

All of these approaches only maintain one clustering across time steps, namely, the one that could not be improved in terms of modularity. This clustering corresponds to the highest level in the dendrogram. In contrast to that, Görke et al. [70] propose to store and possibly modify the whole dendrogram during the course of the algorithm, which leads to the algorithm framework DLOCAL. After all changes have been incorporated into the graph of the lowest level (the original graph), all affected vertices, i.e., vertices that are either new or incident to edge changes, are marked. Additionally, depending on some policy P, some vertices in their close neighbourhood are marked as well. The set of policies evaluated in this study correspond to the aforementioned subset strategies evaluated by Görke et al. [70] for DGLOBAL. This means, P can correspond to freeing vertices within a small hop distance from affected vertices, vertices in the same cluster and vertices found by a bounded breadth first search. Then, vertices on the lowest level are greedily moved until modularity cannot be further improved. Now, affected vertices in the second level of the dendrogram are marked, i.e., subclusters affected by lower level changes. Depending on P, again, some vertices in their neighborhood are marked and possibly moved. This is repeated on all levels. The current clustering can be found at all time steps by considering the projection of the highest to the lowest level. Keeping the whole dendrogram in memory instead of only its highest level yields the additional possibility to merge and split clusters on intermediate levels and therefore increases the search space used for local moving, which leads to possibly better quality. Compared to the algorithm DGLOBAL, the experiments of Görke et al. do not give conclusive results; which of these algorithms performs better depends on the structure of the graph and its communities.

An approach that is very similar to DLOCAL is used in the algorithm A^3CS proposed by Nguyen et al. [44]. The main difference is that the lowest level of the dendrogram is not computed and maintained by a local moving procedure but by an algorithm similar to the static Low-degree Following Algorithm proposed by Dinh and Thai [46]. This algorithm has the nice property to yield approximation guarantees for graphs with a perfect powerlaw degree distribution with a powerlaw exponent larger than 2. This property is inherited by the dynamic algorithm A^3CS. However, the clusters produced by this algorithm are quite small, which is why it makes sense to additionally consider local moving (which includes merging of clusters) on higher levels to further improve its practical performance. Dinh and Thai compare A^3CS to QCA and MIEN, with the result that it dominates both with respect to running time and quality.

The third static algorithm for modularity based clustering, which lends itself especially well to parallelization, is based on the contraction of matchings [124, 131]. To effectively optimize modularity, edges are weighted based on the gain in modularity corresponding to the merge of the two adjacent vertices. Using these edge weights, a weighted matching is computed in a greedy fashion, possibly in parallel [122]. Riedy and Bader [121] propose a dynamic variant of this algorithm especially for the case of larger batch sizes, i.e., many changes between consecutive time steps. Across the time steps, the current clustering together with its community graph is stored. After incorporating the edge changes in the community graph, all (elementary) vertices incident to newly inserted intercluster or deleted intracluster edges are extracted from their community. Then, the matching based agglomeration is restarted from the modified community graph. As the community graph is usually much smaller than the original graph, this potentially saves running time.

Label Propagation/Diffusion. An algorithm that is not based on modularity, but conceptually closely related to the LOUVAIN method is LABEL PROPAGATION [119]. LABEL PROPAGATION can be seen as an algorithm greedily maximizing the number of edges within clusters by moving vertices, very similar to the local moving procedure in the LOUVAIN method. Obviously, the global optimum with respect to the number of edges within clusters is trivial, as it corresponds to assigning all vertices to one cluster. Hence, using local moving in a multilevel scheme, as in the case of modularity maximization, does not make much sense. Instead, one is interested in the local maximum obtained after local moving on the original graph, which corresponds to the lowest level in the dendrogram. LABEL PROPAGATION is very fast and lends itself well to parallelization [139]. Xie and Szymanski propose a modified version of this algorithm called LABEL-RANK [153]. In contrast to the original LABEL PROPAGATION algorithm, each vertex does not maintain one single cluster id or *label*, but instead a vector of containment probabilities for each cluster currently existing in the graph. Label propagation phases alternate with inflation and cutoff steps to decrease the computational complexity and to make the differences in the particular vectors more pronounced. To prevent the algorithm from converging too fast to the (potentially uninteresting) static distribution, only labels of vertices are updated that are sufficiently different from their neighbors. The output of the algorithm can be interpreted as overlapping clusters with the additional information of containment probabilities. This is why, although overlaps are resolved in a preprocessing step by only considering the strongest label for each vertex, we list the algorithm both among overlapping and non overlapping approaches in Table 1. Both LABEL PROPAGATION and LABELRANK have been modified to the dynamic scenario [114,152], roughly based on the idea to only update labels/label vectors of vertices affected by changes in the graph. The dynamic version of LABELRANK is called LABELRANKT.

A concept that is very similar to LABELRANK and has been developed in the context of graph partitioning is *diffusion* [73,97,98]. Similar to the above algorithm, each vertex maintains a vector of size k indicating to which extent it is connected to the vertices of each of the k clusters. The entries of these vectors are called *loads*; loads are distributed through the network along the edges in rounds, which explains the origin of the name diffusion. Based on this concept, Gehweiler and Meyerhenke [61] propose a distributed graph clustering algorithm called DIDIC, which is motivated by the task to cluster nodes of a peer-to-peer based virtual distributed supercomputer. The weight of edges between nodes in this network corresponds to the bandwidth between the associated peers. The idea is to find clusters of highly connected peers that can be exploited to solve a common task in parallel. In contrast to LABELRANKT, they use a second diffusion system drawing the loads associated with cluster i back to the vertices in cluster i, which accelerates the formation of large, connected clusters. In the first time step, the process starts with a random clustering and distributes the load of each cluster to the vertices it contains. After the diffusion process has been run for a certain number of rounds, clusters are reassigned such that each vertex moves to the cluster from which it obtained the highest load value, leading to a complete clustering. The algorithm is made dynamic by initializing the clusters and load vectors with the values obtained in the previous time step, instead of random initialization.

Generative Models. Another popular approach to clustering problems is the use of generative models that assume the graph is randomly generated on the basis of a hidden ground truth clustering. The goal is now to approximately recover the hidden or latent clustering by looking for clusterings that are *likely* given the observed outcome of this random process, which corresponds to the given graph. Given conditional probabilities that estimate this likelihood and a prior distribution over the set of all possible clusterings, the posterior probability of a given clustering can be obtained via Bayes' theorem. For conditional probabilities, a common choice are *stochastic block models* [8,136,149], which generally assume that the probability of a link between two vertices is determined by their cluster membership. If the number of clusters is not provided as an input parameter, a popular choice as a prior distribution for the cluster assignments is the distribution induced by the Chinese restaurant process [7] and its derivations. The goal of maximum a posteriori (MAP) estimations is now to find solutions with high posterior probability.

Among the few approaches based on these concepts that explicitly address the issue of dynamic graphs are FACETNET [93] and the algorithm by Yang et al. [156]. The goal of both approaches is to implicitly enforce smoothness by choosing the prior distribution such that large changes in the clustering between adjacent time steps are assumed to be unlikely. In contrast to traditional stochastic block models, FACETNET builds upon the model proposed by

Yu et al. [157] that assumes "soft community membership", i.e. vertices belong to different clusters to more or less extent. This results in an overlapping clustering. However, these clusters can easily be converted to a complete clustering in a postprocessing step by assigning each vertex to the cluster it participates in to the largest extent. For this reason and the fact that this is often done when comparing complete clusterings to the clusterings produced by FACETNET, we list the algorithm both under overlapping and non overlapping clusterings in Table 1. In the generative model, the probability of a certain cluster assignment at time step t depends on the cluster assignment at step $t-1$. Depending on a parameter ν, the transitions will be more or less smooth. It can be shown that under certain assumptions, the MAP estimation of this model is equivalent to the framework of Chakrabarti [32]. In this context, the Kullback-Leibler divergence [82] between the observed weight matrix and an approximation of it based on cluster assignments is applied as the snapshot cost and the Kullback-Leibler divergence between the clustering at time step t and at time step $t-1$ as history cost. For the inference step, an expectation maximization algorithm is used that is guaranteed to converge towards a locally optimal solution of the corresponding MAP problem. In the FACETNET framework, the number of clusters can change over time. To determine the best number of clusters for each time step, an extension of modularity to soft community memberships is proposed. In the experimental part, synthetic and real world networks are used to evaluate the performance of FACETNET and to compare it to its static counterpart as well as a static and evolutionary (EVOLSPEC) version of spectral clustering [35,133]. With respect to quality, the FACETNET approach compares favorably.

In the algorithm of Yang et al., the number of clusters is given as input. Given the hidden clustering at a certain time step, the conditional probability for a link between two vertices is determined by the linking probabilities associated with their respective clusters. These linking probabilities are in turn random variables such that their prior distribution causes higher linking probabilities for intracluster edges. The whole generative model corresponds to a Bayesian net where the latent variables associated with a certain time step depend on the clustering from the last time step, a matrix A specifying the probability that a vertex moves from a certain cluster to another in the current time step, and the prior distribution for the linking probabilities between clusters. Again, the prior distribution for A biases the moving probabilities in such a way that the probability for each vertex to move to another community k is smaller than the probability to stay in its own cluster, which implicitly biases the model towards smoothness. The model can be generalized to weighted graphs in a straightforward way. For the inference step, the authors evaluate both the online and the offline scenario. In the online scenario, the variables are sampled from time step to time step using the observations seen so far. In the offline scenario, all variables are sampled together by taking both past and future observations into account. In both cases, a Gibbs sampler is used to infer the latent variables. In the offline scenario, additionally, an expectation maximization algorithm is proposed. These two variants are then compared against each other, static statistical blockmodels, and the dynamic algorithms EVOLSPEC and FACETNET and their static counterparts. Experiments on synthetic and real world networks suggest that the approach based on Gibbs sampling in combination with the new generative model yields the best quality. It might be worth to mention that the hyperparameters that influence the priors are tuned by considering the modularity of the resulting clusterings.

Another generative approach has been used by Sun et al. [141] for their algorithm EVO-NETCLUS. Similar to FACETNET, EVO-NETCLUS builds upon soft community membership. In contrast to the models mentioned above, the networks considered are assumed to contain vertices of multiple types, where one of the types is defined as a special "center type". Each edge is incident to exactly one center type vertex; the authors call this star network schema. As an illustrative example that also occurs in the experiments, the authors use publication networks as for example DBLP[1]. Here, vertices of the center type correspond to papers and the other types to authors, conferences and associated terms. Again, the probability of the clustering of time step t is influenced by the clustering at time step $t - 1$, favoring smoothness. The model incorporates changing cluster numbers in each time steps that are not assumed to be input parameters. For the inference step, an online Gibbs sampler is proposed. With respect to quality, the authors compare their model to degenerated models that do not take historical data or only a subset of types into account.

Further Approaches. In this category we list three approaches that do not fit into any of the previous categories. The first approach considers a bicriterial optimization problem while the former approaches focus on a single criterion, the second approach is a general framework that allows to incorporate temporal smoothness into basically every static clustering algorithm, and the third approach claims that considering the input graph as a homogeneous structure where in each region the same criteria for good clusters hold is not appropriate.

The approach of Kim et al. [80] is based on optimizing two different measures simultaneously in a bicriterial approach. Typically, the measures in bicriterial approaches are competing in the sense that one measure tends towards the 1-clustering and the other towards the singleton clustering. The goal is to approximate the *pareto front*, i.e., to find a set of clusterings that are not dominated by other clusterings with respect to both criteria. Kim et al. use as criteria (or *fitness functions*) a global version of MinMaxCut [43], which tends to the 1-clustering, and a global version of the silhouette index [125], which tends to the singleton clustering. They approximate the pareto front by an evolutionary metaheuristic in a dynamic scenario using a *locus-based representation* of clusterings [115], which is a vector of length n storing for each vertex exactly one outgoing edge. The represented clustering then corresponds to the connected components of the induced graph. The locus-based representation has the advantage that different clusterings can be combined (crossed) in a meaningful way by performing *uniform crossover* on the corresponding vectors, which means

---

[1] http://www.informatik.uni-trier.de/~ley/db/.

that each entry in the resulting vector is randomly taken from one of the parent vectors. The dynamization is trivially done by initializing the population of the current time step by the result from the last time step. Different evolutionary metaheuristics are compared with respect to both criteria on a dynamic graph representing YouTube videos.

While the former approach uses evolutionary metaheuristics, which have nothing to do with evolutionary clustering according to Chakrabarti et al. [32], the next approach is again an evolutionary clustering approach. In contrast to other evolutionary clustering approaches, which most often incorporate temporal smoothness into a particular clustering algorithm, the framework introduced by Xu et al. [154] can be applied with any static clustering method. In their publication the authors use the normalized cut spectral clustering approach by Yu and Shi [158]. Although the idea of Xu et al. is inspired by Chakrabarti et al., the main difference is that they do not incorporate temporal smoothness by optimizing a linear combination of snapshot quality and history quality, but adapt the input data for the chosen clustering algorithm based on the community structure found in the previous snapshot. This adaption is done as follows. The adjacency matrices of the snapshots are considered as realizations of a non stationary random process which allows to define an expected adjacency matrix for the current snapshot. Based on this expected matrix a *smoothed adjacency matrix* can be approximated that also takes into account the previous time step. The smoothed adjacency matrix is a convex combination of the smoothed adjacency matrix of the previous time step and the actual adjacency matrix of the current time step. The parameter that balances the two terms of the convex combination is estimated such that it minimizes a mean squared error criterion. The chosen clustering algorithm is then applied to the estimated smoothed adjacency matrix, thus incorporating temporal smoothness to stabilize the variation of the found clusters over time.

All the above clustering approaches use the same objective for the whole graph to get good clusterings. In this way these approaches consider the input graph as homogeneous structure, regardless whether parts of the graph are sparser than others, and thus, possibly require another notion of density for reasonable communities than denser parts. Wang et al. [84] follow Aggarwal et al. [3] who claim that considering networks as homogeneous structures is not an appropriate attempt. This is why Wang et al. introduce patterns describing homogeneous regions that are consolidated in a second step to generate non overlapping clusters. In contrast to density, which depends on the number or the weight of edges within a subgraph or cluster, homogeneity means that all vertices in a pattern have similarly weighted neighbors. In order to efficiently compute the patterns in a dynamic scenario, the authors maintain, by incremental updates, a top-k neighbor list and a top-k candidate list as auxiliary structures. These updates are able to deal with atomic changes as well as with several changes (of vertices and edge weights) in one time step. In comparison with FACETNET [93] and the evolutionary clustering method by Kim and Han [81], experiments on the DBLP, the ACM and the IBM data set prove a better processing rate (number of vertices processed per second) and a better accuracy of the found clusterings in locally heterogeneous graphs.

Summary. To provide a summary of all algorithms described in this section, Table 1 lists some of their basic properties. These include whether the authors aim at running time or smoothness, or both, and if the resulting clusterings are overlapping or not. If applicable, we further give a reference to an existing static algorithm the approach is based upon. Among the algorithms we considered, the majority focuses on the task of finding non overlapping clusters. Interesting is that the number of algorithms aiming at low running time is almost the same as the number of algorithms aiming at smoothness; only very few algorithms take both into account. Note that an entry in the table indicating that an algorithm does not aim at low running time or smoothness does not indicate that the algorithm is slow or produces unsmooth clusterings; it just signifies that this aspect was neither considered in the conception of the algorithm nor evaluated in the experiments. In general, it is expected that smoothness and running time go together quite well, as the use of local updates often improves both of these aspects.

3   Data Sets

Often, researchers developing new or enhanced algorithms are faced with the question which data sets to use to illustrate the advantages and validity of their approach. In the context of clustering evolving graphs, the data needs to have some temporal aspects and, ideally, should come with some well-motivated ground truth clustering. Additionally, using data that has been previously used in the literature makes the comparison to other methods less cumbersome. To simplify the search for suitable test data, this section aims to give an overview on what kind of data sets have been used in current publications regarding the clustering of evolving graphs. In the first part, we concentrate on real world instances, i.e., instances that correspond to data collected from observed relationships between objects or persons. In the second part, we briefly talk about models and generators for evolving networks, with a special focus on synthetic data incorporating a hidden ground truth clustering.

3.1   Real World Instances

Most networks described in this category are based on human interaction and can therefore be classified as *social networks* in the wider sense. We tried to assign them to more fine grained subcategories depending on their structure and interpretation.

Email networks. One of the few publicly available networks corresponding to social interaction and containing both time information and additional metadata is the Enron email dataset[2]. It represents the email exchange of employees of the Enron Corporation and was made public during the legal investigation concerning the Enron corporation. According to the information given on the above mentioned website, the dataset contains about 600000 emails belonging to 158 users. Note that the total number of distinct email addresses in the data is much larger, as also emails to and from non-Enron email addresses are recorded. In most network representations of the dataset, employees are modeled as vertices and two vertices are connected by an edge if and only if the dataset contains an email between the two corresponding employees. Since the data also distinguishes between sender and recipient, edges are sometimes directed. Furthermore, the emails of a certain time period are often aggregated in one snapshot. This may result in multiple edges or in weighted edges representing the frequency. Hence, depending on the concrete modeling, different authors refer to quite different dynamic graphs as "the Enron network", which makes comparisons between the experimental findings rather difficult. This is also the case for static data sets; however, due to even more degrees of freedom, for example the frequency of time steps or the question whether relations age and disappear over time, this is even more immanent in the case of dynamic data. Takaffoli et al. [144] choose monthly snapshots over one year, which considerably decreases the number of vertices and edges in the network. As a kind of ground truth clustering, they identify "persisting topics" based on keyword extraction. Duan et al. [49] and Dinh et al. [47] and Dinh et al. [108] consider emails on a weekly basis and do not consider any metadata. The Enron dataset has also been used in the evaluation of GRAPHSCOPE [140]; however, there the data is considered as a directed and bipartite graph, where one part corresponds to senders and the other to receivers of emails. Görke et al. [70] did not consider Enron data, but an anonymized network of e-mail contacts at the department of computer science at KIT. It comprises about 700000 events collected over a period of about 2.5 years[3]. As metadata, it includes an id for each email address specifying the corresponding chair, which can be considered as ground truth clusters.

Cellphone Data. Very similar to email networks are data about phone calls. Palla et al. [113] cluster a network of phone calls between the customers of a mobile phone company containing data of over 4 million users. They consider edges to be weighted; a phone call contributes to the weight between the participating customers for some time period around the actual time of the call. As metadata to evaluate their community finding approach, they consider zip code and age of customers. Similar data is considered by Green et al. [74], however, they do not consider edge weights. The *Reality MiningDataset* [50] is provided by the MIT Human Dynamics Lab[4] and was collected during a social science experiment in 2004. It includes information about call logs, Bluetooth devices in proximity, cell

---

[2] Available at http://www.cs.cmu.edu/~enron/.

[3] For further details and for downloading the whole dataset, please visit http://i11www.iti.uni-karlsruhe.de/en/projects/spp1307/emaildata.

[4] http://realitycommons.media.mit.edu/realitymining.html.

tower IDs, application usage, and phone status of 94 subjects over the course of an academic year. In the context of dynamic graph clustering, it is possible to extract test data in various ways. Xu et al. [154] construct a dynamic graph where the edge weight between two participants in a time step corresponds to the number of intervals in which they were in close physical proximity. As ground truth clustering, they use the affiliations of the participants. Sun et al. [140] additionally consider the cellphone activity to construct a second dynamic graph.

Online Social Networks and Blogs. Another prime example of social networks are online social networks like Facebook or Flickr. In the context of clustering algorithms, they are particularly interesting due to their size and the fact that friendship links are explicit and not implicitly assumed with the help of other metadata. Viswanath et al. [147] crawled the regional network of Facebook in New Orleans. Only data from public profiles is collected, giving information about approximately 63000 users and 1.5 Mio. friendship links, together with their evolution. Nguyen et al. [108] and Dinh et al. [44] use these data to evaluate their clustering algorithms. Kumar et al. [83] analyze data from Flickr[5] and Yahoo! 360°. Whereas Yahoo! 360° was a typical social network that does not exist anymore, Flickr has a focus on the sharing of photos, although friendship links exist as well. Both datasets are used by the authors in anonymized form and are not publicly available. Closely related to online social networks are networks derived from blogging platforms; here, the edges correspond to entry-to-entry links between different blogs [13,93,111,156]. Angel et al. [9] use sampled data obtained via Twitter's restricted access to its data stream[6]. LiveJournal[7] is somewhere in between a network of blogs and an online social network. Interesting is that users can explicitly create friendship links as well as join groups. In contrast to the usual way dynamic networks are build from blogs, edges do not necessarily correspond to links but can also depend on friendship links. Backstrom et al. [14] study the evolution of communities in LiveJournal using friendship links as edges and group membership as (overlapping) ground truth clustering.

Publication Databases. Publication databases can be used to extract dynamic graphs in several ways. Probably the most common approach is to consider *coauthor graphs*, in which vertices correspond to authors, and two authors are linked if they coauthored at least one publication. Depending on the model, edges are weighted in different ways depending on the number of shared publications and the number of authors on each publication. An orthogonal view on the data yields *copaper networks* where vertices correspond to papers and links exist if papers have at least one author in common. Both of these network types are simplifications of bipartite *author-paper* networks that relate authors to their articles. Another possibility is to not take authorship into account but insert (possibly directed) links between articles if one article cites the other,

---

[5] http://www.flickr.com/.
[6] https://dev.twitter.com/docs/streaming-apis#sampling.
[7] http://www.livejournal.com/.

leading to *citation networks*. It is commonly believed that clusters in all of these networks correspond to different research topics or fields. Due to the fact that publication data is typically not subject to any privacy concerns and their size is reasonably large, they are often used in the evaluation of graph clustering algorithms [11, 13, 14, 16, 44, 47, 49, 70, 93, 108, 141, 143, 152]. Another advantage is that information about conferences and journals the articles appeared in can be used as metadata to evaluate the resulting clusterings. The temporal aspect in the data stems from the fact that each publication has an associated publication year. The two most often considered databases in the context of clustering are DBLP and arXiv. DBLP collects information about publications in the field of computer science; information about how this data can be downloaded as an xml file can be found on the corresponding homepage[8]. The arXiv e-print archive[9] is a repository that stores electronic e-prints, organized in several categories alongside time stamped metadata. To evaluate their dynamic graph clustering framework, Görke et al. [70] used a dataset obtained from this repository, which can be found, together with the source code of the crawler used to extract this data, on the corresponding project page[10]. The KDD cup 2003 also provides further arXiv datasets on its project page[11]; these have been used to evaluate algorithms in the context of modularity based clustering [44, 47, 108].

Small Examples. Many publications about static graph clustering include the analysis of small networks to illustrate some properties of the clusterings produced by their algorithm. A famous example for that is the karate network collected by Zachary in 1977 [159], which describes friendship links between members of a karate club before the club split up due to an internal dispute; a typical question is whether a clustering algorithm is able to predict the split given the network structure. Usually these networks are small enough to be visualized entirely in an article, which enables readers to compare different clusterings of these networks across several publications. The closest evolving analog to Zachary's karate network is the Southern Women data set collected in 1933 by Davis et al. [40]. It contains data on the social activities of 18 women observed over a nine-month period. Within this period, they recorded for 14 informal events whether these women participated or not. It has been used as a test set by Berger et al. [18], Berger-Wolf and Saia [19], and Yang et al. [156]. Another interesting small example is Grevi's zebra data set [142] used by Berger et al. [18]. It consists of information about the spatial proximity of members of a zebra herd observed over three months, corresponding to 44 observations or time steps.

Other Data Sets. In the following, we will list some further sources for dynamic graph data already used to evaluate dynamic graph clustering algorithms. Xie et al. [152] have used graphs representing the topology of the internet at

---

[8] http://www.informatik.uni-trier.de/~ley/db/.

[9] http://arxiv.org/.

[10] http://i11www.iti.uni-karlsruhe.de/en/projects/spp1307/dyneval.

[11] http://www.cs.cornell.edu/projects/kddcup/datasets.html.

the level of autonomous systems (AS Graphs) based on data collected by the University of Oregon Route Views Project [92]. These data are available from the Stanford Large Network Dataset Collection[12]. Xu et al. [154] try to identify communities of spammers in data from Project Honey Pot[13], an ongoing project to identify spammers. Sun et al. [141] use data extracted from the social bookmarking web service Delicious[14], which naturally comes with a plenitude of metadata. Kim et al. [80] use data from youtube crawls[15] in their evaluation. Pang et al. [114] cluster a dynamic network of players of World of Warcraft, where edges are based on the information whether they take part in the same group.

Static Networks with Artificial Dynamics. Apart from real world data with a naturally given temporal evolution, it is also possible to artificially incorporate some dynamics into originally static data. Riedy et al. [121], for example, consider static real world networks that become dynamic by generating random edge deletions and insertions.

3.2 Dynamic Graph Generators

Depending on the aim of designing a certain clustering algorithm, there are good reasons to use synthetic data as well as good reasons to use *not only* synthetic data for the evaluation. Synthetic data refers to graphs that are artificially generated by the help of a graph generator. Given a number of vertices, these generators decide which vertices are connected by an edge based on the probability of such an edge. The edge probabilities are derived for example from a preferential attachment process [17], where vertices that already have a high degree are connected with higher probability than others, or from other rules that are characteristic for the particular generator. In the context of evolving graphs, graph generators usually not only have to decide which vertices are linked but also which vertices or edges are added or deleted. Furthermore, if the generator incorporates a hidden ground truth clustering, this usually evolves randomly as well, which in turn influences edge probabilities.

One reason to include real world instances, i.e., instances that stem from typical applications, in the experimental evaluation is that they frequently exhibit very specific properties and symmetries that are difficult to analyze and rebuild in synthetic data. Hence, to predict the performance of an algorithm in a certain application, using only synthetic data is unrewarding, since experiments involving sample instances stemming from this application are often more accurate.

This raises the question of why to use synthetic data at all. There are some good arguments that justify the use of synthetic data, at least together with real world data:

---

[12] http://snap.stanford.edu/data/index.html.
[13] http://www.projecthoneypot.org/.
[14] https://delicious.com/.
[15] http://netsg.cs.sfu.ca/youtubedata/.

– Tunable characteristics, as for example the density of the generated graphs, allow to evaluate algorithms in detail depending on these characteristics. A scenario where this can be useful is when an algorithm yields good results for some networks but bad results on others. A study on a large set of generated graphs might help to identify characteristics of the graph that are difficult to handle for the algorithm, which in turn might raise some potential for improvements.
– Synthetic graphs can usually be generated in any possible size, even very large networks that might not (yet) exist in practice. This is especially useful in the context of scalability studies.
– Using a graph generator, an unlimited number of different networks with similar properties can be generated, preventing algorithms to focus only on very few benchmark instances. This permits to test algorithms on a representative sample of the graph class one is interested in, ensuring some degree of significance.
– In particular in the context of graph clustering, there is another reason why synthetic networks are quite popular. Since there is no general agreement on a single objective function evaluating the goodness of the clustering, a common approach to evaluate graph clusterings independent of any objective function is the comparison to a known ground truth clustering. The downside of this is that real world graphs with a well-motivated ground truth clustering are still rare. For this reason, synthetic networks incorporating a hidden ground truth clustering that has been used in the generation process are popular.

In the following, we aim to give a short overview of models for synthetic graphs that might be useful in the context of clustering evolving networks. We start with some models especially suited for complex networks that can for example be derived by observing human interaction, with a particular focus on models that try to explain their evolution. In the second part, we give an overview on synthetic benchmark instances that incorporate a hidden ground-truth clustering, together with existing approaches to make these benchmarks dynamic.

Probably the most fundamental model for synthetic graphs are graphs where every edge exists with a fixed, constant probability, as first considered by Gilbert [62] in 1959. Until then, a lot of effort has been put into alternative models that better capture the properties of real world complex networks which typically exhibit characteristics like small diameter, high clustering coefficient and a powerlaw degree distribution [104]. Two classical models are small world networks [151] that explicitly address the first two issues and the Barabási-Albert model [17] that mostly addresses the degree distribution. The latter can be seen as a dynamic model for graph growth according to a preferential attachment process. Numerous variations thereof exist, most of which are targeted in capturing more accurately properties observed in real world social networks [83, 92, 146]. Leskovec et al. [91] determine automatically, among a set of parameterized models, the one fitting a set of four large online networks best based on the associated likelihood values. Similarly, Patro et al. [116] propose to use an evolutionary algorithm to choose among a set of parameterized models of network growth the one fitting a given set of input characteristics best, in order to automatically learn the best model for different graph classes.

Although these models incorporate network growth and already reflect common properties of observed complex networks as for example online social networks very well, they do not come with a well motivated inherent ground truth clustering that can be used to evaluate clustering algorithms. An exception to this is the model by Zheleva et al. [161] that is especially targeted at modeling the growth of social networks where vertices can additionally choose to enter special groups of interest. Here, the assumption is that both the network and the group structure evolve simultaneously, influencing each other. It might be possible to use the group structure chosen by the vertices as a ground truth clustering for overlapping clusters, although the group structure is correlated to the network only to a certain extent. In the model of Bagrow [15], starting from a graph generated according to preferential attachment, edges are randomly rewired to incorporate a given ground truth clustering. While this approach combines a ground truth clustering with a realistic degree distribution, the evolution stemming from the preferential attachment process is lost.

For static graph clustering, two synthetic benchmark sets have been used very frequently in the literature; the GN benchmark introduced by Girvan and Newman [63] and the LFR benchmark introduced by Lancichinetti et al. [87]. The GN benchmark is based on the *planted partition* model [26,38,60], also called *ad hoc* model, that takes as input a given ground truth clustering and two parameters p_{in} and p_{out} that correspond to the linking probabilities between vertices within the same or different clusters. Typically, the GN benchmark is used to determine how well an algorithm is able to recover the ground truth clustering, depending on the gap between p_{in} and p_{out}. The planted partition model has been generalized to weighted [56] and bipartite [75] graphs as well as hierarchical [162] and overlapping [129] ground truth clusterings. Closely related are relaxed caveman graphs [5,150]. Among the dynamic graph clustering algorithms described here, FACETNET [93], the approaches by Yang et al. [156] and Kim and Han [81], and the algorithm framework by Görke et al. [70] have been evaluated with the aid of planted partition graphs. The former two evaluations use graphs from the GN benchmark and introduce dynamics based on vertex moves; in each time step, a constant fraction of vertices leave their cluster and move to a random one. Kim and Han additionally consider a dynamic network that also incorporates the forming and dissolving of clusters and vertex addition and deletion. In contrast to that, Görke et al. use a custom graph generator based on the planted partition model that introduces dynamics by splitting and merging clusters in the ground truth clustering [72]. In each time step, one edge or vertex is added or deleted according to the probabilities prescribed by the current ground truth clustering. Hence, the actual graph structure follows the ground truth clustering with some delay. They also provide an efficient implementation of this generator [69].

In the LFR benchmark, cluster sizes as well as vertex degrees are expected to follow a power law distribution. Similar to the planted partition model, vertices share a certain fraction of their links with other vertices in their cluster and the remaining links with random vertices in other parts of the graph. The LFR benchmark has been generalized to weighted and directed graphs, as well as to overlapping clusters [85]. Among the clustering algorithms described in Sect. 2, Dinh et al. [44] have used a modification of this benchmark to a dynamic setting, whereas Cazabet et al. [30] only use it in a static scenario. Green et al. [74] use dynamic benchmarks based on LFR graphs that incorporate different cluster events, including membership switching, cluster growth, shrinkage, birth and death, and the merge and split of clusters. After the ground truth clustering has been adapted, a new random graph is drawn according to the mechanisms of the LFR benchmark, which results in large differences between adjacent timesteps.

Aldecoa and Marín [6] finally suggest to interpolate between two graphs with a significant clustering structure by rewiring edges at random. This is proposed as an alternative to benchmarks like the GN or LFR benchmark in the context of static clustering algorithms. Here, the assumption is that clusterings of the intermediate states of the graph during the rewiring process should have low distance to both the ground truth clustering of the initial and the final state. The rewiring process could be seen as a model for community evolution. In the context of tracking clusterings over time, Berger et al. [18] do not consider models for dynamic graphs but two scenarios for the evolution of clusters that are more sophisticated than random vertex moves or cluster splits and merges. It remains to mention that, in principle, all generative models used to infer clusterings via a Bayesian approach discussed in Sect. 2 might also be used as benchmark instances, as they naturally come with a dynamic ground truth clustering.

3.3   Summary

Nowadays, a lot of large real world networks have been collected and made available by projects like the Stanford Large Network Dataset Collection[16]. One problem in the context of evaluating clustering algorithms for evolving networks is that even if the original data itself has a temporal aspect, this information is often missing in the thereof constructed networks readily provided in many benchmark sets. On the other hand, the listing in Sect. 3.1 reveals that there is no real lack of dynamic data that is publicly available. A downside of these data is that converting them to dynamic graphs is often laborious and leaves many degrees of freedom. As discussed in the context of the Enron network, data from the same origin can lead to quite different dynamic graphs, depending on the design choices taken. This makes the comparison of results across different publications cumbersome. For static graph clustering, a set of very frequently used networks mostly taken from the websites of Newman[17] and Arenas[18] gained some popularity in the orbit of modularity based methods. It would be nice to have a similar set of common benchmark graphs that are evolving over time.

---

[16] http://snap.stanford.edu/data/.

[17] http://www-personal.umich.edu/~mejn/netdata/.

[18] http://deim.urv.cat/~aarenas/data/welcome.htm.

A related issue arises in the context of synthetic benchmarks that incorporate a ground truth clustering. Although a lot of publications about the static case exist, there is still no general agreement on how to make these data dynamic and what realistic dynamic changes in the ground truth clustering might look like.

4   Conclusion

Clustering evolving networks is at least as difficult as clustering static networks since it inherits all the difficulties from the static case and is further faced with additional problems that arise from the evolution of the considered networks. The difficulties inherited from static graph clustering are the many different ideas of what a good clustering is and what a good clustering algorithm is supposed to do, as well as the absence of approved benchmark instances to evaluate and compare the performance of clustering algorithms. Additional tasks arise whenever we seek for temporal smoothness or want to detect and visualize the evolution of clusters over time. Among the vast number of algorithms designed for detecting clusters in evolving graphs, in this survey we only considered graph clustering approaches in online scenarios with an algorithmic focus on the exploitation of structural information from previous time steps. We presented several state-of-the-art algorithms in different categories and summarized the main features of these algorithms in Table 1. As a first step towards common benchmark sets for the evaluation of clustering algorithms also in evolving networks, we explicitly listed data and graph generators that were used by the authors of the publications presented in this survey. With this list we aim at viewing the variety of available data and providing a collection to other authors in order to help them finding reasonable test instances for their particular algorithm. Furthermore, we discussed tasks like cluster mapping, event detection and visualization, which make the found cluster information beneficial for further analysis. We gave a brief overview on state-of-the-art approaches solving also these problems and gave some further references where the reader can find more information regarding these issues.

References

1. Agarwal, M.K., Ramamritham, K., Bhide, M.: Real time discovery of dense clusters in highly dynamic graphs: identifying real world events in highly dynamic environments. In: Proceedings of the 38th International Conference on Very Large Databases (VLDB 2012), pp. 980–991 (2012)
2. Aggarwal, C.C., Subbian, K.: Evolutionary network analysis: a survey. ACM Comput. Surv. **47**(10), 10:1–10:36 (2014)
3. Aggarwal, C.C., Xie, Y., Yu, P.S.: Towards community detection in locally heterogeneous networks. In: Proceedings of the Fifth SIAM International Conference on Data Mining, pp. 391–402. SIAM (2011)
4. Aggarwal, C.C., Zhao, Y., Yu, P.: A framework for clustering massive graph streams. Stat. Anal. Data Min. **3**(6), 399–416 (2010). http://dx.doi.org/10.1002/sam.10090

5. Aldecoa, R., Marín, I.: Deciphering network community structure by surprise. PLoS ONE **6**, e24195 (2011)
6. Aldecoa, R., Marín, I.: Closed benchmarks for network community structure characterization. Phys. Rev. E **85**, 026109 (2012). http://link.aps.org/doi/10.1103/PhysRevE.85.026109
7. Aldous, D.J.: Exchangeability and related topics. In: Hennequin, P.L. (ed.) École d'Été de Probabilités de Saint-Flour XIII — 1983. LNM, vol. 1117, pp. 1–198. Springer, Heidelberg (1985). doi:10.1007/BFb0099421. http://dx.doi.org/10.1007/BFb0099421
8. Anderson, C.J., Wasserman, S., Faust, K.: Building stochastic blockmodels. Soc. Netw. **14**, 137–161 (1992)
9. Angel, A., Sarkas, N., Koudas, N., Srivastava, D.: Dense subgraph maintenance under streaming edge weight updates for real-time story identification. Proc. VLDB Endow. **5**(6), 574–585 (2012). http://dl.acm.org/citation.cfm?id=2168651.2168658
10. Arenas, A., Duch, J., Fernandez, A., Gomez, S.: Size reduction of complex networks preserving modularity. New J. Phys. **9**(176) (2007). http://dx.doi.org/10.1088/1367-2630/9/6/176
11. Asur, S., Parthasarathy, S., Ucar, D.: An event-based framework for characterizing the evolutionary behavior of interaction graphs. ACM Trans. Knowl. Discovery Data **3**(4), 16:1–16:36 (2009). http://doi.acm.org/10.1145/1631162.1631164
12. Aynaud, T., Fleury, E., Guillaume, J.L., Wang, Q.: Communities in evolving networks definitions detection and analysis techniques. In: Mukherjee, A., Choudhury, M., Peruani, F., Ganguly, N., Mitra, B. (eds.) Dynamics on and of Complex Networks. Modeling and Simulation in Science, Engineering and Technology, vol. 2, pp. 159–200. Springer, New York (2013). http://dx.doi.org/10.1007/978-1-4614-6729-8_9
13. Aynaud, T., Guillaume, J.L.: Static community detection algorithms for evolving networks. In: Proceedings of the 8th International Symposium on Modeling and Optimization in Mobile, Ad Hoc, and Wireless Networks (WiOpt 2010), pp. 513–519. IEEE Computer Society (2010)
14. Backstrom, L., Huttenlocher, D., Kleinberg, J.M., Lan, X.: Group formation in large social networks: membership, growth, and evolution. In: Proceedings of the 12th ACM SIGKDD International Conference on Knowledge Discovery and Data Mining, pp. 44–54. ACM Press (2006). http://doi.acm.org/10.1145/1150402.1150412
15. Bagrow, J.: Evaluating local community methods in networks. J. Stat. Mech.: Theory Exp., P05001 (2008). http://www.iop.org/EJ/abstract/1742-5468/2008/05/P05001/, doi:10.1088/1742-5468/2008/05/P05001
16. Bansal, S., Bhowmick, S., Paymal, P.: Fast community detection for dynamic complex networks. In: F. Costa, L., Evsukoff, A., Mangioni, G., Menezes, R. (eds.) CompleNet 2010. CCIS, vol. 116, pp. 196–207. Springer, Heidelberg (2011). doi:10.1007/978-3-642-25501-4_20. http://dx.doi.org/10.1007/978-3-642-25501-4_20
17. Barabási, A.L., Albert, R.: Emergence of scaling in random networks. Science **286**, 509–512 (1999)
18. Berger-Wolf, T., Kempe, D., Tantipathananandth, C.: A framework for community identification in dynamic social networks. In: Proceedings of the 13th ACM SIGKDD International Conference on Knowledge Discovery and Data Mining. ACM Press (2007). http://compbio.cs.uic.edu/~tanya/research/pubs/TantipathananandhEtal_NetworkCommunities07.pdf

19. Berger-Wolf, T., Saia, J.: A framework for analysis of dynamic social networks. In: Proceedings of the 12th ACM SIGKDD International Conference on Knowledge Discovery and Data Mining, pp. 523–528. ACM Press (2006)

20. Bichot, C.E., Siarry, P. (eds.): Graph Partitioning. Wiley, Hoboken (2011). http://onlinelibrary.wiley.com/book/10.1002/9781118601181

21. Bilgin, C.C., Yener, B.: Dynamic network evolution: models, clustering, anomaly detection. Technical report, Rensselaer University, NY (2008). http://citeseerx. ist.psu.edu/viewdoc/download?rep=rep1&type=pdf&doi=10.1.1.161.6375

22. Blondel, V., Guillaume, J.L., Lambiotte, R., Lefebvre, E.: Fast unfolding of communities in large networks. J. Stat. Mech. Theory Exp. **2008**(10) (2008). http:// dx.doi.org/10.1088/1742-5468/2008/10/P10008

23. Bogdanov, P., Mongiovi, M., Singh, A.K.: Mining heavy subgraphs in time-evolving networks. In: Proceedings of the 2011 IEEE International Conference on Data Mining, pp. 81–90. IEEE Computer Society (2011)

24. Borgwardt, K.M., Kriegel, H.P., Wackersreuther, P.: Pattern mining in frequent dynamic subgraphs. In: Proceedings of the 2006 IEEE International Conference on Data Mining, pp. 818–822. IEEE Computer Society (2006)

25. Brandes, U., Delling, D., Gaertler, M., Görke, R., Höfer, M., Nikoloski, Z., Wagner, D.: On modularity clustering. IEEE Trans. Knowl. Data Eng. **20**(2), 172–188 (2008). http://doi.ieeecomputersociety.org/10.1109/TKDE.2007.190689

26. Brandes, U., Gaertler, M., Wagner, D.: Experiments on graph clustering algorithms. In: Battista, G., Zwick, U. (eds.) ESA 2003. LNCS, vol. 2832, pp. 568–579. Springer, Heidelberg (2003). doi:10.1007/978-3-540-39658-1_52, http://www. springerlink.com/openurl.asp?genre=article&issn=0302-9743&volume=2832& spage=568

27. Brandes, U., Gaertler, M., Wagner, D.: Engineering graph clustering: models and experimental evaluation. ACM J. Exp. Algorithmics **12**(1.1), 1–26 (2007). http://portal.acm.org/citation.cfm?id=1227161.1227162

28. Bron, C., Kerbosch, J.A.G.M.: Algorithm 457: finding all cliques of an undirected graph. Commun. ACM **16**(9), 575–577 (1973)

29. Catalyurek, U., Boman, E., Devine, K., Bozdag, D., Heaphy, R., Riesen, L.A.: Hypergraph-based dynamic load balancing for adaptive scientific computations. In: 21th International Parallel and Distributed Processing Symposium (IPDPS 2007), pp. 1–11. IEEE Computer Society (2007)

30. Cazabet, R., Amblard, F., Hanachi, C.: Detection of overlapping communities in dynamical social networks. In: Proceedings of the 2010 IEEE Second International Conference on Social Computing, pp. 309–314. IEEE (2010)

31. Chakrabarti, D.: AutoPart: parameter-free graph partitioning and outlier detection. In: Proceedings of the 8th European Conference on Principles and Practice of Knowledge Discovery in Databases, pp. 112–124. ACM Press (2004)

32. Chakrabarti, D., Kumar, R., Tomkins, A.S.: Evolutionary clustering. In: Proceedings of the 12th ACM SIGKDD International Conference on Knowledge Discovery and Data Mining, pp. 554–560. ACM Press (2006). http://doi.acm.org/10.1145/ 1150402.1150467

33. Chen, J., Fagnan, J., Goebel, R., Rabbany, R., Sangi, F., Takaffoli, M., Verbeek, E., Zaïane, O.R.: Meerkat: community mining with dynamic social networks. In: Proceedings in the 10th IEEE International Conference on Data Mining - Workshops, pp. 1377–1380. IEEE Computer Society, December 2010

34. Chen, J., Zaïane, O.R., Goebel, R.: Detecting communities in large networks by iterative local expansion. In: Proceedings of the 2009 IEEE International Conference on Computational Aspects of Social Networks, pp. 105–112. IEEE Computer Society (2009)
35. Chi, Y., Song, X., Zhou, D., Hino, K., Tseng, B.L.: Evolutionary spectral clustering by incorporating temporal smoothness. In: Proceedings of the 13th ACM SIGKDD International Conference on Knowledge Discovery and Data Mining, pp. 153–162. ACM Press (2007). http://doi.acm.org/10.1145/1281192.1281212
36. Clauset, A.: Finding local community structure in networks. Phys. Rev. E **72**(2), 026132 (2005). http://link.aps.org/doi/10.1103/PhysRevE.72.026132
37. Clauset, A., Newman, M.E.J., Moore, C.: Finding community structure in very large networks. Phys. Rev. E **70**(066111) (2004). http://link.aps.org/abstract/PRE/v70/e066111
38. Condon, A., Karp, R.M.: Algorithms for graph partitioning on the planted partition model. Randoms Struct. Algorithms **18**(2), 116–140 (2001). http://dx.doi.org/10.1002/1098-2418(200103)18:2⟨116::AID-RSA1001⟩3.0.CO;2-2
39. Cybenko, G.: Dynamic load balancing for distributed memory multiprocessors. J. Parallel Distrib. Comput. **7**(2), 279–301 (1989). http://dx.doi.org/10.1016/0743-7315(89)90021-X
40. Davis, A., Gardner, B., Gardner, M.R.: Deep South. University of Chicago Press, Chicago (1941)
41. Delling, D., Gaertler, M., Görke, R., Wagner, D.: Engineering comparators for graph clusterings. In: Fleischer, R., Xu, J. (eds.) AAIM 2008. LNCS, vol. 5034, pp. 131–142. Springer, Heidelberg (2008). doi:10.1007/978-3-540-68880-8_14
42. Derényi, I., Palla, G., Vicsek, T.: Clique percolation in random networks. Phys. Rev. Lett. **94**, 160202 (2005). http://link.aps.org/abstract/PRL/v94/e160202
43. Ding, C.H.Q., He, X., Zha, H., Gu, M., Simon, H.D.: A min-max cut algorithm for graph partitioning and data clustering. In: Proceedings of the 2001 IEEE International Conference on Data Mining, pp. 107–114. IEEE Computer Society (2001). http://dx.doi.org/10.1109/ICDM.2001.989507
44. Dinh, T.N., Nguyen, N.P., Thai, M.T.: An adaptive approximation algorithm for community detection in dynamic scale-free networks. In: Proceedings of the 32th Annual Joint Conference of the IEEE Computer and Communications Societies (Infocom). IEEE Computer Society Press (2013, to appear)
45. Dinh, T.N., Shin, I., Thai, N.K., Thai, M.T., Znati, T.: A general approach for modules identification in evolving networks. In: Hirsch, M.J., Pardalos, P.M., Murphey, R. (eds.) Dynamics of Information Systems. Springer Optimization and Its Applications, vol. 40, pp. 83–100. Springer, New York (2010). http://dx.doi.org/10.1007/978-1-4419-5689-7_4
46. Dinh, T.N., Thai, M.T.: Community detection in scale-free networks: approximation algorithms for maximizing modularity. IEEE J. Sel. Areas Commun. **31**(6), 997–1006 (2013)
47. Dinh, T.N., Ying, X., Thai, M.T.: Towards social-aware routing in dynamic communication networks. In: Proceedings of the 28th International Performance Computing and Communications Conference (IPCCC), pp. 161–168 (2009)
48. Doll, C., Hartmann, T., Wagner, D.: Fully-dynamic hierarchical graph clustering using cut trees. In: Dehne, F., Iacono, J., Sack, J.-R. (eds.) WADS 2011. LNCS, vol. 6844, pp. 338–349. Springer, Heidelberg (2011). doi:10.1007/978-3-642-22300-6_29
49. Duan, D., Li, Y., Li, R., Lu, Z.: Incremental k-clique clustering in dynamic social networks. Artif. Intell. **38**(2), 129–147 (2012)

50. Eagle, N., Pentland, A.: Reality mining: sensing complex social systems. J. Pers. Ubiquit. Comput. 10(4), 255–268 (2006)
51. Ester, M., Kriegel, H.P., Sander, J., Xu, X.: A density-based algorithm for discovering clusters in large spatial databases with noise. In: Proceedings of the 2nd ACM SIGKDD International Conference on Knowledge Discovery and Data Mining, pp. 226–231. ACM Press (1996)
52. Everett, M.G., Borgatti, S.P.: Analyzing clique overlap. Connections 21(1), 49–61 (1998)
53. Falkowski, T.: Community analysis in dynamic social networks. Ph.D. thesis, Otto-von-Guericke-Universität Magdeburg (2009)
54. Falkowski, T., Bartelheimer, J., Spiliopoulou, M.: Mining and visualizing the evolution of subgroups in social networks. In: IEEE/WIC/ACM International Conference on Web Intelligence, pp. 52–58. IEEE (2006)
55. Falkowski, T., Barth, A., Spiliopoulou, M.: Dengraph: A density-based community detection algorithm. In: IEEE/WIC/ACM International Conference on Web Intelligence, pp. 112–115. IEEE (2007)
56. Fan, Y., Li, M., Zhang, P., Wu, J., Di, Z.: Accuracy and precision of methods for community identification in weighted networks. Phys. A 377(1), 363–372 (2007). http://www.sciencedirect.com/science/article/pii/S0378437106012386
57. Flake, G.W., Tarjan, R.E., Tsioutsiouliklis, K.: Graph clustering and minimum cut trees. Internet Math. 1(4), 385–408 (2004). http://www.internetmathematics. org/volumes/1.htm
58. Fortunato, S.: Community detection in graphs. Phys. Rep. 486(3–5), 75–174 (2010). http://www.sciencedirect.com/science/journal/03701573
59. Fortunato, S., Barthélemy, M.: Resolution limit in community detection. Proc. Natl. Acad. Sci. U.S.A. 104(1), 36–41 (2007). http://www.pnas.org/content/ 104/1/36.full.pdf
60. Gaertler, M., Görke, R., Wagner, D.: Significance-driven graph clustering. In: Kao, M.-Y., Li, X.-Y. (eds.) AAIM 2007. LNCS, vol. 4508, pp. 11–26. Springer, Heidelberg (2007). doi:10.1007/978-3-540-72870-2_2, http://www.springerlink. com/content/nrq6tlm286808887/?p=65f77ccbb2674a16b9a67da6bb370dc7&pi=5
61. Gehweiler, J., Meyerhenke, H.: A distributed diffusive heuristic for clustering a virtual P2P supercomputer. In: Proceedings of the 7th High-Performance Grid Computing Workshop (HGCW 2010) in Conjunction with 24th International Parallel and Distributed Processing Symposium (IPDPS 2010), pp. 1–8. IEEE Computer Society (2010)
62. Gilbert, H.: Random graphs. Ann. Math. Stat. 30(4), 1141–1144 (1959)
63. Girvan, M., Newman, M.E.J.: Community structure in social and biological networks. Proc. Natl. Acad. Sci. U.S.A. 99(12), 7821–7826 (2002)
64. Gloor, P.A., Zhao, Y.: TeCFlow - a temporal communication flow visualizer for social network analysis. In: ACM CSCW Workshop on Social Networks (2004)
65. Gomory, R.E., Hu, T.: Multi-terminal network flows. J. Soc. Ind. Appl. Math. 9(4), 551–570 (1961)
66. Görke, R.: An algorithmic walk from static to dynamic graph clustering. Ph.D. thesis, Fakultät für Informatik, February 2010. http://digbib.ubka.uni-karlsruhe. de/volltexte/1000018288
67. Görke, R., Hartmann, T., Wagner, D.: Dynamic graph clustering using minimum-cut trees. In: Dehne, F., Gavrilova, M., Sack, J.-R., Tóth, C.D. (eds.) WADS 2009. LNCS, vol. 5664, pp. 339–350. Springer, Heidelberg (2009). doi:10.1007/ 978-3-642-03367-4_30. http://dx.doi.org/10.1007/978-3-642-03367-4_30

68. Görke, R., Hartmann, T., Wagner, D.: Dynamic graph clustering using minimum-cut trees. J. Graph Algorithms Appl. **16**(2), 411–446 (2012)
69. Görke, R., Kluge, R., Schumm, A., Staudt, C., Wagner, D.: An efficient generator for clustered dynamic random networks. In: Even, G., Rawitz, D. (eds.) MedAlg 2012. LNCS, vol. 7659, pp. 219–233. Springer, Heidelberg (2012). doi:10.1007/978-3-642-34862-4_16
70. Görke, R., Maillard, P., Schumm, A., Staudt, C., Wagner, D.: Dynamic graph clustering combining modularity and smoothness. ACM J. Exp. Algorithmics **18**(1), 1.5:1.1–1.5:1.29 (2013). http://dl.acm.org/citation.cfm?doid=2444016.2444021
71. Görke, R., Schumm, A., Wagner, D.: Density-constrained graph clustering. In: Dehne, F., Iacono, J., Sack, J.-R. (eds.) WADS 2011. LNCS, vol. 6844, pp. 679–690. Springer, Heidelberg (2011). doi:10.1007/978-3-642-22300-6_58. http://link.springer.com/chapter/10.1007/978-3-642-22300-6_58?null
72. Görke, R., Staudt, C.: A generator for dynamic clustered random graphs. Technical report, ITI Wagner, Faculty of Informatics, Universität Karlsruhe (TH) (2009). http://i11www.iti.uni-karlsruhe.de/projects/spp1307/dyngen, informatik, Uni Karlsruhe, TR 2009-7
73. Grady, L., Schwartz, E.I.: Isoperimetric graph partitioning for image segmentation. IEEE Trans. Pattern Anal. Mach. Intell. **28**(3), 469–475 (2006)
74. Greene, D., Doyle, D., Cunningham, P.: Tracking the evolution of communities in dynamic social networks. In: Proceedings of the 2010 IEEE/ACM International Conference on Advances in Social Networks Analysis and Mining, pp. 176–183. IEEE Computer Society (2010)
75. Guimerà, R., Sales-Pardo, M., Amaral, L.A.N.: Module identification in bipartite and directed networks. Phys. Rev. E **76**, 036102 (2007). http://link.aps.org/doi/10.1103/PhysRevE.76.036102
76. Held, P., Kruse, R.: Analysis and visualization of dynamic clusterings. In: Proceedings of the 46th Hawaii International Conference on System Sciences, pp. 1385–1393 (2013)
77. Hopcroft, J.E., Khan, O., Kulis, B., Selman, B.: Tracking evolving communities in large linked networks. Proc. Natl. Acad. Sci. U.S.A. **101**, 5244–5253 (2004). http://www.pnas.org/content/101/suppl.1/5249.abstract
78. Jaccard, P.: The distribution of flora in the alpine zone. New Phytol. **11**(2), 37–50 (1912)
79. Kannan, R., Vempala, S., Vetta, A.: On clusterings: good, bad, spectral. J. ACM **51**(3), 497–515 (2004)
80. Kim, K., McKay, R.I., Moon, B.R.: Multiobjective evolutionary algorithms for dynamic social network clustering. In: Proceedings of the 12th Annual Conference on Genetic and Evolutionary Computation, pp. 1179–1186. ACM Press (2010)
81. Kim, M.S., Han, J.: A particle-and-density based evolutionary clustering method for dynamic networks. In: Proceedings of the 35th International Conference on Very Large Databases (VLDB 2009), pp. 622–633 (2009)
82. Kullback, S., Leibler, R.A.: On information and sufficiency. Ann. Math. Stat. **22**(1), 79–86 (1951)
83. Kumar, R., Novak, J., Tomkins, A.S.: Structure and evolution of online social networks. In: Proceedings of the 12th ACM SIGKDD International Conference on Knowledge Discovery and Data Mining, pp. 611–617. ACM Press (2006). http://doi.acm.org/10.1145/1150402.1150476
84. Lai, J.H., Wang, C.D., Yu, P.: Dynamic community detection in weighted graph streams. In: Proceedings of the 2013 SIAM International Conference on Data Mining, pp. 151–161. SIAM (2013)

85. Lancichinetti, A., Fortunato, S.: Benchmarks for testing community detection algorithms on directed and weighted graphs with overlapping communities. Phys. Rev. E **80**(1), 016118 (2009)

86. Lancichinetti, A., Fortunato, S., Kertész, J.: Detecting the overlapping and hierarchical community structure of complex networks. New J. Phys. **11**(033015) (2009). http://www.iop.org/EJ/njp

87. Lancichinetti, A., Fortunato, S., Radicchi, F.: Benchmark graphs for testing community detection algorithms. Phys. Rev. E **78**(4), 046110 (2008)

88. Lee, C., Cunningham, P.: Benchmarking community detection methods on social media data. Preprint, arXiv:1302.0739 [cs.SI] (2013)

89. Leicht, E.A., Newman, M.E.J.: Community structure in directed networks. Phys. Rev. Lett. **100**(11), 118703+ (2008). http://dx.doi.org/10.1103/PhysRevLett.100.118703

90. Leighton, F.T., Rao, S.: Multicommodity max-flow min-cut theorems and their use in designing approximation algorithms. J. ACM **46**(6), 787–832 (1999). http://portal.acm.org/citation.cfm?doid=331524.331526

91. Leskovec, J., Backstrom, L., Kumar, R., Tomkins, A.S.: Microscopic evolution of social networks. In: Proceedings of the 14th ACM SIGKDD International Conference on Knowledge Discovery and Data Mining, pp. 462–470. ACM Press (2008)

92. Leskovec, J., Kleinberg, J.M., Faloutsos, C.: Graphs over time: densification laws, shrinking diameters and possible explanations. In: Proceedings of the 11th ACM SIGKDD International Conference on Knowledge Discovery and Data Mining, pp. 177–187. ACM Press (2005). http://portal.acm.org/citation.cfm?id=1081893

93. Lin, Y.R., Chi, Y., Zhu, S., Sundaram, H., Tseng, B.L.: Analyzing communities and their evolutions in dynamic social networks. ACM Trans. Knowl. Discov. Data **3**(2), 8:1–8:31 (2009)

94. Luo, F., Wang, J.Z., Promislow, E.: Exploring local community structures in large networks. In: IEEE/WIC/ACM International Conference on Web Intelligence, pp. 233–239. IEEE (2006). http://ieeexplore.ieee.org/xpl/articleDetails.jsp?arnumber=4061371

95. von Luxburg, U.: A tutorial on spectral clustering. Stat. Comput. **17**(4), 395–416 (2007). http://www.springerlink.com/content/jq1g17785n783661/

96. Meyerhenke, H.: Dynamic load balancing for parallel numerical simulations based on repartitioning with disturbed diffusion. In: 15th International Conference on Parallel and Distributed Systems (ICPADS), pp. 150–157. IEEE (2009)

97. Meyerhenke, H., Monien, B., Sauerwald, T.: A new diffusion-based multilevel algorithm for computing graph partitions. J. Parallel Distrib. Comput. **69**(9), 750–761 (2009). http://dx.doi.org/10.1016/j.jpdc.2009.04.005

98. Meyerhenke, H., Monien, B., Schamberger, S.: Graph partitioning and disturbed diffusion. Parallel Comput. **35**(10–11), 544–569 (2009). http://dx.doi.org/10.1016/j.parco.2009.09.006

99. Mirkin, B.: Eleven ways to look at the chi-squared coefficient for contingency tables. Am. Stat. **55**(2), 111–120 (2001). http://www.jstor.org/stable/2685997

100. Misue, K., Eades, P., Lai, W., Sugiyama, K.: Layout adjustment and the mental map. J. Vis. Lang. Comput. **6**(2), 183–210 (1995). http://www.sciencedirect.com/science/article/pii/S1045926X85710105

101. Moody, J., McFarland, D., Bender-deMoll, S.: Dynamic network visualization. Am. J. Sociol. **110**(4), 1206–1241 (2005)

102. Muelder, C., Ma, K.L.: Rapid graph layout using space filling curves. IEEE Trans. Vis. Comput. Graph. **14**(6), 1301–1308 (2008)

103. Muelder, C., Ma, K.L.: A treemap based method for rapid layout of large graphs. In: Proceedings of IEEE Pacific Visualization Symposium (PacificVis 2008), pp. 231–238 (2008)
104. Newman, M.E.J.: The structure and function of complex networks. SIAM Rev. 45(2), 167–256 (2003). http://dx.doi.org/10.1137/S003614450342480
105. Newman, M.E.J.: Analysis of weighted networks. Phys. Rev. E 70(056131), 1–9 (2004). http://link.aps.org/abstract/PRE/v70/e056131
106. Newman, M.E.J.: Detecting community structure in networks. Eur. Phys. J. B 38(2), 321–330 (2004). http://www.springerlink.com/content/5GTDACX 17BQV6CDC
107. Newman, M.E.J., Girvan, M.: Finding and evaluating community structure in networks. Phys. Rev. E 69(026113), 1–16 (2004). http://link.aps.org/abstract/PRE/v69/e026113
108. Nguyen, N.P., Dinh, T.N., Ying, X., Thai, M.T.: Adaptive algorithms for detecting community structure in dynamic social networks. In: Proceedings of the 30th Annual Joint Conference of the IEEE Computer and Communications Societies (Infocom), pp. 2282–2290. IEEE Computer Society Press (2011)
109. Nicosia, V., Mangioni, G., Carchiolo, V., Malgeri, M.: Extending the definition of modularity to directed graphs with overlapping communities. J. Stat. Mech.: Theory Exp. 2009(03), p03024 (23pp) (2009). http://stacks.iop.org/1742-5468/2009/P03024
110. Ning, H., Xu, W., Chi, Y., Gong, Y., Huang, T.: Incremental spectral clustering with application to monitoring of evolving blog communities. In: Proceedings of the 2007 SIAM International Conference on Data Mining, pp. 261–272. SIAM (2007)
111. Ning, H., Xu, W., Chi, Y., Gong, Y., Huang, T.: Incremental spectral clustering by efficiently updating the eigen-system. Pattern Recogn. 43, 113–127 (2010)
112. Ovelgönne, M., Geyer-Schulz, A.: An ensemble learning strategy for graph clustering. In: Graph Partitioning and Graph Clustering: Tenth DIMACS Implementation Challenge. DIMACS Book, vol. 588, pp. 187–206. American Mathematical Society (2013). http://www.ams.org/books/conm/588/11701
113. Palla, G., Barabási, A.L., Vicsek, T.: Quantifying social group evolution. Nature 446, 664–667 (2007). http://www.nature.com/nature/journal/v446/n7136/abs/nature05670.html
114. Pang, S., Chen, C., Wei, T.: A realtime community detection algorithm: incremental label propagation. In: First International Conference on Future Information Networks (ICFIN 2009), pp. 313–317. IEEE (2009)
115. Park, Y., Song, M.: A genetic algorithm for clustering problems. In: Proceedings of the 3rd Annual Conference on Genetic Programming, pp. 568–575 (1998)
116. Patro, R., Duggal, G., Sefer, E., Wang, H., Filippova, D., Kingsford, C.: The missing models: a data-driven approach for learning how networks grow. In: Proceedings of the 18th ACM SIGKDD International Conference on Knowledge Discovery and Data Mining, pp. 42–50. ACM Press (2012)
117. Pearson, K.: On the criterion that a given system of deviations from the probable in the case of a correlated system of variables is such that it can be reasonably supposed to have arisen from random sampling. Philos. Mag. Ser. 5 50(302), 157–175 (1900)
118. Pons, P., Latapy, M.: Computing communities in large networks using random walks. J. Graph Algorithms Appl. 10(2), 191–218 (2006). http://www.cs.brown.edu/publications/jgaa/

119. Raghavan, U.N., Albert, R., Kumara, S.: Near linear time algorithm to detect community structures in large-scale networks. Phys. Rev. E **76**(3), 036106 (2007). http://link.aps.org/doi/10.1103/PhysRevE.76.036106

120. Rand, W.M.: Objective criteria for the evaluation of clustering methods. J. Am. Stat. Assoc. **66**(336), 846–850 (1971). http://www.jstor.org/stable/2284239?origin=crossref

121. Riedy, J., Bader, D.A.: Multithreaded community monitoring for massive streaming graph data. In: Workshop on Multithreaded Architectures and Applications (MTAAP 2013) (2013, to appear)

122. Riedy, E.J., Meyerhenke, H., Ediger, D., Bader, D.A.: Parallel community detection for massive graphs. In: Wyrzykowski, R., Dongarra, J., Karczewski, K., Waśniewski, J. (eds.) PPAM 2011. LNCS, vol. 7203, pp. 286–296. Springer, Heidelberg (2012). doi:10.1007/978-3-642-31464-3_29. http://dx.doi.org/10.1007/978-3-642-31464-3_29

123. Rissanen, J.: Modeling by shortest data description. Automatica **14**(5), 465–471 (1978)

124. Rotta, R., Noack, A.: Multilevel local search algorithms for modularity clustering. ACM J. Exp. Algorithmics **16**, 2.3:2.1–2.3:2.27 (2011). http://doi.acm.org/10.1145/1963190.1970376

125. Rousseeuw, P.J.: Silhouettes: a graphical aid to the interpretation and validation of cluster analysis. J. Comput. Appl. Math. **20**, 53–65 (1987). http://www.sciencedirect.com/science/article/pii/0377042787901257

126. Saha, B., Mitra, P.: Dynamic algorithm for graph clustering using minimum cut tree. In: Proceedings of the Sixth IEEE International Conference on Data Mining - Workshops, pp. 667–671. IEEE Computer Society, December 2006. http://ieeexplore.ieee.org/xpls/abs_all.jsp?arnumber=4063709

127. Saha, B., Mitra, P.: Dynamic algorithm for graph clustering using minimum cut tree. In: Proceedings of the 2007 SIAM International Conference on Data Mining, pp. 581–586. SIAM (2007). http://www.siam.org/proceedings/datamining/2007/dm07.php

128. Sallaberry, A., Muelder, C., Ma, K.-L.: Clustering, visualizing, and navigating for large dynamic graphs. In: Didimo, W., Patrignani, M. (eds.) GD 2012. LNCS, vol. 7704, pp. 487–498. Springer, Heidelberg (2013). doi:10.1007/978-3-642-36763-2_43. http://dx.doi.org/10.1007/978-3-642-36763-2_43

129. Sawardecker, E.N., Sales-Pardo, M., Amaral, L.A.N.: Detection of node group membership in networks with group overlap. Eur. Phys. J. B **67**, 277–284 (2009). http://dx.doi.org/10.1140/epjb/e2008-00418-0

130. Schaeffer, S.E.: Graph clustering. Comput. Sci. Rev. **1**(1), 27–64 (2007). http://dx.doi.org/10.1016/j.cosrev.2007.05.001

131. Schuetz, P., Caflisch, A.: Efficient modularity optimization by multistep greedy algorithm and vertex mover refinement. Phys. Rev. E **77**(046112) (2008). http://scitation.aip.org/getabs/servlet/GetabsServlet?prog=normal&id=PLEEE8000077000004046112000001&idtype=cvips&gifs=yes

132. Shen, H., Cheng, X., Cai, K., Hu, M.B.: Detect overlapping and hierarchical community structure in networks. Phys. A: Stat. Mech. Appl. **388**(8), 1706–1712 (2009). http://www.sciencedirect.com/science/article/pii/S0378437108010376

133. Shi, J., Malik, J.: Normalized cuts and image segmentation. IEEE Trans. Pattern Anal. Mach. Intell. **22**(8), 888–905 (2000). http://doi.ieeecs.org/10.1109/34.868688

134. Sibson, R.: Slink: an optimally efficient algorithm for the single-link cluster method. Comput. J. **16**(1), 30–34 (1973). http://dx.doi.org/10.1093/comjnl/16.1.30

135. Šíma, J., Schaeffer, S.E.: On the NP-completeness of some graph cluster measures. In: Wiedermann, J., Tel, G., Pokorný, J., Bieliková, M., Štuller, J. (eds.) SOFSEM 2006. LNCS, vol. 3831, pp. 530–537. Springer, Heidelberg (2006). doi:10.1007/11611257_51. http://dx.doi.org/10.1007/11611257_51

136. Snijders, T.A., Nowicki, K.: Estimation and prediction of stochastic blockmodels for graphs with latent block structure. J. Classif. **14**, 75–100 (1997)

137. Spiliopoulou, M., Ntoutsi, I., Theodoridis, Y., Schult, R.: MONIC: modeling and monitoring cluster transitions. In: Proceedings of the 12th ACM SIGKDD International Conference on Knowledge Discovery and Data Mining, pp. 706–711. ACM Press (2006). http://doi.acm.org/10.1145/1150402.1150491

138. Stanton, I., Kliot, G.: Streaming graph partitioning for large distributed graphs. In: Proceedings of the 18th ACM SIGKDD International Conference on Knowledge Discovery and Data Mining, pp. 1222–1230. ACM Press (2012)

139. Staudt, C., Meyerhenke, H.: Engineering high-performance community detection heuristics for massive graphs. In: Proceedings of the 2013 International Conference on Parallel Processing. Conference Publishing Services (CPS) (2013)

140. Sun, J., Yu, P.S., Papadimitriou, S., Faloutsos, C.: Graphscope: parameter-free mining of large time-evolving graphs. In: Proceedings of the 13th ACM SIGKDD International Conference on Knowledge Discovery and Data Mining, pp. 687–696. ACM Press (2007). http://portal.acm.org/citation.cfm?id=1281192.1281266&coll=Portal&dl=GUIDE&CFID=54298929&CFTOKEN=41087406

141. Sun, Y., Tang, J., Han, J., Gupta, M., Zhao, B.: Community evolution detection in dynamic heterogeneous information networks. In: Proceedings of the Eighth Workshop on Mining and Learning with Graphs, pp. 137–146. ACM Press (2010). http://doi.acm.org/10.1145/1830252.1830270

142. Sundaresan, S.R., Fischhoff, I.R., Dushoff, J.: Network metrics reveal differences in social organization between two fission-fusion species, Grevy's zebra and onager. Oecologia **151**(1), 140–149 (2007)

143. Takaffoli, M., Fagnan, J., Sangi, F., Zaïane, O.R.: Tracking changes in dynamic information networks. In: Proceedings of the 2011 IEEE International Conference on Computational Aspects of Social Networks, pp. 94–101. IEEE Computer Society (2011)

144. Takaffoli, M., Rabbany, R., Zaïane, O.R.: Incremental local community identification in dynamic social networks. In: Proceedings of the 2013 IEEE/ACM International Conference on Advances in Social Networks Analysis and Mining, IEEE Computer Society (2013, to appear)

145. Tong, H., Papadimitriou, S., Sun, J., Yu, P.S., Faloutsos, C.: Colibri: fast mining of large static and dynamic graphs. In: Proceedings of the 14th ACM SIGKDD International Conference on Knowledge Discovery and Data Mining, pp. 686–694. ACM Press (2008). http://doi.acm.org/10.1145/1401890.1401973

146. Vázquez, A.: Growing network with local rules: preferential attachment, clustering hierarchy, and degree correlations. Phys. Rev. E **67**, 056104 (2003). http://link.aps.org/doi/10.1103/PhysRevE.67.056104

147. Viswanath, B., Mislove, A., Cha, M., Gummadi, P.K.: On the evolution of user interaction in facebook. In: Proceedings of the 2nd ACM Workshop on Online Social Networks, pp. 37–42. ACM Press (2009). http://doi.acm.org/10.1145/1592665.1592675

148. Wagner, S., Wagner, D.: Comparing clusterings - an overview. Technical report 2006-04, ITI Wagner, Faculty of Informatics, Universität Karlsruhe (TH) (2007). http://digbib.ubka.uni-karlsruhe.de/volltexte/1000011477

149. Wang, Y.J., Wong, G.Y.: Stochastic blockmodels for directed graphs. J. Am. Stat. Assoc. **82**, 8–19 (1987)

150. Watts, D.J.: Networks, dynamics, and the small-world phenomenon. Am. J. Sociol. **105**, 493–527 (1999)

151. Watts, D.J., Strogatz, S.H.: Collective dynamics of 'small-world' networks. Nature **393**(6684), 440–442 (1998)

152. Xie, J., Chen, M., Szymanski, B.K.: LabelRankT: incremental community detection in dynamic networks via label propagation. CoRR abs/1305.2006 (2013). http://arxiv.org/abs/1305.2006

153. Xie, J., Szymanski, B.K.: LabelRank: a stabilized label propagation algorithm for community detection in networks. CoRR abs/1303.0868 (2013). http://arxiv.org/abs/1303.0868

154. Xu, K.S., Kliger, M., Hero, A.O.: Tracking communities in dynamic social networks. In: Salerno, J., Yang, S.J., Nau, D., Chai, S.-K. (eds.) SBP 2011. LNCS, vol. 6589, pp. 219–226. Springer, Heidelberg (2011). doi:10.1007/978-3-642-19656-0_32

155. Xu, X., Yuruk, N., Feng, Z., Schweiger, T.A.J.: Scan: a structural clustering algorithm for networks. In: Proceedings of the 13th ACM SIGKDD International Conference on Knowledge Discovery and Data Mining, pp. 824–833. ACM Press (2007)

156. Yang, T., Chi, Y., Zhu, S., Jin, R.: Detecting communities and their evolutions in dynamic social networks - a Bayesian approach. Mach. Learn. **82**(2), 157–189 (2011)

157. Yu, K., Yu, S., Tresp, V.: Soft clustering on graphs. In: Advances in Neural Information Processing Systems 18, p. 5. MIT Press (2006)

158. Yu, S.X., Shi, J.: Multiclass spectral clustering. In: Proceedings of the 9th IEEE International Conference on Computer Vision, pp. 313–319 (2003)

159. Zachary, W.W.: An information flow model for conflict and fission in small groups. J. Anthropol. Res. **33**, 452–473 (1977)

160. Zhao, Y., Yu, P.S.: On graph stream clustering with side information. In: Proceedings of the Seventh SIAM International Conference on Data Mining, pp. 139–150. SIAM (2013)

161. Zheleva, E., Sharara, H., Getoor, L.: Co-evolution of social and affiliation networks. In: Proceedings of the 15th ACM SIGKDD International Conference on Knowledge Discovery and Data Mining, pp. 1007–1016. ACM Press (2009). http://doi.acm.org/10.1145/1557019.1557128

162. Zhou, H.: Network landscape from a Brownian particle's perspective. Phys. Rev. E **67**, 041908 (2003). http://link.aps.org/doi/10.1103/PhysRevE.67.041908

Integrating Sequencing and Scheduling: A Generic Approach with Two Exemplary Industrial Applications

Wiebke Höhn and Rolf H. Möhring[✉]

Institut Für Mathematik, Technische Universität Berlin, Straße des 17. Juni 136, 10623 Berlin, Germany
{hoehn,moehring}@math.tu-berlin.de

Abstract. A huge number of problems in production planning ask for a somewhat good processing sequence. While all those problems share the common sequencing aspect, the underlying side constraints, and hence, the related scheduling decisions, are highly problem-specific. We propose a generic algorithmic framework which aims at separating the side constraints from the general sequencing decisions. This allows us to only focus on the optimization problem that remains when the processing sequence is already decided—in many cases yet another NP-hard problem. Compared to the overall problem, this subproblem is far more manageable. In cooperation with industry, our approach has been applied to problems of planning (sheet metal) coil coating lines and dairy filling lines, respectively. For the subproblem of the coil coating problem, we could observe a close relation to the maximum weight independent set problem on a special class of multiple-interval graphs. Thorough insights into this problem were finally the basis for achieving a makespan reduction of over 13% on average for the coil coating line of our industrial partner. For the less complex dairy problem, our solutions were on par with the manual production plans in use at the production site of our project partner. We could show that this solution is almost optimal, i.e., within a factor of 1.02 times a lower bound.

1 Introduction

During the industrial revolution in the late 19th century, production processes were widely reorganized from individual manufacturing to assembly line production. The high efficiency of this new concept helped to drastically increase the productivity so that nowadays, it is one of the standard manufacturing concepts in several branches of industry. While over the years, assembly lines improved and got technically more advanced, production planning at the same time became more and more complex. The awareness for the central importance of elaborate schedules, though, raised only slowly. Still, surprisingly many companies do not

Supported by the German Research Foundation (DFG) as part of the Priority Program "Algorithm Engineering.".

L. Kliemann and P. Sanders (Eds.): Algorithm Engineering, LNCS 9220, pp. 330–351, 2016.
DOI: 10.1007/978-3-319-49487-6_10

exploit the full potential of their production lines due to suboptimal planning. It is not unusual that even today production plans are designed manually without any support of planning or optimization tools. Of course, this does not necessarily imply poor plans. On the contrary, due to the long-term experience of the planners, these schedules have very often an impressive quality. However, if the problem becomes too complex it is basically impossible for a human being to oversee the problem as a whole, and hence, to manually produce good plans.

The addressed complex scheduling problems share major problem characteristics. One of the most central components is the sequencing aspect which arises naturally from the architecture of assembly lines. Due to complex side constraints, however, the overall scheduling problem usually goes far beyond basic sequencing problems known from theoretic literature. In most of the cases, the actual processing sequence is just one of many strongly interdependent decisions. Once a sequence is chosen, still certain allocations have to be made—e.g., the allocation of resources or the assignment of time slots for work related to security regulations. Such decisions may depend not only on pairs of adjacently scheduled jobs but also on whole subsequences or even the entire sequence in the worst case. As a consequence, the quest for a good processing sequence can only be answered accurately by taking into account the ensuing allocation problem as well.

Currently, for most of those problems (sufficiently fast) exact algorithms are out of reach. Computationally expensive heuristics are usually also not an option, since planning tools are utilized on a daily basis, and for this reason underlie strict runtime limits. When designing simple heuristics, on the other hand, one is often faced with problems stemming from tangled constraints. Often, the sequencing is tailored to only some of the many side constraints, leading in the end to rather poor results.

In this paper, we investigate a generic integrated approach which aims to avoid these difficulties. Based on a black box sequence evaluation, the algorithm elaborately examines a large number of processing sequences. In this way, the highly problem specific allocation problem is separated from the problem unspecific sequencing task. From a user's perspective, this allows to solely focus on the allocation problem which is usually far more manageable than the overall scheduling problem. This makes the approach very user-friendly, and, as we will see in examples in the following sections, also very successful. As a nice side effect, the approach generates not only one but several good solutions such that expert planners can take the final decision according to criteria possibly not even covered by the actual model. Since it is unlikely to provide a priori performance guarantees, we assess the performance of the algorithms via instance-based lower bounds.

Before addressing related algorithmic challenges, we will first give an idea of what different planning problems and the underlying allocation problems look like.

Automotive Industry. The automotive industry is one well-known example for assembly line production. Cars are usually manufactured while hanging or lying on assembly belts. Due to the constant speed of the lines, and hence, the limited time for each working step, it is not possible to install certain options like sunroofs or special navigation devices at every car in the sequence but only at a certain fraction. Violations of the constraints are measured by a penalty function, representing the cost for additional workers required or the additional time needed. This is a classic problem introduced in [29] already in the 1980s. It has prominently attained attention when being addressed in the 2005 ROADEF Challenge[1] in cooperation with Renault. The results of this competition are summarized in [33]. In this problem, the underlying allocation problem is trivial. Once the processing sequence is chosen, only the penalty cost needs to computed, and no further decisions have to be taken.

Food-Processing Industry. The food industry is characterized by a large and still increasing variety of products which is especially noticeable at the last production stage, the packaging or filling lines. Here, due to special company labels, different destinations or packet sizes, one base product again splits up into several subproducts. This makes the last stage very often the bottleneck of the whole production. As a result of the large number of products, (sequence-dependent) setup operations play a central role in production planning. Focusing only on this aspect, the problem of computing a minimum setup processing sequence can be formulated as Asymmetric Traveling Salesman Problem (ATSP). However, there are usually additional side constraints that make the problem much more complex than ATSP. In most of the cases, this results in non-trivial allocation problems. Classic constraints in this area stem from the limited shelf-life of the products or from hygiene regulations. An overview of typical constraints in food industry and at packaging lines, respectively, can be found in e.g., [2,36]. Additional problem-specific constraints from cheese, ice cream or dairy production are discussed in [8,19,20].

Steel Industry. The final branch of industry we want to use as illustrations is steel industry. It provides complex sequencing problems at almost every stage of production. Some of many examples where such problems occur are the strand casting process as one of the first production stages, the hot and cold rolling mills later on, and the coil coating lines at the very end; see e.g., [7,9,34]. A classic constraint that appears in many of the related scheduling problems is the so called coffin shape. It asks for an ordering of the items so that their width slightly increases for the first few items and decreases afterwards. The reasons for demanding this is that the slabs or coils are grooving the used rollers at their borders, and hence, scheduling the items in the wrong order would lead to visible grooves on subsequent items. Consequently, if the items are not processed in a coffin shape manner, the roller needs to be replaced which delays the further production. There are similar ordering criteria with respect to other characteristics, which in total lead again to an ATSP-like subproblem (if ignoring the initial increase of the width in the coffin shape). Still, similar to the problems from food

---

[1] http://challenge.roadef.org/2005/en/.

industry, there are usually further side constraints that go far beyond ATSP. A class of such constraints is due to the temperature of the processed items which plays a crucial role in many production steps. To avoid the items cooling down too much, it is necessary to limit the waiting times between different processing stages, and hence, it may be necessary to take into account preceding stages when planning.

When implementing the generic algorithmic approach for a concrete problem, the most elaborate part is usually the design of algorithms for the allocation problem which—to serve as a proper subroutine—have to be sufficiently fast. Due to the large number of sequence evaluation calls, we cannot afford allocation algorithms with more than linear or quadratic runtime. In many cases, this permits again only heuristics. Their quality goes very often hand in hand with a good understanding of the problem, not only from the experimental but also from the theoretical perspective. Structural insights gained in the analysis can provide essential ideas for the design of algorithms.

Organization of the Paper. In the remainder of this section, we first formally define the considered class of integrated sequencing and allocation problems, and then discuss related work. In Sect. 2, we present the generic algorithmic framework for solving problems from this class. The approach is applied to two exemplary problems from steel industry and dairy industry in the Sects. 3 and 4, respectively. Finally, in Sect. 5, we give an outlook on a problem setting where the focus is more on social fairness rather than on maximizing the total throughput.

1.1   Abstract Problem Formulation

Encapsulating the structure of the examples given above, we consider basic (ATSP-like) sequencing problems with integrated abstract resource allocation for generated sequences. The latter can be seen as a scheduling problem for a fixed sequence of jobs. However, in order to distinguish it from the overall problem, which we refer to as *scheduling problem*, the term *allocation problem* is used instead. Even though our approach is not limited to a certain objective, the two major problems discussed in this paper address the makespan objective, i.e., the maximum completion time among all jobs which serves as a measure for the throughput. At an abstract level, we can formally describe the considered scheduling problem (with respect to the makespan objective) as follows, and we give an brief outlook on other objectives in Sect. 5.

Integrated Sequencing and Allocation Problem. Consider a set of n jobs $J :=\{1, 2, \ldots, n\}$ with processing time $p_j > 0$ for job $j \in J$, additional job characteristics, and sequence-dependent setups of length $s_{ij} \geq 0$ for any pair $i, j \in J$. Then, given a set of cost functions $c_\pi : \mathcal{A}_\pi \to \mathbb{R}_{\geq 0}$ for any permutation of jobs $\pi \in \Pi_n$ where $\mathcal{A}_\pi$ denotes the set of feasible allocations for π, the task is to compute a minimum makespan *schedule*. A schedule consists of a *sequence* $\pi \in \Pi_n$ and an *allocation* $A(\pi) \in \mathcal{A}_\pi$ for sequence π, and its makespan is given by

$$C_{\max}(\pi, A(\pi)) = \sum_{j \in J} p_j + \sum_{j=1}^{n-1} s_{\pi(j)\pi(j+1)} + c_\pi(A(\pi)).$$

The term $\sum_{j=1}^{n-1} s_{\pi(j)\pi(j+1)} + c_\pi(A(\pi))$ is referred to as *cost* of the schedule.

We typically assume that the allocation problem involves decisions that incorporate not only neighboring jobs but also larger parts of the processing sequence π. Consequently, also the additional allocation cost $c_\pi(A(\pi))$ will be non-local in this sense. In contrast, the term $\sum_{j=1}^{n-1} s_{\pi(j)\pi(j+1)}$ is purely local while $\sum_{j \in J} p_j$ is even independent of the schedule.

1.2 Related Work

Sequencing and allocation problems with setup times and makespan minimization constitute a widely studied field, see [3] for a recent survey. In most of these problems, however, setups are purely *local*, i.e., depend only on two successive jobs on the same machine and thus are closely related to variants of the ATSP, see [5]. Also more general problems with setups typically involve only local setups. One such example are scheduling problems with communication delays and precedence constraints, see [6], where a delay only occurs between pairs of jobs with a non-transitive precedence constraint among them when they are scheduled on different machines.

In contrast, our setup costs need to be calculated in view of several, possibly many preceding jobs. Such *non-local* setups have only sporadically been considered in scheduling, e.g., by [21]. They are more typical in multi product assembly lines. But here one no longer considers sequencing problems with makespan minimization, but aims at finding batches of products that minimize the cycle time of the robotic cells, see e.g., [30].

Setups concurrent with production—as they will occur in the coil coating problem discussed in the following section—also occur in parallel machine scheduling problems with a common server considered by [14]: Before processing, jobs must be loaded onto a machine by a single server which requires some time (the setup time for that job) on that server. Good sequences of jobs on the parallel machines minimize the time jobs have to wait to be setup due to the single server being busy on a different machine. In this model, too, setups are purely local, and once sequences for the machines have been determined, scheduling the server constitutes a trivial task.

Considering the computational complexity of the integrated sequencing and allocation problem, it is in most cases easy to observe that the problem is NP-hard or even inapproximable. Ignoring allocation aspects, the problem is equivalent to the path variant of the ATSP. Straightforward reductions from the Symmetric Travelling Salesman Problem (TSP) allow to transfer results from this classic problem to its path variant, implying that the problem is MAX SNP-hard

---

Algorithm 1.1. Generic algorithmic framework

---

generate initial population of sequences;
repeat
  perform mutation on $m_m \leq m$ randomly chosen sequences from S;
  perform crossover on $m_c \leq m$ randomly chosen pairs of sequences from S;
  add sequences resulting from mutation and crossover to S;
  assess sequences in S utilizing the allocation subroutine;
  delete all but the m best sequences from S;
until *termination criterion fulfilled*;
return S;

---

even in the symmetric case when all edge length are 1 or 2 [28]. Moreover, for the symmetric case with general edge lengths, it follows that there exists no constant factor approximation algorithm unless P $\neq$ NP [31]. Excellent references on TSP include the textbook by Applegate et al. [4] and the more recent collection by Gutin and Punnen [13]. Despite the hardness of ATSP, Helsgaun's efficient implementation of the Lin-Kernighan Heuristic [15] (LKH) can be expected to compute optimal solutions for instances of up to roughly 100 and 1000 cities in approximately one second and one minute, respectively.

2 Generic Algorithmic Framework

In this section, we introduce the generic algorithmic framework for integrated sequencing and allocation problems. For the sequencing part, we utilize a genetic algorithm which by its nature combines solutions, or *individuals*, in such a way that beneficial characteristics of solutions persist, while costly characteristics are eliminated [1]. The set of solutions is commonly referred to as *population*. In a *crossover*, the combination of two parents from the current population brings forth a new individual, while a *mutation* creates a new individual by modifying one already present. In an iteration, or *generation*, the population is first enlarged through crossovers and mutations, before a subset of all these individuals is selected for survival. See [24] for a recent successful application to a different production sequencing problem.

During a run of our algorithm, we maintain a constant population size across generations. For each individual, we (heuristically) solve the allocation subproblem to assess its cost. Individuals with better makespans survive. The algorithm stops if a given termination criterion is fulfilled. The corresponding pseudo code is given in Algorithm 1.1, and its different components are described in more detail below.

Good parameter choices for this approach are usually the result of rigorous testing. Apart from this tuning process, the initial population and the allocation subroutine are the only components which are purely problem-specific, and hence, which need to be fully adapted.

Initial Population. The initial population plays a central role in the success of our approach. Choosing an appropriate and diverse population usually increases the chance to find good solutions and moreover, to do so much faster. In contrast to the final sequence we aim to compute, for the different sequences in the initial population we rather look for sequences that are specifically good with respect to only some of the many constraints. One natural such sequence is the one we obtain when ignoring all side constraints and focusing only on minimizing the total sequence-dependent setup durations s_{ij}. In this restricted form, the problem can be formulated as ATSP with distances s_{ij}, and a solution can be computed utilizing LKH. If necessary, we add further randomly generated sequences. After all, a highly diverse initial population provides a good starting point for later recombinations in the algorithm.

Termination Criterion. An obvious termination criterion is the runtime limit, if given by the customer. From the algorithmic perspective, a more natural criterion is to stop after a certain number of iterations without improvement. This can avoid unnecessary long runtimes on the one hand, and on the other hand, it appears less likely to stop the algorithm right before the very last iterations which might still drastically improve the result. Depending on the problem, one might also consider to restart the algorithm if the solutions are not improving anymore.

Mutation. Mutations are conducted by inverting a random consecutive subsequence of an individual.

Crossover. For crossovers, we implement a classic mechanism for sequencing problems originally proposed by Mühlenbein et al. [27]: Upon the selection of two individuals from the current population, a *donor d* and a *receiver r*, a random consecutive subsequence of random length is taken from the donor to form the basis for the construction of the new individual, or *offspring s*. We complete s by continuing from its last element i with the elements following i in the receiver, until we encounter an element already contained in s. Now we switch back to the donor and continue completing s from d in the same way, going back to r again when encountering an element already added to the offspring. If we can continue with neither r nor d, we add to s the next element from r which is not in s yet, and try again.

3   Integrated Sequencing and Allocation in Steel Industry

In this section, we deal with a particular integrated sequencing and allocation problem as described in Sect. 1.1. The considered problem stems from the final processing step in sheet metal production, the coating of steel coils, which may be seen as a prototype of such an integrated problem: In addition to sequencing coils, a challenging *concurrent setup allocation problem* needs to be solved, and both problems are strongly interdependent.

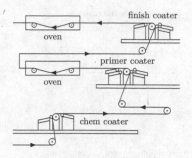

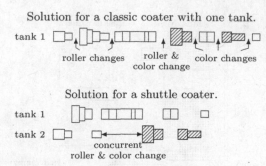

(a) Schematic view of a coil coating line with chem, primer, and finish coater.

(b) Coils of different widths as they are processed in the coating line (from left to right). The example shows the coils' colors at one particular coater.

Fig. 1. Coil coating.

Due to the extremely diverse product portfolio of coils, the coil coating process plays an important role in the overall planning in steel industry. As is typical for paint jobs, it may be subject to long setup times, mainly for the cleaning of equipment, and thus very high setup cost. This problem has been introduced and studied in detail in [18]. However, at this point our focus is rather on illustrating the general approach taken in Algorithm 1.1, and so the description remains high level.

High-Level Problem Formulation. Steel coils are a highly individualized product, and all non-productive time in coil coating depends on certain characteristics of coils. They usually have a length of 1–5 km, and their central attributes are naturally the different coating layers they receive, chosen from a palette of several hundreds, and their width, usually 1.0–1.8 m.

We consider a typical coil coating line as operated by many major steel companies worldwide, consisting of a primer and a finish coater, each comprising two coaters to coat the top and bottom side of a coil, see [10,25]. Before entering production, each coil is unrolled and stapled to the end of its predecessor, so essentially a never-ending strip of sheet metal is continuously running through the coil coating line. After the coating process, the coils are rolled up again, now ready for shipping. A schematic view of a typical coil coating line is depicted in Fig. 1(a).

Each of the coaters may or may not be a *shuttle coater*, that modern coil coating lines are nowadays equipped with in order to reduce the total setup cost. Shuttle coaters possess two separate tanks allowing for holding two different coating materials at the same time. The advantage is twofold: The shuttle can be used to switch between two coatings on the same coater at almost no setup time and cost, or alternatively the unused tank can be set up while coating continues from the other tank in the meantime. We refer to this possibility to perform setup work during production—which is somewhat uncommon in scheduling

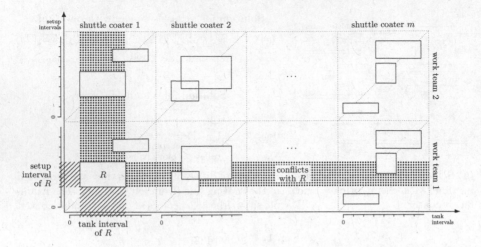

Fig. 2. Interval model for the tank assignment and setup work schedule.

literature—as *concurrent setup*. Setup work is necessary whenever the color of two consecutive coils *on the same tank* changes or when the first of these two coils has a smaller width than the second one. The latter requiring a change of the rubber roller which is used for applying the coating to the coil. See Fig. 1(b) for an illustrative example. The introduction of shuttle coaters significantly changes the flavor and the complexity of production planning: Which tank do we use for which coil, and how do we allocate concurrent setup work without exceeding available work resources?

The optimization goal is to minimize the *makespan* for coating the given set of coils, i.e., the completion time of the last coil in the sequence. Concerning optimal solutions, this is equivalent to minimizing non-productive time, or *cost*, in the schedule.

Technically, once a detailed schedule is chosen, *scrap coils* are inserted in the actual sequence of coils in order to bridge the time required for setup work.

Related Work. Literature regarding optimization in the planning process for coil coating in general is scarce at best: To the best of our knowledge, only Tang and Wang [35] consider planning for a coil coating line. They apply tabu search to a rather basic model without shuttle coaters. The present work is the first incorporating shuttles and concurrent setup work in a thorough mathematical investigation.

Model for the Underlying Allocation Problem. Due to the shuttle coaters, even fixing the coil sequence leaves open the non-trivial question of deciding the tank assignment and the scheduling of concurrent setup work. We develop a representation of solutions of this allocation problem as a family of weighted 2-dimensional intervals or rectangles, where the first dimension is related to a tank assignment and the second dimension to performing concurrent setup work.

More specifically, the x-axis is divided into disjoint segments, one for each shuttle coater. Each of the segments covers the fixed chosen coil sequence, and an "interval" in that sequence corresponds to a maximal sequence of consecutive coils run from the same tank on the corresponding coater. Consequently, intersecting intervals *conflict* in the sense of representing an infeasible tank assignment. The y-axis is similarly divided into segments for each team that can perform setup work. Here, an "interval" in the coil sequence corresponds to a time interval during which setup work is performed concurrently to production. If every team has the ability to perform all possible setup work, we have identical rows of the different teams in the y-direction. In order to properly perform setup work concurrently, the tank on the respective coater must not be changed during this interval, i.e., the rectangle's setup interval must be contained in its tank interval with respect to the segment-based time axis. Intersecting setup intervals conflict since one team can only perform concurrent setup work at one coater at a time. See Fig. 2 for an example.

Finally, we assign weights to the rectangles which represent the (possibly negative) cost savings by the corresponding partial tank assignment and the concurrent setup work performed, compared to running all coils on the same tank without concurrent setup work. The fixed-sequence allocation problem is then equivalent to finding pairwise non-conflicting rectangles with maximum total weight (in a set of properly defined rectangles). This problem is closely related—however, neither being a special case nor a generalization—to the Maximum Weight Independent Set Problem on a restricted class of so called 2-union graphs, which are of interest in their own right. Due to the limited space, here we only focus on the tank assignment and concurrent setup work scheduling. In [18], we provide additional results for the latter problem.

Algorithm for the Underlying Allocation Problem. Studying the allocation problem, we observed that even this fixed-sequence subproblem is strongly NP-hard when the number of shuttle coaters is assumed to be part of the problem input.

Theorem 1 ([18]). *For the number of shuttle coaters m being part of the problem input, the fixed sequence tank assignment and concurrent setup work scheduling problem is strongly NP-hard for any fixed number $r < m$ of work teams for performing setup work.*

This result shows that we cannot assume a polynomial time algorithm for the allocation problem with an arbitrary number of shuttle coaters m. However, in practice, this number is usually rather small so that it is reasonable to assume m to be constant. Under this assumption, we could show that the problem is in P.

Theorem 2 ([18]). *When the number of shuttle coaters m is constant, and the ratio $\max_{j \in C} p_j / \tau$ is polynomial in the size of the input I, denoting by p_j the duration of a coil j and by τ the greatest common divisor of all setup durations, then the fixed sequence tank assignment and concurrent setup work scheduling problem can be solved in polynomial time, even if the number of work teams r is part of the input.*

Even though the designed algorithm has a polynomial runtime, it is far too slow to be used as subroutine in our overall framework. Due to its dynamic programming nature, even a single run of the algorithm for a very small number of coils and shuttle coaters exceeds our overall runtime limit.

Still, the graph theoretical model inspires a fast and good heuristic for the allocation problem which would not have been possible without the above investigations. The complexity of our exact algorithm stems from the need to consider interval selections for all coaters simultaneously in order to ensure that savings from all selected setup intervals can be realized by the scarce work resource(s). Intuitively, the probability that concurrent setup work on different cells can be scheduled feasibly, i.e., one setup at a time, increases with the length of the associated tank interval. This is our core idea for computing good tank assignments heuristically. Instead of considering all coaters at once, we consider them separately. Recall that savings from tank intervals for different coaters can be realized independently in any case. Now, instead of explicitly considering all possible saving rectangles belonging to some tank interval I, we assume that during a certain fraction α of Is length setup work can be performed, and, at this point, we do not care when exactly it is performed and even, if it can be performed that way at all.

Modeling this idea, we define new weights for the intervals. With these modified weights, it suffices to consider tank intervals alone. As a consequence, computing a tank assignment reduces to finding a Maximum Weight Independent Set in an interval graph, which can be dealt with very efficiently; see e.g., [26]. In order to compute a feasible concurrent setup allocation for this tank assignment, we use an earliest-deadline-first strategy as a simple scheduling rule for properly defined deadlines.

Additionally, we also consider the allocation rule which was previously in use at Salzgitter Flachstahl: Whenever subsequent coils have different colors, switch the tank. If the new tank does not contain the required color, a color change on that tank becomes necessary. We refer to this rule as FIFO.

Results. Embedding the above algorithm for the allocation problem into the generic Algorithm 1.1, altogether, we develop a practical heuristic which solves the overall planning problem and computes detailed production schedules for the coil coating line. Besides the allocation algorithm, also the construction of the initial population plays a major role for the overall quality of the algorithm. It generates several very different solutions which are good with respect to only some of the many side constraints. The quality of our schedules is assessed with the help of an integer program which we solve by branch-and-price.

Our algorithm has been added to PSI Metals' planning software suite, and is currently in use for a coil coating line with shuttle coaters at Salzgitter Flachstahl, Germany. At the time of its introduction, it yielded an average reduction in makespan by over 13% as compared to the previous manual planning process. In addition, our lower bounds suggest that the makespan of the solutions computed by our algorithm is within 10% of the optimal makespan

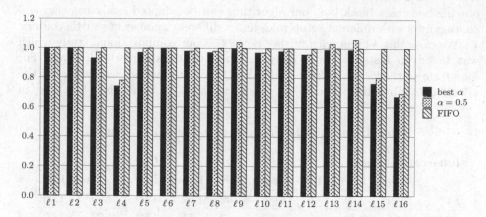

Fig. 3. Comparison of normalized non-productive time (cost) included in our solutions for the long-term instances $\ell 1, \ldots, \ell 16$. From left to right, we show the results of our algorithm using the Independent Set Heuristic with best choice of parameter α, using the same heuristic with fixed choice of $\alpha = 0.5$, and using the FIFO rule.

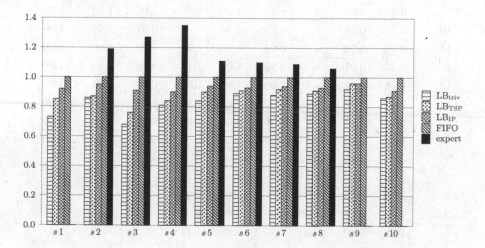

Fig. 4. Comparison of normalized bounds and makespans for representative short-term instances $s 1, \ldots, s 10$ when restricted to FIFO tank assignment. Bounds displayed from left to right are LB_{triv}, the sum of coil processing times, LB_{TSP}, the sum of processing times plus local cost in an optimal, local cost based ATSP solution, and our IP bound LB_{IP}. Makespans are given for our solutions obtained using the FIFO online rule for scheduling, as well as for a reference solution devised by expert human planners if available.

for typical instances[2]. Since most setup cost calculations are incorporated into our methods as a black box, our algorithm can be adapted easily to other coil coating lines with different setup rules and a different number of shuttle coaters.

We close this section with further details on our computational study which was based on instances from daily production at Salzgitter Falchstahl. For long-term instances which cover roughly 72 h of production, we compared the allocation subroutine based on Independent Set for the best α in

Table 1. Normalized results corresponding to Figs. 3 and 4.

Instance	Makespan						Cost		
	LB_{triv}	LB_{TSP}	LB_{IP}	Best α	FIFO	Expert	Best α	$\alpha = 0.5$	FIFO
$\ell 1$	0.71	0.79	–	1.00	1.00	1.22	1.00	1.00	1.00
$\ell 2$	0.69	0.78	–	1.00	1.00	1.11	1.00	1.00	1.00
$\ell 3$	0.71	0.76	–	0.98	1.00	–	0.93	0.97	1.00
$\ell 4$	0.84	0.89	–	0.96	1.00	1.00	0.74	0.78	1.00
$\ell 5$	0.73	0.78	–	0.99	1.00	–	0.97	1.00	1.00
$\ell 6$	0.82	0.88	–	1.00	1.00	1.33	1.00	1.00	1.00
$\ell 7$	0.58	0.79	–	0.99	1.00	–	0.98	0.98	1.00
$\ell 8$	0.71	0.77	–	0.99	1.00	–	0.97	0.98	1.00
$\ell 9$	0.82	0.86	–	1.00	1.00	–	1.00	1.04	1.00
$\ell 10$	0.73	0.78	–	0.99	1.00	–	0.97	0.97	1.00
$\ell 11$	0.71	0.76	–	0.99	1.00	1.29	0.98	1.00	1.00
$\ell 12$	0.75	0.81	–	0.99	1.00	1.28	0.96	0.96	1.00
$\ell 13$	0.78	0.81	–	1.00	1.00	1.31	0.99	1.03	1.00
$\ell 14$	0.78	0.82	–	1.00	1.00	1.23	0.99	1.06	1.00
$\ell 15$	0.83	0.85	–	0.96	1.00	1.00	0.76	0.80	1.00
$\ell 16$	0.82	0.85	–	0.94	1.00	1.00	0.67	0.69	1.00
$s 1$	0.73	0.85	0.92	–	1.00	–	–	–	–
$s 2$	0.86	0.87	0.95	–	1.00	1.19	–	–	–
$s 3$	0.68	0.76	0.91	–	1.00	1.27	–	–	–
$s 4$	0.81	0.84	0.90	–	1.00	1.35	–	–	–
$s 5$	0.84	0.90	0.94	–	1.00	1.11	–	–	–
$s 6$	0.89	0.91	0.93	–	1.00	1.10	–	–	–
$s 7$	0.88	0.92	0.94	–	1.00	1.09	–	–	–
$s 8$	0.89	0.91	0.93	–	1.00	1.06	–	–	–
$s 9$	0.92	0.96	0.96	–	1.00	–	–	–	–
$s 10$	0.86	0.87	0.91	–	1.00	–	–	–	–

---

[2] This success has made this contribution a finalist of the 2009 EURO Excellence in Practice Award.

{0.1, 0.2, . . . , 0.8} and for fixed $\alpha = 0.5$, and the FIFO rule; see Fig. 3 and Table 1. The Independent Set Heuristic outperformed FIFO on 12 of the 16 instances, reducing cost by up to 30% (over FIFO). This translates to makespan savings of up to 6%. When fixing α to 0.5, the Independent Set Heuristic remains similarly superior to FIFO on 8 instances, while incurring an increase in makespan of at most 1% in four cases.

For short-term instances of roughly 24 h, we succeeded in computing lower bounds with respect to the FIFO rule by our IP approach. Yet, we did not solve all instances to integer optimality, so the lower bound is certainly improvable. In Fig. 4 and Table 1 we compare different lower bounds—a trivial bound LB_{triv}, a TSP-based bound LB_{TSP}, and the mention IP bound LB_{IP}, see [18] for further details—with our FIFO results and the solutions we were provided with by Salzgitter Flachstahl. The results show makespan reductions of up to more than 30%. The superiority of the Independent Set Heuristic to FIFO is less significant in short-term planning, so that we focused on FIFO for the short instances.

4   Integrated Sequencing and Allocation in Dairy Industry

After discussing the coil coating problem in the previous section, we now apply our approach from Sect. 2 to a second application from dairy industry. In particular, we consider the planning of filling lines, which are usually the bottleneck in dairy production. The optimization problem incorporates again sequencing and allocation components: For a given set of dairy products—e.g., cream or yogurt—one aims at a filling order with minimum makespan. However, the makespan does not only contain the products' processing times and local sequence-dependent setup times. Additional setup work or waiting periods may become necessary due to different types of side constraints. Several options of performing setup work and inserting waiting periods lead to an allocation problem for the fixed filling sequence. As for the coil coating, we only give a brief description of the problem, further details can be found in [12].

High-Level Problem Formulation. We aim at scheduling a set of dairy products or jobs which are specified by their processing time, their volume, their base (e.g., yogurt or cream), an optional flavor and an article number. Sequence-dependent setups are necessary in between consecutive jobs in the sequence. Roughly speaking, the more parameters of consecutive jobs differ, the more setup work is necessary. Additionally, to meet hygiene regulations, the filling machinery has to be cleaned and sterilized at certain intervals to which we refer to as *cleanings*. Even though some of the sequence-dependent setups may already involve cleanings, further cleanings are usually required to satisfy the regulation. There are two ways of allocating additional cleanings: By job preemption, they can be inserted at any point in the schedule, or, alternatively, they can replace less exhaustive sequence-dependent setups. This yields a tradeoff between keeping the total number of cleanings small by postponing them as long as possible, and saving additional setup time by replacing sequence-dependent setups by cleanings, but

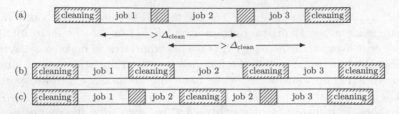

Fig. 5. The schedule (a) contains only jobs and sequence-dependent setups, two of them being cleanings. The time between the cleanings exceeds the maximum time lag Δ_{clean} between cleanings. The schedules (b) and (c) below are feasible with respect to cleaning time lags. The former replaces two existing sequence-dependent setups while the latter preempts the middle job to require only one additional cleaning.

possibly requiring more additional cleanings in total; see Fig. 5. Thus, setups are not only determined by their neighboring jobs, but also by scheduling decisions taken in view of the entire sequence.

Moreover, there are waiting periods which are caused by filling constraints of certain products, or other technical constraints. A typical example for the latter is the production of cream. The pretreated milk is processed in a special tank before the finished cream can be filled into cups. Due to the limited tank size, cream jobs have to be partitioned into different tank fillings. Resulting from the preparation time in the tank and the later cleaning time, the filling machine may be idle, awaiting the first cream job of a new tank filling. We refer to constraints of this type as *capacity constraints*.

Related Work. Previous work on the planning of filling lines in dairy industry mainly focuses on mixed-integer linear programming (MILP) models to minimize weighted cost functions comprising setup cost, storage cost and others. Recently, Kopanos et al. [19] proposed a MILP model for parallel filling machines, taking into account machine- and sequence-dependent setups, due dates, and certain tank capacities. Further MILP models were proposed for different problem settings, e.g., by [11, 22, 23]. For the common planning horizon of one week, the models in [11, 19] could compute optimal solutions at very low computational cost. Based on different relaxations of their model, Marinelli et al. [23] also proposed an algorithm which is heuristically almost optimal, but at a high computational expense. However, different to our problem, regular cleanings of the filling line are performed at the end of the day in all these papers. In combination with their very restricting sequencing constraints, this turns these problems into packing problems rather than into sequencing problems as in our case. The results cannot be adapted to our problem, and, to the best of our knowledge, flexible cleaning scheduling has not been considered in the scheduling literature yet.

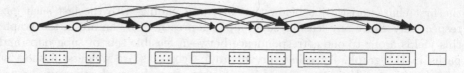

Fig. 6. A sequence of characteristic (dotted) and non-characteristic (white) tasks and chosen subgroups (with some intermediate spacing for clarity). Vertices and arcs are defined as described below. The marked path corresponds to the choice of subgroups as shown in the three boxes.

Model for the Underlying Allocation Problem. The side constraints in the dairy problem exhibit a very similar structure. In fact, we can formulate all of them as *generalized capacity constraints*. Such a constraint is defined in a similar way as normal capacity constraints. For the normal capacity constraints, one is looking for consecutive subgroups of jobs of a certain characteristic, in total not exceeding a given capacity limited—e.g., cream jobs of at most 10000 liters— where *consecutive* means that there are no other jobs of the same characteristic between the jobs of the subgroup. However, jobs of other characteristics may occur inbetween; see the bottom part of Fig. 6. While for the normal capacity constraints there are certain time lags to be observed between the subgroups, in the generalized case, there are general cost. Moreover, generalized capacity constraints do not only take into account jobs but also setup work or waiting periods, jointly referred to as *tasks*.

Note that the dairy problem involves multiple generalized capacity constraints, and a subgroup's cost with respect to a certain constraint strongly depends on the chosen subgroups corresponding to other constraints, since the related costs may represent setup or waiting tasks that have to be added to the sequence. E.g., if inserting cleanings into the schedule for the cleaning constraint, actual waiting times of other constraints may decrease, and so would the cost of the associated subgroups with respect to that constraint. Thus, by optimally satisfying one constraint after the other, in general, we do not obtain an optimum schedule.

Algorithm for the Underlying Allocation Problem. The special case in which the allocation problem consists of a single generalized capacity constraint can be solved with standard dynamic programming. In order to make use of efficient software libraries, we formulate the problem as shortest path problem. For a given sequence of tasks, we define a graph whose vertices lie on the natural time axis of the sequence; see Fig. 6. For each characteristic task, we place a vertex at its start time. Additionally, a vertex is placed at the completion time of the last characteristic task. We insert an arc pointing from one vertex to another if and only if it is a forward arc in the natural sense of time, and the tasks below that arc, i.e., the tasks that lie in the time interval covered by the arc, form a feasible subgroup with respect to the constraint under consideration. More precisely, the subgroup is formed by the subsequence of tasks, starting with the first characteristic job below that arc, and ending with the last such job, respectively. Since any feasible subgroup is represented by an arc in the graph, this yields a one-to-one correspondence between feasible sets of subgroups and paths in the graph. By using the subgroups costs as arc weights, a shortest path in this graph corresponds to a minimum cost choice of subgroups. Since the number of arcs is quadratic in the number of characteristic tasks, Dijkstras shortest path algorithm solves the problem in polynomial time.

If there is more than one generalized capacity constraint, it might as well be possible to solve the allocation problem (under some restrictions) with an adapted dynamic programming approach. However, since we do not expect such an algorithm to be sufficiently fast for being used as subroutine in Algorithm 1.1, our practical focus was more on fast heuristic algorithms. Aiming for at most a quadratic runtime, we examined algorithmic variants which satisfy the different generalized capacity constraints one after the other. The variants differ as well in the order in which the constraints are satisfied as in the subroutine that is used in each constraint step. While in the first setting, we use the above shortest path approach (SP) which satisfies each constraint at minimum cost, in the second variant we use a greedy subroutine (GREEDY) which adds tasks to subgroups as long as the given capacity is not exceeded. The keywords FIRST and LAST indicate that the cleaning constraint is satisfied before and after the generalized capacity constrainst, respectively. In our tests we consider all four variants GREEDYFIRST, GREEDYLAST, SPFIRST and SPLAST.

Currently, the complexity of the allocation problem is still open. For the overall dairy problem with its sequencing and allocation components, already a very restricted special case can be shown to be strongly NP-hard [12].

Results. We evaluate all of the above variants of the allocation algorithm as subroutine in the framework of Algorithm 1.1. For the initial population, we choose a sequence which is optimal with respect to local setup work but disregards all other constraints, and in addition, we choose several random sequences. The former is computed utilizing LKH. Ignoring all cost due to generalized capacity constraints, i.e., accounting only for the job's processing times and local setup times, the LKH solution provides a lower bound for the optimum cost of the general problem. This bound is used to evaluate our solutions.

Our industrial partner Sachsenmilch provided us with actual production data which we used to generate 2880 additional realistic, but in different ways more extreme, problem instances. For the data of Sachsenmilch's current production, our approach achieved an optimality gap of only 2%, and of roughly 15% on average for the generated instances. Due to the presumed weakness of the lower bound, the gap of 2% makes us believe that the former solution is in fact optimal. Comparing the GREEDY and the optimal SP subroutine, it turned out that GREEDY generates in fact better results. A reason for this might be that in contrast to the schedules computed with the shortest path subroutine SP, the GREEDY schedules never preempt jobs. This property seems to allow to adapt to "later" constraints more easily.

In the remainder of this section, we will provide some more detailed computational results. In our evaluations, we filter the instances for one particular parameter, i.e., an instance belongs to such a setting if this parameter is satisfied, no matter how the remaining parameters are set. We examined different base types (all, yoghurt, cream), processing times (original, small, large), volumes of the cream tank (original, small), and lengths of the cleaning intervals (original, small). The setting *original* comprises all instances that were generated according to the actual parameters by Sachsenmilch. The results are shown in Fig. 7 and the corresponding Table 2.

We compared the better result of the two greedy algorithms GREEDYLAST and GREEDYFIRST (BESTGREEDY) with the better result of the two shortest path variants SPLAST and SPFIRST (BESTSP). BESTGREEDY performed always better than BESTSP, up to 4% in makespan average. In fact, for the orginial test setting, BESTSP is never better than BESTGREEDY. If utilizing BESTSP, the worst case gap compared to BESTGREEDY is 6%, whereas conversely this gap may grow up to 210%. As expected, also the runtimes differ greatly; see again Table 2. While BESTSP produces better solutions for the allocation subproblem, the faster running time of BESTGREEDY allows for more iterations of the genetic algorithm and better overall solutions.

If comparing the two greedy and the two shortest path approaches with each other, the difference of the performance is not that striking. The average performance is almost identical. However, considering the worst case behavior, in both cases the LAST-algorithm is only about 6% worse than the FIRST-algorithm, where conversely, the gap can attain 15%. This may be due to the advantage of scheduling the inflexible cleanings as late as possible.

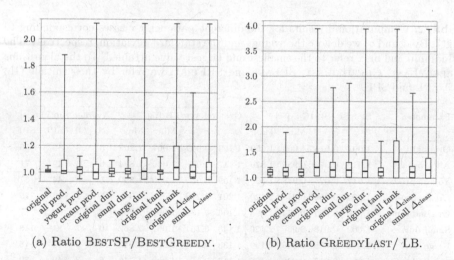

(a) Ratio BESTSP/BESTGREEDY.   (b) Ratio GREEDYLAST/ LB.

Fig. 7. Computational results for the different parameter classes. In the box plots, the average value $\varnothing$ is marked as fat line and the surrounding rectangle shows the interval of the standard deviation σ. The minimum and maximum value is marked at the lower and upper end of the thin vertical line, respectively.

The optimality gap computed with the LKH lower bound is roughly 15%, and it is slightly better for the greedy algorithms than for the shortest path algorithms. Exemplary for our best algorithm GREEDYLAST, the results are shown in Fig. 7(b). For all algorithms, the optimality gap is much worse for a small tank size. However, in this case the lower bound performs very poorly, so that one may assume that this gap is far away from being tight.

5   Outlook on Fairness Objectives

In this final section, we give an outlook on another objective which also plays an important role in production planning. While in makespan minimization, the focus is on maximizing the throughput from a rather global perspective of the machine operator, this new objective aims at schedules which are somehow fair to the jobs. In the related scheduling model, in addition to its actual parameters, a job j has got a weight w_j, which represents its priority. For some non-decreasing cost function f, the goal is then to minimize

$$\sum_{j \in J} w_j f(C_j), \tag{1}$$

where C_j denotes the completion time of job j. In the literature, scheduling problems with this objective are often referred to as *generalized min-sum scheduling*. For instance, setting the cost function to $f(t) := t^p$ for some $p \geq 1$ and applying the p-th square root to the optimal objective function value, we minimize the L_p-norm, a classic fairness measure.

Table 2. Computational results for the different parameter classes corresponding to Fig. 7. By $\varnothing$ and σ, we denote the average and the standard deviation, respectively. The values min and max refer to the smallest and largest value attained. To the algorithms GREEDYLAST, GREEDYFIRST, SPLAST and SPFIRST, we refer by their initials GL, GF, SPL and SPF.

Parameter	BESTSP/BESTGR. [%]				GREEDYLAST/LB [%]				Average runtime [s]			
	$\varnothing$	σ	Min	Max	$\varnothing$	σ	Min	Max	GL	GF	SPL	SPF
Original	1.007	0.015	1.000	1.050	1.105	0.057	1.007	1.203	5	6	39	55
All prod.	1.014	0.079	0.985	1.876	1.117	0.085	1.000	1.894	6	6	49	71
Yogurt prod.	1.017	0.025	0.948	1.120	1.109	0.065	1.010	1.390	5	6	44	67
Cream prod.	1.003	0.058	0.946	2.120	1.201	0.278	1.031	3.943	7	8	53	75
Original dur.	1.010	0.020	0.974	1.093	1.146	0.152	1.007	2.769	7	7	53	77
Small dur.	1.009	0.018	0.956	1.120	1.148	0.151	1.007	2.862	10	11	81	118
Large dur.	1.014	0.098	0.946	2.120	1.133	0.220	1.000	3.943	2	2	13	18
Original tank	1.007	0.017	0.946	1.120	1.117	0.081	1.000	1.716	6	7	49	72
Small tank	1.043	0.155	0.949	2.120	1.321	0.411	1.000	3.943	6	7	49	65
Original Δ_{clean}	1.011	0.045	0.956	1.596	1.121	0.108	1.000	2.671	6	7	49	69
Small Δ_{clean}	1.012	0.069	0.946	2.120	1.164	0.224	1.000	3.943	6	7	49	71

In cooperation with Knorr-Bremse, a producer of train brakes, we considered a problem that involved such a fairness objective: In order to guarantee reliability, different parts of the brakes have to be checked in some measuring and quality control station before the item can be produced in larger quantities. Depending on the type of production line that the item came from (referred to as *machine* henceforth), that machine is idle until the test result is known, or, in the other case, the machine can work on some other job in the meantime. Aiming at avoiding idle times on machines, the job weights (priorities) are chosen accordingly to represent the different machines' behavior. Moreover, the goal was to avoid never ending waiting times for jobs of lower priority. For this reason, a non-linear cost function should be considered. In discussions with the practitioners, the L_p-norm turned out to be a good measure for the quality of schedules for the measuring and quality control station.

Coming back to the general integrated sequencing and allocation problem with objective (1), we started working on a subproblem which plays a similar role as the ATSP for the makespan variant. While the ATSP ignores all global setups, the generalized min-sum variant ignores all side constraints but the prioritized sequencing: We consider a classic single machine scheduling problem in which jobs have only two parameters, processing time and weight. Algorithms for the ATSP were utilized to produce good processing sequences for the initial population as well as for the computation of lower bounds. Likewise, we hope to use efficient algorithms for the classic single machine scheduling problem.

We could already provide a tight analysis of the approximation guarantee of Smith's rule [32]—a simple heuristic which sorts the jobs according to their ratio of weight and processing time—for every fixed convex or concave cost function [17]. For the L_2-norm and the L_3-norm, for instance, the analysis yields approximation factors of only 1.143 and 1.207, respectively. In terms of exact algorithms for $f(t) := t^2$, we showed some properties of the order of the jobs in optimal schedules. Using this property allows us to significantly reduce the search space of algorithms, so that we are also able to compute optimal solutions to the classic single machine scheduling problem, however, at a much higher runtime than Smith's rule [16].

6   Conclusions

Our work shows that good algorithm engineering requires theory and practice to go hand in hand. Problems from industry inspire new interesting theoretical questions which, on the one hand, naturally enhance problems that have been studied in theory, while on the other hand, they provide essential insights for designing better practical algorithms. In the end, algorithm engineering allows to achieve results which would not have been possible with standard techniques and from which in fact both theory and practice benefit.

References

1. Aarts, E., Lenstra, J.K. (eds.): Local Search in Combinatorial Optimization. Wiley, Hoboken (1997)
2. Akkerman, R., van Donk, D.P.: Analyzing scheduling in the food-processing industry: structure and tasks. Cogn. Technol. Work 11, 215–226 (2009)
3. Allahverdi, A., Ng, C.T., Cheng, T.C.E., Kovalyov, M.Y.: A survey of scheduling problems with setup times or costs. Eur. J. Oper. Res. 187, 985–1032 (2008)
4. Applegate, D.L., Bixby, R.E., Chvátal, V., Cook, W.J.: The Traveling Salesman Problem: A Computational Study. Princeton University Press, Princeton (2006)
5. Balas, E., Simonetti, N., Vazacopoulos, A.: Job shop scheduling with setup times, deadlines and precedence constraints. J. Sched. 11, 253–262 (2008)
6. Bampis, E., Guinand, F., Trystram, D.: Some models for scheduling parallel programs with communication delays. Discrete Appl. Math. 72, 5–24 (1997)
7. Bellabdaoui, A., Teghem, J.: A mixed-integer linear programming model for the continuous casting planning. Int. J. Prod. Econ. Theor. Issues Prod. Sched. Control Plan. Control Supply Chains Prod. 104, 260–270 (2006)
8. Claassen, G.D.H., van Beek, P.: Planning and scheduling. Eur. J. Oper. Res. 70, 150–158 (1993)
9. Cowling, P.: A flexible decision support system for steel hot rolling mill scheduling. Comput. Ind. Eng. 45, 307–321 (2003)
10. Delucchi, M., Barbucci, A., Cerisola, G.: Optimization of coil coating systems by means of electrochemical impedance spectroscopy. Electrochim. Acta 44, 4297–4305 (1999)
11. Doganis, P., Sarimveis, H.: Optimal production scheduling for the dairy industry. Ann. Oper. Res. 159, 315–331 (2008)
12. Gellert, T., Höhn, W., Möhring, R.H.: Sequencing, scheduling for filling lines in dairy production. Optim. Lett. 5, 491–504 (2011). Special issue of SEA 2011
13. Gutin, G., Punnen, A.P.: The Traveling Salesman Problem and Its Variations. Springer, New York (2002)
14. Hall, N.G., Potts, C.N., Sriskandarajah, C.: Parallel machine scheduling with a common server. Discrete Appl. Math. 102, 223–243 (2000)
15. Helsgaun, K.: An effective implementation of the Lin-Kernighan traveling salesman heuristic. Eur. J. Oper. Res. 126, 106–130 (2000)
16. Höhn, W., Jacobs, T.: An experimental and analytical study of order constraints for single machine scheduling with quadratic cost. In: Proceedings of the 14th Meeting on Algorithm Engineering and Experiments (ALENEX), pp. 103–117. SIAM (2012)
17. Höhn, W., Jacobs, T.: On the performance of Smith's rule in single-machine scheduling with nonlinear cost. ACM Trans. Algorithms (2014, to appear)
18. Höhn, W., König, F.G., Lübbecke, M.E., Möhring, R.H.: Integrated sequencing and scheduling in coil coating. Manage. Sci. 57, 647–666 (2011)
19. Kopanos, G.M., Puigjaner, L., Georgiadis, M.C.: Optimal production scheduling and lot-sizing in dairy plants: the yogurt production line. Ind. Eng. Chem. Res. 49, 701–718 (2010)
20. Kopanos, G.M., Puigjaner, L., Georgiadis, M.C.: Efficient mathematical frameworks for detailed production scheduling in food processing industries. Comput. Chem. Eng. 42, 206–216 (2012)
21. Koulamas, C., Kyparisis, G.J.: Single-machine scheduling problems with past-sequence-dependent setup times. Eur. J. Oper. Res. 187, 1045–1049 (2008)

22. Lütke Entrup, M., Günther, H.-O., van Beek, P., Grunow, M., Seiler, T.: Mixed-integer linear programming approaches to shelf-life-integrated planning, scheduling in yoghurt production. Int. J. Prod. Res. **43**, 5071–5100 (2005)
23. Marinelli, F., Nenni, M., Sforza, A.: Capacitated lot sizing and scheduling with parallel machines and shared buffers: a case study in a packaging company. Ann. Oper. Res. **150**, 177–192 (2007)
24. Meloni, C., Naso, D., Turchiano, B.: Multi-objective evolutionary algorithms for a class of sequencing problems in manufacturing environments. In: Proceedings of the IEEE International Conference on Systems, Man and Cybernetics, pp. 8–13 (2003)
25. Meuthen, B., Jandel, A.-S.: Coil Coating. Vieweg, Wiesbaden (2005)
26. Möhring, R.H.: Algorithmic aspects of comparability graphs and interval graphs. In: Rival, I. (ed.) Graphs and Order. NATO ASI Series, vol. 147, pp. 41–101. Springer, Heidelberg (1985)
27. Mühlenbein, H., Gorges-Schleuter, M., Krämer, O.: Evolution algorithms in combinatorial optimization. Parallel Comput. **7**, 65–85 (1988)
28. Papadimitriou, C.H., Yannakakis, M.: The traveling salesman problem with distances one and two. Math. Oper. Res. **18**, 1–11 (1993)
29. Perrello, B.D., Kabat, W.C.: Job-shop scheduling using automated reasoning: a case study of the car-sequencing problem. J. Autom. Reason. **2**, 1–42 (1986)
30. Rekieck, B., De Lit, P., Delchambre, A.: Designing mixed-product assembly lines. IEEE Trans. Robot. Autom. **16**, 268–280 (2000)
31. Sahni, S., Gonzalez, T.F.: P-complete approximation problems. J. ACM **23**, 555–565 (1976)
32. Smith, W.E.: Various optimizers for single-stage production. Naval Res. Logist. Q. **3**, 59–66 (1956)
33. Solnon, C., Cung, V.D., Nguyen, A., Artigues, C.: The car sequencing problem: overview of state-of-the-art methods and industrial case-study of the ROADEF'2005 challenge problem. Eur. J. Oper. Res. **191**, 912–927 (2008)
34. Tang, L., Liu, J., Rong, A., Yang, Z.: A review of planning and scheduling systems and methods for integrated steel production. Eur. J. Oper. Res. **133**, 1–20 (2001)
35. Tang, L., Wang, X.: Simultaneously scheduling multiple turns for steel color-coating production. Eur. J. Oper. Res. **198**, 715–725 (2009)
36. van Dam, P., Gaalman, G., Sierksma, G.: Scheduling of packaging lines in the process industry: an empirical investigation. Int. J. Prod. Econ. **30–31**, 579–589 (1993)

Engineering a Bipartite Matching Algorithm in the Semi-Streaming Model

Lasse Kliemann[✉]

Department of Computer Science, Kiel University,
Christian-Albrechts-Platz 4, 24118 Kiel, Germany
lki@informatik.uni-kiel.de

Abstract. We describe the Algorithm Engineering process for designing a pass-efficient semi-streaming algorithm for the bipartite maximum matching problem, using the augmenting paths technique. This algorithm was first published by the author at SEA 2011. This text not only discusses the algorithm, but also describes how Algorithm Engineering helped to invent and refine it.

Outline. The first three sections are an introduction to the matching problem in the semi-streaming model. In Sect. 4, a previous semi-streaming algorithm is described, which grows augmenting paths using the *position limiting* technique [11] and is the basis for all further development. Experimental studies start in Sect. 5, with a description of the experimental setup. In Sect. 6, the algorithm from [11] is analyzed experimentally, which marks the start of the Algorithm Engineering process. The remaining sections describe how this process leads to a new semi-streaming algorithm [21], which experimentally shows much smaller pass counts than the previous one but maintains many of the (good) theoretical properties.

1 Bipartite Matching

Let $G = (A, B, E)$ be a bipartite graph, i.e., $V := A \cup B$ is the set of vertices, $A \cap B = \varnothing$, and $E \subseteq \{\{a, b\}; a \in A \wedge b \in B\}$ are the edges. Denote $n := |V|$ and $N := |E|$ the number of vertices and edges, respectively. The *density* of the graph is $D := \frac{|E|}{|A||B|} \in [0, 1]$, i.e., the ratio of the number of edges to the maximum possible number of edges given the two sets A and B. A *matching* of G is a set $M \subseteq E$ such that $m \cap m' = \varnothing$ for all $m, m' \in M$ with $m \neq m'$. A matching M of G is called *inclusion-maximal* a.k.a. *maximal* if $M \cup \{e\}$ is not a matching for all $e \in E \setminus M$, i.e., if we cannot add another edge without destroying the matching property. A matching M^* of G is called a *cardinality-maximal* a.k.a. *maximum matching* if $|M'| \leqslant |M^*|$ for all matchings M' of G. If M^* is a maximum matching and $\rho \leqslant 1$ is a real number, then a matching M is called a *ρ-approximation* if $|M| \geqslant \rho |M^*|$.

A maximal matching is easy to find algorithmically: just start with $M = \varnothing$ then consider all the edges in an arbitrary order and add an edge to M if this

© Springer International Publishing AG 2016
L. Kliemann and P. Sanders (Eds.): Algorithm Engineering, LNCS 9220, pp. 352–378, 2016.
DOI: 10.1007/978-3-319-49487-6_11

does not destroy the matching property. However, a maximal matching needs not to be a maximum matching, although it is a well-known fact that it is a $\frac{1}{2}$-approximation, i. e., $|M| \geqslant \frac{1}{2} |M^*|$ for all maximal matchings M and a maximum matching M^* of G. Devising polynomial-time algorithms computing maximum matchings in bipartite graphs is a classical and important problem in Combinatorial Optimization.

We introduce some terminology relative to a matching M. A vertex v is called *matched* or *covered* if there is $m \in M$ such that $v \in m$, and v is called *free* otherwise. For $X \subseteq V$, denote free(X) the free vertices in X. We often denote free vertices with lower-case Greek letters. If $v \in V$ is matched, then there is a unique vertex M_v with $\{v, M_v\} \in M$; we call M_v the *mate* of v. An edge $e \in E$ is called a *matching edge* if $e \in M$ and *free* otherwise. Note that the end-vertices of a free edge need not to be free vertices. A path is called *alternating* if it traverses matching edges and free edges alternately. An alternating path where both of the end-vertices are free, is called an *augmenting path*. Clearly, augmenting paths have an odd number of edges and in the bipartite case can be written in the form $(\alpha, e_1, b_1, m_1, a_1, \ldots, m_t, a_t, e_{t+1}, \beta)$, where $\alpha \in$ free(A), $\beta \in$ free(B), $a_1, \ldots, a_t \in A$, $b_1, \ldots, b_t \in B$, $m_1, \ldots, m_t \in M$, and $e_1, \ldots, e_{t+1} \in E \setminus M$ for some $t \in \mathbb{N}$. The length (number of traversed edges) of this path is $2t + 1$. We always denote paths as a sequence of vertices and edges; this notation has redundancy but will be helpful.

By exchanging matching edges for free edges along an augmenting path, a new matching is obtained from M with cardinality $|M| + 1$. Hence a maximum matching does not admit any augmenting paths. By Berge's theorem [5], the converse also holds: when M admits no augmenting paths, then M is a maximum matching. This theorem also holds for general (not necessarily bipartite) graphs.

Finding augmenting paths, or determining that there exist none, is algorithmically easy in the bipartite case. The simplest algorithm starts at an arbitrary free vertex $\alpha \in$ free(A) and does a modified breadth-first search (BFS) which only considers free edges when moving from A to B and only matching edges when moving from B to A.[1] As soon as another free vertex β is found (it will be in B then), an augmenting path $(\alpha, \ldots, \beta)$ is constructed by following the BFS layers. In each such iteration, the matching grows by 1, so in $\mathcal{O}(n)$ iterations a maximum matching is reached, resulting in a bound of $\mathcal{O}(nN) = \mathcal{O}(n^3)$ on the total runtime. If no free vertex β is found, then the next free vertex from free(A) is tried as a starting point for BFS. It can be seen easily that if free(B) is not reachable from free(A) by modified BFS, then there are no augmenting paths. A similar argument cannot be made for general graphs, causing the situation there to be much more complicated. We will not consider general graphs in this work.

---

[1] This can also be achieved by orienting the edges

$$\boldsymbol{E} := \{(a, b) \in A \times B; \{a, b\} \in E \setminus M\} \cup \{(b, a) \in B \times A; \{a, b\} \in M\}$$

and then using normal BFS in this directed bipartite graph $(A, B, \boldsymbol{E})$.

For the bipartite case, the total runtime bound can be reduced to $\mathcal{O}\left(\sqrt{n}N\right) = \mathcal{O}\left(n^{5/2}\right)$ using the algorithm by Hopcroft and Karp [18] (a good description can also be found in [24]). Modified BFS is done starting from free(A), more precisely, we start with the BFS queue containing all vertices from free(A). This BFS constructs a layered graph, with free(A) being one of the end-layers. As soon as a layer with vertices from free(B) is found, depth-first search (DFS) in the layered graph is used to find an *inclusion-maximal set $\mathcal{A}$ of pairwise vertex-disjoint shortest augmenting paths*. We say *disjoint* in the following meaning "pairwise vertex-disjoint". Since the paths in $\mathcal{A}$ are disjoint, they can be used simultaneously to improve the matching, the new matching will be of cardinality $|M| + |\mathcal{A}|$. One such BFS and then DFS is called a *phase* and takes $\mathcal{O}\left(N\right)$ time. The achievement of Hopcroft and Karp lies in recognizing and proving that there are only $\mathcal{O}\left(\sqrt{n}\right)$ phases. An implementation of their algorithm particularly suited for dense graphs having a bound of $\mathcal{O}\left(n^{1.5}\sqrt{N/\log n}\right)$ on the runtime was later given by Alt et al. [3].

There is also a randomized algorithm by Mucha and Sankowski [27], which runs in $\mathcal{O}\left(n^{\omega}\right)$, where ω depends on the running time of the best known matrix multiplication algorithm; currently known is $\omega < 2.38$.

On the experimental side, two early studies by Setubal [30] and Cherkassky et al. [7] are particularly important since they introduce families of bipartite graphs where solving the maximum matching problem is difficult. In practice, the choice of the initial (usually inclusion-maximal) matching can make a difference. Different such initialization heuristics are known and were combined with different algorithms in the more recent study by Langguth et al. [25]. One possible initialization heuristic repeatedly picks a vertex v with minimum degree and matches it to one of its neighbors w, either randomly chosen or also chosen with minimum degree. Then the two matched vertices are removed from all further considerations and the degrees of the remaining vertices updated accordingly, i.e., all remaining neighbors of v and w have their degree reduced.

2   Semi-Streaming Model

Traditionally, random access to the problem instance (here the graph) is assumed to be cheap. For example, the following assumptions are typically made, where $\Delta := \max_{v \in V} \deg(v)$ is the maximum degree in the graph:

- when using an adjacency matrix, whether two vertices are adjacent or not can be tested in $\mathcal{O}\left(1\right)$ time and the neighborhood of a vertex can be collected in $\mathcal{O}\left(n\right)$;
- when using adjacency lists, adjacency can be tested in $\mathcal{O}\left(\Delta\right)$ time (or $\mathcal{O}\left(\log \Delta\right)$ if vertices are ordered) and the neighborhood of a vertex v can be traversed in $\mathcal{O}\left(\deg(v)\right)$.

BFS and DFS, heavily used in the algorithms discussed in the previous section, rely on this. However, when the input becomes very large, perhaps larger than the amount of random access memory (RAM) available on the machine, then those assumptions are no longer realistic. Large data calls for a different access model; a popular class of such models are *streaming models*. In a streaming model, the input is given as a sequence of items, e. g., numbers or pairs of numbers (which could represent edges in a graph). Such a sequence is called a *stream*. An algorithm can request to see the stream once or multiple times, and each time each of the items is presented to the algorithm, one by one. Seeing the stream once is called a *pass* or a *pass over the input*. No assumption is made on the order in which the items are presented, and often algorithms are designed so that the order of items is allowed to change from one pass to the other. A pass is assumed to be costly, and a *bound on the number of passes*, the *pass guarantee*, is an important characteristic of a streaming algorithm. It is generally accepted that the pass guarantee should be *independent of the input size*, but is allowed to depend on approximation parameters.

The first streaming algorithms were devised starting in the late 1970s (see, e. g., [15,28]), with one of the most influential works published in 1996 by Alon et al. [2]. The term "streaming" was coined shortly after by Henzinger et al. [17].

Besides the pass guarantee, another important characteristic is the amount of RAM an algorithm requires. RAM should be substantially smaller than the input size for the streaming model to make any sense. On the other hand, RAM should be large enough in order that something useful can be done. For graph problems, the graph is given as a stream of its edges, i. e., as a sequence $e_1, \ldots, e_N$ where each e_i is a pair (interpreted as an unordered pair) of numbers from $[n] = \{1, \ldots, n\}$ when n is the number of vertices. Feigenbaum et al. [13] showed that $\mathcal{O}\,(\text{poly} \log n)$ bits[2] is not enough to even determine whether a path of length 2 exists between two given vertices, i. e., if their neighborhoods are disjoint or not, unless an input-size-dependent number of passes is allowed. The argument is based on the fact that set disjointness has $\Omega\,(n)$ communication complexity [19], and a $p(n)$-pass streaming algorithm with $\mathcal{O}\,(\text{poly} \log n)$ bits of RAM, say $\mathcal{O}\,(\log^c n)$ for some $c > 0$, would allow the problem to be solved with only $\mathcal{O}\,(p(n) \cdot \log^c n)$ bits of communication. So any pass guarantee of $p(n) = o(n/\log^c n)$ is ruled out, in particular $p(n) = \mathcal{O}\,(1)$ is impossible. It follows that logarithmic space makes not much sense for graph problems.

Shortly before the work by Feigenbaum et al., in 2003, Muthukrishnan had proposed the *semi-streaming model* [29], where RAM is restricted to $\mathcal{O}\,(n \cdot \text{poly} \log n)$ bits, meaning that we can store a *linear (in n) number of edges* at a time.[3] In this model, they investigate [13] several graph problems, in particular they devise a semi-streaming $\left(\frac{2}{3} - \varepsilon\right)$-approximation algorithm, $0 < \varepsilon < \frac{1}{3}$, for the bipartite maximum matching problem with a pass guarantee of $\mathcal{O}\left(\varepsilon^{-1} \log \varepsilon^{-1}\right)$. This is an impressive bound, but on the other hand a

---

[2] By $\text{poly}\,x$ we denote a polynomial in x, another way to write $\mathcal{O}\,(\text{poly}\,x)$ is $x^{\mathcal{O}(1)}$.

[3] The "semi" attribute was chosen since the term "streaming model" is generally associated with *logarithmic* space. The semi-streaming model is considered "between" logarithmic and quadratic space, the latter being equivalent to the RAM model since a graph can be stored in $\mathcal{O}\,(n^2)$ bits of space.

$\frac{1}{2}$-approximation, not that far from $\frac{2}{3}$, can already be obtained in just one pass, since in one pass a maximal matching can be computed.

The exact algorithms from the previous section cannot be directly applied to the streaming situation. If the order of edges in unfortunate, a naively implemented BFS can require as many passes as the number of layers. Feigenbaum et al. [14] prove that indeed, any BFS algorithm computing the first k layers with probability at least 2/3, requires more than $k/2$ passes if staying within the limits of the semi-streaming model (see Guruswami and Onak [16] for improved lower bounds). In the next section, we will see that in order to obtain a $(1 + \varepsilon)^{-1}$-approximation, it is sufficient to consider augmenting paths of length $\mathcal{O}(\varepsilon^{-1})$, so BFS can be concluded when reaching this depth. But then we still need a bound on the number of augmentation steps (for the simple algorithm), but even for Hopcroft-Karp we only know the input-size-dependent $\mathcal{O}(\sqrt{n})$ bound on the number of phases. It is worth noting that even the simple initialization heuristic described at the end of Sect. 1 cannot be carried over to the streaming situation, since each degree update takes one pass.

Now suppose that in Hopcroft-Karp, we could bring the number of phases down to $\mathcal{O}(\text{poly}\,\varepsilon^{-1})$. This still leaves us with a problem, namely how to do the DFS in a pass-efficient way. The breakthrough idea for this was given by McGregor [26], namely to perform a "blend" of BFS and DFS: depending on which edge comes along in the stream, we either grow in breadth or in depth. Using this technique, McGregor gave a randomized $(1 + \varepsilon)^{-1}$-approximation algorithm for *general* graphs, but with an exponential dependence on ε^{-1} in the pass guarantee. Eggert et al. [11] showed that this dependence remains even if restricting to bipartite input, namely we have a worst-case lower bound of $\Omega\left(\varepsilon^{-(\varepsilon^{-1})}\right)$ on the number of passes required by McGregor's algorithm. This is due mainly to the randomized nature of the algorithm, requiring a large number of iterations in order to attain a useful success probability. Using the concept of *position limiting*, Eggert et al. [11] gave a new BFS/DFS blend for the bipartite case, with a pass guarantee of $\mathcal{O}(\varepsilon^{-5})$. Subsequently, Algorithm Engineering was performed on this algorithm yielding an experimentally much faster derivative [21]. This algorithm and the engineering process that led to its creation will be described in Sect. 7 and experimentally analyzed in Sect. 8.

In a different line of research, Ahn and Guha devised linear-programming-based algorithms for a variety of matching-type graph problems [1]. For the bipartite matching problem, an algorithm with pass guarantee $\mathcal{O}(\varepsilon^{-2}\log\log\varepsilon^{-1})$ is presented (with the number of passes being a factor in the RAM requirement). An experimental evaluation of these techniques is still an open task. Konrad et al. [23] gave algorithms for the bipartite maximum matching problem which work in one or two passes with approximation guarantees slightly above the known $\frac{1}{2}$ of an inclusion-maximal matching. For further work on graph streams (connectivity, spanning trees, weighted matching, cuts) the work by Zelke [31] is a good starting point.

More recent results include the following; we restrict the discussion to upper bounds. In *dynamic graph streams*, edges that have been announced can also be removed from the graph at a later point in the stream, and edges may be inserted multiple times. Assadi et al. [4] give a one-pass randomized algorithm for approximating a maximum matching in a bipartite dynamic graph stream, with a parameter controlling a trade-off between approximation and required memory. Konrad [22] gives a similar result for the case of a slightly more general dynamic streaming model. A constant-factor randomized approximation for the *size* of a maximum matching in *planar* graphs is achieved in sub-linear memory size and with one pass with the algorithm by Esfandiari et al. [12]. They generalize to graphs with bounded arboricity. For extensions to dynamic graph streams, see Bury and Schwiegelshohn [6]. Chitnis et al. [8,9], for dynamic graph streams, give a one-pass algorithm for maintaining an *inclusion-maximal* matching using $\mathcal{O}\left(k \cdot \operatorname{poly} \log n\right)$ bits, provided that at any point in time, no inclusion-maximal matching of size greater than k exists. For computing a polylogarithmic approximation to the *size* of a maximum matching, Kapralov et al. [20] give a one-pass algorithm requiring only polylogarithmic space, requiring the stream to be presented in random order. Extensions are given in [6]. Much has been done for *weighted* matching, and a recent approximation algorithm was presented by Crouch and Stubbs [10]. Again, for extensions, see [6].

3   Approximation Theory for Matching

Since BFS covering the whole graph can take too many passes, we restrict to augmenting paths up to an input-size-independent length. This section presents a general framework for this.

A *DAP algorithm* ("DAP" standing for "disjoint augmenting paths") is one that finds, given a matching M, a set of disjoint augmenting paths. For $\lambda \in \mathbb{N}$, we call a path a λ *path* if it is of length at most $2\lambda + 1$; the length of a path being the number of its edges. For $\lambda_1, \lambda_2 \in \mathbb{N}$, $\lambda_1 \leqslant \lambda_2$, a set $\mathcal{D}$ of paths is called a (λ_1, λ_2) *DAP set* if:

(i) All paths in $\mathcal{D}$ are augmenting λ_2 paths.
(ii) Any two paths in $\mathcal{D}$ are vertex-disjoint.
(iii) We cannot add another augmenting λ_1 path to $\mathcal{D}$ without violating condition (ii).

We call $s := \frac{\lambda_2}{\lambda_1}$ the *stretch*, since it specifies how far paths may stretch beyond λ_1. Given $\delta \in [0,1]$, a DAP algorithm is called a $(\lambda_1, \lambda_2, \delta)$ *DAP approximation algorithm* if it always delivers a result $\mathcal{A}$ of disjoint augmenting λ_2 paths such that there exists a (λ_1, λ_2) DAP set $\mathcal{D}$ so that $|\mathcal{D}| \leqslant |\mathcal{A}| + \delta\,|M|$. Let $\delta_{\mathrm{inn}}, \delta_{\mathrm{out}} \in [0,1]$ and DAP be a $(\lambda_1, \lambda_2, \delta_{\mathrm{inn}})$ DAP approximation algorithm. All our algorithms utilize the loop shown in Algorithm 1. When this loop terminates, clearly there exists a (λ_1, λ_2) DAP set $\mathcal{D}$ with $|\mathcal{D}| \leqslant |\mathcal{A}| + \delta_{\mathrm{inn}}\,|M| \leqslant \delta_{\mathrm{out}}\,|M| + \delta_{\mathrm{inn}}\,|M| = (\delta_{\mathrm{inn}} + \delta_{\mathrm{out}})\,|M|$, where M denotes the matching before the last augmentation. Let $k \in \mathbb{N}$; this will control approximation, $\frac{1}{k}$ takes the role of ε as it was used in the previous section. Moreover, let $k \leqslant \lambda_1 \leqslant \lambda_2$ and

$$\delta(k, \lambda_1, \lambda_2) := \frac{\lambda_1 - k + 1}{2k\lambda_1\,(\lambda_2 + 2)} > 0. \tag{1}$$

Algorithm 1. Outer Loop

---
1 $M :=$ any inclusion-maximal matching;
2 **repeat**
3 | $c := |M|$;
4 | $\mathcal{A} := \mathtt{DAP}(M)$;
5 | augment M using $\mathcal{A}$;
6 **until** $|\mathcal{A}| \leqslant \delta_{\mathrm{out}} c$;

Following the pattern of [11, Lemmas 4.1 and 4.2] we can prove:

Lemma 1. *Let M be an inclusion-maximal matching. Let $\mathcal{D}$ be a (λ_1, λ_2) DAP set such that $|\mathcal{D}| \leqslant 2\delta\,|M|$ with $\delta = \delta(k, \lambda_1, \lambda_2)$. Then M is a $\left(1 + \frac{1}{k}\right)^{-1}$-approximation.*

The lemma yields the $\left(1 + \frac{1}{k}\right)^{-1}$ approximation guarantee for Algorithm 1 when $\delta_{\mathrm{inn}} = \delta_{\mathrm{out}} = \delta(k, \lambda_1, \lambda_2)$. What are desirable values for λ_1 and λ_2? The DAP approximation algorithms presented in later sections (the path-based and the tree-based one) can work with any allowable setting for λ_1 and λ_2, so we have some freedom of choice. We assume that constructing longer paths is more expensive, so we would like to have those values small and in particular $\lambda_1 = \lambda_2$. (We will later encounter situations where it is conceivable that $\lambda_1 < \lambda_2$, in other words $s > 1$, may be beneficial.) On the other hand, we would like to have δ large in order to terminate quickly. The function $\lambda \mapsto \delta(k, \lambda, \lambda)$ climbs until $\lambda = k - 1 + \sqrt{k^2 - 1} \leqslant 2k - 1$ and falls after that. Since we only use integral values for λ_1, the largest value to consider is $\lambda_1 = \lambda_2 = 2k - 1$. The smallest one is $\lambda_1 = \lambda_2 = k$. We parametrize the range in between by defining

$$\lambda(\gamma) := \lceil k\,(1 + \gamma) \rceil - 1 \quad \text{for each} \quad \gamma \in [{}^1\!/k, 1]. \tag{2}$$

Consider the setting $\lambda_1 := \lambda_2 := \lambda(\gamma)$ and $\delta_{\mathrm{inn}} := \delta_{\mathrm{out}} := \delta(k, \lambda_1, \lambda_2)$. Then increasing γ increases path length, but also increases δ_{inn} and δ_{out}, which means that we are content with a less good approximation from the DAP algorithm and also relax the stopping condition of the outer loop. So γ controls a trade-off between path length and stopping criterion, and different choices should be compared experimentally.

Now we have a general framework, but a major building block, a $(\lambda_1, \lambda_2, \delta_{\mathrm{inn}})$ DAP approximation algorithm, is still missing. We will give two different algorithms, one we call *path-based* and the other *tree-based*, the latter being the result of Algorithm Engineering.

4 Path-Based DAP Approximation

The following paragraphs describe how we find a $(\lambda_1, \lambda_2, \delta_{\mathrm{inn}})$ DAP approximation with $\lambda_1 = \lambda_2$, following [11]. Since both length parameters are the same, we write $\lambda = \lambda_1 = \lambda_2$. Pseudocode is given as Algorithm 2.

Algorithm 2. Path-Based DAP Approximation

Input: inclusion-maximal matching M, parameters $\lambda \in \mathbb{N}$ and $\delta_{\mathrm{inn}} \in \mathbb{R}_{>0}$

Output: $(\lambda, \lambda, \delta_{\mathrm{inn}})$ DAP approximation $\mathcal{A}$

1 **foreach** $m \in M$ **do** $\ell(m) := \lambda + 1$;

2 **foreach** $\alpha \in \mathrm{free}(A)$ **do** $P(\alpha) := (\alpha)$;

3 **repeat**

4   **foreach** $e = \{a, b\} \in E$ *with* $a \in A$ *and* $b \in B$ **do**

5   **if** $a, b \in \mathrm{remain}(V)$ *and* $\exists P(\alpha^*) = (\alpha^*, e^*, b^*, m_1, \ldots, m_t, a)$ **then**

6   **if** $b \in \mathrm{free}(B)$ **then**

7             $P(\alpha^*) := (\alpha^*, \ldots, m_t, a, e, b)$;

8             $\mathcal{A} := \mathcal{A} \cup \{P(\alpha^*)\}$;

9   mark vertices on $P(\alpha^*)$ as used;

10   **else if** $t + 1 < \ell(m)$ *with* $m := \{b, M_b\}$ **then**

11   **if** $\exists P(\tilde{\alpha}) = (\tilde{\alpha}, \ldots, \tilde{a}, \tilde{e}, b, m, M_b, e_1, b_1, \tilde{m}_1, \ldots, \tilde{m}_j, a_j)$ **then**

12                $P(\tilde{\alpha}) := (\tilde{\alpha}, \ldots, \tilde{a})$;

13                $P(\alpha^*) := (\alpha^*, \ldots, m_t, a, e, b, m, M_b, \ldots, a_j)$;

14                $\ell(m) := t + 1$;

15   **foreach** $i \in \{1, \ldots, j\}$ **do** $\ell(\tilde{m}_i) := t + 1 + i$;

16   **else**

17                $P(\alpha^*) := (\alpha^*, \ldots, m_t, a, e, b, m, M_b)$;

18                $\ell(m) := t + 1$;

19    $c := |\{\alpha \in \mathrm{free}(A) \cap \mathrm{remain}(A) \; ; \; P(\alpha) > 0\}|$;

20   **foreach** $\alpha \in \mathrm{free}(A) \cap \mathrm{remain}(A)$ **do**

21   **if** $P(\alpha)$ *was not modified in the previous pass* **then**

22   backtrack by removing last two edges from $P(\alpha)$;

23 **until** $c \leqslant \delta_{\mathrm{inn}} |M|$;

24 **return** $\mathcal{A}$;

Fix an inclusion-maximal matching M; such a matching can be constructed in just one pass by starting with the empty matching and adding edges as they come along in the stream as long as they do not destroy the matching property. We construct disjoint alternating paths starting at vertices of $\mathrm{free}(A)$, the *constructed paths*, and we index them by their starting vertices: $(P(\alpha))_{\alpha \in \mathrm{free}(A)}$. During construction, these paths have an even length. When we find augmenting paths, they are stored in a set $\mathcal{A}$ and their vertices marked as *used*; a vertex not being used is called *remaining*. Denote $\mathrm{remain}(X)$ the remaining vertices in a set $X \subseteq V$. Suppose $P(\alpha) = (\alpha, e_1, b_1, m_1, a_1, \ldots, m_t, a_t)$ is a path with free vertex $\alpha \in \mathrm{free}(A)$, vertices $a_1, \ldots, a_t \in A$ and $b_1, \ldots, b_t \in B$, free edges $e_1, \ldots, e_t \in E$ and matching edges $m_1, \ldots, m_t \in M$. Then we say that matching edge m_i has *position* i, $i \in [t]$. Each matching edge m has a *position limit* $\ell(m)$, initialized to $\ell(m) := \lambda + 1$, which will be ensured to be an impossible position to take in any constructed path. We perform *position limiting*, i.e., a matching edge m will only be inserted into a constructed path if its new position is strictly smaller than its position limit (so in particular $\lambda + 1$ is an impossible position to take). When a matching edge is inserted, its position limit is decremented to its position in the constructed path.

Insertion takes place the following way. Let $e = \{a, b\}$ come along in the stream with $a, b \in \text{remain}(V)$. Then we check whether a is the end of any constructed path. If so, say $P(\alpha^*) = (\alpha^*, \ldots, m_t, a)$ with t matching edges, then we check whether b is free. If so, then we have found an augmenting path $(\alpha^*, \ldots, m_t, a, e, b)$, which is immediately stored away into the result set $\mathcal{A}$, and its vertices are marked used. Otherwise, if b is matched, let $m := \{b, M_b\}$ and we check if $t + 1 < \ell(m)$. If so, we extend the path to $(\alpha^*, \ldots, m_t, a, e, b, m, M_b)$ if m is in no constructed path already. If on the other hand there is $P(\tilde{\alpha}) = (\tilde{\alpha}, \ldots, \tilde{a}, \tilde{e}, b, m, M_b, \ldots)$, we move m and everything behind it from $P(\tilde{\alpha})$ to $P(\alpha^*)$, so that the latter looks like: $(\alpha^*, \ldots, m_t, a, e, b, m, M_b, \ldots)$. Position limits of all inserted or migrated matching edges are updated, i. e., $\ell(m) := t + 1$, and so on for all matching edges behind m, if any. For example, if the next matching edge is called $\tilde{m}$, then $\ell(\tilde{m}) := t + 2$. After migration, we have $P(\tilde{\alpha}) = (\tilde{\alpha}, \ldots, \tilde{a})$.

After each pass, we backtrack conditionally: each constructed path that was not modified during that preceding pass has its last two edges removed. This gives other edges the chance to be inserted there in subsequent passes. When the number of constructed paths of positive length (considered before backtracking) falls on or below $\delta_{\text{inn}} |M|$, we terminate and deliver all augmenting paths found so far. Note that although constructed paths are initialized with length zero (one vertex and no edges), during the first pass, every possible extension is considered and hence the set of positive-length paths can be large after the first pass, when the termination criterion is considered for the first time. It was proved in [11]:

(i) The procedure described above is a $(\lambda, \lambda, \delta_{\text{inn}})$ DAP approx. algorithm.
(ii) Termination occurs after at most $2\lambda\delta_{\text{inn}}^{-1} + 1$ passes.

Position limiting is important for bounding the number of passes [11, Lemma 7.1]. Finally we consider the whole matching algorithm, which in general means multiple invocations of the DAP algorithm. By the stopping criterion of the outer loop, there can be at most $\delta_{\text{out}}^{-1} + 1$ such invocations [11, Theorem 7.2]. Hence, with (2), we have the following bound on the number of passes conducted in total:

$$\left(\delta_{\text{out}}^{-1} + 1\right) \left(2\lambda\delta_{\text{inn}}^{-1} + 1\right) = \mathcal{O}\left(\gamma^{-2} k^5\right). \tag{3}$$

Let us specify γ by $\tilde{\gamma} \in [0, 1]$ via the relation $\gamma = k^{-\tilde{\gamma}}$. Then for $\tilde{\gamma} = 0$ the bound is $\mathcal{O}\left(k^5\right)$, for $\tilde{\gamma} = \frac{1}{2}$ it is $\mathcal{O}\left(k^6\right)$, and for $\tilde{\gamma} = 1$ it is $\mathcal{O}\left(k^7\right)$. We will compare these three values for $\tilde{\gamma}$ in experiments in Sect. 6.

Concluding this section, we give a series of figures demonstrating possible operations of the path-based DAP approximation algorithm for $k = 7$. Constructed paths are drawn starting at the left and growing to the right. In the state drawn in the following figure, there are 5 constructed paths of positive length and 1 constructed path of length 0. If the dotted edge comes along in the stream, having a free vertex at its right end, then the 4th path (counting from the top) is completed to an augmenting one and immediately stored away in the set $\mathcal{A}$:

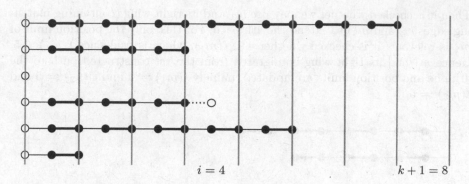

On the other hand, if the other end of the dotted edge is matched, so there is a matching edge m there, then it is checked whether we can use m to extend that constructed path. Since we are talking about position $i = 4$, it is checked whether $4 < \ell(m)$. If so, the dotted edge and the matching edge m are appended to the path, and the position limit of m is updated to $\ell(m) := 4$:

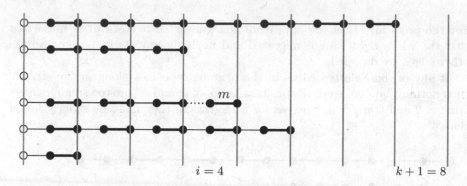

If m is no part of any constructed path, then this is all that has to be done. On the other hand, let us consider that an edge comes along in the stream connecting the end of the 4th constructed path to some matching edge inside of the 1st constructed path:

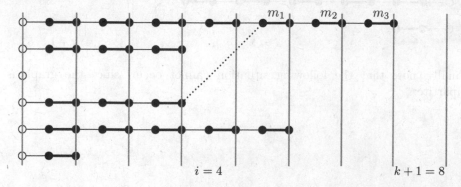

Then it is checked whether we may use m_1 and its right wing (containing matching edges m_2 and m_3) to extend the 4th path. For this, only the position limit of m_1 is relevant. It is checked whether $4 < \ell(m_1)$, which is true since $\ell(m_1) = 5$. Hence m_1 and its right wing is migrated from the 1st constructed path to the 4th one and position limits are updated, namely $\ell(m_1) := 4$ and $\ell(m_2) := 5$ and $\ell(m_3) := 6$:

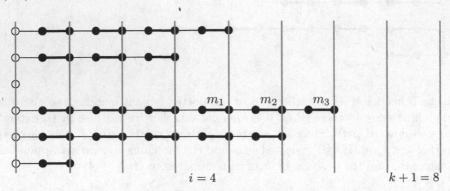

For the proof in [11] of the approximation guarantee to work, it is important that the whole right wing is migrated and not only, say, its first edge and the other edges are dropped.

If any of those dotted edges in the next figure comes along in the stream, then nothing happens since the position limits of m and m' are too small, namely $\ell(m) = 2$ and $\ell(m') = 4$, whereas for a migration, they must be strictly larger than 4:

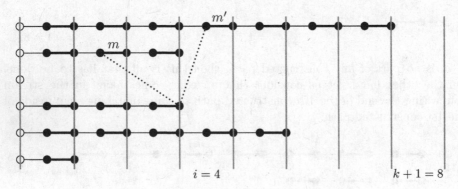

Finally, note that the following situation cannot occur, since the graph is bipartite:

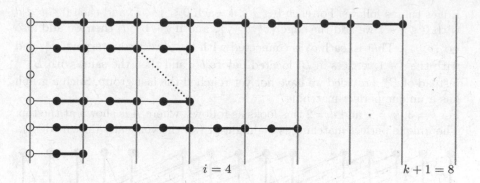

$$i = 4 \qquad\qquad\qquad\qquad k + 1 = 8$$

5 Experimental Setup

In order to perform Algorithm Engineering, we need an implementation and a way to generate difficult test instances. All algorithms and instance generators were implemented in C++. Each instance is kept completely in RAM while the algorithms are working on it, as an array of pairs of 32 bit integers. So the streaming situation is only simulated (although one might argue that even in RAM, sequential access can be beneficial). When randomizing the order of edges for the sake of experiments, it is important to really randomize the array and not just access it using randomized indices. If s [] is the array holding the stream, it should be first randomized by permuting its entries so that during a pass we access it using s[i] with $i = 1, \ldots, N$. Another possibility would be to use an array p [] being a random permutation of $[N]$ and then obtain a random permutation of the stream by accessing it using s[p[i]] with $i = 1, \ldots, N$. However, the latter is substantially less efficient and would lead to much fewer experiments conducted per time.

We generate instances with various structure:

rand: Random bipartite graph; each edge in $\{\{a, b\} ; \ a \in A \land b \in B\}$ occurs with probability $p \in [0, 1]$, which is a parameter.

degm: The degrees in one partition, say A, are a linear function of the vertex index, which runs from 1 to $|A|$. The neighbors in partition B are chosen uniformly at random. A parameter $p \in [0, 1]$ is used to scale degrees.

The following three classes were introduced in [7,30], see also [25]. The constructions work by dividing vertices into groups of equal size and connect them following certain rules.

hilo: Parameters are $l, k, d \in \mathbb{N}$, with $d \leqslant k$ and $|A| = |B| = lk$. Denote $A = \{a_0, \ldots, a_{lk-1}\}$ and $B = \{b_0, \ldots, b_{lk-1}\}$. Define the *groups* by

$$A^i := \{a_j; \ ki \leqslant j < k\,(i+1)\} \quad \text{and} \quad B^i := \{b_j; \ ki \leqslant j < k\,(i+1)\}$$

for each $0 \leqslant i < l$. This makes l groups in each partition, each group being of size k. Denote $A^i = \{a_0^i, \ldots, a_{k-1}^i\}$ und $B^i = \{b_0^i, \ldots, b_{k-1}^i\}$ for each $0 \leqslant i < l$.

Edges run as follows. For each $0 \leqslant i < l$, each $0 \leqslant j < k$, and each $0 \leqslant t < d$ with $0 \leqslant j - t$ we add the edge $\{a_j^i, b_{j-t}^i\}$, and if $i + 1 < l$, then we add also $\{a_j^i, b_{j-t}^{i+1}\}$. That is, each a_j^i is connected with its "direct counterpart" b_j^i, and with the $d - 1$ vertices in B^i located before b_j^i; and then the same with B^{i+1} instead of B^i, provided we have not yet reached the last group. Such a graph has a unique perfect matching.

For $l = 3$, $k = 5$, and $d = 2$ this looks as follows, where A is shown at the top. The unique perfect matching is highlighted (all the vertically drawn edges).

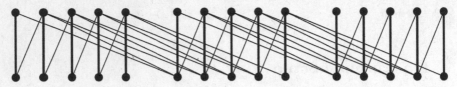

rbg: Parameters are $l, k \in \mathbb{N}$ and $p \in [0, 1]$, where again $|A| = |B| = lk$. Groups $(A^i)_{i=1}^l$ and $(B^i)_{i=1}^l$ are defined as for hilo. For each $0 \leqslant i < l$ and each $j \in \{i - 1, i, i + 1\}$ (where the arithmetic is modulo l, hence $-1 = l - 1$ and $l = 0$) and each vertex $v \in A^i$ and each vertex $w \in B^j$, we add $\{v, w\}$ with probability p. That is, we have a random bipartite graph between each group of A and its three "nearest" groups in B, with wrap-around. This class is also known as fewg and manyg, depending on the size of parameter l.

rope: Parameters and definition of groups is as in rbg. Edges run as follows. For each $0 \leqslant i < l$, we add a perfect matching between A^i and B^i. For each $1 \leqslant i < l$, we add each possible edge between A^i and B^{i-1} with probability p. Such a graph has a unique perfect matching. The following picture gives an example for $l = 3$ and $k = 4$, with the unique perfect matching highlighted. From left to right, we have $A^0, B^0, A^1, B^1, A^2, B^2$.

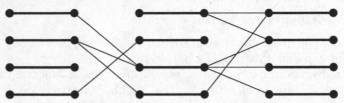

We impose a hard limit of 1×10^9 on $|E|$, meaning about 7.5 GiB (each vertex is stored as a 32 bit unsigned integer). A *series* is specified by a density limit $D_{\max}$ and a set of values for n. For each n of a series and for each class, we generate 256 instances on n vertices. For hilo, rbg, and rope, parameter l is chosen randomly from the set of divisors of $|A| = \frac{n}{2}$. For all classes, a parameter controlling the (expected) number of edges (e.g., p for rand) is being moved through a range such that we start with very few (expected) edges and go up to (or close to) the maximum number of edges possible, given the hard limit, the limit $D_{\max}$ on the density (allowing some overstepping due to randomness), and any limit resulting from structural properties (e.g., number of groups l). This way we produce instances of different densities. For rand and degm, we use 16 different densities and generate 16 instances each. For hilo, rbg, and rope, we use 64 random choices of l and for each 4 different densities. This amounts to 256 instances per n and class. After an instance is generated, its edges are brought into random order. Then each algorithm is run on it once, and then again with partitions A and B swapped. During one run of an algorithm, the order of edges in the stream is kept fix. We use the term *pass count* to refer to the number of passes occurring until the algorithm terminates; clearly the pass guarantee is an upper bound on any pass count.

6   Experimental Results for Path-Based DAP Approximation

The starting point for the Algorithm Engineering process is an implementation of the algorithm explained in Sect. 4. We give concrete theoretical bounds for this algorithm for $k = 9$, which means a 90%-approximation (i.e., $|M| \geqslant 0.9 \cdot |M^*|$ for the constructed matching M and a maximum matching M^*). Recall that $\tilde{\gamma}$ controls the length parameters $\lambda_1 = \lambda_2 = \lambda(\gamma) = \lambda(k^{-\tilde{\gamma}})$, given by (2), and results in different pass guarantees, given by (3). Recall also that we use the same value for δ_{inn} and δ_{out}, given by (1). Writing $\lambda = \lambda_1 = \lambda_2$ and $\delta = \delta_{\text{inn}} = \delta_{\text{out}}$ we obtain the following expression as a bound on the number of passes:

$$(\delta^{-1} + 1)\,(2\lambda\delta^{-1} + 1),$$

where

$$\delta = \frac{\lambda - k + 1}{2k\lambda\,(\lambda + 2)} \quad \text{and} \quad \lambda = \left\lceil k\,(1 + k^{-\tilde{\gamma}}) \right\rceil - 1.$$

Concrete numbers for $k = 9$ are given in the following table.

Table 1. Theorerical pass guarantees

	$\tilde{\gamma} = 0$	$\tilde{\gamma} = \frac{1}{2}$	$\tilde{\gamma} = 1$
How pass guarantee depends on k:	$\mathcal{O}\left(k^5\right)$	$\mathcal{O}\left(k^6\right)$	$\mathcal{O}\left(k^7\right)$
Length parameter $\lambda(k^{-\tilde{\gamma}})$ for $k = 9$:	17	11	9
Termination parameter δ^{-1} for $k = 9$:	646	858	1 782
Concrete pass guarantee for $k = 9$:	14 211 355	16 215 343	57 193 291

Despite the relatively weak dependence on k, these are daunting numbers. It is therefore good to see that actual pass counts are much lower, as shown in the following table.

Numbers state the maximum and rounded mean pass counts, respectively, that were observed for the different choices of parameters and instance classes. This series uses $n = 40\,000, 41\,000, \ldots, 50\,000$ and a density limit of $D_{\max} = \frac{1}{10}$. Number of edges ranges up to about $|E| = 62 \times 10^6$.

Table 2. Experimentally observed pass counts (taken from [21])

$\tilde{\gamma}$		Maximum					Mean				
		rand	degm	hilo	rbg	rope	rand	degm	hilo	rbg	rope
0	$\mathcal{O}\left(k^5\right)$	11 886	14 180	7 032	4 723	2 689	107	145	3 337	257	378
$\frac{1}{2}$	$\mathcal{O}\left(k^6\right)$	7 817	31 491	7 971	4 383	3 843	80	127	2 071	500	541
1	$\mathcal{O}\left(k^7\right)$	7 121	32 844	9 106	5 687	5 126	74	166	2 033	844	790

It is not only interesting to note that these numbers are much smaller than the pass guarantees, but also that there is no best $\tilde{\gamma}$ setting. When only looking at how the pass guarantee depends on k and also when looking at the concrete pass guarantees (as per Table 1), the setting $\tilde{\gamma} = 0$ is superior. However, in experiments it shows to be inferior to $\tilde{\gamma} = 1$ in the following cases: for the maximum and mean for rand, and for the mean for hilo. Especially for hilo this is interesting since this instance class shows by far the highest mean. This is yet another reminder that the performance of an algorithm observed in practice will not necessarily be predicted by theoretical analysis, even if constants otherwise hidden in $\mathcal{O}\left(\cdot\right)$ notation are taken into account.

7   Tree-Based DAP Approximation

A general observation during the experiments with the path-based DAP approximation was that the vast majority of edges come along in the stream during a pass without anything happening. Recall the series of examples at the end of Sect. 4 and consider that an edge $\{a, b\}$ comes along in the stream with b being free but a is not at the end of any constructed path but somewhere in the middle. For example, this could look like this, the edge $\{a, b\}$ drawn dotted:

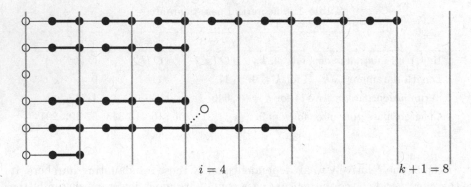

Then nothing will happen, the edge will be ignored. Moreover assume that the 5th constructed path as shown above has a "dead end", i.e., there is no way to complete it to an augmenting path, not until the last two matching edges have been removed by backtracking. (Recall that after each pass, we backtrack conditionally: each constructed path that was not modified during that preceding pass has its last two edges removed.) After backtracking was performed twice on the 5th path – which will take at least 2 more passes – the dotted edge shown above can finally become effective and complete the path to an augmenting path. The question arises: would it not be a good idea to complete a path as soon as it is possible instead?

Another question is raised if the completion is not imminent, but we could complete for example using one intermediate matching edge m:

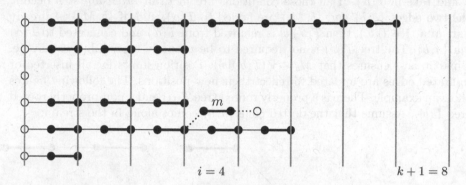

To benefit from this without having to go through all the backtracking, we would have to first remember m when the first dotted edge comes along in the stream and later complete when the second dotted edge comes along. So in fact, we would *not be growing paths, but growing trees*. These considerations give rise to the *first version* of our tree-based DAP approximation algorithm, described formally in the following.

First Version

An *alternating tree* is a pair consisting of a tree T that is a subgraph of G, and a vertex $r \in V(T)$, called its *root*, so that each path from r to any other vertex of T is an alternating path. For $v \in V(T)$ the subtree induced by all vertices reachable from r via v is called the *subtree below v* and denoted $T[v]$. An *alternating forest* consists of one or more alternating trees being pairwise vertex-disjoint. Our tree-based DAP algorithm maintains an alternating forest with trees indexed by their roots; we write $T(r) = (V(r), E(r))$ for the tree rooted at $r \in V$. The *forest* $\mathcal{F}$ consists of all such trees that are rooted at a remaining vertex, i.e., $\mathcal{F} = \{T(r);\ r \in \text{remain}(V)\}$. We call a tree *properly rooted* if its root is a free vertex. A properly rooted tree $T(\alpha)$ together with an edge $\{a, \beta\}$ with β being free and $a \in V(T)$ at an even distance from α, yield an augmenting path.

We initialize by setting $T(\alpha) := (\{\alpha\}, \varnothing)$ for each $\alpha \in \text{free}(A)$ and $T(r) := (\varnothing, \varnothing)$ for each $r \in V \setminus \text{free}(A)$. So we have empty trees and one-vertex trees with a free vertex of A. Position limits are initialized $\ell(m) := \lambda_1 + 1$ for each $m \in M$ as usual. If $(\alpha, e_1, b_1, m_1, a_1, \ldots, m_t, a_t)$ is a path in the properly rooted tree $T(\alpha)$, then we say that matching edge m_i, $i \in [t]$, has *position i*. Results (i.e., the augmenting paths found) will be stored into a set $\mathcal{A}$, that is initialized to $\mathcal{A} := \varnothing$.

Trees grow over time, and there may also emerge non-properly rooted trees. When a free edge $\{a, b\}$ between two remaining vertices goes by in the stream with b being covered, the algorithm checks whether to extend any of the trees. Conditions are: the tree has to be properly rooted, say $T(\alpha)$, it must contain a, and $i < \ell(\{b, M_b\})$, where i is the position that the matching edge $\{b, M_b\}$ would take in $T(\alpha)$. If all those conditions are met, an *extension step* occurs: the two edges $\{a, b\}$ and $\{b, M_b\}$ are added to $T(\alpha)$, and, if $\{b, M_b\}$ is already part of a tree $T(b')$, then $T(b')[b]$ is removed from $T(b')$ and connected to $T(\alpha)$ via $\{a, b\}$. The tree $T(b')$ is not required to be properly rooted, but it may be. Bipartiteness ensures that $M_b \in V(T(b')[b])$. Position limits for all inserted or migrated edges are updated to reflect their new positions. The following figures show an example. There is a properly rooted tree $T(\alpha)$ and a non-properly rooted tree $T(b')$. Assume that the dotted edge $\{a, b\}$ comes along in the stream:

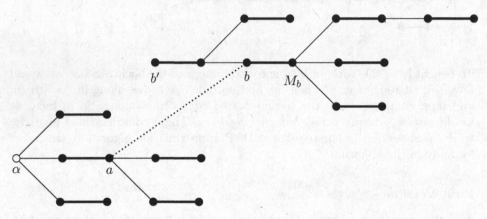

Provided that position limits allow, then part of $T(b')$, namely the subtree $T(b')[b]$, is migrated to $T(\alpha)$. The migrated edges will receive new position limits, e.g., $\ell(\{b, M_b\}) := 2$. There are only 3 edges left in tree $T(b')$, two matching edges and one free edge:

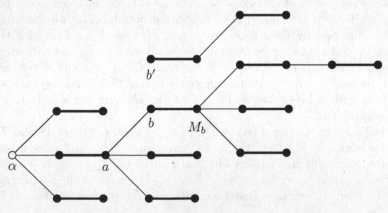

When a free edge $\{a, \beta\}$ with $a, \beta \in \text{remain}(V)$ goes by in the stream with β being *free*, then we check whether we can build an augmenting path. If there is a properly rooted tree $T(\alpha)$ with $a \in V(\alpha)$, the path P in $T(\alpha)$ from α to β is augmenting. In that case, a *completion step* occurs: we store P into the result set $\mathcal{A}$, and mark all vertices on P as used. Also, we adjust our forest as follows. For each $a \in V(P) \cap A$ and each of its neighbors in $T(\alpha)$ and not in P, i.e., for each $b \in N_{T(\alpha)}(a) \setminus V(P)$, we set $T(b) := T(\alpha)[b]$. In other words, we "cut" P out of $T(\alpha)$ and make each of the resulting subtrees that "fall off" a new tree of its own. None of those is properly rooted, and also they are rooted at vertices of partition B, not A as the properly rooted ones. However, they – or parts of them – can subsequently be connected to remaining properly rooted trees by an extension step as described before.

After each pass, it is checked whether it is time to terminate and return the result $\mathcal{A}$. We terminate when any of the following two conditions is met:

(T1) During the last pass, no extension or completion occurred. In other words, the forest did not change. (It then would not change during further passes.)

(T2) The number of properly rooted trees (which is also the number of remaining free vertices of A) is on or below $\delta_{\text{inn}} |M|$.

A backtracking step as for the path-based algorithm makes no sense here since its purpose was to free up the ends of constructed paths in order that other edges can be attached there – which obviously is not necessary for the tree-based algorithm.

Experiments and Proof Attempt

An implementation of this algorithm was substantially more involved than for the path-based one, but experiments were rewarding since they showed astonishing low pass counts, far below 100, even for heavy instances like hilo. Consequently, the time required to run a series of experiments dropped to a fraction of that time needed when using the path-based algorithm. Next, a theoretical analysis was attempted, first for the approximation guarantee. Recall that for the path-based algorithm, we have the following termination criterion: when the number of constructed paths of positive length falls on or below $\delta_{\text{inn}} |M|$, we terminate. The proof of its approximation guarantee (given in [11]) works by showing that

(i) if that threshold is reduced to 0, then a (λ_1, λ_2) DAP set is returned (where $\lambda_1 = \lambda_2$ by the way we choose these parameters for the path-based algorithm);[4]

(ii) and by using any threshold $\tau > 0$, at most τ augmenting paths are missed.

---

[4] The statement (i) can also be formulated as the algorithm with threshold 0 being a $(\lambda_1, \lambda_2, 0)$ DAP approximation algorithm.

It immediately follows that for $\tau = \delta_{\text{inn}} |M|$, there exist a (λ_1, λ_2) DAP set $\mathcal{D}$ with $|\mathcal{D}| \leqslant |\mathcal{A}| + \delta_{\text{inn}} |M|$, namely we can take for $\mathcal{D}$ the set that would have been constructed for threshold 0. By definition, we thus have a $(\lambda_1, \lambda_2, \delta_{\text{inn}})$ DAP approximation algorithm, as required. The most complicated step in this proof is (i). An attempt to follow through the same program for the first version of the tree-based algorithm failed, and it must fail due to the following simple example. Let the following graph and matching be given:

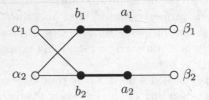

Assume $\{\alpha_1, b_1\}$ and $\{\alpha_1, b_2\}$ come first in the stream. Then the two matching edges $\{b_1, a_1\}$ and $\{b_2, a_2\}$ are built into $T(\alpha_1)$ and their position limits set to 1, so they can never migrate to $T(\alpha_2)$. Thus, at most one of the four augmenting paths is found (namely either $(\alpha_1, b_1, a_1, \beta_1)$ or $(\alpha_1, b_2, a_2, \beta_2)$, depending on whether (a_1, β_1) or (a_2, β_2) comes next in the stream), leaving one behind that is disjoint to the found one.

This does not only spoil the proof attempt, but it is a serious flaw in the algorithm, which can lead to poor approximation. To see this, we generalize the above example to $|\text{free}(A)| = t$ for some $t \in \mathbb{N}$.

Example 1. This is a family of examples, parameterized by $t \in \mathbb{N}$. For each t, a graph and an ordering of its edges in the stream is specified. The graph looks like this:

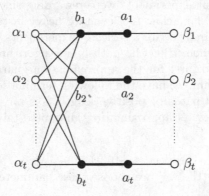

The set of edges is

$$E = \{\{\alpha_i, a_j\}\,;\ i, j \in [t]\} \cup \{\{a_i, b_i\}\,;\ i \in [t]\} \cup \{\{b_i, \beta_i\}\,;\ i \in [t]\}.$$

The stream is ordered like so (for each of the passes):

$$\{a_1, b_1\}, \{a_2, b_2\}, \ldots, \{a_t, b_t\},$$
$$\{\alpha_1, b_1\}, \{\alpha_1, b_2\}, \ldots, \{\alpha_1, b_t\},$$
$$\{\alpha_2, b_1\}, \{\alpha_2, b_2\}, \ldots, \{\alpha_2, b_t\},$$
$$\ldots,$$
$$\{\alpha_t, b_1\}, \{\alpha_t, b_2\}, \ldots, \{\alpha_t, b_t\},$$
$$\{b_1, \beta_1\}, \{b_2, \beta_2\}, \ldots, \{b_t, \beta_t\}$$

This concludes the definition of the example.

Due to the order in the stream, the initial matching will be $\{\{a_i, b_i\}; \; i \in [t]\}$ as shown in the picture above, since all these edges come first and are hence picked. This is just a $\frac{1}{2}$-approximation, not better. During the first pass of the DAP approximation, the tree $T(\alpha_1)$ will first grab all the matching edges and when $\{a_1, \beta_1\}$ comes along in the stream, we will have the augmenting path $(\alpha_1, b_1, a_1, \beta_1)$. Due to position limits, nothing more will happen in this phase, so the DAP approximation terminates, delivering just one augmenting path. It will be used to improve the matching from size t to size $t + 1$, but then the whole algorithm, which is Algorithm 1, will terminate since $|\mathcal{A}| = 1$. Strictly, this requires t to be sufficiently large compared to k, but these requirements are easily met. For example, let $k = 9$ and $\lambda_1 = \lambda_2 = k$ (i.e., $\tilde{\gamma} = 1$). Then $\delta(k, \lambda_1, \lambda_2) = \frac{1}{1782}$ as per (1). So if $t \geqslant 1782$ and $k = 9$ then the algorithm will terminate after one invocation of the DAP approximation and as a result it will miss its goal of a 90%-approximation by far.

Second and Third Version

In order to remedy the flaw that has become evident above, we add a feature to the completion step. Recall that in the completion step, an augmenting path is "cut" out of a properly rooted tree, perhaps leaving some non-properly rooted trees behind. The second version introduces *position limit release*: we reset position limits to $\lambda_1 + 1$ on edges of the new (non-properly rooted) trees; we say that the position limits on those edges are *released*. This can be considered an implicit form of backtracking.

Pass counts in experiments went up only moderately after position limit release was implemented. However, something else unexpectedly happened: the batch system at the Computing Center where the experiments were run killed many of the jobs after some time because they exceeded their memory limit. The batch system requires to give a bound on the memory requirements of a job and will kill the job if this bound is exceeded. One way is to just give the total amount of memory available on the desired compute nodes, but this would be a waste of computing power since the nodes have multiple cores and the memory limit set to total memory would mean that only one (single-threaded) job would run on it. So the memory requirement of the program was estimated by looking at its data structures and this estimate was given to the batch system. With this estimate, jobs were killed, even after introduction of an extra safety margin. Additional assert () statements finally revealed that the data structure used for storing trees grew beyond all expectations, in particular paths much longer than $2\lambda_2 + 1$ were constructed (i.e., they were not λ_2 paths anymore).[5] Indeed, a review of the algorithm showed that this is to be expected after position limit release was introduced, explained in the following.

In an extension step, although position limits at first are not higher than $\lambda_1 + 1$, edges can be included in a tree at positions beyond λ_1. Assume $m = \{b, M_b\}$ is inserted at position $i \leqslant \lambda_1$ into a properly rooted tree $T(\alpha)$ and subsequently, more edges are inserted behind m. Then an augmenting path is found in $T(\alpha)$ not incorporating m, hence the position limit of m is released. Later m can be inserted at a position j with $\lambda_1 \geqslant j > i$ in another properly rooted tree $T(\alpha')$. When m carries a sufficiently deep subtree with it, then $T(\alpha')$ could grow beyond λ_1, even though $j \leqslant \lambda_1$. This is no good since we expect a DAP approximation algorithm to deliver λ_2 paths for a parameter λ_2 (which so far was chosen to be equal to λ_1); cf. Sect. 3.

As a solution, the second length parameter λ_2 takes on a special role. The third version of the tree-based DAP approximation includes the following feature: when the migrated subtree is too deep, we trim its branches just so that it can be migrated without making the destination tree reach beyond λ_2. The trimmed-off branches become non-properly rooted trees of their own. We control a trade-off this way: higher λ_2 means fewer trimming and hence that we destroy fewer of our previously built structure. But higher λ_2 reduces $\delta(\lambda_1, \lambda_2)$ and so may prolong termination. Choosing $\lambda_2 := \lambda_1$ is possible, so we may stick to a single length parameter as before, but can also experiment with larger λ_2. Recall that the *stretch* $s = \frac{\lambda_2}{\lambda_1}$ is used as a measure how far beyond λ_1 our structures may stretch.

This experience shows that appropriate technical restrictions during experimentation, such as memory limits, can be helpful not only to find flaws in the implementation *but also to find flaws in the algorithm design*.

We are finally able to give an approximation guarantee:

Lemma 2. *The third version of the tree-based algorithm (with position limit release and trimming) is a $(\lambda_1, \lambda_2, \delta_{\text{inn}})$ DAP approximation algorithm.*

Proof. Recall the termination conditions on page 18. When the algorithm terminates via condition (T2), it could have, by carrying on, found at most $\delta_{\text{inn}} |M|$ additional augmenting paths. We show that when we restrict to termination condition (T1), we have a $(\lambda_1, \lambda_2, 0)$ DAP approximation algorithm. Clearly, by trimming, only λ_2 paths can be returned. It is also obvious that all paths returned are augmenting paths and disjoint.

It remains to show that we cannot add an augmenting λ_1 path to $\mathcal{A}$ without hitting at least one of the paths already included. Suppose there is an augmenting

---

[5] Recall that so far we have always used $\lambda_1 = \lambda_2$. A distinction between the two parameters will be made shortly.

path $(\alpha, e_1, b_1, m_1, a_1, e_2, b_2, m_2, a_2, \ldots, a_t, e_{t+1}, \beta)$ with $t \leqslant \lambda_1$, $\alpha \in \text{free}(A)$ and $\beta \in \text{free}(B)$ that is disjoint to all paths in $\mathcal{A}$. We show that when the algorithm terminates, then a_t is in a properly rooted tree T. This is a contradiction: first, by the stopping criterion, a_t was there for the whole pass, since an extension step would have made termination impossible. But then the algorithm would have pulled out an augmenting path from T when $e_{t+1} = \{a_t, \beta\}$ came along in the stream during the last pass and so it would not have been allowed to terminate.

We proceed by induction, denoting $a_0 := \alpha$. We show that when the algorithm terminates, a_i is in a properly rooted tree for each $i \in \{0, \ldots, t\}$ with a distance of at most $2i$ from the root (for $i > 0$ this means that m_i is at position at most i). As argued above, the vertex being there at the end of the last pass means that it was there during the whole last pass.

The induction base is clear: $a_0 = \alpha$ is in the properly rooted tree $T(\alpha)$ with distance 0 from the root. Let $i \in [t]$ and assume the statement to be true for $i-1$, with $a_{i-1} \in V(T(\alpha'))$ for some α' during the last pass. If a_i is in no properly rooted tree, then $m_i = \{b_i, a_i\}$ has position limit $\lambda_1 + 1$. If a_i is in a properly rooted tree at distance more than $2i$ from the root, then m_i has position limit more than i. In both cases, m_i would have been inserted into $T(\alpha')$ when the edge $\{a_{i-1}, b_i\}$ came by during the last pass. Since this did not happen (as it would contradict termination), the only remaining alternative holds: a_i is in a properly rooted tree at distance at most $2i$ from the root. $\qquad\square$

Pass Guarantee

By the previous lemma, we have an approximation guarantee for the third version. What about a pass guarantee? Unfortunately, no input-size-independent pass guarantee can be made. This is seen by Example 1. First the tree $T(\alpha_1)$ will grab all the matching edges, then one augmenting path is found and position limits on all the matching edges not in that path are released. During the next pass, tree $T(\alpha_2)$ will grab all the matching edges, and so on. There will be $\Omega(t) = \Omega(n)$ passes.

On the upside, this is easily fixed: we simply let an algorithm using the path-based DAP approximation run in parallel, feeding it the same edges from the stream. We terminate when one of the two terminates. Since both have an approximation guarantee of $(1 + \frac{1}{k})^{-1}$, we know that we have a good approximation no matter from which of the two algorithms we take the solution. Since the path-based algorithm has a pass guarantee of $\mathcal{O}(k^{\mathcal{O}(1)})$, we know that termination will certainly occur after that many passes – in practice it is of course reasonable to expect that termination will occur much earlier than the pass guarantee predicts and also that termination will be triggered by the tree-based algorithm.

On the other hand, in the special case of Example 1, the path-based algorithm would require at most 4 passes: one pass to establish the initial matching, one pass to construct the paths (α_i, b_i, a_i) in the order $i = 1, \ldots, t$, one pass to complete each such path to $(\alpha_i, b_i, a_i, \beta_i)$, then the DAP approximation terminates and an augmentation step occurs, and then there is at most one more final pass to realize that nothing more can be done, the DAP approximation returns $\mathcal{A} = \varnothing$. The result is the perfect matching $\{\{\alpha_i, b_i\}, \{a_i, \beta_i\}; \ i \in [t]\}$.

8   Experimental Results for Tree-Based DAP Approximation

We review detailed experimental results (as they were stated in [21]) for the third and final version of the tree-based DAP approximation algorithm. For the sake of a simpler implementation, we did not combine it with the path-based algorithm as described in the previous section, since the pass counts we observed did not indicate that this would be necessary.

The following table is based on the same instances as Table 2, in particular we have $n = 40\,000, 41\,000, \ldots, 50\,000$ and $D_{\max} = \frac{1}{10}$. As before, we use $\tilde{\gamma} \in \{0, \frac{1}{2}, 1\}$. For the stretch, we use $s = 1$ (i.e., $\lambda_1 = \lambda_2$) and $s = 2$ (i.e., $2\lambda_1 = \lambda_2$). Except for one case (maximum value for hilo and $\tilde{\gamma} = 0$), there is no improvement of $s = 2$ over $s = 1$; on the contrary, the higher stretch lets maximum for rope and $\tilde{\gamma} = 0$ jump from 79 to 94. Among the $s = 1$ results, $\tilde{\gamma} = 1$ is the best except for the maximum for rbg, which is one less for $\tilde{\gamma} = \frac{1}{2}$. But $\tilde{\gamma} = \frac{1}{2}$ shows inferior results for several other classes.

Table 3. Pass counts for the tree-based algorithm

$\tilde{\gamma}$	s	Maximum					Mean				
		rand	degm	hilo	rbg	rope	rand	degm	hilo	rbg	rope
0	1	6	9	75	41	79	3	3	51	5	22
0	2	6	9	74	52	94	3	3	51	5	26
$\frac{1}{2}$	1	6	9	59	37	63	3	3	38	5	20
$\frac{1}{2}$	2	6	9	59	44	70	3	3	38	5	22
1	1	6	9	54	38	61	3	3	35	5	20
1	2	6	9	55	40	67	3	3	36	6	21

All the following experiments are done with the good (and almost always best) settings $\tilde{\gamma} = 1$ and $s = 1$. The highest pass count we have seen for this setting in Table 3 is 61. We increase number of vertices up to a million and first keep the density limit at $D_{\max} = \frac{1}{10}$. The following table shows development for growing n. The number of edges ranges up to about the hard limit of $|E| = 1 \times 10^9$, which takes about 7.5 GiB of space (Table 4).

The previous highest pass count of 61 is exceeded, for $n = 1\,000\,000$ and hilo we observe 65 in this new series. However, this is only a small increase, and moreover the mean values show no increase. The linear worst-case dependence on n, as seen by Example 1, is not reflected by these results.

Next, we lower the density limit. The following table is based on two series: one with $D_{\max} = 1 \times 10^{-3}$ and the other with $D_{\max} = 1 \times 10^{-4}$ (Table 5).

Table 4. Pass counts for $\tilde{\gamma} = 1$ and $s = 1$ and higher n.

n	Maximum					Mean				
	rand	degm	hilo	rbg	rope	rand	degm	hilo	rbg	rope
100 000	3	8	53	30	62	2.5	3.2	35.0	5.1	19.8
200 000	3	7	56	31	63	2.5	2.8	37.6	4.7	19.1
300 000	3	7	55	29	64	2.5	2.9	38.6	3.9	18.2
400 000	3	8	56	33	63	2.5	2.9	36.3	5.3	15.6
500 000	3	7	58	34	64	2.5	3.0	36.7	4.4	19.4
600 000	3	9	58	30	64	2.5	3.5	38.4	3.3	18.1
700 000	6	9	56	35	62	2.5	3.6	37.4	3.9	18.5
800 000	3	8	58	31	63	2.5	3.5	37.9	3.1	16.2
900 000	7	8	61	32	62	2.6	3.3	37.0	3.7	14.5
1 000 000	6	9	60	34	65	2.5	3.1	33.4	4.6	18.2

Table 5. Pass counts for lower densities

n	Maximum					Mean				
	rand	degm	hilo	rbg	rope	rand	degm	hilo	rbg	rope
100 000	40	41	53	48	46	10.3	12.0	28.3	18.9	24.0
200 000	43	43	54	48	46	8.3	11.1	28.5	14.6	21.9
300 000	41	42	56	52	55	6.5	8.4	29.1	11.9	21.2
400 000	44	42	56	48	55	6.1	8.2	29.6	8.6	18.2
500 000	48	45	59	41	56	4.9	7.1	28.9	8.2	18.5
600 000	48	40	58	42	56	5.3	8.3	29.2	6.0	19.1
700 000	40	42	57	32	55	4.3	6.6	29.4	4.7	16.4
800 000	30	42	57	34	57	3.8	6.3	30.9	4.8	16.7
900 000	46	45	58	48	60	4.3	6.6	30.0	4.6	16.6
1 000 000	48	45	59	42	60	4.1	7.5	31.3	4.5	17.3

For several classes, in particular rand and degm, lower density elicits higher pass counts. But still the previous maximum of 65 is not exceeded. This remains true even for the following series going up to two million vertices and $D_{\max} = 1 \times 10^{-4}$ (Table 6).

Practical Considerations. This section has focused on the setting $\tilde{\gamma} = 1$ and stretch $s = 1$. But for any practical application, the author would recommend to experiment with different settings. Given the rise in multicore processors, it may be conceivable to have multiple instances of the algorithm perform in parallel with different parameters, being fed from the same stream. The six combinations of $\tilde{\gamma}$ and s in Table 3 may be a good place to start. Also the combination with the path-based algorithm in parallel should be tested, with different $\tilde{\gamma}$ parameters.

Table 6. Pass counts for up to two million vertices.

n	Maximum					Mean				
	rand	degm	hilo	rbg	rope	rand	degm	hilo	rbg	rope
1 000 000	48	43	62	41	48	5.5	8.8	29.6	4.5	17.7
1 100 000	47	49	60	42	50	5.1	8.6	29.6	4.4	17.0
1 200 000	45	51	60	33	52	4.4	8.4	29.0	5.2	14.5
1 300 000	31	30	59	41	47	3.9	8.7	30.0	3.9	15.5
1 400 000	32	35	61	35	51	4.5	7.6	28.5	4.6	14.9
1 500 000	28	29	57	33	51	3.9	8.5	28.7	4.5	15.5
1 600 000	25	27	58	34	52	4.1	6.9	26.7	4.5	15.9
1 700 000	28	42	60	35	52	3.6	7.7	28.8	4.6	16.2
1 800 000	31	29	60	35	54	4.1	6.8	28.1	3.2	15.2
1 900 000	23	26	56	34	50	3.2	6.3	27.7	4.6	14.0
2 000 000	32	21	60	35	49	3.4	6.4	28.9	4.5	15.7

Acknowledgments. I thank Peter Munstermann for helpful discussions. Financial support through DFG Priority Program "Algorithm Engineering" (Grants Sr7/12-2, Sr7/12-3, and KL 2078/1-1) is also gratefully acknowledged.

References

1. Ahn, K.J., Guha, S.: Linear programming in the semi-streaming model with application to the maximum matching problem. Inf. Comput. **222**, 59–79 (2013). Conference version at ICALP 2011
2. Alon, N., Matias, Y., Szegedy, M.: The space complexity of approximating the frequency moments. J. Comput. Syst. Sci. **58**, 137–147 (1999). Conference version at STOC 1996
3. Alt, H., Blum, N., Mehlhorn, K., Paul, M.: Computing a maximum cardinality matching in a bipartite graph in time $O(n^{1.5}\sqrt{m/\log n})$. Inf. Process. Lett. **37**, 237–240 (1991)
4. Assadi, S., Khanna, S., Li, Y., Yaroslavtsev, G.: Tight bounds for linear sketches of approximate matchings (2015). http://arxiv.org/abs/1505.01467
5. Berge, C.: Two theorems in graph theory. Proc. Natl. Acad. Sci. United States Am. **43**(9), 842–844 (1957). http://www.pnas.org/content/43/9/842.short
6. Bury, M., Schwiegelshohn, C.: Sublinear estimation of weighted matchings in dynamic data streams. In: Bansal, N., Finocchi, I. (eds.) ESA 2015. LNCS, vol. 9294, pp. 263–274. Springer, Heidelberg (2015). doi:10.1007/978-3-662-48350-3_23
7. Cherkassky, B.V., Goldberg, A.V., Martin, P.: Augment or push: a computational study of bipartite matching and unit-capacity flow algorithms. ACM J. Exp. Algorithms **3**, Article No. 8 (1998). http://www.jea.acm.org/1998/CherkasskyAugment/
8. Chitnis, R., Cormode, G., Esfandiari, H., Hajiaghayi, M.T., Monemizadeh, M.: Brief announcement: new streaming algorithms for parameterized maximal matching and beyond. In: Proceedings of the 27th ACM Symposium on Parallel Algorithms and Architectures, Portland, Oregon, USA, June 2015 (SPAA 2015) (2015)

9. Chitnis, R., Cormode, G., Hajiaghayi, M.T., Monemizadeh, M.: Parameterized streaming: maximal matching and vertex cover. In: Proceedings of the 26th Annual ACM-SIAM Symposium on Discrete Algorithms, San Diego, California, USA, January 2015 (SODA 2015) (2015)
10. Crouch, M., Stubbs, D.M.: Improved streaming algorithms for weighted matching, via unweighted matching. In: Proceedings of the International Workshop on Approximation Algorithms for Combinatorial Optimization Problems and Randomization and Computation, Barcelona, Spain, September 2014 (APPROX RANDOM 2014) (2014)
11. Eggert, S., Kliemann, L., Munstermann, P., Srivastav, A.: Bipartite matching in the semi-streaming model. Algorithmica 63, 490–508 (2012). Conference version at ESA 2009
12. Esfandiari, H., Hajiaghayi, M.T., Liaghat, V., Monemizadeh, M., Onak, K.: Streaming algorithms for estimating the matching size in planar graphs and beyond. In: Proceedings of the 26th Annual ACM-SIAM Symposium on Discrete Algorithms, San Diego, California, USA, January 2015 (SODA 2015) (2015)
13. Feigenbaum, J., Kannan, S., McGregor, A., Suri, S., Zhang, J.: On graph problems in a semi-streaming model. Theoret. Comput. Sci. 348, 207–217 (2005). Conference version at ICALP 2004
14. Feigenbaum, J., Kannan, S., McGregor, A., Suri, S., Zhang, J.: Graph distances in the data-stream model. SIAM J. Comput. 38, 1709–1727 (2008)
15. Flajolet, P., Martin, G.N.: Probabilistic counting algorithms for data base applications. J. Comput. Syst. Sci. 31(2), 182–209 (1985)
16. Guruswami, V., Onak, K.: Superlinear lower bounds for multipass graph processing. In: Electronic Colloquium on Computational Complexity (2013)
17. Henzinger, M.R., Raghavan, P., Rajagopalan, S.: Computing on data streams. External Memory Algorithms. DIMACS Series in Discrete Mathematics and Theoretical Computer Science, vol. 50, pp. 107–118 (2000)
18. Hopcroft, J.E., Karp, R.M.: An $n^{5/2}$ algorithm for maximum matchings in bipartite graphs. SIAM J. Comput. 2(4), 225–231 (1973)
19. Kalyansundaram, B., Schnitger, G.: The probabilistic communication complexity of set intersection. SIAM J. Discrete Math. 5, 545–557 (1992)
20. Kapralov, M., Khanna, S., Sudan, M.: Approximating matching size from random streams. In: Proceedings of the 25th Annual ACM-SIAM Symposium on Discrete Algorithms, Portland, Oregon, USA, January 2014 (SODA 2014) (2014)
21. Kliemann, L.: Matching in bipartite graph streams in a small number of passes. In: Pardalos, P.M., Rebennack, S. (eds.) SEA 2011. LNCS, vol. 6630, pp. 254–266. Springer, Heidelberg (2011). doi:10.1007/978-3-642-20662-7_22
22. Konrad, C.: Maximum matching in turnstile streams. In: Bansal, N., Finocchi, I. (eds.) ESA 2015. LNCS, vol. 9294, pp. 840–852. Springer, Heidelberg (2015). doi:10.1007/978-3-662-48350-3_70
23. Konrad, C., Magniez, F., Mathieu, C.: Maximum matching in semi-streaming with few passes. In: Gupta, A., Jansen, K., Rolim, J., Servedio, R. (eds.) APPROX/RANDOM -2012. LNCS, vol. 7408, pp. 231–242. Springer, Heidelberg (2012). doi:10.1007/978-3-642-32512-0_20
24. Kozen, D.C.: The Design and Analysis of Algorithms. Springer, Heidelberg (1992)
25. Langguth, J., Manne, F., Sanders, P.: Heuristic initialization for bipartite matching problems. ACM J. Exp. Algorithmics 15, 1.3:1.1–1.3:1.22 (2010). http://doi.acm.org/10.1145/1712655.1712656

26. McGregor, A.: Finding graph matchings in data streams. In: Chekuri, C., Jansen, K., Rolim, J.D.P., Trevisan, L. (eds.) APPROX/RANDOM -2005. LNCS, vol. 3624, pp. 170–181. Springer, Heidelberg (2005). doi:10.1007/11538462_15

27. Mucha, M., Sankowski, P.: Maximum matchings via Gaussian elimination. In: Proceedings of the 45th Annual IEEE Symposium on Foundations of Computer Science, Rome, Italy, (FOCS 2004), pp. 248–255 (2004). http://doi.ieeecomputersociety.org/10.1109/FOCS.2004.40, http://www.mimuw.edu.pl/mucha/pub/mucha_sankowski_focs04.pdf

28. Munro, J.I., Paterson, M.: Selection and sorting with limited storage. Theoret. Comput. Sci. **12**, 315–323 (1980). Conference version at FOCS 1978

29. Muthukrishnan, M.: Data streams: algorithms and applications. Found. Trends Theoret. Comput. Sci. **1**(2), 1–67 (2005). http://algo.research.googlepages.com/eight.ps

30. Setubal, J.C.: Sequential and parallel experimental results with bipartite matching algorithms. Technical report IC-96-09, Institute of Computing, University of Campinas, Brazil (1996). http://www.dcc.unicamp.br/ic-tr-ftp/1996/96-09.ps.gz

31. Zelke, M.: Algorithms for streaming graphs. Ph.D. thesis, Mathematisch-Naturwissenschaftliche Fakultät II, Humboldt-Universität zu Berlin (2009). http://www.tks.informatik.uni-frankfurt.de/getpdf?id=561

Engineering Art Galleries

Pedro J. de Rezende[1]([✉]), Cid C. de Souza[1], Stephan Friedrichs[2],
Michael Hemmer[3], Alexander Kröller[3], and Davi C. Tozoni[1]

[1] Institute of Computing, University of Campinas, Campinas, Brazil
{rezende,cid}@ic.unicamp.br, davi.tozoni@gmail.com
[2] Max Planck Institute for Informatics, Saarbrücken, Germany
sfriedri@mpi-inf.mpg.de
[3] TU Braunschweig, IBR, Algorithms Group, Braunschweig, Germany
mhsaar@gmail.com, kroeller@perror.de

Abstract. The Art Gallery Problem (AGP) is one of the most well-known problems in Computational Geometry (CG), with a rich history in the study of algorithms, complexity, and variants. Recently there has been a surge in experimental work on the problem. In this survey, we describe this work, show the chronology of developments, and compare current algorithms, including two unpublished versions, in an exhaustive experiment. Furthermore, we show what core algorithmic ingredients have led to recent successes.

Keywords: Art gallery problem · Computational geometry · Linear programming · Experimental algorithmics

1 Introduction

The Art Gallery Problem (AGP) is one of the classic problems in Computational Geometry (CG). Originally it was posed forty years ago, as recalled by Ross Honsberger [37, p. 104]:

> "At a conference in Stanford in August, 1973, Victor Klee asked the gifted young Czech mathematician Václav Chvátal (University of Montreal) whether he had considered a certain problem of guarding the paintings in an art gallery. The way the rooms in museums and galleries snake around with all kinds of alcoves and corners, it is not an easy job to keep an eye on every bit of wall space. The question is to determine the minimum number of guards that are necessary to survey the entire building."

It should be noted that a slightly different definition is used today, where not only the walls of the gallery have to be guarded, but also the interior (this is indeed a different problem, see Fig. 1a). AGP has received enormous attention from the CG community, and today no CG textbook is complete without a treatment of it. We give an overview on the most relevant developments in Sect. 2, after introducing the problem more formally.

L. Kliemann and P. Sanders (Eds.): Algorithm Engineering, LNCS 9220, pp. 379–417, 2016.
DOI: 10.1007/978-3-319-49487-6_12

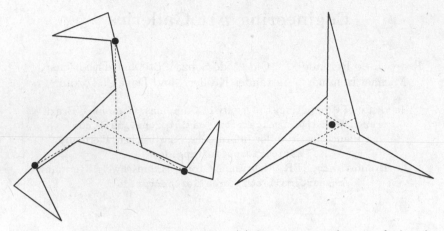

(a) Three guards suffice to cover the walls, but not the interior.

(b) One point guard covers the interior, but a vertex guard cannot

Fig. 1. Edge cover and vertex guard variants have better and worse solutions than the classic AGP, respectively.

Besides theoretical interest, there are practical problems that turn out to be AGP. Some are of these are straightforward, such as guarding a shop with security cameras, or illuminating an environment with few lights. For another example, consider a commercial service providing indoors laser scanning: Given an architectural drawing of an environment, say, a factory building, a high-resolution scan needs to be obtained. For that matter, the company brings in a scanner, places it on a few carefully chosen positions, and scans the building. As scanning takes quite a while, often in the range of several hours per position, the company needs to keep the number of scans as low as possible to stay competitive — this is exactly minimizing the number of guards (scan positions) that still survey (scan) the whole environment.

In this paper, we provide a thorough survey on experimental work in this area, i.e., algorithms that compute optimal or good solutions for AGP, including some problem variants. We only consider algorithms that have been implemented, and that underwent an experimental evaluation. During the past seven years, there have been tremendous improvements, from being able to solve instances with tens of vertices with simplification assumptions, to algorithm implementations that find optimal solutions for instances with several thousands of vertices, in reasonable time on standard PCs. We avoid quoting experimental results from the literature, which are difficult to compare to each other due to differences in benchmark instances, machines used, time limits, and reported statistics. Instead, we conducted a massive unified experiment with 900 problem instances with up to 5000 vertices, comparing six different implementations that were available to us. This allows us to pinpoint benefits and drawbacks of each implementation, and to exactly identify where the current barrier in problem complexity lies.

Given that all benchmarks are made available, this allows future work to compare against the current state. Furthermore, for this paper, the two leading implementations were improved in a joint work ·between their respective authors, using what is better in each. The resulting implementation significantly outperforms any previous work, and constitutes the current frontier in solving AGP.

The remainder of this paper is organized as follows. In the next section, we formalize the problem and describe related work. In Sect. 3, we turn our attention to the sequence of experimental results that have been presented in the past few years, with an emphasis on the chronology of developments. This is followed by an experimental cross-comparison of these algorithms in Sect. 4, showing speedups over time, and the current frontier. In Sect. 5, we take an orthogonal approach and analyze common and unique ingredients of the algorithms, discussing which core ideas have been most successful. This is followed by a discussion on closely related problem variants and current trends in Sect. 6, and a conclusion in Sect. 7.

2 The Art Gallery Problem

Before discussing Art Gallery Problem (AGP) in detail, let us give a formal definition and introduce the necessary notation.

2.1 Problem and Definitions

We are given a polygon P, possibly with holes, in the plane with vertices V and $|V| = n$. P is *simple* if and only if its boundary, denoted by ∂P, is connected. For $p \in P$, $\mathcal{V}(p) \subseteq P$ denotes all points *seen* by p, referred to as the *visibility region* of p, i.e., all points $p' \in P$ that can be connected to p using the line segment $\overline{pp'} \subset P$. We call P *star-shaped* if and only if $P = \mathcal{V}(p)$ for some $p \in P$, the set of all such points p represents the *kernel* of P. For any $G \subseteq P$, we denote by $\mathcal{V}(G) = \bigcup_{g \in G} \mathcal{V}(g)$. A finite $G \subset P$ with $\mathcal{V}(G) = P$ is called a *guard set* of P; $g \in G$ is a guard. We say that g *covers* all points in $\mathcal{V}(g)$. The AGP asks for such a guard set of minimum cardinality.

Note that visibility is symmetric, i.e., $p \in \mathcal{V}(q) \iff q \in \mathcal{V}(p)$. The inverse of $\mathcal{V}(\cdot)$ describes all points that can see a given point p. This is easily confirmed to be

$$\mathcal{V}^{-1}(p) := \{q \in P : p \in \mathcal{V}(q)\} = \mathcal{V}(p).$$

We use two terms to refer to points of P, making the discussion easier to follow. We call a point a *guard position* or *guard candidate* when we want to stress its role to be selected as part of a guard set. The second term comes from the fact that in a feasible solution, every point in $w \in P$ needs to be covered by some visibility polygon. We refer to such a point as *witness* when we use it as certificate for coverage.

Let $G, W \subseteq P$ be sets of guard candidates and witnesses such that $W \subseteq \mathcal{V}(G)$. The AGP variant were W has to be covered with a minimum number of guards, which may only be picked from G, can be formulated as an Integer Linear Program (ILP):

$$\mathrm{AGP}(G, W) := \quad \min \quad \sum_{g \in G} x_g \tag{1}$$

$$\mathrm{s.t.} \quad \sum_{g \in \mathcal{V}(w) \cap G} x_g \geq 1, \ \forall w \in W, \tag{2}$$

$$x_g \in \{0, 1\}, \qquad \forall g \in G. \tag{3}$$

Essentially the model above casts the AGP variant in terms of a Set Covering Problem (SCP). But note that, depending on the choice of G and W, $\mathrm{AGP}(G, W)$ may have an infinite number of variables and/or constraints, i.e., be a semi- or doubly-infinite ILP. We discuss three major variants of AGP:

- The classic AGP definition, allowing for arbitrary *point guards*, i.e., allowing to place guards anywhere within P. It requires that all of P, boundary and interior, is guarded. This corresponds to $\mathrm{AGP}(P, P)$. We refer to this variant as "the" AGP.
- In $\mathrm{AGP}(V, P)$, all of P has to be guarded, but guards are restricted to be placed on vertices of P only. We refer to such guards as *vertex guards*. Trivially, a vertex guard solution is a solution for AGP as well, but the reverse is not necessarily true, see Fig. 1b.
- The variant that Victor Klee actually described, i.e., where only the polygon's boundary needs to be guarded, is described by $\mathrm{AGP}(P, \partial P)$. A solution for $\mathrm{AGP}(P, P)$ also solves $\mathrm{AGP}(P, \partial P)$, but not vice versa (see Fig. 1a).

There are many more AGP variants that deserve (and received) attention, however, these are the three versions that are mostly relevant for this paper.

In the following, unless explicitly stated otherwise, we use G and W to indicate discretized versions of the AGP. For example $\mathrm{AGP}(G, P)$ may refer to a (sub-)problem where all of P needs to be guarded, but a *finite* set of guard candidates is already known. Analogously, $\mathrm{AGP}(P, W)$ is the version where only a finite set of points needs to be covered, and $\mathrm{AGP}(G, W)$ is the fully discretized version.

The semi-infinite case $\mathrm{AGP}(G, P)$ provides some structure that can be exploited in algorithms. Consider Fig. 2. We denote by $\mathcal{A}(G)$ the arrangement obtained by overlaying all visibility polygons $\mathcal{V}(g)$ for every $g \in G$. Every feature (face, edge, or vertex) of $\mathcal{A}(G)$ has a well-defined set of guards that completely sees it. Hence, any of those guards covers the entire feature, and we refer to them as Atomic Visibility Polygon (AVPs). We define a partial order on them as follows: For two faces f_1 and f_2, we define $f_1 \succ f_2$ if they are adjacent in $\mathcal{A}(G)$ and the set of guards seeing f_1 is a superset of those seeing f_2. The maximal (minimal) elements in the resulting poset are called light (shadow) AVPs. They can be exploited to solve the two semi-infinite cases: For given finite G, any subset of G that covers all shadow AVPs also covers P, hence is feasible for $\mathrm{AGP}(G, P)$. For finite W, there is always an optimal solution for $\mathrm{AGP}(P, W)$ that uses only guards in light AVPs of $\mathcal{A}(W)$, with at most one guard per AVP.

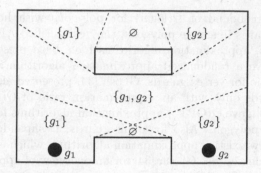

Fig. 2. The visibility arrangement $\mathcal{A}(\{g_1, g_2\})$ induced by two guards g_1 and g_2 in a polygon with one hole.

2.2 Related Work

Chvátal [13] was the first to prove the famous "Art Gallery Theorem", stating that $\lfloor n/3 \rfloor$ guards are sometimes necessary and always sufficient for polygons with n vertices. Later Fisk [32] came up with a simple proof for this theorem, beautiful enough to be included in the BOOK [1]. It also translates directly into a straightforward algorithm to compute such a set of guards. Note that for every n, there exist polygons that can be guarded by a single point (i.e., star-shaped polygons). So any algorithm producing $\lfloor n/3 \rfloor$ guards is merely a $\Theta(n)$-approximation. There are excellent surveys on theoretical results, especially those by O'Rourke [49] and Urrutia [61] should be mentioned.

Many variants of the problem have been studied in the past. For example, Kahn et al. [38] established a similar theorem using $\lfloor n/4 \rfloor$ guards for orthogonal polygons. There are variants where the characteristics of the guards have been changed. For example, *edge guards* are allowed to move along an edge and survey all points visible to some point on this edge. Instead of patrolling along an edge, *diagonal guards* move along diagonals, *mobile guards* are allowed to use both. See Shermer [56] for these definitions. Alternatively, variations on the guard's task have been considered, for example, Laurentini [44] required visibility coverage for the polygon's edges only. Another relevant problem related to the coverage of polygons considers watchman routes. A watchman route is a path in the interior of a polygon P such that every point of P is seen by at least one point in the path. Therefore, a mobile guard moving along this path can do the surveillance of the entire polygon's area. Results on this problem can be found, for example, in Mitchell [47,51].

AGP and its variants are typically hard optimization problems. O'Rourke and Supowit [50] proved AGP to be NP-hard by a reduction from 3SAT, for guards restricted to be located on vertices and polygons with holes. Lee and Lin [45] showed NP-hardness also for simple polygons. This result was extended to point guards by Aggarwal [49]. Schuchardt and Hecker [55] gave NP-hardness proofs for rectilinear simple polygons, both for point and vertex guards. Eidenbenz et al. [25] established lower bounds on the achievable approximation ratio.

They gave a lower bound of $\Omega(\log n)$ for polygons with holes. For vertex, edge and point guards in simple polygons, they established APX-hardness. For restricted versions, approximation algorithms have been presented. Efrat and Har-Peled [24] gave a randomized approximation algorithm with logarithmic approximation ratio for vertex guards. Ghosh [34] presented algorithms for vertex and edge guards only, with an approximation ratio of $O(\log n)$. For point guards Nilsson [48] gave $O(\text{OPT}^2)$-approximation algorithms for monotone and simple rectilinear polygons. Also for point guards, Deshpande et al. [23] proposed one of the few existing approximation algorithms which is not constrained to a few polygon classes. See Ghosh [34] for an overview of approximation algorithms for the AGP. The first known exact algorithm for point guard problem was proposed by Efrat and Har-Peled [24] and has complexity $O((nc)^{3(2c+1)})$, where c is the size of the optimal solution. No experimental results with this algorithm have been reported so far. The exponential grow of the running time with c probably makes it useless to solve large non-trivial instances.

3   Timeline

After receiving mainly a theoretical treatment for over thirty years, several groups have started working on solving the AGP using the Algorithm Engineering methodology, aiming at providing efficient implementations to obtain optimal, or near-optimal, solutions.

Especially two groups, the Institute of Computing at the University of Campinas, Brazil, and the Algorithms Group at TU Braunschweig, Germany, developed a series of algorithms that substantially improve in what kind of instances can be solved efficiently. In this section, we give a chronological overview on these efforts, and describe the algorithms that were developed. It should be noted that all these approaches follow similar core ingredients, e.g., the AGP is treated as an infinite Set Covering Problem (SCP). As finite SCP instances can be solved reasonably fast in practice, the AGP is reduced to finite sets, and different techniques are employed to connect the finite and infinite cases.

3.1   Stony Brook 2007:
$AGP(P, P)$ Heuristics, Tens of Vertices

Amit et al. [2] were among the first to experiment with a solver for $AGP(P, P)$, see the journal version [3] and the PhD thesis by Packer [52] for extended presentations.

In this work, greedy algorithms are considered, following the same setup: A large set G of guard candidates is constructed, with the property that P can be guarded using G. Algorithms pick guards one after the other from G, using a priority function μ, until P is fully guarded. Both G and μ are heuristic in nature. The authors present 13 different strategies (i.e., choices for G and μ), and identify the three that are the best: In A_1, G consists of the polygon vertices, and of one additional point in every face of the arrangement obtained by adding edge extensions to the polygon. Priority is given to guards that can see the most of the currently unguarded other positions in G. The second strategy, A_2 follows the same idea. Additionally, after selecting a guard g, it adds $\mathcal{V}(g)$ to the arrangements and creates additional candidate positions in the newly created faces. Finally, A_{13} employs a weight function ω on G, used as a random distribution. In each step, a point from G is selected following ω. Then, a random uncovered point p is generated, and all guard candidates seeing p get their weight doubled.

To produce lower bounds, greedy heuristics for independent witnesses (i.e., witnesses whose visibility regions do not overlap) are considered. Using a pool of witness candidates, consisting of the polygon's convex vertices and points on reflex-reflex edges, a witness set is constructed iteratively. In every step, the witness seeing the fewest other witness candidates is added, and dependent candidates are removed.

The authors conducted experiments with 40 input sets, including randomly generated as well as hand-crafted instances, with up to 100 vertices. Both simple polygons and ones with holes are considered. By comparing upper and lower bounds, it was found that the three algorithms mentioned above always produced solutions that are at most a factor 2 from the optimum. Algorithm A_1 was most successful in finding optimal solutions, which happened in 12 out of 37 reported cases.

3.2 Campinas 2007:
$AGP(V, P)$ for Orthogonal Simple Polygons, Hundreds of Vertices

In 2007, Couto et al. [20,21] focused on the development of an exact algorithm for the AGP with vertex guards, AGP(V, P), restricted to orthogonal polygons without holes. To the best of our knowledge, these works were the first in the literature to report extensive experimentation with an exact algorithm for a variant of the AGP. Early attempts to tackle the orthogonal AGP(V, P) also involved reductions to the SCP [27,28] and aimed either to obtain heuristic solutions or to solve it exactly [57,58]. However, experiments in these works only considered a few instances of limited sizes. In contrast, in the work of Couto et al., thousands of instances, some of which with 1000 vertices, were tested and later assembled into a benchmark, made publicly available for future comparisons [16], containing new classes of polygons including some very hard problem instances.

Moreover, in [21], the group in Campinas derived theoretical results that were later extended and proved to be instrumental to obtain exact solutions for more general variants of the AGP. They showed that AGP(V, P) can be solved through a single instance of the SCP by replacing the infinite set of points P by a finite set of suitably chosen witnesses from P.

The basic idea of the algorithm is to select a discrete set of points W in P and then solve the AGP variant whose objective consists in finding the minimum number of vertices sufficient to cover all points in W. This discretized

AGP is then reduced to an SCP instance and modeled as an ILP. The resulting formulation is subsequently solved using an ILP solver, in their case, XPRESS. If the solution to the discretized version covers the whole polygon, then an optimal solution has been found. Otherwise, additional points are added to W and the procedure is iterated. The authors prove that the algorithm converges in a polynomial number of iterations, $O(n^3)$ in the worst case.

An important step of this exact algorithm is to decide how to construct the set of witnesses W. Couto et al. study various alternatives and investigated the impact on the performance of the algorithm. In the first version [20], a single method for selecting the initial discretization is considered, which is based on the creation of a regular grid in the interior of P. In the journal version [21], four new discretizations are proposed: *Induced grid* (obtained by extending the lines of support of the edges of the polygon), *just vertices* (comprised of all vertices of P), *complete AVP* (consisting of exactly one point in the interior of each AVP), and *reduced AVP* (formed by one point from each shadow AVP). The authors prove that, with the *shadow AVP* discretization, that it takes the algorithm only one iteration to converge to an optimal solution for the orthogonal AGP.

The first experimental results were largely surpassed by those reported in the journal version. Besides introducing new discretizations, the *shadow AVP* discretization increased the polygon sizes fivefold (to 1000 vertices). In total, almost 2000 orthogonal polygons were tested, including von Koch polygons, which give rise to high density visibility arrangements and, as a consequence, to larger and harder to solve SCP instances.

The authors highlight that, despite the fact that the visibility polygons and the remaining geometric operations executed by the algorithm can be computed in polynomial time, in practice, the preprocessing phase (i.e. geometric operations such as visibility polygon computation) is responsible for the majority of the running time. At first glance, this is surprising since the SCP is known to be NP-hard and one instance of this problem has to be solved at each iteration. However, many SCP instances are easily handled by modern ILP solvers, as is the case for those arising from the AGP. Furthermore, the authors also observe that, when reasonable initial discretizations of the polygon are used, the number of iterations of the algorithm is actually quite small.

Knowing that the *reduced AVP* discretization requires a single iteration, albeit an expensive one timewise, the authors remark that a trade-off between the number of iterations and the hardness of the SCP instances handled by the ILP solver should to be sought. Extensive tests lead to the conclusion that the fastest results were achieved using the *just vertices* discretization since, although many more iterations may be required, the SCP instances are quite small.

3.3 Torino 2008: $AGP(P, \partial P)$, Hundreds of Vertices

In 2008, Bottino and Laurentini [9] proposed a new algorithm for the AGP variant whose objective consists in only covering the edges of a polygon P, denoted AGP$(P, \partial P)$. Hence, in this version, coverage of the interior of P is not required. Despite being less constrained than the original AGP, the $AGP(P, \partial P)$ was proven to be NP-hard [44]. In this context, the authors presented an algorithm capable of optimally solving $AGP(P, \partial P)$ for polygons with and without holes provided the method converges in a finite number of steps. This represents a significant improvement in the search for optimal solutions for the AGP.

The algorithm by Bottino and Laurentini works iteratively. First, a lower bound specific for P is computed. The second step consists of solving an instance of the so called Integer Edge Covering Problem (IEC). In this problem, the objective is also to cover the whole boundary of the polygon with one additional restriction: each edge must be seen entirely by at least one of the selected guards. It is easy to see that a solution to the IEC is also viable for $AGP(P, \partial P)$ and, consequently, its cardinality is an upper bound for the latter. After obtaining a viable solution, the gap between the upper and lower bounds is checked. If it is zero (or less than a predefined threshold) the execution is halted. Otherwise, a method is used to find *indivisible* edges, which are edges that are entirely observed by one guard in some or all optimal solutions of $AGP(P, \partial P)$. The identification of these edges can be done in polynomial time from the visibility arrangement. After identifying them, those classified as not indivisible are split and the process starts over.

Tests were performed on approximately 400 random polygons with up to 200 vertices. The instances were divided into four classes: simple, orthogonal, random polygons with holes and random orthogonal polygons with holes. Reasonable optimality percentages were obtained using the method. For instance, on random polygons with holes, optimal results were achieved for 65% of the instances with 60 vertices. In cases where the program did not reach an optimal solution (due to the optimality gap threshold or to timeout limits), the final upper bound was, on average, very close to the lower bound computed by the algorithm. On average, for all classes of polygons, the upper bound exceeded the lower bound by $\sim 7\%$.

3.4 Campinas 2009:
$AGP(V, P)$ for Simple Polygons, Thousands of Vertices

Couto et al. went on to study how to increase the efficiency of the algorithm for $AGP(V, P)$ proposed in the works discussed in Sect. 3.2. A complete description of their findings can be found in a 2011 paper [19], with a preliminary version available as a 2009 technical report [15]. The basic steps of the algorithm are explained in [17], and are illustrated in the companion video [18].

Compared to the previous works by the same authors, the new algorithm was extended to cope with more general classes of polygons, still without holes, but now including non-orthogonal polygons. Experiments on thousands of instances confirmed the robustness of the algorithm. A massive amount of data was subsequently made publicly available containing the entire benchmark used for these tests, see also Sect. 4.1.

Essentially, some implemented procedures were improved relative to the approach in [21] to enable handling non-orthogonal polygons. Moreover, two new initial discretization techniques were considered. The first one, called *single vertex*, consists in the extreme case where just one vertex of the polygon forms the initial discretized set W. As the second strategy, named *convex vertices*, W comprises all convex vertices of P.

The authors made a thorough analysis of the trade-off between the number and nature of the alternative discretization methods and the number of iterations. Their tests were run on a huge benchmark set of more than ten thousand polygons with up to 2500 vertices. The conclusion was that the decision over the best discretization strategy deeply depends on the polygon class being solved. As anticipated, the fraction of time spent in the preprocessing phase was confirmed to be large for sizable non-orthogonal polygons and even worse in the case of von Koch and random von Koch polygons. Moreover, while using shadow AVPs as the initial discretization produces convergence after just one iteration of the algorithm, the resulting discretization set can, in this case, be so large that the time cost of the preprocessing phase overshadows the solution of the ensued SCP. For this reason, the *just vertices* strategy lead to the most efficient version of the algorithm, in practice, as the small SCP instances created counterbalanced the larger number of iterations for many polygon classes.

3.5   Braunschweig 2010: Fractional Solutions for $AGP(P, P)$, Hundreds of Vertices

In 2010, Baumgartner et al. [7] (see Kröller et al. [41] for the journal version) presented an exact algorithm for the fractional variant of AGP(P, P). In it, solutions may contain guards $g \in P$ with a fractional value for x_g. This corresponds to solving a Linear Program (LP), namely the LP relaxation of AGP(P, P) which is obtained by replacing Constraint (3) of AGP(G, W) with

$$0 \leq x_g \leq 1 \quad \forall g \in G. \tag{4}$$

We denote by AGP$_{\mathbb{R}}(G, W)$ the LP relaxation of AGP(G, W). Note that this is the first exact algorithm where neither guard nor witness positions are restricted.

The authors present a primal-dual approach to solve the problem. They notice that AGP$_{\mathbb{R}}(G, W)$ can be easily solved using an LP solver, provided G and W are finite and not too large. The proposed algorithm picks small, carefully chosen sets for G and W. It then iteratively extends them using cutting planes and column generation:

- *cutting planes:* If there is an uncovered point $w \in P \setminus W$, this corresponds to a violated constraint of AGP$_{\mathbb{R}}(P, P)$, so w is added to W. Otherwise the current solution is feasible for AGP$_{\mathbb{R}}(G, P)$, and hence an upper bound of AGP$_{\mathbb{R}}(P, P)$. We also refer to this part as *primal separation*.
- *column generation:* A violated constraint of the dual of AGP$_{\mathbb{R}}(P, P)$ corresponds to a guard candidate $g \in P \setminus G$ that improves the current solution, and g is added to G. Otherwise the current solution optimally guards the witnesses in W, i.e. is optimal for AGP$_{\mathbb{R}}(P, W)$, and hence provides a lower bound for AGP$_{\mathbb{R}}(P, P)$. We also refer to this part as *dual separation*.

It can be shown that, if the algorithm converges, it produces an optimal solution for $AGP_R(P, P)$. Furthermore, the authors use the algorithm for the integer AGP, but only insofar that LP solutions sometimes are integer by chance, but without any guarantee.

The algorithm has many heuristic ingredients, e.g., in the choice of initial G and W and the placement strategy for new guards and witnesses. The authors conducted an exhaustive experiment, with 150 problem instances with up to 500 vertices. They compared different strategies for all heuristic ingredients of the algorithm. There were four separation strategies: (1) Focusing on upper bounds by always running primal separation, but dual only when primal failed. (2) Focusing on lower bounds, by reversing the previous one. (3) Always running both separators, in the hope of quickly finishing. (4) Alternating between foci, by running primal separation until an upper bound is found, then switching to running dual separation until a lower bound is found, and repeating. There were four different separators, i.e., algorithms to select new candidates for G resp. W; these included selecting the point corresponding to a maximally violated constraint, selecting points in all AVPs, a greedy strategy to find independent rows (columns) with a large support, and placing witnesses on AVP edges. For initial choice of G and W, four heuristics were used: (1) Using all polygon vertices, (2) starting with an empty set (for implementation reasons, a single point had to be used here), (3) selecting half the vertices to keep the set smaller but still allowing full coverage, and finally two strategies based on the work by Chwa et al. [14]. Here, G is initialized to use all reflex vertices, and W is initialized to have a witness on every polygon edge that is incident to a reflex vertex.

The trial consisted of over 18,000 runs, allowing for a direct comparison of individual parameter choices. It was found that many instances could be solved optimally within 20 min. This happened for 60% of the 500-vertex polygons, and 85% of the 100-vertex polygons. Other findings included the importance of the initial solution, where the best strategy (the Chwa-inspired one) led to an overall speedup factor of 2. The best primal and dual separators were identified in a similar way. Furthermore, the authors were first to observe the bathtub-shaped runtime distribution that is still prominent in today's algorithms: Either the algorithm finishes very quickly, usually in the first few seconds, with an optimal solutions, or it engages in an excruciatingly slow process to find good guards and witnesses, often failing to finish within time.

3.6 Torino 2011: $AGP(P, P)$, Tens of Vertices

In 2011, Bottino et al. [10] improved their previous work (see Sect. 3.3) by applying similar ideas to solve the original AGP rather than the $AGP(P, \partial P)$. The objective was to develop an algorithm capable of finding nearly-optimal solutions for full polygon coverage, since, at that time, there was a lack of practical methods for this task.

The first step of the presented technique consists of using the algorithm discussed in Sect. 3.3, which allows for obtaining a lower bound and also multiple optimal solutions for the AGP$(P, \partial P)$. These are then tested in the search for a coverage of the entire polygon. According to the authors, if such solution exists, it is automatically a nearly optimal one for AGP(P, P). If a viable solution is not among those, guards are then added using a greedy strategy until a feasible solution is found.

It should be noted that there are worst-case instances for AGP(P, P) that only possess a single optimal solution, where no characterization of the guard positions is known. Therefore this algorithm, and none of the subsequent ones presented in this paper, can guarantee to find optimal solutions. This common issue is discussed in more detail in Sect. 6.2.

For the experiments presented in [10], 400 polygons with sizes ranging from 30 to 60 vertices were examined. As in the previous work, the following classes were tested: simple, orthogonal, random polygons with holes and also random orthogonal polygons with holes. Guaranteed optimal solutions were found in about 68% of the polygons tested. Note that, in about 96% of the cases, the solution found for AGP$(P, \partial P)$ in the first step of this algorithm was also viable for AGP(P, P). In addition, the authors also implemented the most promising techniques by Amit et al. (see Sect. 3.1), in order to enable comparison between both works. As a result, this technique was more successful than the method by Amit et al. considering the random polygons tested.

3.7 Braunschweig 2012: $AGP(P, P)$, Hundreds of Vertices

In 2012, the primal-dual method introduced by the Braunschweig group was extended. The previous version could find optimal point guards, but only for the LP relaxation which allows fractional guards. Integer solutions could only be found by chance. Now, two ingredients were added to find integer solutions: An ILP-based routine and cutting planes. See Friedrichs [33] for a detailed discussion on the cutting planes, and Fekete et al. [30,31] for the combined approach. As it turned out, this algorithm could solve the classic problem AGP(P, P) on instances of several hundreds of vertices with holes, a factor 10 more than in previous work.

The 2012 algorithm switches between primal and dual phases. In the primal phase, feasible solutions are sought, i.e., upper bounds. Unlike the 2010 version, now only integer solutions are considered. For the current set G of guards and W of witnesses, AGP(G, W) is solved optimally using an ILP formulation. The visibility overlay $\mathcal{A}(G)$ is scanned for insufficiently covered spots, and additional witnesses are generated accordingly. The primal phase ends when no new witnesses are generated, with a feasible integer solution for AGP(G, P), and hence an upper bound for AGP(P, P). In the dual phase, new guard positions are found using the dual arrangement $\mathcal{A}(W)$. For that, a dual solution is needed, which is provided by solving the LP relaxation AGP$_{\mathbb{R}}(G, W)$. The dual phase ends with an optimal solution for AGP$_{\mathbb{R}}(P, W)$, which is a lower bound for AGP$_{\mathbb{R}}(P, P)$, and hence also AGP(P, P). The procedure computes a narrowing sequence of upper bounds for AGP(P, P) and lower bounds for AGP$_{\mathbb{R}}(P, P)$, leaving the issue of closing the integrality gap between them. This may lead to terminating with a suboptimal solution, however with a provided lower bound. As a leverage against this shortcoming, cutting planes are employed to raise the lower bounds [33]. Two classes of facet-defining inequalities for the convex hull of all feasible integer solution of $AGP(G, W)$ are identified. While the NP-hardness of AGP indicates that it is hopeless to find a complete polynomial-size facet description, it is shown that the new inequalities contain a large set of facets, including all with coefficients in $\{0, 1, 2\}$, see also [6]. The dual phase is enhanced with separation routines for the two classes, consequently improving the lower bounds, and often allowing the algorithm to terminate with provably optimal solutions.

To evaluate this work, experiments were conducted on four different polygon classes, sized between 60 and 1000 vertices. These included both orthogonal and non-orthogonal instances, both with and without holes, and polygons where optimal solutions cannot use vertex guards. Different parametrizations of the algorithms were tested, and it was found that the ILP-based algorithm itself (without applying cutting planes) could identify good integer solutions, sometimes even optimal ones, and considerably surpassed the previous 2010 version. The algorithm was able to find optimal solutions for 500-vertex instances quite often. Instances with 1000 vertices were out of reach though.

3.8 Campinas 2013:
$AGP(P, P)$, Hundreds of Vertices

The work by Tozoni et al. [60] generalizes to $AGP(P, P)$ the ideas developed for $AGP(V, P)$ by Couto et al. (see Sect. 3.4). The paper proposes an algorithm that iteratively generates upper and lower bounds while seeking to reach an exact solution. Extensive experiments were carried out which comprised 1440 simple polygons with up to 1000 vertices from several classes, all of which were solved to optimality in a matter of minutes on a standard desktop computer. Up to that point in time, this was the most robust and effective algorithm available for $AGP(P, P)$, for simple polygons. The restriction to simple polygons in this version as well as earlier versions of the Campinas branch originates from the fact that no visibility algorithm for general polygons was available to the group in Campinas, yet.

The algorithm generates, through a number of iterations, lower and upper bounds for the $AGP(P, P)$ through the resolution of the two semi-infinite discretized variants of the original AGP, namely $AGP(P, W)$ (asking for the minimum number of guards that are sufficient to cover the finite set W of witnesses) and $AGP(G, P)$ (computing the minimum number of guards from G that are sufficient to cover P). Notice that in these variants, either the witness or the guard candidate set is infinite, preventing the formulation of these problem variants as an ILP. However, remarkable results [60] show that both variants can be reduced to a compact set covering problem.

To solve $AGP(P, W)$ instance, the algorithm constructs $\mathcal{A}(W)$, and chooses the vertices of the light AVPs to become part of the guard candidates set G.

Assuming that $|W|$ is bounded by a polynomial in n, the same holds for $|G|$. Therefore, the SCP instance corresponding to $AGP(G, W)$ admits a compact ILP model. Tozoni et al. showed that an optimal solution for $AGP(P, W)$ can be obtained by solving $AGP(G, W)$. Thus, the algorithm computes a lower bound for $AGP(P, P)$ using an ILP solver.

Now, to produce an upper bound for $AGP(P, P)$, an idea similar to the one developed by Couto et al. [19] to solve the $AGP(V, P)$ is used. The procedure starts with the same sets G and W used for the lower bound computation. The $AGP(G, W)$ is solved as before. If the optimal solution found in this way covers P, then it is also feasible for $AGP(P, P)$ and provides an upper bound for the problem. Otherwise, new witnesses are added to the set W and the procedure iterates. The upper bound procedure is known to converge in a number of iterations that is polynomial in n.

The lower and upper bound procedures are repeated until the gap between the two bounds reaches zero or a predefined time limit is reached. For certain initial discretization sets and strategies for updating the witness set, one can construct fairly simple instances that lead the algorithm to run indefinitely. Therefore, it remains an important open question whether there exists a discretization scheme that guarantees that the algorithm always converges, see also Sect. 6.2.

An important step of this algorithm, which greatly affects the performance of the final program, is how the initial witness set should be chosen and updated throughout the iterations. Two initial discretizations were tested in practice and are worth noting. The first one, called *Chwa-Points*, is based on the work by Chwa et al. [14] and chooses the middle points of reflex-reflex edges and the convex vertices that are adjacent to reflex vertices. This is similar to the initialization used in [7,41]. The second, called *Convex-Vertices*, comprises all convex vertices of P.

The computational results obtained by this algorithm confirmed its robustness. A total of 1440 instances were tested from different polygon classes, including simple, orthogonal and von Koch ones. Optimal solutions were found for all of them. Also, comparisons with previous published papers showed that the algorithm was effective and far more robust than its competitors. Experiments with different initial witness sets revealed that, on average, *Chwa Points* attained the best results. However, on von Koch Polygons, *Convex Vertices* performed better.

3.9   Campinas 2013 (Journal Version): $AGP(P, P)$, Thousands of Vertices

After presenting an algorithm for AGP with point guards in spring 2013 (see Sect. 3.8), the research group from Campinas continued working on the subject. In this context, improvements were implemented, including the development of their own visibility algorithm that was also able to handle polygons with holes, giving rise to a new version of the algorithm [59]. The resulting implementation is able to solve polygons with thousands of vertices in a few minutes on a standard computer.

Several major improvements were introduced in order to reach this redesigned version. Among them, a Lagrangian Heuristic method (see Sect. 5.2) was implemented to help the ILP solver expedite the computation of optimal solutions for SCP instances. Moreover, a procedure for removing redundant variables and constraints from the SCP formulation was also used to speed up the ILP resolution process.

One of the most effective changes consisted in reversing the point of view of visibility testing from the perspective of the guards to that of the witnesses. Since these are fewer in number and their arrangement has already been computed, much of the geometric computation is simplified.

In the end, 2440 instances were tested and optimal solutions were found for more than 98% of them. The test bench included several different classes of polygons, with and without holes, with up to 2500 vertices. Besides the classes tested in the previous version [60], the authors also used a newly created benchmark instance for polygons with holes (see also Sect. 4.1) and the spike polygons presented by Kröller et al. [41]. Also, comparisons were made with the work of Kröller et al. and an analysis of the effects obtained from different discretizations for the initial witness set were presented. Moreover, the authors evaluated the impact of using a Lagrangian heuristic on the overall performance of the method and concluded that it reduces the average execution time in most of the cases.

3.10   Braunschweig 2013 (Current Version): $AGP(P, P)$, Thousands of Vertices

A deeper runtime analysis of the former code from 2012 revealed that the main bottlenecks where the geometric subroutines, namely (i) the computation of visibility polygons (an implementation of a $O(n \log n)$ rotational sweep as in Asano [4]), (ii) the overlays of these visibility polygons to form $\mathcal{A}(G)$ and $\mathcal{A}(W)$ $(O(n^2 m^2 \log(nm)))$, where m is the size of G resp. W), and (iii) point location algorithms to determine membership in AVPs. This was somewhat surprising as all of these algorithms have fairly low complexity, especially when compared to LP solving (worst-case exponential time when using the Simplex algorithm) and ILP solving (NP-hard in general, and used to solved the NP-hard Set Cover problem). Still the geometric routines made up for over 90% of the runtime.

The group in Braunschweig focused on the improvement of these geometric subroutines: (i) A new Computational Geometry Algorithms Library [12] (CGAL) package for visibility polygon computation was developed in Braunschweig [36], which contains a new algorithm named triangular expansion [11]. Though the algorithm only guarantees an $O(n^2)$ time complexity, it usually performs several magnitudes faster than the rotational sweep. (ii) The code now uses the lazy-exact kernel [54], which delays (or even avoids) the construction of exact coordinates of intersection points as much as possible. The impact is most evident in the construction of the overlays, which contain many intersection points. (iii) The algorithm was restructured to allow a batched point location [62, Sect. 3][1] of all already existing guards (or witnesses) with respect to a new visibility polygon at once.

The new code now runs substantially faster, allowing it to solve much larger instances than the previous one. This paper contains the first experimental evaluation of this new algorithm. Section 4 contains results from running the algorithm and comparing it to the other approaches presented here. Section 5 discusses the speedup obtained by the new subroutines.

3.11   Campinas and Braunschweig 2013 (Current Version): $AGP(P, P)$, Thousands of Vertices

This implementation is the result of a joint effort by the Braunschweig and the Campinas groups. With the intent of achieving robustness, its core is the algorithm from Campinas (Sect. 3.9), refitted with optimizations from Braunschweig that greatly improved its efficiency.

The new code now also uses the lazy exact kernel (cf. Sect. 3.10) of CGAL and the triangular expansion algorithm [11] of the new visibility package [36] of CGAL. While the impact of the new visibility polygon algorithm was huge for both approaches the usage of the lazy kernel was also significant since the overlays in the approach of Campinas contain significantly more intersection points. To see more about how changes in kernel and visibility affect the solver, consult Sect. 5.

Moreover, the current version of Campinas also includes new approaches on the algorithm side. One of the ideas developed was to postpone the computation of an upper bound (solving $AGP(G, P)$) to the time that a good lower bound, and, consequently, a "good" set of guard candidates is obtained. This can be done by repeatedly solving only $AGP(P, W)$ instances until an iteration where the lower bound is not improved is reached. This situation possibly means that the value obtained will not change much in the next iterations. It also increases the chances that the first viable solution found is also provably optimal, which automatically reduces the number of $AGP(G, P)$ instances which must be resolved.

Other changes that are the inclusion of a new strategy for guard positioning, where only one interior point from each light AVP is chosen to be part of the guard candidate set (instead of all its vertices), and the possibility of using IBM ILOG CPLEX Optimization Studio [22] (CPLEX) solver instead of XPRESS.

This new version was tested in experiments conducted for this paper, using 900 problem instances ranging from 200 to 5000 vertices. Section 4 presents the obtained results in detail. The implementation proved to be efficient and robust for all classes of polygons experimented.

---

[1] This is an an $O((n+m) \log n)$ sweep line algorithm, where n is the number of polygon vertices and m the number of query points.

4   Experimental Evaluation

To assess how well the AGP can be solved using current algorithms, and how their efficiency has developed over the last years, we have run exhaustive experiments. The experiments involve all algorithms for which we could get working implementations, and were conducted on the same set of instances and on the same machines, see Sect. 4.2. We refrain from providing comparisons based on numbers from the literature.

We had several snapshots from the Braunschweig and Campinas code available, these are:

- For Braunschweig, the versions from 2010 (Sect. 3.5), 2012 (Sect. 3.7), and 2013 (Sect. 3.10). These will be referred to as BS-2010, BS-2012, and BS-2013, respectively.
- For Campinas, the version from 2009 (Sect. 3.4), and the two snapshots from 2013 (Sects. 3.8 and 3.9). These will be referred to as C-2009, C-2013.1 and C-2013.2, respectively.
- The latest version is the combined approach from Campinas and Braunschweig that was obtained during a visit of Davi C. Tozoni to Braunschweig (Sect. 3.11), which we refer to as C+BS-2013.

The older versions have already been published, for these we provide a unified evaluation. The versions BS-2013 and C+BS-2013 are, as of yet, unpublished.

4.1   AGPLib

For the performed experiments, several classes of polygons were considered. The majority of them were collected from AGPLib [16], which is a library of sample instances for the AGP, consisting of various classes of polygons of multiple sizes. They include the test sets from many previously published papers [7, 19–21, 41, 59, 60].

To find out more about how each of the classes was generated, see [20] and [59]. Below, we show a short description of the six classes of instances considered in this survey; all of them are randomly generated:

"simple": Random non-orthogonal simple polygons as in Fig. 3a.

"simple-simple": Random non-orthogonal polygons as in Fig. 3b. These are generated like the "simple" polygon class, but with holes. The holes are also generated like the first class and randomly scaled and placed until they are in the interior of the initial polygon.

"ortho": Random floorplan-like simple polygons with orthogonal edges as in Fig. 3c.

"ortho-ortho": Random floorplan-like orthogonal polygons as in Fig. 3d. As the simple-simple class, these polygons are generated by using one polygon of the ortho class as main polygon, and then randomly scaling and translating smaller ortho polygons until they are contained within the main polygon's interior, where they are used as holes.

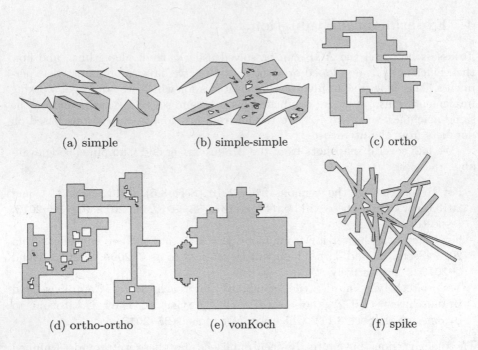

(a) simple   (b) simple-simple   (c) ortho

(d) ortho-ortho   (e) vonKoch   (f) spike

Fig. 3. Example instances of different polygon classes.

"von Koch": Random polygons inspired by randomly pruned Koch curves, see Fig. 3e.

"spike": Random polygons with holes as in Fig. 3f. Note that this class is specifically designed to provide polygons that encourage placing point guards in the intersection centers. It has been published along with the BS-2010 algorithm (Sect. 3.5), which was the first capable of placing point guards.

4.2   Experimental Setup

The experiments were run on identical PCs with eight-core Intel Core i7-3770 CPUs at 3.4 GHz, 8 MB cache, and 16 GB main memory running a 64-bit Linux 3.8.0 kernel. All algorithms used version 4.0 of Computational Geometry Algorithms Library [12] (CGAL) and IBM ILOG CPLEX Optimization Studio [22] (CPLEX) 12.5. The only component using concurrency is the ILP solver CPLEX, everything else was single-threaded. For each polygon-class/complexity combination, we tested 30 different polygons. Each test run had a runtime limit of 20 min.

4.3   Results

Historically, the two lines of algorithms have been working towards AGP from different angles. Campinas focused on binary solutions, which initially came at the expense of being limited to given guard discretization, like vertex guards: $AGP(V, P)$. The Braunschweig work started with point guards, but the price were fractional solutions: $AGP_R(P, P)$. It only became possible to reliably solve the binary AGP with point guards, $AGP(P, P)$, with the BS-2012 and C-2013.1 algorithms.

Therefore, we first sketch progress for the AGP with vertex guards and the fractional AGP before discussing the experimental results for AGP with point guards itself.

Vertex Guards. $AGP(V, P)$ is one of the semi-infinite variants. We believe it to be a considerably simpler problem than $AGP(P, P)$ for two reasons: (1) Both variants are NP-hard, but we know $AGP(V, P)$ is in NP, which is uncertain for $AGP(P, P)$ as it is unknown if there is a polynomial-size representation of guard locations. (2) Experience and experimental results indicate that finding good guard candidates is the hardest part of the problem and leads to many iterations; but for $AGP(V, P)$ we only have to place witnesses and solve SCP instances, which is usually possible in a comparably short time frame with a good ILP solver. The first experimental work on $AGP(V, P)$ was Campinas 2007, but unfortunately, the implementation is no longer available.

Table 1. Optimality rates for vertex guards. Notice that BS-2010 finds fractional vertex guard solutions, whereas the others find integer ones.

	Polygons without holes					Polygons with holes				
	200	500	1000	2000	5000	200	500	1000	2000	5000
C-2009	100.0	100.0	100.0	100.0	63.3	–	–	–	–	–
C-2013.1	100.0	100.0	100.0	100.0	63.3	–	–	–	–	–
BS-2013	100.0	100.0	100.0	100.0	88.9	100.0	97.8	67.8	77.8	66.7
C-2013.2	100.0	100.0	100.0	100.0	37.8	100.0	100.0	66.7	65.6	0.0
C+BS-2013	100.0	100.0	100.0	100.0	100.0	100.0	100.0	77.8	88.9	66.7
BS-2010 *	100.0	100.0	100.0	100.0	33.3	100.0	100.0	100.0	98.9	0.0

Table 1 shows optimality rates, i.e., how many of the instances each implementation could solve, given a 20 min time limit per instance. The polygons were grouped in two categories: those without holes, including the instances classes **simple**, **ortho** and **von Koch**, and those with holes composed by the instances in the classes **simple-simple**, **ortho-ortho** and **spikes**. The Campinas versions prior to C-2013.2 could not deal with holes in input polygons, so these entries are empty. It should also be noted that BS-2010 solves the easier case of fractional vertex guards. It is clearly visible how all algorithms (including the five-year-old C-2009) can solve all simple polygons with up to 2000 vertices as well as most simple 5000-vertex polygons. For instances with holes, however, the solution percentages of all algorithms (except BS-2010 which solves an easier problem) start significantly dropping at 1000 vertices. This demonstrates two effects: First, for smaller sizes, the problem is easier to solve as the search for good guard candidates is unnecessary. Second, for larger sizes, finding optimal solutions to large instances of the NP-hard SCP dominate, resulting in a computational barrier. The difficulty to handle large SCP instances also shows up when we consider the results of the Campinas codes C-2013.1 and C-2013.2. As the size of the polygons increases and the SCPs to be solved grow in complexity, the Lagrangian heuristic employed by C-2013.2 version uses more computational time but does not help the ILP solver to find optimal solutions for the AGP(G,W) instances, due to the deterioration of the primal bounds. This inefficiency causes a decrease in the solver's performance, as can be seen in the optimality rate shown in Table 1 for simple polygons with 5000 vertices. In this case, if C-2013.2 did not use the Lagrangian heuristic by default, a result at least similar to that obtained by C-2013.1 would be expected.

The high solution rates allow us to directly analyze the speedup achieved over time. Table 2 shows how much faster than C-2009 later algorithms could solve the problem. The shown numbers are log-averages over the speedup against C-2009 for all instances solved by both versions. As C-2009 cannot process holes, this analysis is restricted to simple polygons. It is clearly visible that

Table 2. Speedup for vertex guards. Numbers indicate how many times faster than C-2009 later implementations became, computed as log-average. The comparison is only possible when there is at least one instance of the group that was solved by all considered solvers. This table is restricted to simple polygons, since C-2009 does not support polygons with holes.

Class	n	Speedup factor					
		C-2009	BS-2010	C-2013.1	C-2013.2	BS-2013	C+BS-2013
Simple	200	1.00	0.66	1.03	1.21	7.54	**6.75**
	500	1.00	0.66	1.01	1.02	7.79	**10.21**
	1000	1.00	0.66	1.02	0.95	8.03	**14.65**
	2000	1.00	0.68	1.00	0.90	10.24	**18.97**
	5000	–	–	–	–	–	–
Orthogonal	200	1.00	0.64	1.01	1.05	**6.46**	6.15
	500	1.00	0.63	1.01	0.98	6.67	**10.82**
	1000	1.00	0.65	1.00	0.92	7.75	**15.67**
	2000	1.00	0.65	0.98	0.82	9.57	**19.52**
	5000	1.00	0.72	1.00	0.75	12.63	**28.64**
von Koch	200	1.00	0.38	1.02	1.33	2.09	**3.45**
	500	1.00	0.44	1.09	1.37	1.86	**4.27**
	1000	1.00	0.60	0.95	1.40	1.95	**4.75**
	2000	1.00	0.92	1.34	1.39	2.67	**6.18**
	5000	–	–	–	–	–	–

BS-2013 is about five times faster then C-2009, and the changes from C-2013.2 to C+BS-2013 led to a speedup factor of about seven. These stem from a number of changes between versions, however, roughly a factor 5 can be attributed to improvements in geometric subroutines — faster visibility algorithms, lazy-exact CGAL kernel, reduced point constructions. We discuss the influence of geometry routines in Sect. 5.1.

Fractional Guards. The Braunschweig line of work started with solving the fractional point guard variant $AGP_R(P, P)$ and all Braunschweig versions, even those designed for binary solutions, still support the fractional AGP. Table 3 shows how often the three implementations could find optimal solutions, and how often they achieved a 5% gap by the end of the 20-min runtime limit. Here again, the polygons have been grouped: those with holes and those without holes.

Table 3. Optimality rates for fractional point guards.

		Polygons without holes					Polygons with holes				
		200	500	1000	2000	5000	200	500	1000	2000	5000
OPT	BS-2010	55.6	27.8	14.4	3.3	0.0	54.4	32.2	28.9	30.0	0.0
	BS-2012	53.3	30.0	11.1	4.4	0.0	54.4	32.2	27.8	31.1	0.0
	BS-2013	56.7	24.4	12.2	1.1	0.0	50.0	31.1	27.8	31.1	33.3
5% gap	BS-2010	93.3	100.0	100.0	100.0	33.3	96.7	100.0	98.9	75.6	0.0
	BS-2012	93.3	100.0	100.0	100.0	33.3	97.8	98.9	98.9	72.2	0.0
	BS-2013	91.1	100.0	100.0	100.0	98.9	97.8	98.9	98.9	98.9	33.3

Unsurprisingly, there is no significant difference between the BS-2010 and BS-2012 versions, the development between these snapshots focused on the integer case. The improvements from BS-2012 to BS-2013 stem from improved geometry subroutines which are beneficial to both, the binary and the fractional mode. It can be seen that near-optimal solutions are obtained almost every time, but the gap is not always closed. Furthermore, with the 20-min time limit, there is an barrier between 2000 and 5000 vertices, where the success rate drops sharply, indicating that the current frontier for input complexity lies roughly in this range.

Point Guards. We turn our attention to the classic AGP, $AGP(P, P)$: Finding integer solutions with point guards. We report optimality in three different ways: Which percentage of instances could be solved optimally with a matching lower bound (i.e., proven optimality) is reported in Table 4; we show in how many percent of the cases an instance could be solved optimally, whether or not a matching bound was found in Table 5; Table 6 reports how many percent of the solutions were no more than 5% away from the optimum. This allows to

Table 4. Optimality Rate for point guards.

Class	n	Optimality rate (%)				
		BS-2012	C-2013.1	C-2013.2	BS-2013	C+BS-2013
Simple	200	**100.0**	**100.0**	**100.0**	96.7	**100.0**
	500	76.7	**100.0**	**100.0**	96.7	**100.0**
	1000	70.0	96.7	**100.0**	90.0	**100.0**
	2000	36.7	6.7	50.0	60.0	**100.0**
	5000	0.0	0.0	0.0	26.7	**100.0**
Orthogonal	200	96.7	**100.0**	**100.0**	96.7	96.7
	500	86.7	**100.0**	96.7	93.3	93.3
	1000	70.0	**100.0**	**100.0**	86.7	**100.0**
	2000	46.7	70.0	90.0	70.0	**100.0**
	5000	0.0	0.0	0.0	40.0	93.3
Simple-simple	200	93.3	–	**100.0**	86.7	**100.0**
	500	76.7	–	83.3	60.0	**100.0**
	1000	3.3	–	0.0	13.3	**100.0**
	2000	0.0	–	0.0	0.0	46.7
	5000	0.0	–	0.0	0.0	0.0
Ortho-ortho	200	83.3	–	96.7	86.7	**100.0**
	500	53.3	–	83.3	53.3	**100.0**
	1000	16.7	–	3.3	16.7	96.7
	2000	0.0	–	0.0	0.0	33.3
	5000	0.0	–	0.0	0.0	0.0
von Koch	200	**100.0**	**100.0**	**100.0**	**100.0**	**100.0**
	500	**100.0**	96.7	**100.0**	93.3	**100.0**
	1000	**100.0**	46.7	**100.0**	96.7	**100.0**
	2000	83.3	0.0	0.0	86.7	**100.0**
	5000	0.0	0.0	0.0	0.0	0.0
Spike	200	**100.0**	–	**100.0**	96.7	**100.0**
	500	**100.0**	–	**100.0**	**100.0**	**100.0**
	1000	3.3	–	96.7	**100.0**	**100.0**
	2000	0.0	–	96.7	**100.0**	**100.0**
	5000	0.0	–	0.0	96.7	**100.0**

distinguish between cases where BS-2013 does not converge, and cases where the integrality gap prevents it from detecting optimality.

The C+BS-2013 implementation solves the vast majority of instances from our test set to proven optimality, the only notable exception being some classes of very large polygons with holes and the 5000-vertex Koch polygons. Given how the best known implementation by 2011, the Torino one from Sect. 3.6, had an optimality rate of about 70% for 60-vertex instances, it is clearly visible

Table 5. Optimality Rate without proof for point guards.

Class	n	Optimality rate (%) without proof				
		BS-2012	C-2013.1	C-2013.2	BS-2013	C+BS-2013
Simple	200	**100.0**	**100.0**	**100.0**	96.7	**100.0**
	500	80.0	**100.0**	**100.0**	**100.0**	**100.0**
	1000	73.3	**100.0**	**100.0**	**100.0**	**100.0**
	2000	50.0	50.0	80.0	93.3	**100.0**
	5000	0.0	0.0	0.0	83.3	**100.0**
Orthogonal	200	96.7	**100.0**	**100.0**	96.7	96.7
	500	86.7	**100.0**	**100.0**	93.3	93.3
	1000	70.0	**100.0**	**100.0**	90.0	**100.0**
	2000	50.0	96.7	93.3	90.0	**100.0**
	5000	0.0	0.0	0.0	50.0	93.3
Simple-simple	200	96.7	–	**100.0**	90.0	**100.0**
	500	93.3	–	96.7	80.0	**100.0**
	1000	33.3	–	20.0	73.3	**100.0**
	2000	0.0	–	0.0	33.3	50.0
	5000	0.0	–	0.0	0.0	0.0
Ortho-ortho	200	93.3	–	**100.0**	**100.0**	**100.0**
	500	80.0	–	93.3	90.0	**100.0**
	1000	70.0	–	30.0	70.0	96.7
	2000	0.0	–	0.0	30.0	43.3
	5000	0.0	–	0.0	0.0	0.0
von Koch	200	**100.0**	**100.0**	**100.0**	**100.0**	**100.0**
	500	**100.0**	**100.0**	**100.0**	93.3	**100.0**
	1000	**100.0**	70.0	**100.0**	96.7	**100.0**
	2000	83.3	0.0	30.0	90.0	**100.0**
	5000	0.0	0.0	0.0	0.0	0.0
Spike	200	**100.0**	–	**100.0**	96.7	**100.0**
	500	**100.0**	–	**100.0**	**100.0**	**100.0**
	1000	3.3	–	**100.0**	**100.0**	**100.0**
	2000	0.0	–	96.7	**100.0**	**100.0**
	5000	0.0	–	0.0	96.7	**100.0**

how the developments in the last years pushed the frontier. With C+BS-2013, instances with 2000 vertices are usually solved to optimality, showing an increase in about two orders of magnitude. The success of C+BS-2013 is multifactorial: It contains improved combinatorial algorithms as well as faster geometry routines, most notably a fast visibility implementation. Section 5 discusses its key success factors.

Table 6. Rate of upper bound within 5% distance to lower bound.

Class	n	5% gap Rate in (%)				
		BS-2012	C-2013.1	C-2013.2	BS-2013	C+BS-2013
Simple	200	100.0	100.0	100.0	100.0	100.0
	500	100.0	100.0	100.0	100.0	100.0
	1000	100.0	100.0	100.0	100.0	100.0
	2000	100.0	100.0	96.7	100.0	100.0
	5000	0.0	0.0	0.0	100.0	100.0
Orthogonal	200	100.0	100.0	100.0	100.0	100.0
	500	100.0	100.0	100.0	100.0	100.0
	1000	100.0	100.0	100.0	100.0	100.0
	2000	100.0	100.0	100.0	100.0	100.0
	5000	0.0	0.0	0.0	100.0	100.0
Simple-simple	200	100.0	— —	100.0	100.0	100.0
	500	100.0	— —	93.3	100.0	100.0
	1000	100.0	— —	33.3	100.0	100.0
	2000	0.0	— —	0.0	96.7	80.0
	5000	0.0	— —	0.0	0.0	0.0
Ortho-ortho	200	100.0	— —	100.0	100.0	100.0
	500	100.0	— —	100.0	100.0	100.0
	1000	100.0	— —	40.0	96.7	100.0
	2000	56.7	— —	0.0	76.7	86.7
	5000	0.0	— —	0.0	0.0	0.0
von Koch	200	100.0	100.0	100.0	100.0	100.0
	500	100.0	100.0	100.0	100.0	100.0
	1000	100.0	73.3	100.0	100.0	100.0
	2000	100.0	0.0	56.7	100.0	100.0
	5000	0.0	0.0	0.0	3.3	0.0
Spike	200	100.0	— —	100.0	96.7	100.0
	500	100.0	— —	100.0	100.0	100.0
	1000	3.3	— —	96.7	100.0	100.0
	2000	0.0	— —	96.7	100.0	100.0
	5000	0.0	— —	0.0	96.7	100.0

It can be seen from Table 6 that many algorithms are able to find near-optimal solutions (5% gap) for most instances, indicating that for practical purposes, all 2013 algorithms perform very well. The frontier on how large instances can be solved with small gap is between 2000 and 5000 vertices for most polygons with holes and beyond 5000 vertices for simple polygons.

Comparing Tables 4, 5 and 6, it can be seen that the primal-dual approach (BS-2010 and BS-2012) produces decent upper bounds, often optimal ones, but does have an issue with finding matching lower bounds. This drawback has been much improved in BS-2012 but is still measurable.

Finally, we analyze how difficult the individual instance classes are. In Tables 4, 5 and 6, we group them by size and based on whether they feature holes. Table 7 shows optimality rates for each class. We restrict presentation to BS-2013 here, for the simple reason that it has the highest variation in reported rates. In each class, we see a continuous decline with increasing input complexity, indicating that local features of an instance play a major role in how hard it is to solve it, rather than this being an intrinsic property of the generator. The only generator that produces "easier" instances than the others is Spike. These are instances tailored for showing the difference between vertex and point guards, requiring few guards to be placed in the middle of certain free areas. We include the Spike instances in our experiments because they are an established class of test cases, being aware that all of the current implementations are able to identify good non-vertex positions for guards, and that this class has to be considered easy.

Table 7. Optimality rates for BS-2013 on different instance classes.

	200	500	1000	2000	5000	Avg
Simple	96.7	96.7	90.0	60.0	26.7	74.0
Orthogonal	96.7	93.3	86.7	70.0	40.0	77.3
simple-simple	86.7	60.0	13.3	0.0	0.0	32.0
ortho-ortho	86.7	53.3	16.7	0.0	0.0	31.3
von Koch	100.0	93.3	96.7	86.7	0.0	75.3
Spike	96.7	100.0	100.0	100.0	96.7	98.7

5   Success Factors

As seen in Sect. 3, the most effective algorithms for the AGP can be decomposed into four elements:

- Geometric subroutines dealing with computing visibility relations, determining feasibility,
- Set Cover subroutines computing (near-)optimal solutions for finite cases,
- Routines to find candidates for discrete guard and witness locations, and
- An outer algorithm combining the three parts above.

In this section, we focus on these techniques.

5.1   Geometric Subroutines

Both groups use the 2D Arrangements package [62] of CGAL which follows the *generic programming paradigm* [5]. For instance, in the case of arrangements it is possible to change the curve type that is used to represent the planar subdivisions or the kernel that provides the essential geometric operations and also determines the number type used. In the context of this work, it is clear that the used curves are simply segments[2]. However, the choice of the geometric kernel can have a significant impact on the runtime.

First of all, it should be noted that among the different kernels that CGAL offers only kernels that provide exact constructions should be considered as any inexact construction is likely to induce inconsistencies in the data structure of the arrangements package. This already holds for seemingly simple scenarios as the code of the arrangement package heavily relies on the assumption that constructions are exact.

This essentially leaves two kernels: The Cartesian kernel and the lazy-exact kernel. For both kernels it is possible to exchange the underlying exact rational number type, but CGAL :: Gmpq [35] is the recommended one[3].

The **Cartesian kernel**, is essentially the naive application of exact rational arithmetic (using the one that it is instantiated with, in this case CGAL :: Gmpq). Thus, coordinates are represented by a numerator and denominator each being an integer using as many bits as required. This implies that even basic geometric constructions and predicates are not of constant cost, but depend on the bit-size of their input. For instance, the intersection point of two segments is likely to require significantly more bits than the endpoints of the segments. And this is even more relevant in case of cascaded constructions as the bit growth is cumulative. This effect is very relevant in both approaches due to there iterative nature, e.g., when such a point is chosen to be a new guard or witness position.

The **lazy-exact kernel** [54] tries to attenuate all these effects by using exact arithmetic only when necessary. Every arithmetic operation and construction is first carried out using only double interval arithmetic, that is, using directed rounding, an upper and a lower of the exact value is computed. The hope is that for most cases this is already sufficient to give the correct and certified answer, for instance whether a point is above or below a line. However, for the case when this is not sufficient, each constructed object also knows its history, which makes it possible to carry out the exact rational arithmetic as it is done in the Cartesian kernel in order to determine the correct result. This idea is implemented by the lazy kernel not only on the number type level[4], but also for predicates and constructions, which reduces the overhead (memory and time) that is induced by maintaining the history.

---

[2] In the context of fading [43] circular arcs may also be required.

[3] Other options are, for instance, leda::rational [46] or CORE::BigRat [39], but, compared to Gmpq, both imply some overhead and are only recommended in case the usage of the more complex number types of these libraries is required.

[4] This can be achieved by the instantiation of the Cartesian kernel with CGAL::Lazy_exact_nt<CGAL::Gmpq>.

Table 8. The speedup factor of C+BS-2013 using the Cartesian kernel and the lazy-exact kernel. Similar numbers were obtained for BS-2012. The lazy-exact kernel is now the standard configuration in BS-2013 and C+BS-2013.

Class size	200	500	1000	2000	5000
Simple	1.27	1.46	1.55	1.49	1.35
Orthogonal	1.44	1.60	1.66	1.69	1.65
Simple-simple	2.15	1.72	1.44	1.37	-
Ortho-ortho	1.54	1.30	1.21	1.20	-
von Koch	1.02	1.06	1.10	1.16	-
Spike	1.15	1.61	1.76	2.10	2.56

By the genericity of CGAL it is possible to easily exchange the used geometric kernel. Table 8 shows the speedup factors by using the Cartesian kernel vs the lazy-exact kernel for the different instances for C-2013.1. It should be noted that all Braunschweig and Campinas implementations since 2007 use a complexity reduction step together with the Cartesian kernel: Whenever a point in a face is generated, it is rounded to a nearby point of lower bit complexity. Without this, neither implementation would be able to solve any instance of substantial size. This speedup technique is missing in the variant with the lazy-exact kernel, as it requires to actually compute the point coordinates before rounding, which would defeat the purpose of the kernel. Therefore the table compare the lazy-exact kernel against the Cartesian kernel with explicit complexity reduction.

For the random polygons, as well as for the spike ones, it can be observed that the lazy-exact kernel is usually almost twice as fast as the Cartesian kernel. However, for the von Koch polygons the lazy-exact kernel only gives a mild speedup. We explain this by two effects. First, the bit-size of the input polygons is not very large and also the bit-size of intermediate constructions do not grow as much, as the horizontal and vertical lines dominate the scene. Second, the instance induces degenerate situations in which the lazy-exact kernel must fall back to the exact arithmetic in which cases effort for interval arithmetic and maintaining the history is a real overhead. The lazy-exact kernel is now the standard configuration in BS-2013 and C+BS-2013.

Visibility Computation. One of the most significant improvements with respect to speed is due to the new upcoming visibility package [36] of CGAL. This package was developed by the group in Braunschweig with this project being the main motivation. Of course, this packages was also made available to the group in Campinas prior to its actual integration in CGAL. Figure 4 illustrates the tremendous impact on the runtime for both approaches. The left side shows the split up of total runtime for the code from Braunschweig in 2012 and 2013. While in 2012 the update time (dominated by visibility computation) used about two third of the time for visibility computation is now almost negligible. The same holds for improvements achieved in the code from Campinas, see right side of Fig. 4. It can be noticed in the latter graph that the time spent by C+BS-2013

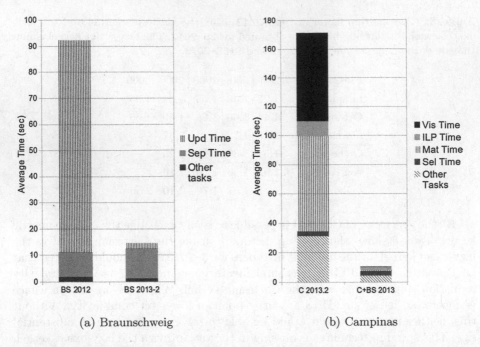

(a) Braunschweig   (b) Campinas

Fig. 4. Split up of average total time for different configurations on all simple instances with 1000 vertices. (left) The update time which is dominated by the visibility polygon computation almost vanishes in BS-2013 compared to the BS-2012. (right) The time spent on visibility in C+BS-2013 is almost negligible compared to the time spend in C-2013.2.

in building the constraint matrices for the ILPs, denoted by Mat Time, also suffered a huge reduction relative to C-2013.2. As commented in Sect. 3.11, this was mostly due to the execution of the visibility testing from the perspective of the witnesses rather than the guards.

5.2   Set Cover Optimization

Many AGP algorithms rely on repeatedly solving $AGP(G, W)$ (Eqs. (1)–(3)) for finite G and W as a subroutine, corresponding to the NP-hard SCP. Therefore improving the solutions times for these SCP instances can benefit the overall algorithm.

Lagrangian Relaxation. In the algorithm developed by the research group in Campinas subsequent to the journal version from 2013 [59] (Sect. 3.9), attempts were made to reduce the time spent by the ILP solver through the implementation of some known techniques, such as ILP matrix reduction and Lagrangian heuristic.

A standard method for reducing constraints and variables was used, which is based on inclusion properties among columns (guard candidates) and rows (witnesses) of the Boolean constraint matrix of the ILP that models the SCP instance.

Furthermore, their algorithm employs a *Lagrangian Heuristic* in order to obtain good, hopefully optimal, feasible starting solutions for the SCP to speedup the convergence towards an optimum. See [8] for a comprehensive introduction to this technique. The heuristic implemented is based on the work presented in [8]. Figure 5 shows how the use of this technique positively influenced the average run time of the approach.

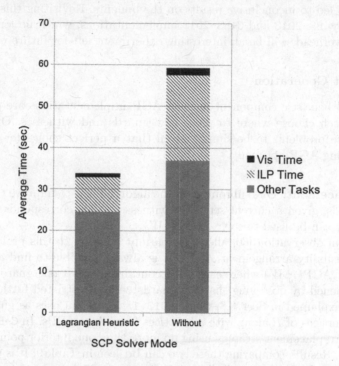

Fig. 5. Average Time needed for the current Campinas version to solve von Koch polygons with 1000 vertices with and without the Lagrangian Heuristic.

DC Programming. A different solution method for the Braunschweig approach was discussed in Kröller et al. [42]. Here, the ILP representing the SCP for $AGP(G, W)$ was rewritten as

$$\min_{x \in \mathbb{R}^G} F(x) \text{ , where } F(x) := \sum_{g \in G} x_g - \theta \sum_{g \in G} x_g(x_g - 1) + \chi(x). \quad (5)$$

Here, θ is a sufficiently large constant used to penalize fractional values for x_g, and $\chi \colon \mathbb{R}^G \to \{0, \infty\}$ is an indicator function with

$$\chi(x) = 0 \quad :\Longleftrightarrow \quad \begin{cases} \sum_{g \in \mathcal{V}(w)} x_g \geq 1 \ \forall w \in W \\ 0 \leq x_g \leq 1 \qquad \forall g \in G \end{cases}. \quad (6)$$

It is easy to see that F can be expressed as $F(x) := f_1(x) - f_2(x)$, where

$$f_1(x) := \sum_{g \in G} x_G + \chi(x), \quad \text{and} \quad f_2(x) = \theta \sum_{g \in G} x_g(x_g - 1), \tag{7}$$

i.e., the SCP instance is reduced to minimizing the difference of two non-linear convex functions. For such optimization problems, the DCA algorithm [53] can be used. In experiments, it was shown that solutions for $AGP(G, W)$ could be found very quickly, however, at the time, the large runtime overhead of the geometric subroutines led to inconclusive results on the benefits. Revisiting this approach with the new BS-2013 and C+BS-2013 implementations, which no longer suffer from this overhead, will be an interesting experiment left for future work.

5.3   Point Generation

The central heuristic component in most AGP implementations are point generators, which choose where to place new guards and witnesses. One cannot expect these problems to be simple, given that a perfect choice for G and W equals solving AGP optimally.

Guard Placement. One subroutine in the algorithms is to improve the current set of guards, given a current set W of witnesses. This corresponds to finding guards that can be used to solve $AGP(P, W)$.

A critical observation [60] allows for elegant solution to this problem: Consider the visibility arrangement $\mathcal{A}(W)$. It is always possible to find an optimal solution for $AGP(P, W)$ where each AVP contains at most one guard. This can be strengthened by observing that the guards can be restricted further to light AVPs. As explained in Sect. 3.8, the C-2013.1 algorithm uses as guard candidates the vertices of P along with all vertices from light AVPs. In C+BS-2013, a second guard placement strategy using no more than one interior point per AVP is available. Results comparing these two can be seen in Table 9. It is possible to conclude that the latest guard placement strategy, which consists of using only one point within each light AVP, is often the best option. The explanation for

Table 9. Percentage of instances solved to binary optimality by the current implementation from Campinas with guard candidates on vertices or inside light AVPs.

	Vertices of light AVPs	Interior of light AVPs
simple 2000	100.0	100.0
ortho 2000	100.0	100.0
simple-simple 2000	6.7	33.3
ortho-ortho 2000	13.3	46.7
von Koch 2000	100.0	100.0
spike 2000	100.0	100.0

this success is probably related to the fact that, with the winning strategy, there is a reduced number of visibility tests between witnesses and guard candidates, as well as a smaller size of SCP instances to be solved.

For $AGP_R(P, W)$, as solved by the Braunschweig line of algorithms, this observation can be extended further: If an optimal dual solution for $AGP_R(G, W)$ is available, selecting additional guards corresponds to a column generation process. Therefore, the BS algorithms place guards only in light AVPs where the dual solution guarantees an improvement in the objective function. To avoid cycling in the column generation process, G is monotonically growing, leading over time to a large number of guard positions.

Witness Placement. The choice of witnesses is as important as that of the guards. In principle, the same reasoning as for guards can be used: Given guard candidates G, creating W with one witness in every shadow AVP of $\mathcal{A}(G)$ guarantees that a solution for $AGP(G, W)$ is also a solution for $AGP(G, P)$. A naïve placement algorithm based on this observation would simply create witnesses in shadow AVPs. However, this leads to the problem of creeping shadows at reflex vertices, see Fig. 6: Placing witness in the interior of the AVP adjacent to the polygon boundary creates an infinite chain of guard/witness positions that converges towards a witness on the boundary, but not reaching it. Both the Braunschweig and the Campinas algorithms therefore can create additional witnesses on the edges of shadow AVPs.

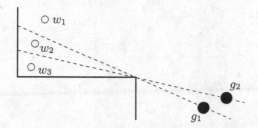

Fig. 6. Creeping shadow effect.

Initial Set. The selection of the first candidates for guards and witnesses, i.e., the initial choice of G and W can have tremendous impact on algorithm runtime. In principle, a good heuristic here could pick an almost optimal set for G and a matching W to prove it, and reduce the algorithm afterwards to a few or even no iterations.

Chwa et al. [14] provide a partial answer to this problem: They attempt to find a finite set of witnesses with the property that guarding this set guarantees guarding the whole polygon. If such a set exists, the polygon is called *witnessable*. Unfortunately this is not always the case. However, for a witnessable polygon, the set can be characterized and quickly computed. Current algorithms do not bother checking for witnessability (although Chwa et al. provide an algorithm), but rather directly compute this set and use it for initial witnesses. Should the polygon be witnessable, the algorithm automatically terminates in the first iteration.

Considering the current version from Campinas, two initial discretizations are used: Convex Vertices (CV) and Chwa Points (CP). The first one includes only the convex vertices of the polygon in the initial set, while the second chooses the middle points of reflex-reflex edges and the convex vertices that are adjacent to reflex vertices.

The two charts in Fig. 7 show the average run time necessary to find optimal solutions when using CV and CP strategies on simple-simple and spike polygons. From these charts, one can perceive that there is an advantage in using the CP discretization for polygons from the simple-simple class. On the other hand, the chart corresponding to the spike polygons shows that the implementation works much better when the strategy chosen is the CV one. In this last case, the program required four times less time to solve the same set of polygons when using the CV strategy as opposed to CP.

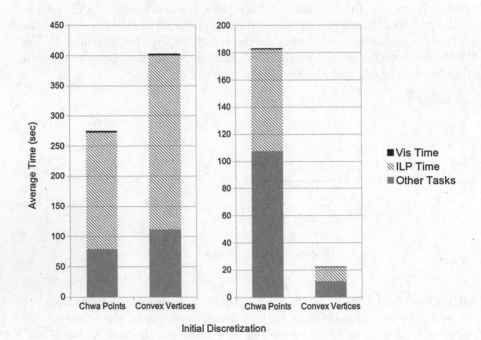

Fig. 7. Average Time needed to solve ortho-ortho (left) and spike (right) polygons with 1000 vertices using the Convex Vertices and the Chwa Points discretization.

For BS-2010, several strategies were implemented, see Table 10 which is extracted from the corresponding paper [41]: Leaving G and W empty (for implementation reasons, both contained one arbitrary point), putting guards

Table 10. Speedup factors in BS-2010 obtained by varying initial guards and witnesses [41].

Initial G	Initial W	Speedup
Single Point	Single Point	1.00
Every other vertex	Every other vertex	1.59
All vertices	All vertices	1.64
All vertices	Reflex edges	1.74
Reflex vertices	Reflex edges	2.02

and witnesses on every (or every other) vertex of the polygon, putting guards on all reflex vertices, and putting a witness on every edge adjacent to a reflex vertex. The Chwa-inspired combination allowed for a speedup of around two.

5.4   Lower Bounds

A crucial success factor for solving the binary AGP variants is the quality of the lower bounds. This is especially visible in BS-2013, which was tested with and without the cutting planes, i.e., with and without the features published in [31,33] and outlined in Sect. 3.7. Table 11 compares the solution rates for the different classes of instances with 500 vertices and clearly shows that using cutting planes greatly improves solution rates. Cutting planes increase the lower bounds and improve the solution rates for all classes of instances.

Table 11. Percentage of instances solved to binary optimality comparing two variants of code from Braunschweig 2013, one with and without cutting planes, for 500-vertex instances.

Class/Technique	With cutting planes	Without cutting planes
ortho	80.0%	63.3%
simple	86.7%	40.0%
von Koch	100.0%	70.0%
ortho-ortho	63.3%	13.3%
simple-simple	70.0%	6.7%
spike	100.0%	96.7%

For the Campinas approach, the quality of the lower bound computed is a very important issue. For AGP(P, P), the lower bound is obtained by solving an AGP(P, W) instance, where W is a discretized set of witnesses points within P. Therefore, it is fair to say that the quality of the value computed is directly dependent on the strategy applied to select the points that comprise the set W. For more information on how the witness set is managed and how it affects convergence of Campinas method, see Sect. 5.3.

6   Variants and Open Problems

6.1   Fading

An interesting variant for the AGP was proposed by Joe O'Rourke in 2005: What if visibility suffers from fading effects, just like light in the real world does? To be precise, we assume that for a guard g with intensity x_g, a witness $w \in \mathcal{V}(g)$ is illuminated with a value of $\varrho(d(g, w))x_g$, where $d(g, w)$ is the Euclidean distance between g and w, and ϱ is a fading function, usually assumed to be

$$\varrho(d) := \begin{cases} 1 & \text{if } d < 1 \\ d^{-\alpha} & \text{if } 1 \le d < R \\ 0 & \text{if } d \ge R \end{cases}. \tag{8}$$

Here, α is a constant (2 for natural light in 3D space), and R is a maximal radius beyond which illumination is neglected. Fixing $\varrho(d)$ to 1 for small d is necessary to keep the problem well-defined. Otherwise, an infinitesimally small light can illuminate a small circle around it. Then, no finite solution can exist, because it can always be improved by creating additional guards between the existing ones, and reducing intensity for all. This converges towards the setup of $G = P$, with all $x_g = 0$, which is not feasible.

Very little is known about this variant. A restricted case has been discussed by Eisenbrand et al. [26], where a 1-dimensional line segment is illuminated from a fixed set of guards. It is shown how to solve this problem exactly and approximatively using techniques from mathematical programming.

The primal-dual Braunschweig algorithm was shown to apply to this variant as well: Kröller et al. [43] have modified the ILP formulation (1)–(3) to use the constraint

$$\sum_{g \in \mathcal{V}(w)} \varrho(d(g, w))x_g \ge 1 \quad \forall w \in W \tag{9}$$

instead of (2). Two algorithms for vertex guards were proposed and tested [29], based on the BS-2013 implementation. The first approximates ϱ with a step function, and uses updated primal and dual separation routines that operate on overlays of visibility polygons and circular arcs, resulting in an FPTAS for the fractional AGP(V, P). The other is based on continuous optimization techniques, namely a simplex partitioning approach. In an experimental evaluation using polygons with up to 700 vertices, it was found that most polygons can be solved (to an 1.2-approximation in case of the discrete approach) within 20 min on a standard PC. The continuous algorithm turned out to be much faster, and very often finishing with an almost-optimal solution with a gap under 0.01%. In an experimental work by Kokemüller [40], AGP(P, P) with fading was analyzed. It was found that placing guards makes the problem substantially more difficult. This is mainly due to an effect where moving one guard requires moving chains of other guards as well to cover up for decreased illumination. It was also found that scaling an input polygon has an impact on the structure of solutions and number of required guards, resulting in a dramatic runtime impact.

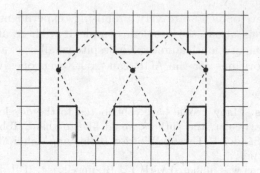

Fig. 8. A simple orthogonal polygon possessing only a single optimal solution.

6.2 Degeneracies

The experiments conducted by different groups as well as the results shown in Sect. 4 indicate that practically efficient algorithms exist, and a growing number of input instances can be solved to optimality. This raises the question whether it can be expected that all instances can be solved, given sufficient time.

Unfortunately, the answer to this question is "no". As a counterexample, consider the polygon depicted in Fig. 8. The three indicated guard positions form the only optimal solution. There is no variation allowed—shifting any guard by any $\varepsilon > 0$, in an arbitrary direction, will create a shadow, requiring a fourth guard and thereby losing optimality.

None of the currently known algorithms can solve such problems, as no way to characterize these points is known. To see this, consider perturbations of the shown polygon: It is possible to slightly move all vertices in a way that keeps the dashed lines intact. It is not clear how to find the shadow alignment points on the boundary, which in turn define the optimal guard positions. It should be noted, however, that it remains an open question whether there are polygons given by rational coordinates that require optimal guard positions with irrational coordinates.

To summarize, after forty years of research on AGP, it is still not known whether there exist finite-time algorithms for it. Even membership in NP is unclear, as it is not known if guard locations can be encoded in polynomial size.

7   Conclusion

In this paper, we have surveyed recent developments on solving the Art Gallery Problem (AGP) in a practically efficient manner. After over thirty years of mostly theoretical work, several approaches have been proposed and evaluated over the last few years, resulting in dramatic improvements. The size of instances for which optimal solutions can be found in reasonable time has improved from tens to thousands of vertices in just a few years.

While these developments are very promising, experimental findings have led to new questions about the problem complexity. There are bad instances that current implementations cannot solve despite small size, and it is not clear whether exact algorithms for the AGP can exist, even ones with exponential runtime.

Acknowledgments. Many people have contributed to the developments described in this paper. In particular, the authors would like to thank Tobias Baumgartner, Marcelo C. Couto, Sándor P. Fekete, Winfried Hellmann, Mahdi Moeini, Eli Packer, and Christiane Schmidt.

Stephan Friedrichs was affiliated with TU Braunschweig, IBR during most of the research.

This work was partially supported by the Deutsche Forschungsgemeinschaft (DFG) under contract number KR 3133/1-1 (Kunst!), by Fundação de Amparo à Pesquisa do Estado de São Paulo (FAPESP, #2007/52015-0, #2012/18384-7), Conselho Nacional de Desenvolvimento Científico e Tecnológico (CNPq, grants #311140/2014-9, #477692/2012-5 and #302804/2010-2), and FAEPEX/UNICAMP. Google Inc. supported the development of the Computational Geometry Algorithms Library [12] (CGAL) visibility package through the 2013 Google Summer of Code.

References

1. Aigner, M., Ziegler, G.M.: Proofs from THE BOOK, 4th edn. Springer Publishing Company Incorporated, Heidelberg (2009)
2. Amit, Y., Mitchell, J.S.B., Packer, E.: Locating guards for visibility coverage of polygons. In: ALENEX, pp. 120–134 (2007)
3. Amit, Y., Mitchell, J.S.B., Packer, E.: Locating guards for visibility coverage of polygons. Int. J. Comput. Geom. Appl. **20**(5), 601–630 (2010)
4. Asano, T.: An efficient algorithm for finding the visibility polygon for a polygonal region with holes. IEICE Trans. **68**(9), 557–559 (1985)
5. Austern, M.H.: Generic Programming and the STL. PUB-AW (1999)
6. Balas, E., Ng, S.M.: On the set covering polytope: II. lifting the facets with coefficients in $\{0, 1, 2\}$. Math. Program. **45**, 1–20 (1989). doi:10.1007/BF01589093. http://dx.doi.org/10.1007/BF01589093
7. Baumgartner, T., Fekete, S.P., Kröller, A., Schmidt, C.: Exact solutions and bounds for general art gallery problems. In: Proceedings of the SIAM-ACM Workshop on Algorithm Engineering and Experiments, ALENEX 2010, pp. 11–22. SIAM (2010)
8. Beasley, J.E.: Lagrangian relaxation. In: Reeves, C.R. (ed.) Modern Heuristic Techniques for Combinatorial Problems, pp. 243–303. Wiley, New York (1993). http://dl.acm.org/citation.cfm?id=166648.166660
9. Bottino, A., Laurentini, A.: A nearly optimal sensor placement algorithm for boundary coverage. Pattern Recogn. **41**(11), 3343–3355 (2008)
10. Bottino, A., Laurentini, A.: A nearly optimal algorithm for covering the interior of an art gallery. Pattern Recogn. **44**(5), 1048–1056 (2011). http://www.sciencedirect.com/science/article/pii/S0031320310005376
11. Bungiu, F., Hemmer, M., Hershberger, J., Huang, K., Kröller, A.: Efficient computation of visibility polygons. CoRR abs/1403.3905 (2014). http://arxiv.org/abs/1403.3905

12. CGAL (Computational Geometry Algorithms Library). http://www.cgal.org/
13. Chvátal, V.: A combinatorial theorem in plane geometry. J. Comb. Theory Ser. B **18**, 39–41 (1974)
14. Chwa, K., Jo, B., Knauer, C., Moet, E., van Oostrum, R., Shin, C.: Guarding art galleries by guarding witnesses. Int. J. Comput. Geom. Appl. **16**(2–3), 205–226 (2006). http://dx.doi.org/10.1142/S0218195906002002
15. Couto, M.C., de Rezende, P.J., de Souza, C.C.: An exact algorithm for an art gallery problem. Technical report IC-09-46, Institute of Computing, University of Campinas, November 2009
16. Couto, M.C., de Rezende, P.J., de Souza, C.C.: Instances for the art gallery problem (2009). http://www.ic.unicamp.br/~cid/Problem-instances/Art-Gallery
17. Couto, M.C., de Rezende, P.J., de Souza, C.C.: An IP solution to the art gallery problem. In: SoCG 2009: Proceedings of the 25th Annual Symposium on Computational Geometry, pp. 88–89. ACM, New York (2009)
18. Couto, M.C., de Rezende, P.J., de Souza, C.C.: Video: an IP solution to the art gallery problem. In: 18th Video Review of Computational Geometry at the 25th Annual Symposium on Computational Geometry, June 2009. www.computational-geometry.org/SoCG-videos/socg09video/video1-couto.mov
19. Couto, M.C., de Rezende, P.J., de Souza, C.C.: An exact algorithm for minimizing vertex guards on art galleries. Int. Trans. Oper. Res. **18**, 425–448 (2011)
20. Couto, M.C., de Souza, C.C., de Rezende, P.J.: An exact and efficient algorithm for the orthogonal art gallery problem. In: SIBGRAPI 2007: Proceedings of the XX Brazilian Symposium on Computer Graphics and Image Processing, pp. 87–94. IEEE Computer Society, Washington, DC (2007)
21. Couto, M.C., Souza, C.C., Rezende, P.J.: Experimental evaluation of an exact algorithm for the orthogonal art gallery problem. In: McGeoch, C.C. (ed.) WEA 2008. LNCS, vol. 5038, pp. 101–113. Springer, Heidelberg (2008). doi:10.1007/978-3-540-68552-4_8
22. IBM ILOG CPLEX Optimization Studio. http://www.ibm.com/software/integration/optimization/cplex-optimizer/
23. Deshpande, A., Kim, T., Demaine, E.D., Sarma, S.E.: A pseudopolynomial time $O(\log n)$-approximation algorithm for art gallery problems. In: Dehne, F., Sack, J.-R., Zeh, N. (eds.) WADS 2007. LNCS, vol. 4619, pp. 163–174. Springer, Heidelberg (2007). doi:10.1007/978-3-540-73951-7_15
24. Efrat, A., Har-Peled, S.: Guarding galleries and terrains. Inf. Process. Lett. **100**(6), 238–245 (2006)
25. Eidenbenz, S., Stamm, C., Widmayer, P.: Inapproximability results for guarding polygons and terrains. Algorithmica **31**(1), 79–113 (2001)
26. Eisenbrand, F., Funke, S., Karrenbauer, A., Matijevic, D.: Energy-aware stage illumination. In: Proceedings of the 21st ACM Symposium on Computational Geometry (SCG 2005), pp. 336–345 (2005). http://portal.acm.org/citation.cfm?doid=1064092.1064144
27. Erdem, U.M., Sclaroff, S.: Optimal placement of cameras in floorplans to satisfy task requirements and cost constraints. In: Proceedings of the Fifth International Workshop on Omnidirectional Vision, Camera Networks and Non-classical Cameras, pp. 30–41 (2004)
28. Erdem, U.M., Sclaroff, S.: Automated camera layout to satisfy task-specific and floor plan-specific coverage requirements. Comput. Vis. Image Underst. **103**(3), 156–169 (2006)
29. Ernestus, M., Friedrichs, S., Hemmer, M., Kokemüller, J., Kröller, A., Moeini, M., Schmidt, C.: Algorithms for art gallery illumination. arXiv e-prints, October 2014

30. Fekete, S.P., Friedrichs, S., Kröller, A., Schmidt, C.: Facets for art gallery problems. In: Du, D.-Z., Zhang, G. (eds.) COCOON 2013. LNCS, vol. 7936, pp. 208–220. Springer, Heidelberg (2013). doi:10.1007/978-3-642-38768-5_20

31. Fekete, S.P., Friedrichs, S., Kröller, A., Schmidt, C.: Facets for art gallery problems. Algorithmica **73**(2), 411–440 (2014)

32. Fisk, S.: A short proof of Chvátal's watchman theorem. J. Comb. Theory Ser. B **24**(3), 374–375 (1978)

33. Friedrichs, S.: Integer solutions for the art gallery problem using linear programming. Master's thesis, TU Braunschweig (2012)

34. Ghosh, S.K.: Approximation algorithms for art gallery problems in polygons and terrains. In: Rahman, M.S., Fujita, S. (eds.) WALCOM 2010. LNCS, vol. 5942, pp. 21–34. Springer, Heidelberg (2010). doi:10.1007/978-3-642-11440-3_3

35. GNU Multiple Precision Arithmetic Library (2013). http://gmplib.org

36. Hemmer, M., Huang, K., Bungiu, F.: 2D visibility. In: CGAL User and Reference Manual. CGAL Editorial Board (2014, to appear)

37. Honsberger, R.: Mathematical Gems II. Mathematical Association of America, Washington, DC (1976)

38. Kahn, J., Klawe, M., Kleitman, D.: Traditional art galleries require fewer watchmen. SIAM J. Algebr. Discrete Methods **4**(2), 194–206 (1983)

39. Karamcheti, V., Li, C., Pechtchanski, I., Yap, C.: A core library for robust numeric and geometric computation. In: Proceedings of the 15th Annual ACM Symposium of Computational Geometry (SCG), pp. 351–359 (1999)

40. Kokemüller, J.: Variants of the art gallery problem. Master's thesis, TU Braunschweig (2014)

41. Kröller, A., Baumgartner, T., Fekete, S.P., Schmidt, C.: Exact solutions and bounds for general art gallery problems. ACM J. Exp. Algothmmics 17, Article ID 2.3 (2012)

42. Kröller, A., Moeini, M., Schmidt, C.: A novel efficient approach for solving the art gallery problem. In: Ghosh, S.K., Tokuyama, T. (eds.) WALCOM 2013. LNCS, vol. 7748, pp. 5–16. Springer, Heidelberg (2013). doi:10.1007/978-3-642-36065-7_3

43. Kröller, A., Schmidt, C.: Energy-aware art gallery illumination. In: Proceedings of the 28th European Workshop on Computational Geometry (EuroCG 2012), pp. 93–96 (2012)

44. Laurentini, A.: Guarding the walls of an art gallery. Vis. Comput. **15**(6), 265–278 (1999)

45. Lee, D.T., Lin, A.K.: Computational complexity of art gallery problems. IEEE Trans. Inf. Theory **32**(2), 276–282 (1986)

46. Mehlhorn, K., Näher, S.: LEDA: A Platform for Combinatorial and Geometric Computing. PUB-CAMB, Cambridge (2000)

47. Mitchell, J.S.B.: Approximating watchman routes. In: Proceedings of the Twenty-Fourth Annual ACM-SIAM Symposium on Discrete Algorithms, SODA 2013, NewOrleans, Louisiana, USA, 6-8 January 2013, pp. 844–855 (2013)

48. Nilsson, B.J.: Approximate guarding of monotone and rectilinear polygons. In: Caires, L., Italiano, G.F., Monteiro, L., Palamidessi, C., Yung, M. (eds.) ICALP 2005. LNCS, vol. 3580, pp. 1362–1373. Springer, Heidelberg (2005). doi:10.1007/11523468_110

49. O'Rourke, J.: Art Gallery Theorems and Algorithms. International Series of Monographs on Computer Science. Oxford University Press, New York (1987)

50. O'Rourke, J., Supowit, K.: Some NP-hard polygon decomposition problems. IEEE Trans. Inf. Theory **29**(2), 181–190 (1983)

51. Packer, E.: Computing multiple watchman routes. In: McGeoch, C.C. (ed.) WEA 2008. LNCS, vol. 5038, pp. 114–128. Springer, Heidelberg (2008). doi:10.1007/978-3-540-68552-4_9

52. Packer, E.: Robust geometric computing and optimal visibility coverage. Ph.D. thesis, SUNY Stony Brook (2008)

53. Tao, P.D., An, L.T.H.: Convex analysis approach to D.C. programming: theory, algorithms and applications. Acta Mathematica Vietnamica **22**(1), 289–355 (1997)

54. Pion, S., Fabri, A.: A generic lazy evaluation scheme for exact geometric computations. In: 2nd # WOR-LCSD (2006). http://www.citebase.org/abstract?id=oai:arXiv.org:cs/0608063

55. Schuchardt, D., Hecker, H.D.: Two NP-hard art-gallery problems for orthopolygons. Math. Log. Q. **41**, 261–267 (1995)

56. Shermer, T.C.: Recent results in art galleries (geometry). Proc. IEEE **80**(9), 1384–1399 (1992)

57. Tomás, A.P., Bajuelos, A.L., Marques, F.: Approximation algorithms to minimum vertex cover problems on polygons and terrains. In: Sloot, P.M.A., Abramson, D., Bogdanov, A.V., Dongarra, J.J., Zomaya, A.Y., Gorbachev, Y.E. (eds.) ICCS 2003. LNCS, vol. 2657, pp. 869–878. Springer, Heidelberg (2003). doi:10.1007/3-540-44860-8_90

58. Tomás, A.P., Bajuelos, A.L., Marques, F.: On visibility problems in the plane - solving minimum vertex guard problems by successive approximations. In: Proceedings of the 9th International Symposium on Artificial Intelligence and Mathematics (AI & MATH 2006) (2006, to appear)

59. Tozoni, D.C., de Rezende, P.J., de Souza, C.C.: Algorithm 966: a practical iterative algorithm for the art gallery problem using integer linear programming. ACM Trans. Math. Softw. **43**(2), 16:1–16:27 (2016). doi:10.1145/2890491. Article no. 16

60. Tozoni, D.C., Rezende, P.J., Souza, C.C.: The quest for optimal solutions for the art gallery problem: a practical iterative algorithm. In: Bonifaci, V., Demetrescu, C., Marchetti-Spaccamela, A. (eds.) SEA 2013. LNCS, vol. 7933, pp. 320–336. Springer, Heidelberg (2013). doi:10.1007/978-3-642-38527-8_29

61. Urrutia, J.: Art gallery and illumination problems. In: Sack, J.R., Urrutia, J. (eds.) Handbook on Computational Geometry, pp. 973–1026. Elsevier Science Publishers, Amsterdam (2000)

62. Wein, R., Berberich, E., Fogel, E., Halperin, D., Hemmer, M., Salzman, O., Zukerman, B.: 2D arrangements. In: CGAL User and Reference Manual, 4.0 edn., CGAL Editorial Board (2012)

Author Index

Printed in the United States
By Bookmasters